Carbon Dioxide and Terrestrial Ecosystems

Physiological Ecology

A Series of Monographs, Texts, and Treatises

Series Editor
Harold A. Mooney
Stanford University, Stanford, California

Editorial Board
Fakhri Bazzaz   F. Stuart Chapin   James R. Ehleringer
Robert W. Pearcy   Martyn M. Caldwell   E.-D. Schulze

T. T. KOZLOWSKI. Growth and Development of Trees, Volumes I and II, 1971

D. HILLEL. Soil and Water: Physical Principles and Processes, 1971

V. B. YOUNGER and C. M. McKELL (Eds.). The Biology and Utilization of Grasses, 1972

J. B. MUDD and T. T. KOZLOWSKI (Eds.). Responses of Plants to Air Pollution, 1975

R. DAUBENMIRE. Plant Geography, 1978

J. LEVITT. Responses of Plants to Environmental Stresses, Second Edition
Volume I: Chilling, Freezing, and High Temperature Stresses, 1980
Volume II: Water, Radiation, Salt, and Other Stresses, 1980

J. A. LARSEN (Ed.). The Boreal Ecosystem, 1980

S. A. GAUTHREAUX, JR. (Ed.). Animal Migration, Orientation, and Navigation, 1981

F. J. VERNBERG and W. B. VERNBERG (Eds.). Functional Adaptations of Marine Organisms, 1981

R. D. DURBIN (Ed.). Toxins in Plant Disease, 1981

C. P. LYMAN, J. S. WILLIS, A. MALAN, and L. C. H. WANG. Hibernation and Torpor in Mammals and Birds, 1982

T. T. KOZLOWSKI (Ed.). FLooding and Plant Growth, 1984

E. I. RICE. Allelopathy, Second Edition, 1984

M. L. CODY (Ed.). Habitat Selection in Birds, 1985

R. J. HAYNES, K. C. CAMERON, K. M. GOH, and R. R. SHERLOCK (Eds.). Mineral Nitrogen in the Plant–Soil System, 1986

T. T. KOZLOWSKI, P. J. KRAMER, and S. G. PALLARDY. The Physiological Ecology of Woody Plants, 1991

H. A. MOONEY, W. E. WINNER, and E. J. PELL (Eds.). Response of Plants to Multiple Stresses, 1991

The list of titles in this series continues at the end of this volume.

Carbon Dioxide and Terrestrial Ecosystems

Edited by

George W. Koch
Department of Biological Sciences
Northern Arizona University
Flagstaff, Arizona

Harold A. Mooney
Department of Biological Sciences
Stanford University
Stanford, California

Academic Press

San Diego   New York   Boston   London   Sydney   Tokyo   Toronto

Cover photograph: Central Californian annual grassland at Stanford University's Jasper Ridge Biological Preserve showing open-top chambers for ecosystem CO_2 experiments. (Photograph courtesy of Chris Field.)

This book is printed on acid-free paper. ∞

Copyright © 1996 by ACADEMIC PRESS, INC.

All Rights Reserved.
No part of this publication may be reproduced or transmitted in any form or by any means, electronic or mechanical, including photocopy, recording, or any information storage and retrieval system, without permission in writing from the publisher.

Academic Press, Inc.
A Division of Harcourt Brace & Company
525 B Street, Suite 1900, San Diego, California 92101-4495

United Kingdom Edition published by
Academic Press Limited
24-28 Oval Road, London NW1 7DX

Library of Congress Cataloging-in-Publication Data

Carbon dioxide and terrestrial ecosystems / edited by
  George W. Koch, Harold A. Mooney
    p.   cm. -- (Physiological ecology series)
  Includes bibliographical references and index.
  ISBN 0-12-505295-2 (alk. paper)
  1. Plants, Effect of atmospheric carbon dioxide on. 2. Trees-
-Effect of atmospheric carbon dioxide on. 3. Atmospheric carbon-
dioxide--Environmental aspects. I. Mooney, Harold A. II. Series:
Physiological ecology.
  QK753.C3C37   1995
  581.5'222--dc20   95-31377
                                                      CIP

PRINTED IN THE UNITED STATES OF AMERICA
95 96 97 98 99 00 BB 9 8 7 6 5 4 3 2 1

Contents

Contributors xiii
Preface xvii

1. Tree Responses to Elevated CO_2 and Implications for Forests
Richard J. Norby, Stan D. Wullschleger, and Carla A. Gunderson

I. Introduction 1
II. Experimental Approach 3
III. Results 5
IV. Discussion 15
V. Conclusions 19
References 19

2. Effects of CO_2 and N on Growth and N Dynamics in Ponderosa Pine: Results from the First Two Growing Seasons
Dale W. Johnson, P. H. Henderson, J. Timothy Ball, and R. F. Walker

I. Introduction 23
II. Site and Methods 24
III. Results and Discussion 26
IV. Perspective: Applicability of Results to Mature Forests 32
V. Summary and Conclusions 37
References 38

3. Linking Above- and Belowground Responses to Rising CO_2 in Northern Deciduous Forest Species
Peter S. Curtis, Donald R. Zak, Kurt S. Pregitzer, John Lussenhop, and James A. Teeri

I. Introduction 41
II. Study Site and Experimental Methods 44
III. Results and Discussion 47
IV. Conclusions 49
References 50

4. The Effects of Tree Maturity on Some Responses to Elevated CO_2 in Sitka Spruce (*Picea sitchensis* Bong. Carr.)
Helen S. J. Lee and Paul G. Jarvis

    I. Introduction   53
   II. Materials and Methods   57
  III. Results and Discussion   59
  IV. Conclusions   68
      References   69

5. Growth Strategy and Tree Responses to Elevated CO_2: A Comparison of Beech (*Fagus sylvatica*) and Sweet Chestnut (*Castanea sativa* Mill.)
Marianne Mousseau, Eric Dufrêne, Asmae El Kohen, Daniel Epron, Denis Godard, Rodolphe Liozon, Jean Yves Pontailler, and Bernard Saugier

    I. Introduction   71
   II. Materials and Methods   72
  III. Results   73
  IV. Discussion   82
      References   85

6. Litter Quality and Decomposition Rates of Foliar Litter Produced under CO_2 Enrichment
Elizabeth G. O'Neill and Richard J. Norby

    I. Introduction   87
   II. Litter Quality and the Decomposition Process   89
  III. CO_2 Enrichment Effects on Litter Quality and Decomposition   92
  IV. A Few Words to the Wise Decomposer   96
   V. Summary   100
      References   101

7. CO_2-Mediated Changes in Tree Chemistry and Tree–Lepidoptera Interactions
Richard L. Lindroth

    I. Introduction   105
   II. Forest Lepidoptera   106
  III. Carbon–Nutrient Balance Theory: A Predictive Tool   106
  IV. Effects of Elevated CO_2 on Tree Chemistry   107

V. Effects on Insect Herbivores   112
VI. Potential Community and Ecosystem Responses   115
VII. Future Research Directions   117
References   118

8. The Jasper Ridge CO_2 Experiment: Design and Motivation
Christopher B. Field, F. Stuart Chapin III, Nona R. Chiariello, Elisabeth A. Holland, and Harold A. Mooney

I. Introduction   121
II. The Challenge   122
III. Jasper Ridge   125
IV. The Suite of Experiments   129
V. Experimental Facilities   131
VI. Results   138
VII. Concluding Remarks   141
References   142

9. Ecosystem-Level Responses of Tallgrass Prairie to Elevated CO_2
Clenton E. Owensby, Jay M. Ham, Alan Knapp, Charles W. Rice, Patrick I. Coyne, and Lisa M. Auen

I. Introduction   147
II. Study Site and Experimental Design   150
III. Results and Discussion   151
IV. Summary and Conclusions   157
References   160

10. Direct Effects of Elevated CO_2 on Arctic Plant and Ecosystem Function
Walter C. Oechel and George L. Vourlitis

I. Introduction   163
II. Individual Plant Response to Elevated CO_2 165
III. Ecosystem-Level Response to Elevated CO_2 167
IV. Long-Term Ecosystem Response to Elevated CO_2 171
V. Summary and Conclusions   172
References   174

11. Response of Alpine Vegetation to Elevated CO_2
Christian Körner, Matthias Diemer, Bernd Schäppi, and Lukas Zimmerman

I. Introduction   177
II. Methods   179

III. Results of Two Years of Field Experimentation   185
IV. Conclusions   194
    References   195

12. Long-Term Elevated CO_2 Exposure in a Chesapeake Bay Wetland: Ecosystem Gas Exchange, Primary Production, and Tissue Nitrogen
Bert G. Drake, Gary Peresta, Esther Beugeling, and Roser Matamala

I. Introduction   197
II. Results   199
III. Discussion   203
IV. Conclusions   210
V. Summary   211
   References   212

13. Free-Air CO_2 Enrichment: Responses of Cotton and Wheat Crops
Paul J. Pinter, Jr., Bruce A. Kimball, Richard L. Garcia, Gerard W. Wall, Douglas J. Hunsaker, and Robert L. LaMorte

I. Introduction   215
II. Materials and Methods   218
III. Results and Discussion   221
IV. Summary and Future Investigations   239
   Appendix   241
   References   245

14. Response of Growth and CO_2 Uptake of Spring Wheat and Faba Bean to CO_2 Concentration under Semifield Conditions: Comparing Results of Field Experiments and Simulations
Paul Dijkstra, Sanderine Nonhebel, Cees Grashoff, Jan Goudriaan, and Siebe C. van de Geijn

I. Introduction   251
II. Materials and Methods   252
III. Results   254
IV. Discussion   259
V. Summary and Conclusions   261
   References   262

15. Assessment of Rice Responses to Global Climate Change: CO_2 and Temperature
Jeffrey T. Baker, L. Hartwell Allen, Jr., Kenneth J. Boote, and Nigel B. Pickering

    I. Introduction   265
    II. Materials and Methods: Outdoor, Sunlit, Controlled-Environment Chambers   265
    III. Results and Discussion   267
    IV. Conclusions and Research Recommendations   279
    V. Summary   280
    References   281

16. Interactions between CO_2 and Nitrogen in Forests: Can We Extrapolate from the Seedling to the Stand Level?
Dale W. Johnson and J. Timothy Ball

    I. Introduction   283
    II. Nature of Nitrogen Cycling in Forests   284
    III. Potential Effects of Elevated CO_2 on Nitrogen Cycling   289
    IV. What Do Seedling–Sapling Studies Tell Us about Ecosystem-Level Response?   291
    V. Conclusions   294
    References   295

17. Protection from Oxidative Stress in Trees as Affected by Elevated CO_2 and Environmental Stress
Andrea Polle

    I. Introduction   299
    II. Detoxification of Reactive Oxygen Species   300
    III. Interactions of Environmental Stresses and Elevated CO_2 303
    IV. Summary   309
    References   311

18. Integrating Knowledge of Crop Responses to Elevated CO_2 and Temperature with Mechanistic Simulation Models: Model Components and Research Needs
Jeffrey S. Amthor and Robert S. Loomis

    I. Introduction   317
    II. Models   319
    III. Processes That Should Be Included in Models Used to Predict Crop

Responses to Elevated CO_2　326
　IV. Knowledge Needs　337
　V. Conclusions　339
　　　References　340

19. Progress, Limitations, and Challenges in Modeling the Effects of Elevated CO_2 on Plants and Ecosystems
James F. Reynolds, Paul R. Kemp, Basil Acock, Jia-Lin Chen, and Daryl L. Moorhead

　　I. Introduction　347
　 II. Leaf-Level Models　349
　III. Plant-Level Models　351
　IV. Population, Community, and Stand Models　355
　 V. Ecosystem Models　359
　VI. Regional and Global Models　366
　VII. Future Challenges　369
　　　References　372

20. Stimulation of Global Photosynthetic Carbon Influx by an Increase in Atmospheric Carbon Dioxide Concentration
Yiqi Luo and Harold A. Mooney

　　I. Introduction　381
　 II. The Model　383
　III. Results and Discussion　384
　IV. Summary　393
　　　Appendix　394
　　　References　394

21. Biota Growth Facter β: Stimulation of Terrestrial Ecosystem Net Primary Production by Elevated Atmospheric CO_2
Jeffery S. Amthor and George W. Koch

　　I. Introduction　399
　 II. Modeling and Measuring Plant and Ecosystem Responses to Elevated CO_2　400
　III. Information Needs　409
　IV. Where Do We Stand with Respect to β?　409
　　　References　411

22. Response of Terrestrial Ecosystems to Elevated CO_2: A Synthesis and Summary
George W. Koch and Harold A. Mooney

    I. Cross-System Comparisons   415
   II. Future Research Needs   423
  III. Conclusions   427
       References   428

Index   431

Contributors

Numbers in parentheses indicate the pages on which the authors' contributions begin.

Basil Acock (347), Systems Research Laboratory, USDA, Beltsville, Maryland 20705

L. Hartwell Allen, Jr. (265), United States Department of Agriculture, Agricultural Research Service, University of Florida, Gainesville, Florida 32611

Jeffrey S. Amthor (317, 399), Global Climate Research Division, Lawrence Livermore National Laboratory, Livermore, California 94550

Lisa M. Auen (147), Department of Agronomy, Kansas State University, Manhattan, Kansas 66506

Jeffrey T. Baker (265), Agronomy Department, Institute of Food and Agriculture Sciences, University of Florida, Gainesville, Florida 32611

J. Timothy Ball (23, 283), Desert Research Institute and University of Nevada, Reno, Nevada 89506

Esther Beugeling (197), Smithsonian Environmental Research Center, Edgewater, Maryland 21037

Kenneth J. Boote (265), Agronomy Department, Institute of Food and Agricultural Sciences, University of Florida, Gaineville, Florida 32611

F. Stuart Chapin III (121), Department of Integrative Biology, University of California, Berkeley, California 94720

Jia-Lin Chen (347), Department of Botany, Duke University, Durham, North Carolina 27708

Nona R. Chiariello (121), Jasper Ridge Biological Preserve, Stanford University, Stanford, California 94305

Patrick I. Coyne (147), Ft. Hays Branch Experiment Station, Kansas State University, Hays, Kansas 67601

Peter S. Curtis (41), Department of Plant Biology, The Ohio State University, Columbus, Ohio 43210

Matthias Diemer (177), Botanishces Institut der Universität Basel, CH-4056 Basel, Switzerland

Paul Dijkstra[1] (251), Research Institute for Agrobiology and Soil Fertility (AB-DLO), 6700 AA Wageningen, The Netherlands

---

[1] Present Address: C/O Ross Hinkle, Dynamac Corporation, Kennedy Space Center, Florida 32899.

Bert G. Drake (197), Smithsonian Environmental Research Center, Edgewater, Maryland 21037

Eric Dufrêne (71), Laboratoire d'Ecologie Végétale, Université Paris-Sud, 91405 Orsay, France

Asmae El Kohen (71), Laboratoire d'Ecologie Végétale, Université Paris-Sud, 91405 Orsay, France

Daniel Epron (71), Institute des Sciences et Techniques de l'Environment, Pôle Universitaire du pay de Montbéliard, 25211 Montbéliard, France

Christopher B. Field (121), Department of Plant Biology, Carnegie Institution of Washington, Stanford, California 94305

Richard L. Garcia (215), U. S. Water Conservation Lab, Phoenix, Arizona 85040

Denis Godard (71), Laboratoire d'Ecologie Végétale, Université Paris-Sud, 91405 Orsay, France

Jan Goudriaan (251), Department of Theoretical Production Ecology, Wageningen Agricultural University, 6708 PD Wageningen, The Netherlands

Cees Grashoff (251), Research Institute for Agrobiology and Soil Fertility (AB-DLO), 6700 AA Wageningen, The Netherlands

Carla A. Gunderson (1), Environmental Sciences Division, Oak Ridge National Laboratory, Oak Ridge, Tennessee 37831

Jay M. Ham (147), Department of Agronomy, Kansas State University, Manhattan, Kansas 66506

P. H. Henderson (23), Desert Research Institute and University of Nevada, Reno, Nevada 89506

Elisabeth A. Holland (121), National Center for Atmospheric Research, Boulder, Colorado 80307

Douglas J. Hunsaker (215), U. S. Water Conservation Lab, Phoenix, Arizona 85040

Paul G. Jarvis (53), Institute of Ecology & Resource Management, Darwin Building, University of Edinburgh, Edinburgh EH9 3JU, Scotland

Dale W. Johnson (23, 283), Desert Research Institute and University of Nevada, Reno, Nevada 89506

Paul R. Kemp (347), Department of Botany, Duke University, Durham, North Carolina 27708

Bruce A. Kimball (215), U. S. Water Conservation Lab, Phoenix, Arizona 85040

Alan Knapp (147), Division of Biology, Kansas State University, Manhattan, Kansas 66506

George W. Koch (399, 415), Department of Biological Sciences, Northern Arizona University, Flagstaff, Arizona 86011

Christian Körner (177), Botanishces Institut der Universität Basel, CH-4056 Basel, Switzerland

Robert L. LaMorte (215), U. S. Water Conservation Lab, Phoenix, Arizona 85040

Helen S. J. Lee (53), Institute of Ecology & Resource Management, Darwin Building, University of Edinburgh, Scotland EH9 3JU

Richard L. Lindroth (105), Department of Entomology, University of Wisconsin, Madison, Wisconsin 53706

Rodolphe Liozon (71), Laboratoire d'Ecologie Végétale, Université Paris-Sud, 91405 Orsay, France

Robert S. Loomis (317), Department of Agronomy and Range Science, University of California, Davis, California 95616

Yiqi Luo (381), Biological Sciences Center, Desert Research Institute, Reno, Nevada 89506

John Lussenhop (41), Biology Department, University of Illinois, Chicago, Illinois 60680

Roser Matamala (197), Smithsonian Environmental Research Center, Edgewater, Maryland 21037

Harold A. Mooney (121, 381, 415), Department of Biological Sciences, Stanford University, Stanford, California 94305

Daryl L. Moorhead (347), Department of Biological Sciences, Texas Tech University, Lubbock, Texas 79409

Marianne Mousseau (71), Laboratoire d'Ecologie Végétale, Université Paris-Sud, 91405 Orsay, France

Sanderine Nonhebel (251), Department of Theoretical Production Ecology, Wageningen Agricultural University, 6708 PD Wageningen, The Netherlands

Richard J. Norby (1, 87), Environmental Sciences Division, Oak Ridge National Laboratory, Oak Ridge, Tennessee 37831

Elizabeth G. O'Neill (87), Environmental Sciences Division, Oak Ridge National Laboratory, Oak Ridge, Tennessee 37831

Walter C. Oechel (163), Global Change Research Group and Systems Ecology Research Group, San Diego State University, San Diego, California 92182

Clenton E. Owensby (147), Department of Agronomy, Kansas State University, Manhattan, Kansas 66506

Gary Peresta (197), Smithsonian Environmental Research Center, Edgewater, Maryland 21037

Nigel B. Pickering (265), Agricultural Engineering Department, University of Florida, Gainesville, Florida 32611

Paul J. Pinter, Jr. (215), U. S. Water Conservation Lab, Phoenix, Arizona 85040

Andrea Polle (299), Institut für Forstbotanik und Baumphysiologie, Albert-Ludwigs Universität Freiburg, D-79085 Freiburg, Germany

Jean Yves Pontailler (71), Laboratoire d'Ecologie Végétale, Université Paris-Sud, 91405 Orsay, France

Kurt S. Pregitzer (41), School of Forestry, Michigan Technological University, Houghton, Michigan 49931

James F. Reynolds (347), Department of Botany, Duke University, Durham, North Carolina 27708

Charles W. Rice (147), Department of Agronomy, Kansas State University, Manhattan, Kansas 66506

Bernard Saugier (71), Laboratoire d'Ecologie Végétale, Université Paris-Sud, 91405 Orsay, France

Bernd Schäppi (177), Botanishces Institut der Universität Basel, CH-4056 Basel, Switzerland

James A. Teeri (41), Biological Station, University of Michigan, Pellston, Michigan 49769

Siebe C. van de Geijn (251), Research Institute for Agrobiology and Soil Fertility (AB-DLO), 6700 AA Wageningen, The Netherlands

George L. Vourlitis (163), Global Change Research Group and Systems Ecology Research Group, San Diego State University, San Diego, California 92182

R. F. Walker (23), University of Nevada, Reno, Nevada 89506

Gerard W. Wall (215), U. S. Water Conservation Lab, Phoenix, Arizona 85040

Stan D. Wullschleger (1), Environmental Sciences Division, Oak Ridge National Laboratory, Oak Ridge, Tennessee 37831

Donald R. Zak (41), School of Natural Resources and Environment, University of Michigan, Ann Arbor, Michigan 48109

Lukas Zimmerman (177), Botanishces Institut der Universität Basel, CH-4056 Basel, Switzerland

Preface

In the area of global change there are certainties as well as uncertainties. In the case of atmospheric CO_2 the certainty is that the concentration is increasing and has been doing so since the beginning of the industrial revolution. We know that the increased emissions due to industrialization and deforestation are accounted for only in part by increases in the atmospheric CO_2. The remainder of the CO_2 is taken up by the oceans and by terrestrial ecosystems. The question is how much is partitioned into these two realms and, in turn, partitioned within them. In this book we examine the terrestrial part of the equation. We ask a very specific question: "Are different terrestrial ecosystems equally effective at absorbing increased atmospheric carbon dioxide?" The first tentative answer was "probably yes," based on early physiological measurements and on initial experiments with crops. The fact that most natural ecosystems are resource (water, nutrient, or light) limited, however, has led to the suggestion that they may in fact be differentially responsive to increasing CO_2 (Mooney *et al.,* 1991).

This book represents a first synthesis of the major ecosystem-level CO_2 experiments, many of which are ongoing. It is apparent from the number of studies presented here that research on ecosystem effects of CO_2 has increased greatly since the last major syntheses (Lemon, 1983; Strain and Cure, 1985), albeit from a meager baseline. Worldwide there are now, or have been, CO_2 experiments in about a dozen different ecosystems, with more coming on-line in the near future. This dramatic shift in focus from photosynthesis and growth of plants raised in environmental chambers to primary production, biogeochemical cycling, and plant–herbivore interactions of intact ecosystems has occurred in parallel with the increasing realization that CO_2 effects may not scale linearly across levels of ecological organization, that complex environmental interactions may strongly modify CO_2 responses, and that ecosystem structure and composition may change in response to elevated CO_2. Overarching these issues is the concern about the potential climatic effects of rising CO_2 and the question of the extent to which a stimulation of biosphere carbon sequestration might mitigate this rise.

This volume summarizes much of what has been and is being learned from ecosystem CO_2 studies and suggests where the major gaps in our understanding lie, both conceptually and in terms of important ecosystems yet to be studied. Portions of some studies presented here have appeared elsewhere, while other results are from recently completed research. The emphasis in this volume is on "natural" (i.e., nonagricultural) ecosystems, although several chapters on key crop experiments have been included because it is felt that the unique features of these systems extend the continuum of system response along important resource axes and thereby provide additional explanatory power in understanding variation in CO_2 responses among different ecosystems.

Chapters are grouped into three sections according to general plant life form: woody ecosystems (Chapters 1–7), unmanaged herbaceous systems (Chapters 8–12), and crop systems (Chapters 13–15). Each chapter summarizes the project's principal findings to date and synthesizes what the authors feel is and is not known about the effects of CO_2 on the particular ecosystem. A fourth section of the book (Chapters 16–22) serves several important purposes: to highlight arising issues in CO_2 research, including the key role of the nitrogen cycle in regulating responses to CO_2 and potential interactions of CO_2 and air pollutants; to discuss strategies for developing useful models of ecosystem function in elevated CO_2; to summarize our current understanding of the biosphere's sensitivity to CO_2 in terms of photosynthetic and biomass responses; and to synthesize patterns of response to elevated CO_2 among ecosystems differing in life form and abiotic stress factors.

This book was developed in cooperation with the Global Change and Terrestrial Ecosystem project of the International Geosphere Biosphere Program and funded by the U.S. Department of Energy and the Electric Power Research Institute. We are grateful to Drs. Antonio Raschi and Franco Miglietta and the Centro Studi "I Cappuccini" della Cassa di Risparmio di San Miniato S.p.A. for providing an environment for the stimulating and enriching discussions that led to the production of this book.

References

Lemon, E. R. (ed.) (1983). "CO_2 and Plants: The Response of Plants to Rising Levels of Atmospheric Carbon Dioxide." Westview Press, Boulder, CO.
Strain, B. R., and Cure J. D. (eds.) (1985). "Direct Effects of Increasing Carbon Dioxide on Vegetation." DOE/ER-0238, U.S. Department of Energy, Washington, DC.
Mooney, H. A., Drake, B. G., Luxmoore, R. J., Oechel, W. C., and Pitelka, L. F. (1991). Predicting ecosystem responses to elevated CO_2 concentrations. *BioScience* 41(2), 96–104.

HAROLD A. MOONEY
GEORGE W. KOCH

1
Tree Responses to Elevated CO_2 and Implications for Forests

Richard J. Norby, Stan D. Wullschleger, and Carla A. Gunderson

I. Introduction

The responses of forest ecosystems to elevated concentrations of atmospheric CO_2 cannot be measured directly in experiments that approximate environmental conditions of the future. Even if large and expensive facilities are constructed for manipulation of the abiotic environment in a forest, the edaphic and biotic adjustments that will occur over decade time frames cannot be reproduced in an experimental system. Nevertheless, there is a compelling need to understand the responses of forests to elevated CO_2 because of the prominent role of forests in the global carbon cycle and the potential for CO_2 fertilization to alter the relationship between the carbon cycle and the climate system (Solomon and Cramer, 1993).

Our ability to measure forest response is constrained by two prominent features of forests that do not similarly constrain research on other ecosystems. The first constraint is that the dominant organisms of forests—trees—are large and have a long life span, precluding simple pot experiments for assessing growth responses to elevated CO_2, as well as many of the important physiological responses. Such studies have been invaluable with crop plants and grasses, where the plant can be grown to its full size and life span. With trees we must instead rely on extrapolation from relevant physiological and morphological indicators of response, and multiyear studies must be conducted in the field so that conclusions are not confounded by the artifactual restrictions of root growth in pots (Eamus and Jarvis, 1989).

The second important constraint is that the ecological complexity of forest ecosystems greatly increases the importance of considerations beyond single tree responses. Of course, assessment of the response of any ecosystem must encompass considerations of many biotic and abiotic influences besides the response of a single dominant plant species, but the larger spatial and temporal scales of forest ecosystems increase the importance of these considerations—and the difficulty in addressing them—manyfold. In systems of smaller stature, intact ecosystems can be enclosed in chambers and ecosystem-level responses to CO_2 measured directly (Oechel *et al.*, 1991; Drake, 1992; Owensby *et al.*, 1993), but the scale of forests precludes similar studies. Instead, the CO_2 responses of critical components of the forest, including interactions and feedback between trees and multiple environmental resources, must be studied separately in a manner conducive to reassembly in an integrated analysis. Scale is an especially important consideration in a complex and diverse forest type such as the eastern North American deciduous forest, raising the issue of differential responses of competing tree species to changes in the atmospheric CO_2 concentration (Bazzaz and Miao, 1993).

Although these problems limiting our ability to obtain data on tree responses that are relevant to ecosystem and global-scale questions seemed daunting, we began a series of experiments in 1981 designed to investigate the responses of trees, especially deciduous species, to elevated CO_2 (Norby *et al.*, 1994). The overriding question was whether nutrient limitation, as commonly occurs in unmanaged forests, would preclude growth enhancement in response to CO_2 enrichment. The experiments focused on mechanisms of CO_2 × nutrient interactions, especially on belowground processes such as exudation, rhizosphere microbial activity, and symbiotic relationships that could alter nutrient availability. The results showed that belowground processes are often especially stimulated by CO_2 enrichment and that nutrient deficiency does not necessarily preclude growth responses to elevated CO_2, at least during experiments lasting several months. The mechanisms sustaining growth response under nutrient-limited conditions can include increased nutrient availability in the plant–soil system, as well as decreased physiological demand for nutrients by the plant (Norby *et al.*, 1986, 1994). Although questions at a finer scale of resolution (e.g., biochemical mechanisms of response) naturally occurred during the course of these experiments, the most important new questions that developed concerned whether short-term responses could be sustained over several growing seasons under field conditions. These are key issues that control whether the growth chamber studies are at all useful in predicting forest responses. The questions that arose included the following: Does feedback between growth and photosynthesis limit the enhancement of photosynthesis over time? How does growth in elevated CO_2 alter the processes of nutrient and carbon storage and retranslocation? Do CO_2-induced changes in tissue chemistry influence decomposition and nutrient turnover?

Here we describe the field experiment that was designed to test and extend the concepts developed in the series of growth chamber experiments. We will discuss the results of the field experiment in relation to the larger issue of forest ecosystem response to elevated CO_2, and finally we will suggest how our new insights should logically lead to new experimental approaches.

II. Experimental Approach

In order to begin to address these questions, it was necessary to grow trees in elevated CO_2 for more than one growing season under conditions more closely resembling the forest environment than is possible in a growth chamber experiment with potted plants. Hence, an open–top chamber experiment was initiated in 1989. There were three primary objectives of this field experiment:

1. To determine whether the short-term responses of tree seedlings to elevated CO_2 are sustained over several growing seasons under field conditions.
2. To compare the responses to elevated CO_2 of *Liriodendron tulipifera* L. (yellow poplar or tulip tree, family Magnoliaceae) and *Quercus alba* L. (white oak, family Fagaceae). These species are important, cooccurring components of the deciduous forest of eastern North America that have many contrasting physiological, morphological, and ecological features. For example, yellow poplar grows faster than white oak, initiates new leaves throughout the growing season, is more nutrient demanding, and is less drought resistant, and its litter decomposes more rapidly.
3. To provide data and insights relevant for predicting forest ecosystem responses to elevated CO_2. Because of the experimental limitations discussed earlier, simulation models will be necessary to predict forest ecosystem responses. These modeling approaches will be much more useful and believable if they incorporate the best evidence about how forest processes, resource interactions, and biogeochemical feedback will be influenced by elevated CO_2.

Seeds of the two species were collected from single trees in Fall 1988, germinated, and planted in pots. The seedlings initially were raised in growth chambers containing CO_2 concentrations of 380, 500, or 650 μmol mol^{-1}. After a dormant period at 0–4°C, the seedlings were planted in soil within six open-top chambers in May 1989. The chambers were of standard design (Rogers *et al.*, 1983), 3 m in diameter and 2.4 m tall (later extended to 3.6 m tall). Ten seedlings of each species were planted in each chamber, and later the plots were thinned to five of each species. The faster growing yellow poplar plants were in the northern half of the chamber so as not to shade the white oak plants. Four chambers were provided with CO_2 continuously to maintain CO_2 concentrations of 150 and 300 μmol mol^{-1} higher than ambient (two replicate chambers per treatment, designated +150 and +300). In addition, there were two chambers with no added CO_2 (ambient CO_2 or +0). Carbon dioxide enrichment was not continued during the winter when the trees were leafless and dormant (mid-November to mid-April). No fertilization or supplemental irrigation was provided during the experiment. Nitrogen mineralization at this site is estimated to be about 100 kg ha^{-1} yr^{-1}. Foliar N concentrations of the yellow poplar trees (Table I) were lower than the range considered typical for the species (Leaf, 1973), suggesting possible N limitation. A period of low rainfall in spring of the second year was reflected in lower stomatal conductance (Gunderson *et al.*, 1993), but otherwise rainfall was plentiful, there were no morphological manifestations of drought stress, and growth was vigorous.

Measurements made during the course of the experiment included stem height and basal diameter, from which stem dry mass could be estimated,

Table I Nitrogen Content and Distribution in Yellow Poplar Saplings after Three Growing Seasons in Ambient or Elevated CO_2 Concentrations[a]

CO_2 enrichment (μmol mol^{-1})	Leaf	Stem	Woody root	Fine and small roots	Whole plant
		N Concentration (mg g^{-1})			
+0	15.2 ± 1.2	3.3 ± 0.3	4.8 ± 0.4	8.0 ± 1.7	6.7 ± 0.3
+150	11.5 ± 1.4	2.8 ± 0.2	3.9 ± 0.4	6.1 ± 1.3	5.3 ± 0.3
+300	12.5 ± 1.1	2.9 ± 0.2	3.4 ± 0.3	6.4 ± 2.1	5.2 ± 0.2
P	0.001	0.008	0.134	0.032	0.001
		N Amount (g plant^{-1})			
+0	8.1 ± 3.8	3.2 ± 1.4	3.2 ± 1.1	1.0 ± 0.5	15.5 ± 6.5
+150	7.1 ± 4.0	3.6 ± 2.2	2.9 ± 1.5	1.4 ± 0.5	15.0 ± 8.1
+300	6.8 ± 2.7	3.5 ± 1.5	2.7 ± 1.1	1.6 ± 0.9	14.6 ± 5.9
P	ns	ns	ns	0.096	ns
		N Distribution (%)			
+0	51.9 ± 2.7	20.4 ± 1.9	20.9 ± 2.9	6.7 ± 0.8	
+150	47.1 ± 5.3	22.8 ± 4.1	20.4 ± 4.6	9.7 ± 2.3	
+300	47.1 ± 4.9	23.3 ± 2.4	18.4 ± 2.5	11.2 ± 2.4	
P	0.010	0.016	0.128	0.001	

[a] Data are the means (±SD) of five plants per chamber in two replicate chambers per CO_2 concentration. Nitrogen concentrations of fine (<2 mm) and small (2–7 mm) roots were determined on samples collected from soil cores; these data were converted to grams of N per plant as described in Fig. 2.

light-saturated photosynthesis, and stomatal conductance. Leaves were occasionally sampled for constituent analysis. At the end of each growing season, leaves were collected as they abscised. These senescent leaves were measured for determination of final leaf area of the plants and for litter quality and decomposition experiments [O'Neill and Norby, 1996 (this volume)]. Yellow poplar saplings were harvested in August of the third growing season (1991); the slower growing oak saplings were not harvested until the end of the fourth growing season (November 1992). Leaves, branches, and bole were separately dried and weighed. The tap roots were excavated from the soil and weighed. Lateral root mass was estimated by regression analysis of the diameters of lateral roots at the point of attachment to the tap root. These regressions were established from 8 yellow poplar lateral roots ($R^2 = 0.94$) and 46 white oak lateral roots ($R^2 = 0.91$) that were completely excavated. Fine roots were estimated in 1991 and 1992 in five 10-cm-diameter soil cores taken from the half of the chamber where the trees were harvested. Fine roots of the other species were easily excluded on the basis of obvious differences in morphology and odor.

A second experiment was initiated in 1990 following the same experimental design, except with three replications of the CO_2 treatments (Wullschleger et al., 1992b). One-year-old, bare-rooted seedlings from a commercial tree nursery were planted in the chambers rather than plants grown from seed in regulated CO_2 concentrations as in the first experiment.

III. Results

A. Gas Exchange

Light-saturated photosynthesis of individual leaves of both species was higher in elevated CO_2 and remained higher throughout the experiment (Fig. 1; Gunderson et al., 1993). The relative effect of CO_2 was unchanged as leaves aged and were longer exposed to elevated CO_2. The relative effect of CO_2 also was similar throughout the canopy, and there was no effect of CO_2 on the timing of leaf senescence or abscission, which would alter the seasonal duration of photosynthesis. Analysis of the relationship between internal CO_2 concentration and assimilation rate (A–C_i analysis) indicated that there was no loss of photosynthetic capacity in the plants grown in elevated CO_2 (Gunderson et al., 1993). Photosynthesis increased with CO_2 enrichment despite lower concentrations of chlorophyll and (in yellow poplar) nitrogen (Wullschleger et al., 1992b; Gunderson et al., 1993).

Stomatal conductance, which is expected to be lower in elevated CO_2, did not respond consistently to CO_2 enrichment in this experiment (Gunderson et al., 1993). Conductance generally was lower in elevated CO_2, but the differences between treatments were rarely significant and were much

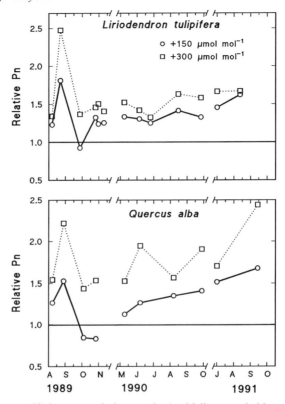

Figure 1 The rate of light-saturated photosynthesis of fully expanded leaves of plants growing in elevated CO_2 expressed relative to that in ambient CO_2. The mean ±SD values of photosynthesis in ambient CO_2 were 8.6 ± 1.9 μmol m^{-2} s^{-1} for *Liriodendron* and 7.6 ± 2.9 μmol m^{-2} s^{-1} for *Quercus*. The original data were the means of 3–5 leaves per chamber in two replicate chambers per CO_2 concentration (Gunderson *et al.*, 1993).

smaller than seasonal variations. Nevertheless, instantaneous water-use efficiency was always significantly higher in elevated CO_2.

The dark respiration rate of leaves was measured during the last growing season for each species (Wullschleger *et al.*, 1992a; Wullschleger and Norby, 1992) and in the repeat experiment (Wullschleger *et al.*, 1992b). Respiration rate declined with increasing CO_2 concentration in both species, and the reductions were attributed to reductions in both the maintenance and growth components of respiration. The lower respiration rate of CO_2-enriched yellow poplar was correlated with a reduction in foliar N concentration.

B. Metabolites

Carbohydrates were measured one time in October in the leaves of both species in the evening and the next morning (Wullschleger et al., 1992b). Starch accumulated in leaves during the day, and at the end of the day starch concentrations were significantly higher (by 22 mg g^{-1} in yellow poplar and by 15 mg g^{-1} in white oak) in leaves of CO_2-enriched trees. Starch concentrations declined during the night, but there was no difference with CO_2 treatment in the rate of nocturnal starch decomposition, and the concentrations were still significantly higher in leaves growing in elevated CO_2 in the morning. Sucrose concentrations showed the opposite trend, declining with increasing CO_2 concentration, and there was no difference in glucose or fructose concentrations. Thus, the total of the four major nonstructural carbohydrates changed much less in response to CO_2 enrichment than has been reported in other systems [e.g., Körner and Arnone, 1992; Körner et al., 1996 (this volume)], and they accounted for only a small part of the significant increase in leaf mass per unit area.

Preliminary measurements of metabolites in the roots of the yellow poplar and white oak saplings indicated no differences in carbohydrate concentration, but substantially lower concentrations of free amino acids in CO_2-enriched trees (T. J. Tschaplinski, Oak Ridge National Laboratory, personal communication).

All components of the yellow poplar saplings were analyzed for nutrient content, and a complete nitrogen budget is shown in Table I. Foliar N concentration was significantly lower in plants grown in elevated CO_2 than in plants grown in ambient CO_2. Although leaf mass per unit area was higher in elevated CO_2, N concentrations were lower whether expressed on a leaf area or leaf mass basis. The reduced concentration cannot be attributed to a decline in N uptake during the study since whole-plant N content was not significantly different between treatments. There were significant differences in the distribution of the N pool among tissues, primarily reflecting differences in biomass distribution. Relatively less of the total N pool was allocated to leaves and more was allocated to stems and fine roots, but all tissues had a lower N concentration in elevated CO_2. Phosphorus was the only other nutrient with a significantly lower concentration in leaves in elevated CO_2. There was no significant difference in the whole-plant content of any nutrient.

C. Dry Mass

The effect of CO_2 concentration on dry mass at final harvest was substantially different in yellow poplar and white oak (Fig. 2). There was no significant effect of CO_2 concentration on yellow poplar dry mass, despite the sustained increase in photosynthesis and reduced foliar respiration

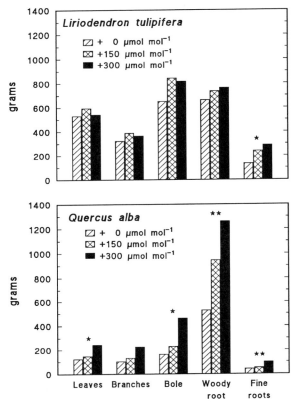

Figure 2 Dry mass of components of trees after nearly three growing seasons (*Liriodendron*) or four full growing seasons (*Quercus*) of exposure to ambient or elevated CO_2. Fine root mass was measured in five 81-cm^2 soil cores per chamber; the data on fine root density were converted to a per tree basis by assuming that 4.5 trees contributed to each core and that each tree had a 3-m-diameter rooting zone. Data on other plant components are the means of five trees in each of two replicate chambers. Coefficients of variation were about 38% (*Liriodendron*) and 64% (*Quercus*). The effect of CO_2 concentration was significant at $P < 0.05$ (*) or $P < 0.01$ (**). Data are from Norby *et al.* (1992, 1995).

(Norby *et al.*, 1992). The dry mass of the oak trees, on the other hand, was 59% higher in +150 than in ambient-grown trees and 134% higher in +300 ($P < 0.04$). The leaf area of yellow poplar was 10% lower in +300 (not significant) and that of white oak was 95% higher ($P < 0.05$) in +300, but the leaf area ratio (leaf area divided by whole-plant mass) was significantly lower in elevated CO_2 in both species. There was no effect of CO_2 concentration on any index of canopy structure (e.g., branch-to-bole mass ratio) in either species, but wood density increased by 7% in yellow poplar

($P < 0.02$). Fine root mass increased significantly in both species, but this component composed a relatively small percentage of the total tree mass.

D. Growth Dynamics

The large difference in CO_2 responses of the two species in dry matter accumulation must be explained. If different species respond to CO_2 enrichment so differently in the same soil and atmosphere, the prediction of forest responses to elevated CO_2 will be much more difficult. Differences between yellow poplar and white oak might be associated with inherent attributes of the species with respect to nutrient use or patterns of leaf initiation (Norby and O'Neill, 1991). Differences in carbon allocation patterns, such as the large difference between the species in root-to-shoot ratio (see Fig. 2), could alter sink strength for excess photosynthate. These possibilities cannot be evaluated appropriately on the basis of a static measure of dry mass at one point in time. Instead, the dynamics of growth should be examined.

The time course of aboveground growth of white oak was made possible by our ability to estimate stem mass from periodic nondestructive measures of stem height and diameter. The white oak plants were in an exponential growth phase throughout the experiment, with a seasonal growth pattern imposed on top of the exponential pattern (Fig. 3). The absolute effect of CO_2 enrichment increased through time, although the proportional response to CO_2 remained fairly constant. This trend is most apparent

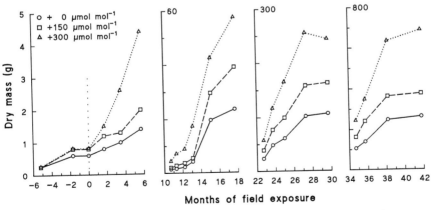

Figure 3 Stem dry mass of *Quercus alba* seedlings during four growing seasons in open-top chambers. Stem mass was estimated from periodic data on stem height and basal diameter by using regression equations. Seedlings were planted in the field at time 0; prior to that they were grown in pots in CO_2-controlled growth chambers.

when the annual stem mass data are plotted on a log scale (Fig. 4). The slopes of the lines on a log scale are proportional to the mean relative growth rates of the plants, and they are essentially equal across treatments except during the first growing season. Aboveground growth of yellow poplar (Fig. 4) was virtually identical in the three treatments throughout the experiment.

The initial growth enhancement of white oak seedlings by elevated CO_2 began in the growth chambers prior to planting in the field (Fig. 3). The dry mass difference between seedlings in +0 and +150 at the time of planting was only 200 mg, but this proportionate difference remained constant for the next 4 yr, implying that the initial response to CO_2 enrichment was not sustained. Seedlings in the +300 treatment continued to respond to CO_2 enrichment for the first several months in the field, but there was not a sustained response. This analysis suggests that, even if the elevated CO_2 had been removed after 2 months, the growth curves would have been the same as those in Figs. 3 and 4. Some verification of this inference is provided by an analysis of covariance. With the seedling mass after 2 months in the field used as a covariate, there were no significant differences in stem mass at the end of the experiment—a strikingly different conclusion from that stating a doubling of whole-plant mass in elevated CO_2.

A consideration of leaf area dynamics in the white oak plants in this experiment and in a developing forest stand is critical to understanding

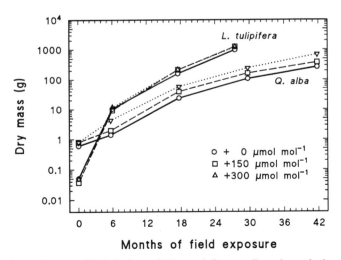

Figure 4 Stem dry mass of *Liriodendron tulipifera* and *Quercus alba* at the end of each growing season. The data are presented on a log scale such that the slopes of the lines are equal to the mean relative growth rate.

this system. The initial stimulation of growth in elevated CO_2 was associated with increased leaf area, and increased leaf area provides greater growth potential and subsequent leaf area production and so on (Blackman, 1919). Hence, the absolute difference between CO_2 treatments increases with time, even without a sustained CO_2 effect on growth rate. This compound interest effect cannot continue indefinitely, however. In a developing forest, leaf area eventually reaches a maximum determined primarily by the availability of light, water, nutrients, and other resources (Waring and Schlesinger, 1985). The only lasting effect of the early stimulation in growth by elevated CO_2 may be a shortening of the time required to reach canopy closure. The growth curves for white oak in Fig. 4 are shifted along the x-axis by less than 1 yr. Although an effect of CO_2 on seedling establishment could have important consequences for forest stand development (Eamus and Jarvis, 1989), this does not appear to be a robust conclusion of this experiment, which was not designed to investigate seedling establishment. There was no effect of CO_2 concentration on the dry mass of white oak saplings in the repeat experiment (unpublished data). In that experiment, 1-yr-old seedlings from a nursery were planted in the field, and these seedlings probably started with more stored carbon than the seedlings grown in an accelerated first-year growth cycle. Stored carbon may buffer the plant against the benefits of an increased supply of carbon from the atmosphere.

Whole-tree dry mass seemingly is the measure most relevant to questions about the effects of CO_2 concentration on carbon storage in forests, and many short-term studies with tree seedlings have drawn conclusions about forest tree growth on the basis of differences in biomass at the end of the experiment. However, the analysis of oak response illustrates why a static measure such as difference in tree mass is not a good predictor of long-term response. Differences in biomass established during exponential growth are unlikely to be sustained after internal and environmental constraints impose limits on growth. Furthermore, mass during exponential growth is highly dependent on prior leaf area development, which may be affected by factors other than CO_2, and is subject to small, perhaps uncontrollable differences in initial experimental conditions.

Projections of tree growth in a forest should be based on fundamental plant processes that are not dependent on leaf area or initial conditions. A useful, integrating measure is the growth efficiency index—the annual increment in biomass (usually stem mass) per unit leaf area (Waring and Schlesinger, 1985), which is similar to the concept of net assimilation rate used in agriculture. High values of this index represent a favorable balance in the allocation of photosynthate, whereas lower values might suggest a limitation in stored reserves. The growth efficiency index has been used to interpret the relative constraints of various environmental factors on tree growth (Waring and Schlesinger, 1985).

Growth efficiency of the yellow poplar and white oak saplings was determined by plotting stem mass increment (estimated from diameter and height measurements) vs leaf area at the end of each growing season for each individual plant (Fig. 5). Growth efficiency, which is the slope of the linear regressions, was similar for the two species: 106 g m^{-2} yr^{-1} for yellow poplar and 122 g m^{-2} yr^{-1} for white oak in ambient CO_2. The response to +300 μmol mol^{-1} CO_2 was also similar for the two species: increases of 35 and 37%, respectively. The only clear difference between yellow poplar and white oak was the relative response to the intermediate CO_2 concentration. Hence, despite the large difference between species in the response of biomass to elevated CO_2, at a more fundamental level the two species

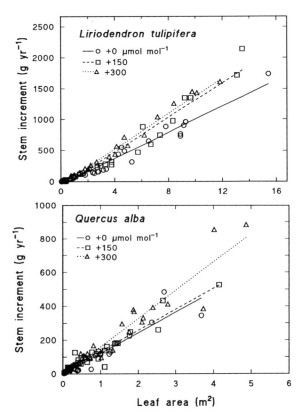

Figure 5 Annual stem mass increment, determined as in Fig. 3, vs leaf area at the end of the year for each of 10 plants per CO_2 concentration during three growing seasons. The slopes of the lines, which represent growth efficiency (Waring and Schlesinger, 1985), are significantly different at $P < 0.05$.

responded very similarly. It is also noteworthy that the 35–37% increase in growth efficiency closely corresponds to the average short-term response of biomass in many CO_2 enrichment experiments with annual plants (Kimball, 1983) and woody plants (Wullschleger et al., 1994). Unlike the response of biomass production, which could not continue to the same degree after leaf area reaches a maximum, there is no obvious reason to assume that the relative effect of CO_2 on growth efficiency would decline after canopy closure.

E. Carbon Cycling

Carbon sequestration by forests includes components besides tree growth, but the construction of a complete carbon budget is highly problematic. It is difficult to measure accurately all of the carbon pools and fluxes, even in a simplified system. Hence, we must resort to a qualitative analysis of the effects of CO_2 enrichment on carbon budget. The problem is made apparent by the discrepancy between the large increase in photosynthesis (coupled with a reduction in foliar respiration), the relatively smaller increase in growth efficiency, and the absence of a response in relative growth rate. Since growth efficiency is based only on aboveground production, "excess" carbon could reside in root mass. The root-to-stem mass ratio changes during plant development, so that a static measure of root-to-stem ratio based on the root mass at the end of the experiment (the only data available) is potentially misleading if there are CO_2 effects on ontogeny (Norby, 1994). Allometric analysis of root mass vs stem mass indicated similar dry matter allocation between stems and roots in yellow poplar in the different CO_2 concentrations, but in CO_2-enriched oak there was a small but significant increase in the relative dry matter allocation to woody roots (Norby, 1994).

Carbon cycling in the yellow poplar and white oak systems may have been influenced by changes in carbon allocation between leaves and roots. The reason that the relative growth rate did not increase in either species with CO_2 enrichment even though annual stem production per unit leaf area (growth efficiency) was higher is that there was a relative reduction in leaf area production or leaf area ratio (leaf area divided by whole-plant mass). In yellow poplar, the actual leaf area was slightly lower (not statistically significant) in elevated CO_2, similar to the response previously observed in a growth chamber experiment (Norby and O'Neill, 1991). In white oak, leaf area was much greater in elevated CO_2, but the leaf area ratio was still significantly lower. Hence, after the initial stimulation in growth in elevated CO_2, leaf area did not increase as fast relative to plant mass in elevated CO_2 as in ambient CO_2, and this trend offset the increase in growth efficiency. A general prediction of the process-based plant–soil simulation model G'DAY described by Comins and McMurtrie (1993) was that the optimal allocation of carbon to leaf production for maximal long-term stem production is less in elevated CO_2. Compensatory responses between process rates per unit leaf area and leaf area production may be a general reaction to CO_2 enrichment and one that must be understood when attempting to scale responses to a forest stand.

The consequence of a reduction in carbon allocation to leaves may be a compensatory increase in carbon allocation to fine roots. Fine root density (mass per unit ground area) at the end of the experiments was significantly greater in elevated CO_2. There are no data on fine root turnover in this system, however, from which to calculate fine root production. The relative increase in fine root density in yellow poplar greatly exceeded the relative increase in whole-plant mass (Norby et al., 1992). The relative increases in fine roots and whole-plant mass were of similar magnitude in white oak, but these estimates of fine root density do not include the carbon allocated to extramatrical hyphae of mycorrhizal roots, and the mycorrhizal density in white oak (but not yellow poplar) was significantly greater in elevated CO_2 (G. M. Berntson, Harvard University, personal communication). The increase in fine root density was associated with a higher CO_2 efflux from the soil, which was first suggested by a single measurement (Hanson et al., 1993) made in 1991 under the yellow poplar (Norby et al., 1992), and then was shown to be a consistent response in 1992 under the oak. Although the data set supports only qualitative conclusions, it appears that CO_2 enrichment of this experimental system increased the carbon flux through the system much more than it increased carbon storage per unit leaf area.

Fine roots represented a small percentage of the total dry mass of the trees in this experiment, so that their response to CO_2 enrichment had little direct impact on the total amount of carbon stored in the plant. Furthermore, the mean residence time of live fine roots is usually about 1 yr, although longer times are also reported (Joslin and Henderson, 1987). Hence, increased fine root production in elevated CO_2 is unlikely to be an important long-term mechanism for increasing carbon sequestration by plants. Nevertheless, this response could be a critical component of the integrated response of a forest ecosystem. Fine roots undoubtedly are physiologically important for water and nutrient uptake, yet it is difficult to make a direct connection between increased fine root production and increased resource acquisition. Indeed, the yellow poplar saplings in elevated CO_2 did not have increased nutrient content despite the greater fine root density. Fine roots are also the platform for interactions with soil microbes, including those effecting nutrient turnover, and a CO_2 effect on root production has been associated with increased N mineralization (Zak et al., 1993). Finally, fine roots can account for at least 25% of soil organic matter in hardwood forests (McClaugherty et al., 1984); increased fine root production in elevated CO_2 could be a route to long-term carbon sequestration by the ecosystem, even if not by the plant.

IV. Discussion

One of the primary objectives of this experiment was to compare the responses of *Liriodendron tulipifera* and *Quercus alba* and determine whether differences in response to elevated CO_2 could be attributed to differences in species characteristics. In the first analysis, there was a substantial difference between the species—there was no significant increase in the mass of the yellow poplar in response to CO_2 enrichment, whereas white oak saplings more than doubled in mass. A more detailed analysis with a process orientation, however, showed the species' responses to be essentially similar. In both species, growth efficiency increased by about 36% in ambient +300 μmol mol^{-1} CO_2, but this response was offset by a decline in leaf area ratio such that there was no effect of CO_2 concentration on the relative growth rate after the first several months in the field. The absence of an effect on relative growth rate is not to say that the trees did not respond to CO_2 enrichment. Photosynthesis was enhanced, and it remained enhanced throughout the course of the experiment. The evidence suggests that the additional carbon was cycled through the plant rather than remaining in biomass. These responses to elevated CO_2 occurred in soil without supplemental fertilization or irrigation, and they were similar in yellow poplar and in white oak despite the many differences between the species.

It is premature to evaluate whether the responses of these two species can be considered to be general reactions of deciduous trees to elevated CO_2. The other long-term study of a broadleaf tree in elevated CO_2 reported that sour orange trees more than doubled in growth with CO_2 enrichment (Idso and Kimball, 1993). It was concluded from that study that forest productivity would be similarly enhanced across the earth. However, the responses of the sour orange trees were remarkably similar to those of the oak trees described here, and a similar analysis of the longer term implications applies. That is, when the aboveground production data are plotted on a log scale, it is apparent that there was not a sustained growth response to elevated CO_2 and that the large difference in tree size between treatments was the result of an apparent stimulation in growth early in the study. Whether this initial difference was a specific result of CO_2 enrichment or a condition prior to the onset of CO_2 exposure cannot be determined from the published reports. Although the assumption that the 2.8-fold increase in productivity would be sustained might be applicable to an orchard managed with spacing to minimize canopy interactions, it cannot be extended to a forest with a closed canopy. However, in remarkable agreement with the responses of yellow poplar and white oak, the growth efficiency of the sour orange trees increased by 33% in elevated CO_2, on the basis of the increase in aboveground volume (including fruit rinds) during 1990 (Idso and Kimball, 1993) and the 101% difference in leaf area during this period (Idso et al., 1993).

Ultimately, an objective of research on forest responses to CO_2 will be to predict the response of net primary productivity and carbon sequestration of a forest ecosystem. While the number of longer term studies of tree responses under field conditions is increasing, we are still studying the responses of individual trees, not forest ecosystems. This experiment highlights some of the key uncertainties that are hindering our ability to extend these tree responses to the scale of a forest:

1. How do the responses to CO_2 concentration change as a stand of trees approaches canopy closure? No data address whether the leaf area index (LAI) of a closed stand will be different in a CO_2-enriched atmosphere, and from what is known about CO_2 effects on resource availability and utilization, multiple hypotheses can be put forth. Responses to CO_2 of individual trees that are driven by differences in leaf area, however, are not pertinent to a closed canopy with a constrained leaf area. Responses on a unit leaf area basis such as growth efficiency are more likely to be pertinent, but we do not know whether the response of growth efficiency to CO_2 concentration is independent of canopy development. The question is critical, since production per unit ground area is the product of growth efficiency and LAI (Waring and Schlesinger, 1985).

2. Similar to the concept of canopy closure, how do tree responses to CO_2 change when rooting space reaches capacity? Although increased fine root production in an expanding stand might be associated with increased capacity for nutrient or water uptake, this response may be unimportant if the soil is already fully exploited by roots [see also Johnson et al., 1996 (this volume)]. However, the larger scale question becomes focused on the relationships between CO_2 effects on fine root production, fine root activity, and longer term carbon sequestration in soil organic matter (Norby, 1994).

3. Does a shift in allocation from leaf to fine root production imply an increased capacity to withstand resource shortages or environmental stresses? At the physiological level, an increase in the ratio of fine roots to leaf area was associated with increased whole-plant water-use efficiency in yellow poplar seedlings (Norby and O'Neill, 1991). At the level of forest stand dynamics, a simulation with the stand competition model LINKAGES (Post et al., 1992) suggested that increased fine root productivity in elevated CO_2 could result in increased N availability over decade time scales. Over the long term, the response of a forest ecosystem to global change may be more dependent on indirect effects of CO_2 on the capacity of trees to

withstand infrequent environmental stresses than on the direct effects of CO_2 on growth.

An important product of open-top chamber studies on tree responses to CO_2 is the identification of key uncertainties to be addressed on a larger scale. Questions 1 and 2 will probably require CO_2 enrichment facilities that can accommodate much larger experimental systems, such as FACE experiments (Hendrey, 1992). This large-scale, expensive approach will be more efficient if the experiments are designed around specific, testable hypotheses generated by smaller scale experiments. For example, observations on the canopy development of a young forest or tree plantation exposed to elevated CO_2 concentrations beginning at the seedling stage could answer the question of whether CO_2 alters maximum LAI or decreases the time required to attain maximum LAI. The issue of indirect effects of CO_2 and stress interactions suggested in question 3, on the other hand, does not necessarily require large experiments. Alternative approaches to increasing the scale of the investigation include the addition of interacting environmental variables or the replication of experiments over larger geographical areas.

A new study on the interaction of CO_2 and temperature is an example of the increase in scope needed to address the more complex questions of resource interactions that will modify forest ecosystem response. Models of forest landscape dynamics that are based on simple phenomenological relationships between tree growth processes and local environmental feedback suggest that climatic warming will alter the structure, composition, and biomass of forests (Prentice *et al.*, 1993). A major uncertainty in such projections, however, is the potential for increasing atmospheric CO_2 concentrations to ameliorate or compensate for the stresses associated with climate change (Solomon and Cramer, 1993). A general hypothesis to be tested is that elevated CO_2 concentrations will alter the response of trees to temperature increases, and the magnitude of the interaction will vary with species, genotypic adaptations, and geographical variations in environmental conditions. Plants in warm climates usually have a temperature optimum similar to the prevailing temperature, while plants in cooler climates respond positively to increased temperature (Brock, 1970). For example, growth of *Acer saccharum* (sugar maple) is more likely to be adversely affected by temperature increases (and associated changes in water status) in Tennessee, near the southern limit of the species' North American range, than in Wisconsin, near the northern limit. The geographic limit of a species in the equatorial direction is probably related to a decline in its competitive ability rather than to any direct lethal effect of high temperature (Woodward, 1987). The range of *Fagus grandifolia* (American beech) partially overlaps that of sugar maple, but since the northern and southern range limits of the two species do not coincide, differential responses to temperature and a changing competitive balance might be expected near the margins of their ranges.

Initial experiments on the CO_2 × temperature interaction have been conducted in Oak Ridge, TN, with sugar maple seedlings in new open-top chambers with temperature control. The physiological responses of the sugar maples in an unreplicated set of four of these chambers, comprising ambient and elevated temperatures combined with ambient and elevated CO_2 concentrations, were measured during the first year of exposure. Photosynthesis initially was higher in elevated CO_2 (Fig. 6), but the response was not sustained. Several plants, however, had a second flush of growth that was much more responsive to CO_2 enrichment. In marked contrast to the response of yellow poplar and white oak, stomatal conductance was

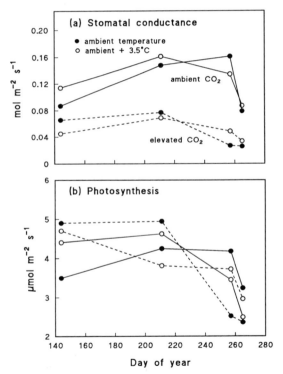

Figure 6 Stomatal conductance and light-saturated photosynthesis of first-flush leaves of *Acer saccharum* seedlings in temperature-controlled open-top chambers: solid symbols, chambers maintained at ambient temperature; open symbols, chambers maintained at ambient + 3.5°C; solid lines, ambient CO_2; solid lines, ambient CO_2 dashed lines, chambers at ambient + 300 μmol mol^{-1} CO_2. Data are means of five plants per chamber; chambers were unreplicated.

substantially lower in elevated CO_2 (Fig. 6). There was no effect of growth temperature on these responses. The dry mass of sugar maple stems at elevated temperatures was less than half that of plants growing in ambient temperature; root mass, however, was higher at elevated temperatures. Both root and stem mass increased with CO_2 enrichment. Consistent with the working hypotheses, growth of the American beech was greater at elevated temperatures but did not respond to CO_2 enrichment. These preliminary results suggest a substantial difference from the previously measured responses of yellow poplar and white oak at this site, and they are presented as a cautionary note. It is important to measure the CO_2 responses of more tree species under realistic field conditions before accepting too many generalizations about tree and ecosystem responses.

V. Conclusions

Experimental approaches to the prediction of forest ecosystem responses to global changes must recognize the constraints imposed by the basic characteristics of forests and trees. Experiments must be designed to study critical components of forest ecosystems, even if entire intact ecosystems cannot be manipulated directly. There also must be great care in choosing the appropriate measures of response from which to infer the responses on a larger scale. Some common measures such as dry mass or leaf area of individual plants at a single point in time can be highly misleading. Measures of response with more of a process orientation, such as productivity per unit leaf area, are much more informative. The CO_2 enrichment experiments with tree species that are currently being conducted are identifying some key uncertainties about forest response, including questions pertaining to canopy development, belowground processes, and interactions with other environmental variables and stresses. Answers to these questions will require increases in the scope of the experimental approach, both through the exposure of large systems and networks of open-top chamber experiments for systematic inclusion of interacting variables.

Acknowledgments

This research was sponsored by the Global Change Research Program of the Environmental Sciences Division, U. S. Department of Energy, and the Laboratory Directed Research and Development Program of the Oak Ridge National Laboratory, managed for the U.S. Department of Energy by Lockheed Martin Energy Systems, Inc. under Contract No. DE-AC05-84OR21400. This is Publication No. 4292, Environmental Sciences Division, Oak Ridge National Laboratory, Oak Ridge, TN.

References

Bazzaz, F. A., and Miao, S. L. (1993). Successional status, seed size, and responses of tree seedlings to CO_2, light, and nutrients. *Ecology* 74, 104–112.

Blackman, V. H. (1919). The compound interest law and plant growth. *Ann. Bot.* 33, 353–360.
Brock, T. D. (1970). High temperature systems. *Annu. Rev. Ecol. Syst.* 1, 191–220.
Comins, H. N., and McMurtrie, R. E. (1993). Long-term response of nutrient-limited forests to CO_2 enrichment: equilibrium behavior of plant-soil models. *Ecol. Appl.* 3, 666–681.
Drake, B. G. (1992). The impact of rising CO_2 on ecosystem production. *Water Air Soil Pollut.* 64, 5–44.
Eamus, D., and Jarvis, P. G. (1989). The direct effects of increase in the global atmospheric CO_2 concentration on natural and commercial temperate trees and forests. *Adv. Ecol. Res.* 19, 1–55.
Gunderson, C. A., Norby, R. J., and Wullschleger, S. D. (1993). Foliar gas exchange responses of two deciduous hardwoods during 3 years of growth in elevated CO_2: No loss of photosynthetic enhancement. *Plant Cell Environ.* 16, 797–807.
Hanson, P. J., Wullschleger, S. D., Bohlman S. A., and Todd, D. E. (1993). Seasonal and topographic patterns of forest floor efflux from an upland oak forest. *Tree Physiol.* 13, 1–15.
Hendrey, G. R. (1992). Global greenhouse studies: need for a new approach to ecosystem manipulation. *Crit. Rev. Plant Sci.* 11, 61–74.
Idso, S. B., and Kimball, B. A. (1993). Tree growth in carbon dioxide enriched air and its implications for global carbon cycling and maximum levels of atmospheric CO_2. *Global Biogeochem. Cycles* 7, 537–555.
Idso, S. B., Wall, G. W., and Kimball, B. A. (1993). Interactive effects of atmospheric CO_2 enrichment and light intensity reductions on net photosynthesis of sour orange tree leaves. *Environ. Exp. Bot.* 33, 367–375.
Johnson, D. W., Henderson, P. H., Ball, J. T., and Walker, R. F. (1996). Effects of CO_2 and N on growth and N dynamics in ponderosa pine: Results from the first two growing seasons. *In* "Carbon Dioxide and Terrestrial Ecosystems" (G. W. Koch and H. A. Mooney, eds.). Academic Press, San Diego.
Joslin, J. D., and Henderson, G. S. (1987). Organic matter and nutrients associated with fine root turnover in a white oak stand. *Forest. Sci.* 33, 330–346.
Kimball, B. A. (1983). Carbon dioxide and agricultural yield: an assemblage and analysis of 330 prior observations. *Agron. J.* 75, 779–788.
Körner, C., and Arnone, J. A., III. (1992). Responses to elevated carbon dioxide in artificial tropical ecosystems. *Science* 257, 1672–1675.
Körner, C., Dimer, M., Shäppi, B., and Zimmermann, L. (1996). Responses of alpine vegetation to elevated CO_2. *In* "Carbon Dioxide and Terrestrial Ecosystems" (G. W. Koch and H. A. Mooney, eds.). Academic Press, San Diego.
Leaf, A. L. (1973). Plant analysis as an aid in fertilizing forests. *In* "Soil Testing and Plant Analysis" (L. M. Walsh and J. D. Beaton, eds.), pp. 427–454. Soil Science Society of America, Madison, WI.
McClaugherty C. A., Aber, J. D., and Melillo, J. M. (1984). Decomposition dynamics of fine roots in forested ecosystems. *Oikos* 42, 378–386.
Norby, R. J. (1994). Issues and perspective for investigating root responses to elevated atmosphere carbon dioxide. *Plant Soil* 165, 9–20.
Norby, R. J., and O'Neill, E. G. (1991). Leaf area compensation and nutrient interactions in CO_2-enriched yellow poplar (*Liriodendron tulipifera* L.) seedlings. *New Phytol.* 117, 515–528.
Norby, R. J., O'Neill, E. G., and Luxmoore, R. J. (1986). Effects of atmospheric CO_2 enrichment on the growth and mineral nutrition of *Quercus alba* seedlings in nutrient-poor soil. *Plant Physiol.* 82, 83–89.
Norby, R. J., Gunderson, C. A., Wullschleger, S. D., O'Neill, E. G., and McCracken, M. K. (1992). Productivity and compensatory responses of yellow-poplar trees in elevated CO_2. *Nature* 357, 322–324.
Norby, R. J., O'Neill, E. G., and Wullschleger, S. D. (1995). Belowground responses to atmospheric carbon dioxide in forests. *In* "Carbon Forms and Functions in Forest Soils" (W. W. McFee and J. M. Kelly, eds.), pp. 397–418. Soil Science Society of America, Madison, WI.

Norby, R. J., Wullschleger, S. D., Gunderson, C. A., and Nietch, C. T. (1995). Increased growth efficiency of *Quercus alba* trees in a CO_2-enriched atmosphere. *New Phytol.* 131, (in press).

Oechel, W. C., Riechers, G., Lawrence, W. T., Prudhome, T. T., Grulke, N., and Hastings, S. J. (1991). Long-term *in situ* manipulation and measurement of CO_2 and temperature. *Funct. Ecol.* 6, 86–100.

O'Neill, E. G., and Norby, R. J. (1996). Litter quality and decomposition rates of foliar litter produced under CO_2 enrichment. *In* "Carbon Dioxide and Terrestrial Ecosystems" (G. W. Koch and H. A. Mooney, eds.). Academic Press, San Diego.

Owensby, C. E., Coyne, P. I., Ham, J. M., Auen, L. M., and Knapp, A. K. (1993). Biomass production in a tallgrass prairie ecosystem exposed to ambient and elevated carbon dioxide. *Ecol. Appl.* 3, 644–653.

Post, W. M., Pastor, J., King, A. W., and Emanuel, W. R. (1992). Aspects of the interaction between vegetation and soil under global change. *Water Air Soil Pollut.* 64, 345–363.

Prentice, I. C., Sykes, M. T., and Cramer, W. (1993). A simulation model for the transient effects of climate change on forest landscapes. *Ecol. Model.* 65, 51–70.

Rogers, H. H., Heck, W. W, and Heagle, A. S. (1983). A field technique for the study of plant responses to elevated carbon dioxide concentrations. *J. Air Pollut. Control Assoc.* 33, 42–44.

Solomon, A. M., and Cramer, W. (1993). Biospheric implications of global environmental change. *In* "Vegetation Dynamics & Global Change" (A. M. Solomon and H. H. Shugart, eds.), pp. 25–52. Chapman & Hall, New York.

Waring, R. H., and Schlesinger, W. H. (1985). "Forest Ecosystems. Concepts and Management." Academic Press, Orlando.

Woodward, F. I. (1987). "Climate & Plant Distribution." University Press, Cambridge, UK.

Wullschleger, S. D., and Norby, R. J. (1992). Respiratory cost of leaf growth and maintenance in white oak saplings exposed to atmospheric CO_2 enrichment. *Can. J. Forest Res.* 22, 1717–1721.

Wullschleger, S. D., Norby, R. J., and Gunderson, C. A. (1992a). Growth and maintenance respiration in leaves of *Liriodendron tulipifera* L. saplings exposed to long-term carbon dioxide enrichment in the field. *New Phytol.* 121, 515–523.

Wullschleger, S. D., Norby, R. J., and Hendrix, D. L. (1992b). Carbon exchange rates, chlorophyll content, and carbohydrate status of two forest tree species exposed to carbon dioxide enrichment. *Tree Physiol.* 10, 21–31.

Wullschleger, S. D., Post, W. M., and King, A. W. (1995). On the potential for a CO_2 fertilization effect in forest trees—an assessment of 58 controlled-exposure studies and estimates of the biotic growth factor. *In* "Biotic Feedbacks in the Global Climatic System: Will the Warming Feed the Warming?" (G. M. Woodwell and F. T. Mackenzie, eds.), pp. 85–107. Oxford University Press, New York.

Zak, D. R., Pregitzer, K. S., Curtis, P. S., Teeri, J. A., Fogel, R., and Randlett, D. L. (1993). Elevated atmospheric CO_2 and feedback between carbon and nitrogen cycles in forested ecosystems. *Plant Soil* 151, 105–117.

2

Effects of CO_2 and N on Growth and N Dynamics in Ponderosa Pine: Results from the First Two Growing Seasons

Dale W. Johnson, P. H. Henderson, J. Timothy Ball, and R. F. Walker

I. Introduction

Forest ecosystems throughout the world have experienced and will probably continue to experience a significant increase in atmospheric carbon dioxide (CO_2) concentrations (Strain and Thomas, 1992). The potential growth response to these increases is highly dependent upon the interactions between CO_2 and nitrogen (N), the most commonly limiting nutrient to forest growth in the Northern Hemisphere (Gessel et al., 1973; Aber et al., 1989; Johnson, 1992). In polluted regions, increases in N deposition likely have caused increased forest growth in some cases (e.g., Kauppi et al., 1992), whereas excessive N deposition has apparently resulted in the appearance of other nutrient deficiencies in other cases (Schulze et al., 1989). Both atmospheric CO_2 concentrations and N deposition rates are projected to increase over the next few decades, and thus it is important to understand the interactions between CO_2 and N in order to make reasonable forecasts of forest response to these nutrients and of the potential for carbon sequestration as a result of such responses.

There are several ways in which CO_2 and N can interact so as to allow growth response to CO_2 even under N limitation. Many seedling studies have reported reduced tissue N concentrations under elevated CO_2 regimes, facilitating growth increases under suboptimal N conditions [Brown, 1991; Campagna and Margolis, 1989; Norby et al., 1986a,b, 1996 (this volume); Samuelson and Seiler, 1993]. One explanation for the often-observed decreases in foliar N concentration with increased CO_2 is that plants may produce lower concentrations of the enzymes of the photosynthetic carbon reduction (PCR) cycle, particularly the carboxylating enzyme ribulose-1,5-bisphosphate-carboxylase/oxygenase (rubisco). If this response occurs under field conditions, it implies that the N deficiencies common to many forest ecosystems will not preclude a growth increase in response to CO_2.

Another factor that may allow N-deficient forests to respond to CO_2 is increased uptake. Elevated CO_2 might facilitate increased uptake by stimulating root growth. Several studies have shown that increased CO_2 causes greater carbohydrate allocation to roots and mycorrhizae, causing disproportionate growth increases in root growth and thereby facilitating greater soil exploration (Norby *et al.*, 1986a,b; van Veen *et al.*, 1991; Rogers *et al.*, 1992; Walker *et al.*, 1994). In addition, studies have suggested that stimulation of rhizosphere activity by elevated CO_2 can cause increased N availability. Körner and Arnone (1992) reported increased soil respiration and soil solution NO_3^- concentrations with increased CO_2 in artificial tropical ecosystems. Zak et al. (1993) showed that elevated CO_2 caused increases in labile C and N in rhizosphere soil from *Populus grandidentata* seedlings. The authors posed a conceptual model whereby elevated CO_2 creates positive feedback on soil C and N dynamics and tree growth because of increased carbohydrate allocation and, consequently, increased N availability in the rhizosphere. Curtis *et al.* (1996, volume) report data from studies of *P. grandidentata* and *Populus euramericana* supporting this model. On the other hand, Diaz *et al.* (1993) reported that increased root exudation from native herbaceous plants caused microbial immobilization of nutrients, potentially causing nutrient feedback in the opposite direction of that posed by Zak *et al.* (1993). Indeed, the addition of labile organic C with a low C/N ratio is known to cause the immobilization of available N (Paul and Clark, 1989; Johnson, 1992).

In this paper, we report on CO_2–N interactions during the first two growing seasons in ponderosa pine seedlings treated with three levels of N and three levels of CO_2 in open-top chambers. On the basis of the literature reviewed earlier, we hypothesized that CO_2 would result in a growth increase with or without N addition due to a combination of greater biomass production per unit N uptake (lower tissue N concentrations) and greater N uptake resulting from the soil due to greater root biomass and N mineralization.

II. Site and Methods

The open-top chamber site was located at the Institute of Forest Genetics in Placerville, CA. The soil is Aiken clay loam, a xeric haplohumult derived from andesite. Soils were sampled intensively prior to chamber establishment and found to be very uniform. Some average chemical and physical properties of the soils from the site are shown in Table I.

During February–April 1991, 24 hexagonal open-top chambers (3.6 m in diameter) were established on the site. The basic experimental design consisted of three levels of nitrogen (0, 10, and 20 g m^{-2} yr^{-1} of N as ammonium sulfate, applied in early spring) and four CO_2 treatments (ambient, no chamber; ambient, chambered; 525 μl l^{-1} CO_2; and 700 μl l^{-1} CO_2). Water was delivered to each plot via a timed stand pipe to a looped, 1-in.-diameter manifold and low pressure spray heads. Each of the chambered treatments was replicated three times, and each of the unchambered treatments was replicated twice. Only the results from the chambered measurements will be reported here. Due to cost limitations, the 10 g m^{-2} yr^{-1} N, 525 μl l^{-1} treatment was excluded. Treatments were begun in May 1991.

In May of 1991, ponderosa pine (*Pinus ponderosa*) was planted in each chamber. Seedlings were grown from seed (21 planting locations per chamber) and seedlings (21 per chamber), the latter being a backup in the event of excessive mortality. Seed-grown seedling survival was very good, and the seedling-grown stock was removed in October 1991.

In October 1991, three trees from each chamber were harvested, including complete root systems. In October 1992, three trees from each chamber were harvested again, but only one complete root system per chamber was obtained because of the increased size of the seedlings and concern for

Table I Some Chemical and Physical Properties of the Placerville Site Soils (Aiken Clay Loam, Xeric Haplohumult Derived from Andesite)

Horizon (depth, cm)	Db (g cm^{-3})	% > 2 mm	C[a] (mg g^{-1})	N[b] (mg g^{-1})	C/N	Bray P^2 (mg kg^{-1})
Ap (0–18)	1.14	1	22.0	0.9	24	12.1
Bw (18–30)	1.24	1	18.0	0.9	21	10.9
Bt (30+)			7.1	0.4	16	1.6

| Horizon (depth, cm) | pH[c] | CEC[d] (cmol$_c$ kg^{-1}) | Exchangeable cations (cmol$_c$ kg^{-1}) | | | | | % BS[e] |
			Ca	Mg	K	Na	Al	
Ap (0–18)	5.1	11.24	4.37	0.62	0.74	0.04	0.68	51
Bw (18–30)	5.1	9.39	4.26	0.62	0.74	0.03	0.78	65
Bt (30+)	5.5	14.89	6.11	1.18	0.90	0.04	0.02	57

[a] Perkin-Elmer 2400 CHN analyzer.
[b] 0.5 M HCl + 1 M NH$_4$F (Olsen and Sommers, 1982).
[c] 0.1 M CaCl$_2$.
[d] Cation exchange capacity and exchangeable cations by 1 M NH$_4$Cl extraction followed by 1 M KCl.
[e] Percent base saturation.

excessive plot disturbance. Root biomass by size class and mycorrhizal infection were analyzed in each case and will be reported later (R. F. Walker, unpublished data). Only total root biomass will be reported here. Seedlings were dried, weighed by major component (foliage, branch, stem, roots), and analyzed for N on a Perkin-Elmer 2400 CHN analyzer.

In May of 1993, ceramic cup lysimeters were installed in the unfertilized chambers (two per chamber for a total of six replicates) at an 18-cm depth (just beneath the Ap horizon), and soil solutions were collected on a monthly basis for NH_4^+ and NO_3^- analyses (automated colorimetric analysis at the Water Analysis Laboratory at DRI). Twice per year, the full suite of cations and anions were also analyzed in soil solution (Ca^{2+}, Mg^{2+}, K^+, and Na^+ by atomic absorption; Cl^- and SO_4^{2-}, and orthophosphate by a Dionex ion chromatograph; pH; and HCO_3^- by titration to pH 5.0).

In the October 1993 harvest, rhizosphere soil was collected and analyzed for microbial populations, microbial biomass and N content, and total N (P. Henderson, unpublished data). Results reported here are NH_4^+ and NO_3^- production during a 10-day incubation (25°C, 25% moisture content) of Ap horizons. At the end of the incubation, samples were extracted with $2\ M$ KCl and the extracts were analyzed for NH_4^+ and NO_3^- as described earlier.

Statistical analyses consisted of ANOVA for fractional factorial designs using SYSTAT software (Wilkinson *et al.*, 1992). The standard error of each mean is included in the presentation of results that follows to provide an indication of the variation among sample means.

III. Results and Discussion

Both CO_2 and N produced statistically significant, positive growth responses at 6 and 18 months (Fig. 1). The N treatment effect was most pronounced at 6 months, whereas the CO_2 effect was more pronounced at 18 months. As noted in the previous controlled environment study (Johnson *et al.*, 1994a), there was a tendency for the 525 $\mu l\ l^{-1}$ CO_2 treated seedlings to have greater biomass than the 700 $\mu l\ l^{-1}$ seedlings; however, this effect was transient and disappeared at 30 months according to all indications from monthly diameter and height measurements (J. T. Ball, unpublished data). Further aspects of this transient growth response will be discussed in later papers.

Elevated CO_2 caused statistically significant reductions in N concentration in all tissues at 6 months (one growing season) but only in needles and fine roots at 18 months (two growing seasons) (Figs. 2 and 3). There was no overall decline in foliar N concentrations between 6 and 18 months, but the CO_2 effect attenuated somewhat because of increases in the

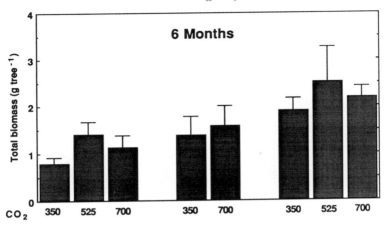

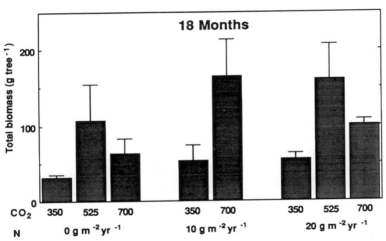

Figure 1 Seedling biomass at 6 and 18 months in the Placerville study. For CO_2 treatments, concentrations are given in microliters per liter (standard errors are given).

700 μl l^{-1} CO_2 treatment (Fig. 2). In a previous controlled environment study with ponderosa pine growing on N-poor soil, the foliar N concentration declined sharply with time overall and CO_2 effects were transient, as was the case here (Johnson *et al.*, 1994a). Also similar to the previous controlled environment study was the fact that the initial (6 month) reduction in tissue N concentration with elevated CO_2 occurred with or without N fertilization and, thus, cannot be interpreted as an attempt by the seed-

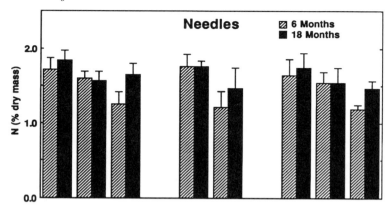

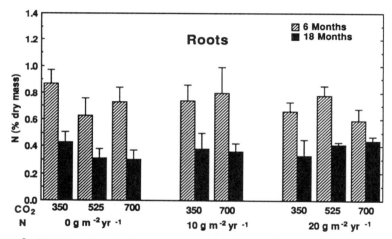

Figure 2 Needle and root nitrogen concentrations in ponderosa pine seedlings at 6 and 18 months in the Placerville study. Root concentrations are weighted average values for all size fractions. For CO_2 treatments, concentrations are given in microliters per liter (total root N content divided by total root biomass × 100) (standard errors are given).

lings to optimize scarce N supplies. In contrast to the previous controlled environment study, however, foliar N concentrations in the Placerville study were quite high, even in the unfertilized treatment, and there was no significant N fertilization effect on foliar N concentration. As shown here, however, there was a significant growth response to N fertilization.

Both stem and weighted average root N concentrations declined substantially between 6 and 18 months, as woody tissues began to accumulate in both of these components (Fig. 3). The effects of CO_2 and N treatments on stems and root N concentration were sporadic, however. At 6 months, there were significant N and CO_2 treatment effects on stem N concentration (the CO_2 effect being evident only in the fertilized treatments), with the expected pattern of increase with N and decrease with CO_2 (Figs. 2 and 3). By 18 months, however, the N effect had disappeared and the CO_2 effect had reversed: there was a significant increase in stem N concentration

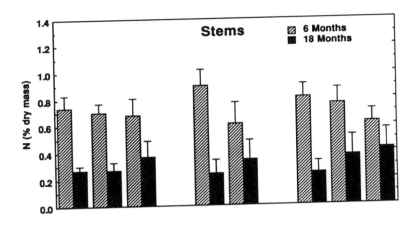

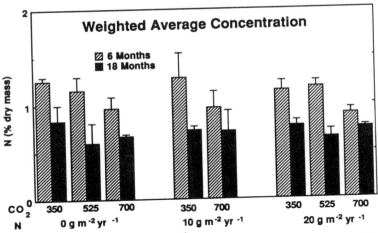

Figure 3 Stem and seedling weighted average N concentrations at 6 and 18 months in the Placerville study. For CO_2 treatments concentrations are given in microliters per liter (seedling N content divided by biomass $\times$ 100) (standard errors are given).

with CO_2 at all N fertilization levels. The reasons for this are not known, but they may include a CO_2-induced change in N translocation and storage with the plant.

The effects of CO_2 on weighted average root N concentration (total root N content divided by root biomass × 100; not shown) were inconsistent, occurring mainly in the unfertilized treatment at 6 months and largely disappearing at 18 months. There was a decline in weighted average root N concentration between 6 and 18 months due to the increase in woody root biomass. The pattern of reduced root N concentration with increasing CO_2 was maintained in the fine roots at 18 months (Fig. 2), whereas the woody (tap) root concentrations were not affected by CO_2 (not shown). Because most root biomass at 18 months was in larger, more woody roots, CO_2 treatment effects on weighted average root N concentration generally were not significant.

Weighted average seedling N concentration (seedling N content × 100 divided by biomass) was significantly lower with elevated CO_2 at 6 months, but not at 18 months (Fig. 3). The disappearance of CO_2 effects on weighted average N concentration at 18 months was due to two factors: (1) increasing importance of woody tissues with low and unresponsive N concentrations and (2) generally smaller effects of CO_2 on foliar N concentrations. In any event, the observed increases in growth with elevated CO_2 were accompanied by a more economical use of N only in the initial 6 months of growth. This is easily seen from seedling N content, which was significantly greater with elevated CO_2 at all three levels of N (Fig. 4). Thus, elevated CO_2 facilitated the acquisition and uptake of additional N from the soil, even without fertilization.

The increased N uptake with elevated CO_2 was due to the greater root biomass and soil exploration rather than increases in N mineralization. N mineralization rates in the rhizosphere soils were quite low and unaffected by either CO_2 or N treatment in 1993 (Table II). Soil solution NH_4^+ and NO_3^- concentrations in the unfertilized chambers were very low and unaffected by CO_2. (Soil solutions were not sampled in the fertilized chambers). The only statistically significant effects of CO_2 on soil solution were in the cases of pH and HCO_3^- (Table III). As noted in a previous paper (Johnson et al., 1994b), the elevated CO_2 treatments (especially the 525 $\mu l\, l^{-1}$ treatment) caused large, significant increases in soil atmospheric CO_2 concentrations (soil pCO_2), which in turn caused increases in the rate of carbonic acid leaching (the latter of which is manifested as increased soil solution HCO_3^- after dissociation of H^+ and exchange for base cations on the cation exchange complex).

The role of root growth in the acquisition of additional N is strongly suggested by the root/shoot ratio data. At 6 months, there was no significant CO_2 or N effect on the root/shoot ratio (Fig. 5). At 18 months, however,

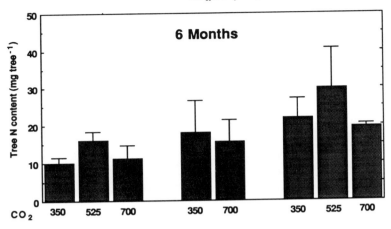

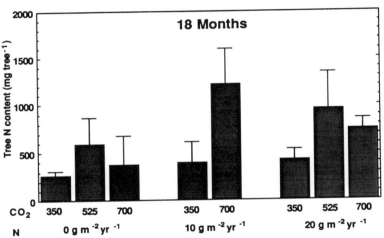

Figure 4 Seedling N content at 6 and 18 months in the Placerville study. For CO_2 treatments, concentrations are given in microliters per liter (standard errors are given).

there was a large positive CO_2 effect on the root/shoot ratio in the unfertilized treatment, as would be expected in response to increased need for N. Indeed, the patterns of the root/shoot ratio are very similar to those of total seedling biomass in the unfertilized treatment. Root/shoot ratios were significantly lower and the CO_2 effects were less in fertilized than in unfertilized treatments, as would be expected when N supplies are more abundant.

Table II Ammonium and Nitrate Nitrogen in Rhizosphere Soils after 7 Days of Incubation[a]

Nitrogen	CO_2 treatment (μg N g soil^{-1})		
	Ambient	525 μl l^{-1}	700 μl l^{-1}
Unfertilized			
NH_4^+–N	2.69 + 0.76	2.24 + 0.26	2.32 + 0.31
NO_3^-–N	2.22 + 0.20	1.79 + 0.36	0.60 + 0.54
10 g N m^{-2} yr^{-1}			
NH_4^+–N	2.23 + 0.81		2.84 + 0.55
NO_3^-–N	2.98 + 1.23		0.93 + 0.58
20 g N m^{-2} yr^{-1}			
NH_4^+–N	2.61 + 0.35	3.44 + 1.07	3.29 + 0.74
NO_3^-–N	2.72 + 1.26	3.13 + 2.65	2.06 + 0.84

[a] P. Henderson, unpublished data; sampled in October 1993.

IV. Perspective: Applicability of Results to Mature Forest Ecosystems

As noted in detail in a companion paper [Johnson and Ball, 1996 (this volume)], there are many differences in the nature of nutrient cycling between seedlings and mature, closed-canopy forests. The results of this study, while preliminary, do have some significant, process-level implications for mature forest ecosystems, even though N cycling processes in the seedling study are immature and incomplete relative to mature forests. First, the fact that we found no stimulation of N mineralization by elevated CO_2 is highly significant if it holds in mature forests, because it implies that the only possibility for increased N uptake is through greater root growth and soil exploration. Whether this can occur in mature forests, where "root closure" may have occurred, is an open question. Vogt *et al.*

Table III Soil Solution Concentrations (μmol_c l^{-1}, 20-cm Depth) at the Placerville Site (Unfertilized Chambers Only)

	CO_2 level		
	350 ppm	525 ppm	700 ppm
pH	6.7 ± 0.4	6.7 ± 0.4	6.3 ± 0.1
NO_3^-	2 ± 1	<1	1 ± 1
NH_4^+	4 ± 5	<1	13 ± 15
HCO_3^-	162 ± 38	200 ± 38	234 ± 21

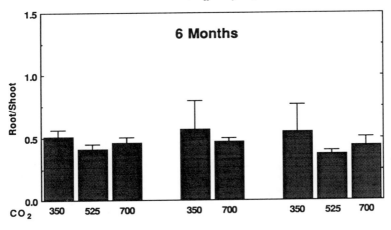

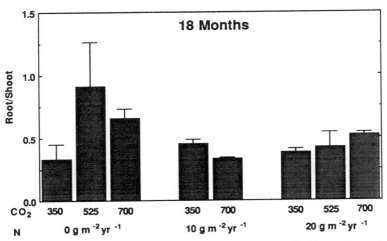

Figure 5 Root/shoot ratios in seedlings at 6 and 18 months in the Placerville study. For CO_2 treatments, concentrations are given in microliters per liter (standard errors are given).

(1983) found that fine root biomass in Douglas fir stands peaked at about the same time as folial biomass, that is, at the time of canopy closure. After that, fine root biomass in high site quality stands decreased (perhaps as nutrient demands declined), whereas root biomass in low site quality stands remained approximately stable. Thus, there is evidence for "root closure" at about the time of canopy closure, but there is also evidence that root biomass is elevated in low nutrient and xeric sites, suggesting a response to nutrient or water limitations [see also Keyes and Grier (1981); Santantonio and Hermann, 1985; Comerau and Kimmins, 1989].

Second, the changes (if any) in woody tissue N concentration with elevated CO_2 can be of considerable significance in mature forests. Wood contained only 30–40% of total tree C and 10–20% of total tree N in the Placerville seedling studies (Fig. 6). However, wood contains the greatest biomass and tree N in most mature forests, despite its relatively low N

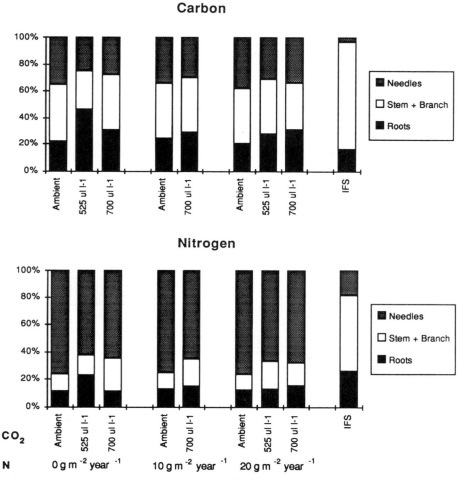

Figure 6 Distribution of biomass (top) and N (bottom) in Placerville seedlings at 18 months and average values for the Integrated Forest Study (IFS, right-hand bar; Johnson and Lindberg, 1991).

concentration. This is illustrated in the far right-hand bar of Fig. 6, which shows the average organic matter distribution in trees for the Integrated Forest Study sites (IFS; Johnson and Lindberg, 1991). Foliage contains a disproportionate amount of tree N ($\approx$10–40%) compared to biomass ($\approx$1–10%), but wood contains the greatest proportion of tree N ($\approx$40–70%) because it constitutes a large proportion of total biomass ($\approx$70–90%). As noted in the companion paper [Johnson and Ball, 1996 (this volume)], foliage biomass (and, presumably, fine root biomass) reaches an approximate steady state after crown closure, and only woody plant biomass continues to accumulate. The N increment in this woody tissue must originate from either the soil or the atmosphere, whereas the annual uptake of N for building new foliage and fine roots can be obtained from decomposition of litterfall, root turnover, and internal translocation. The effects of CO_2 on the N concentration and, consequently, the N increment in woody tissues obviously are of considerable importance and merit much more study.

Third, these results, in combination with those of the first two controlled environment studies, provide a spectrum of ponderosa pine growth responses to both N and CO_2 that relates to the spectrum of N responses under field conditions. In the first controlled environment study, we found no growth response to CO_2 in the lowest (unfertilized) N treatment over a 13-month period. The contrast between these results and those obtained over the first 18 months in the field study can be explained in terms of the degree of N deficiency: the soil for the first controlled environment study was created by diluting very N-poor soil by 5-fold (with a combination of sand, perlite, and peat moss) in order to create an extremely N-deficient condition. The total N concentration of the soil mix used in that study (0.02%) was only 20% of that in the Aiken soil (Table I). (The same N amendments were used in both the controlled environment and field studies.) The comparison of field and controlled environment studies results demonstrates what should have been obvious: N limitation is not an either/or situation but a continuum, and only at the extreme end of the deficiency scale (such as in a new landslide or a mine spoil, where soils are nearly devoid of N) is growth response to CO_2 precluded. It can thus be concluded that, since all degrees of N limitation are present in the field, forest response to CO_2 will not, on an average or a global scale, be precluded by N limitation.

Finally, some perspective concerning the relative importance of soil vs vegetation C and N pools can be gained by comparing the Placerville data with those typical of mature stands. At 18 months, the seedlings at the Placerville site contained less than 1% of the total ecosystem C and N capital (Fig. 7). Among the IFS sites, trees contain a much larger proportion of total ecosystem C capital (20–70%), but still less than 25% of total ecosystem N capital (Fig. 7, right-hand bar). The continuing importance

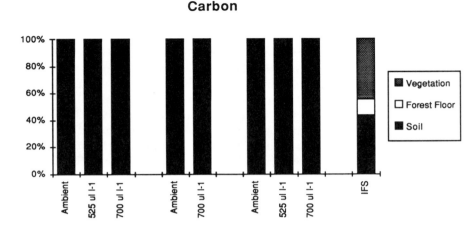

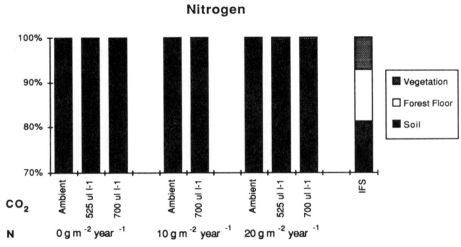

Figure 7 Distribution of organic matter (top) and N (bottom) in vegetation and soils in the Placerville study at 18 months and average values for the Integrated Forest Study (IFS, right-hand bar; Johnson and Lindberg, 1991).

of soil as a major N pool in mature forests is due to its lower C/N ratio (approximately 20) compared to that of the vegetation (approximately 230) (Johnson, 1994). Thus, while soils certainly are important components of ecosystem C capital in all cases, the relative importance of tree growth response (especially growth in wood volume) increases very substantially as forests mature. In the case of N, soils clearly are the dominant pool in all successional steps.

The forest floor can be a very important C and N pool in forest ecosystems with cool climates. In seedling stage studies, a forest floor is often absent unless it is artificially introduced (e.g., Körner and Arnone, 1992) or left over from a previous clearcut; in either event, seedling stage studies can offer no information on the effects of CO_2 on forest floor development in mature forests. Introduction of an artificial forest floor was not entertained as a possibility in the Placerville studies for fear of introducing a significant artifact of unknown dimensions into the study.

V. Summary and Conclusions

The combination of initial results from the Placerville study and the first controlled environment study at DRI have shown that seedling growth response to CO_2 can occur in the absence of N fertilization if the soil has sufficient unexplored N reserves to support such a response. This response can occur even if growth response to N fertilization indicates N limitation, but it is precluded in extreme cases of N deficiency (e.g., newly exposed soil nearly devoid of organic C and N pools). Thus, N limitation should be thought of as a continuum rather than an either/or situation, with the potential for growth response to CO_2 probably increasing as N limitation decreases.

Increased soil exploration through greater root biomass was the major mechanism by which the seedlings were able to respond to elevated CO_2 under N limitation in our study. There was no evidence of increased N mineralization due to elevated CO_2 in our study, in contrast to some other studies [Körner and Arnone, 1992; Zak *et al.*, 1993; Curtis *et al.*, 1996 (this volume)]. The often-observed reduction in tissue N concentration was rather transient in our study and was not sufficient to facilitate a large growth response without substantial additional N uptake. Furthermore, there was evidence that CO_2 could cause increases in N concentrations in woody tissues. The potential importance of CO_2 effects upon litter quality and decomposition has been well-recognized [Strain, 1985; Norby *et al.*, 1986b; Couteaux *et al.*, 1991, O'Neill and Norby, 1996 (this volume)] and is under investigation as part of our study at this time. However, the potential importance of CO_2 effects on woody tissue N concentration has not been adequately considered and will require longer term studies that allow more full development of woody biomass.

If our results from seedlings are applicable to mature forests, they imply that the response of mature forests to elevated CO_2 will require additional

N uptake and that this additional N will have to be obtained through increased root growth and soil exploration. We found little evidence for potential growth response to CO_2 without additional N uptake (i.e., little reduction in tissue N concentrations) and no evidence for increased rhizosphere soil N mineralization due to elevated CO_2. Given the prolonged opportunities for root exploration of the soil over a stand's lifetime, one might expect that the possibilities for further tapping of N reserves would be minimal. However, many studies have shown greater root biomass under infertile rather than fertile soil conditions after "root closure," suggesting that trees do attempt to exploit soils to a greater degree when nutrients or water is in greater demand. The questions of whether elevated CO_2 can induce such increases in root exploration and whether such root increases are successful in taking more N can only be resolved through stand-level CO_2 manipulations in mature forests.

Acknowledgments

Research was supported by the Electric Power Research Institute (RP3041-02) and the Nevada Agricultural Experiment Station, University of Nevada, Reno, NV. Technical assistance by Valerie Yturiaga, Carol Johnson, Peter Ross, and Greg Ross is greatly appreciated.

References

Aber, J. D., Nadelhoffer, K. J., Streudler, P., and Melillo, J. M. (1989). Nitrogen saturation in northern forest ecosystems. *Bioscience* 39, 378–386.

Brown, K. R. (1991). Carbon dioxide enrichment accelerates the decline in nutrient status and relative growth rate of *Popullus tremuloides* Michx. seedlings. *Tree Physiol.* 8, 161–173.

Campagna, M. A., and Margolis, H. A. (1989). Influence of short-term atmospheric CO_2 enrichment on growth, allocation patterns, and biochemistry of black spruce seedlings at different stages of development. *Can. J. Forest Res.* 19, 773–782.

Comerau, P. G., and Kimmins, J. P. (1989). Above- and below-ground biomass and production of lodgepole pine on sites with differing soil moisture regimes. *Can. J. Forest Res.* 19, 447–454.

Couteaux, M.-M., Mousseau, M., Celkerier, M.-L., and Bottner, P. (1991). Increased atmospheric CO_2 and litter quality decomposition of sweet chestnut litter with animal food webs of different complexities. *Oikos* 61, 54–64.

Curtis, P. S., Zak, D. R., Pregitzer, K. S., Lussnehop, J., and Teeri, J. A. (1996). Linking above- and below-ground responses to rising CO_2 in northern deciduous forest species. *In* "Carbon Dioxide and Terrestrial Ecosystems" (G. W. Koch and H. A. Mooney, eds.). Academic Press, San Diego.

Diaz, S., Grime, J. P. Harris, J., and McPherson, E. (1993). Evidence of feedback mechanism limiting plant response to elevated carbon dioxide. *Nature* 364, 616–617.

Gessel, S. P., Cole, D. W., and Steinbrenner, E. C. (1973). Nitrogen balances in forest ecosystems of the Pacific Northwest. *Soil Biol. Biochem.* 5, 19–34.

Johnson, D. W. (1992). Nitrogen retention in forest soils. *J. Environ. Qual.* 21, 1–12.

Johnson, D. W. (1994). Role of carbon in the cycling of other nutrients in forest ecosystems. *In* "Carbon: Forms and Functions in Forest Soils" (J. M. Kelly and W. M. McFee, eds.). 8th North American Forest Soil Conference, Special publication, Soil Science Society of America (in press).

Johnson, D. W., and Ball, J. T. (1996). Interactions between CO_2 and Nitrogen in Forests: Can We Extrapolate from the Seedling to the Stand Level? *In* "Carbon Dioxide and Terrestrial Ecosystems" (G. W. Koch and H. A. Mooney, eds.). Academic Press, San Diego.

Johnson, D. W., and Lindberg, S. E. (eds.) (1991). "Atmospheric Deposition and Forest Nutrient Cycling: A Synthesis of the Integrated Forest Study." Ecological Series 91, Springer-Verlag, New York.

Johnson, D. W., Ball, J. T., and Walker, R. F. (1994a). Effects of CO_2 and nitrogen on nutrient uptake in ponderosa pine seedlings. *Plant Soil* (in press).

Johnson, D. W., Geisinger, D. R., Walker, R. F., Vose, J., Elliot, K., and Ball, J. T. (1994b). Soil pCO_2, soil respiration, and root activity in CO_2-fumigated ponderosa pine. *Plant Soil* (in press).

Kauppi, P. E., Mielikäinen, K., and Kuusela, K. (1992). Biomass and carbon budget of European Forests, 1971 to 1990. *Science* 256, 70–74.

Keyes, M. R., and Grier, C. C. (1981). Above- and below-ground net production in 40-year-old Douglas-fir stands on low and high productivity sites. *Can. J. Forest Res.* 8, 265–279.

Körner, C., and Arnone, J. A. (1992). Responses to elevated carbon dioxide in artificial tropical ecosystems. *Science* 257, 1672–1675.

Norby, R. J., O'Neill, E. G., and Luxmoore, R. J. (1986a). Effects of atmospheric CO_2 enrichment on the growth and mineral nutrition of *Quercus alba* seedlings in nutrient-poor soil. *Plant Physiol.* 82, 83–89.

Norby, R. J., Pastor, J., and Melillo, J. M. (1986b). Carbon-nitrogen interactions in CO_2-enriched white oak: physiological and long-term perspectives. *Tree Physiol.* 2, 233–241.

Norby, R. J., Wullschlezer, S. D., and Gunderson, C. A. (1996). Tree Responses to Elevated CO_2 and Implications for Forests. *In* "Carbon Dioxide and Terrestrial Ecosystems" (G. W. Koch and H. A. Mooney, eds.). Academic Press, San Diego.

Olsen S. R., and Sommers, L. E. (1982). Phosphorus. *In* "Methods of Soil Analysis. Part 2. Chemical and Microbiological Properties," Second Edition, Number 9 in the Series, Agronomy (A. L. Page, R. H. Miller, and D. R. Keeney, eds.), pp. 403–430. American Society of Agronomy, Madison, WI.

O'Neill, E. G., and Norby, R. J. (1996). Litter Quality and Decomposition Rates of Foliar Litter Produced under CO_2 Enrichment. *In* "Carbon Dioxide and Terrestrial Ecosystems" (G. W. Koch and H. A. Mooney, eds.). Academic Press, San Diego.

Paul, E. A., and Clark, F. E. (1989). "Soil Microbiology and Biochemistry." Academic Press, New York.

Rogers, H. H., Peterson, C. M., McCrimmon, J. N., and Cure, J. D. (1992). Response of plant roots to elevated atmospheric carbon dioxide. *Plant, Cell, Environ.* 15, 749–752.

Samuelson, L. J., and Seiler, J. R. (1993). Interactive role of elevated CO_2, nutrient limitations, and water stress in the growth responses of red spruce seedlings. *Forest Sci.* 39, 348–358.

Santantonio, D., and Hermann, R. K. (1985). Standing crop, production, and turnover of fine roots on dry, moderate, and wet sites of mature Douglas-fir in western Oregon. *Ann. Sci. For.* 42, 113–142.

Schulze, E.-D., De Vries, W., Hauhs, M., Rosen, K., Rasmussen, L., Tamm, C. O., and Nilsson, J. (1989). Critical loads for nitrogen deposition on forest ecosystems. *Water, Air, Soil Pollut.* 48, 451–456.

Strain, B. R. (1985). Physiological and ecological controls on carbon sequestering in terrestrial ecosystems. *Biogeochemistry* 1, 219–232.

Strain B. R., and Thomas, R. B. (1992). Field measurements of CO_2 enhancement and climate change in natural vegetation. *Water, Air, Soil Pollut.* 64, 45–60.

van Veen, J. A., Liljeroth, E., and Lekkerkerk, L. J. A. (1991). Carbon fluxes in plant-soil systems at elevated CO_2 levels. *Ecol. Appl.* 1, 175–181.

Vogt, K. A., Moore, E. E., Vogt, D. J., Redlin, M. J., and Edmonds, R. L. (1983). Conifer fine root and mycorrhizal root biomass within the forest floors of Douglas-fir stands of different ages and site productivities. *Can. J. Forest Res.* 13, 429–437.

Walker, R. F., Geisinger, D. R., Johnson, D. W., and Ball, J. T. (1994). Interactive effects of atmospheric CO_2 enrichment and soil N on growth and ectomycorrhizal colonization of ponderosa pine seedlings. *Forest Sci.* (in press).

Wilkinson, L., Hill, M., and Vang, E. (1992). "SYSTAT Statistics," pp. 281–283. SYSTAT Inc., Evanston, IL.

Zak, D. R., Pregitzer, K. S., Curtis, P. S., Teeri, J. A., Fogel, R., and Randlett, D. L. (1993). Elevated atmospheric CO_2 and feedback between carbon and nitrogen cycles. *Plant Soil* 151, 105–117.

3

Linking Above- and Belowground Responses to Rising CO_2 in Northern Deciduous Forest Species

Peter S. Curtis, Donald R. Zak, Kurt S. Pregitzer, John Lussenhop, and James A. Teeri

I. Introduction

Rising atmospheric CO_2 partial pressure has many potential consequences for terrestrial vegetation, ranging from short-term physiological responses to long-term changes in ecosystem structure and function. The prediction of forest ecosystem responses will require an understanding of the degree to which dominant tree species show sustained increases in photosynthesis and growth as CO_2 levels rise (Graham et al., 1990). The likelihood of such increases is open to question on several grounds: (1) the potential reduction or elimination of positive photosynthetic responses to elevated CO_2 following prolonged exposure to high CO_2 [reviewed by Gunderson and Wullschleger (1994)], (2) the potential reduction or elimination of positive growth responses to elevated CO_2 under conditions where other environmental factors, particularly nutrients, strongly limit growth (Thomas et al., 1991; Norby et al., 1992; Wong et al., 1992), and (3) reduction in soil nitrogen availability and plant productivity through microbial N immobilization as a result of increased litter C:N or lignin:N ratios (Diaz et al., 1993). These processes, operating alone or together, could significantly reduce plant responses to elevated CO_2 and, hence, moderate potential direct effects of CO_2 on ecosystems.

A. The UMBS Conceptual Model of Tree Responses to Elevated CO_2

Our research program at the University of Michigan Biological Station is organized around a conceptual model linking above- and belowground responses to rising atmospheric CO_2 in forest ecosystems (Fig. 1; Zak et al., 1993). In this model, soil nitrogen (N) availability is linked to net carbon (C) assimilation through changes in fine root and mycorrhizal biomass, soil labile C pools, soil microbial biomass, and N mineralization rates. Both positive and negative feedback is possible within this system. A basic assumption of this model is that net C assimilation and plant productivity determine the magnitude of C input to the soil and, all other factors being equal, the soil microbial biomass. It is therefore important to understand the degree to which whole plant assimilatory capacity (that is, both leaf

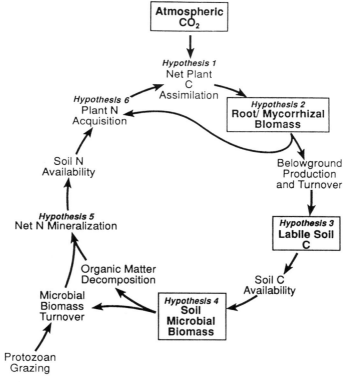

Figure 1 The UMBS conceptual model of tree responses to elevated CO_2. Soil N availability is linked to net C assimilation through changes in fine root and mycorrhizal biomass, soil labile C pools, soil microbial biomass, and N mineralization rates. The model's time frame is from weeks to months, and factors operating at longer time intervals, such as leaf and stem litter input are not included.

area duration and single leaf photosynthetic capacity) is affected by elevated CO_2 and the efficacy with which C is transported belowground. We also make the assumption that C input to the root zone via fine root and/or mycorrhizal turnover, and rhizodeposition, will have a more immediate effect on rhizosphere microbial dynamics than will C input via leaf or stem litter deposited on the soil surface.

The amount and form of C available for microbial growth and maintenance under elevated CO_2 are important regulators of soil N dynamics in our model. Nitrogen mineralization will increase with greater rhizosphere labile C input if microbial populations both increase and cycle more quickly. A number of studies have shown that N mineralization increases in the presence of plant roots, presumably via increased protozoan grazing on soil bacteria (Clarholm, 1985). However, N mineralization could slow if microbial growth were not balanced by turnover or the quality of C input were very low (e.g., high lignin:N ratio; Melillo et al., 1984). The latter should be most important at the soil surface or where surface litter is rapidly incorporated into lower soil horizons.

Nitrogen acquisition by trees will be a function of both soil N availability, via mineralization or exogenous input, and the ability of the tree to forage for N, via root and mycorrhizal growth. Growth at high CO_2 generally results in proportionate increases in both root and shoot biomasses, except under low nutrient or drought conditions where the relative increase in root biomass may be greater than the increase in shoot biomass (Stulen and den Hertog, 1993). This increase in root mass should promote N acquisition, even with little net change in N mineralization, in situations where available soil N is not fully exploited (e.g., in many aggrading systems) or where roots compete for spatially variable or relatively immobile N resources. Other factors to be considered regarding N acquisition are CO_2 effects on root architecture, the relative energetic and nutrient costs of different root systems, and the timing of root growth relative to N availability.

Increased C flux belowground and greater N acquisition could affect photosynthetic C assimilation through at least two mechanisms. First, greater C allocation to roots and mycorrhizae should mitigate negative feedback on photosynthetic capacity associated with carbohydrate buildup in source leaves. Second, greater N acquisition should limit the degree to which N is reallocated away from rubisco in source leaves and should reduce the decline in photosynthetic capacity associated with lowered leaf N at high CO_2.

We are testing a series of hypotheses directed at each causal link in our model. Our underlying hypothesis is that there is positive feedback between net C assimilation, root and mycorrhizal growth, microbial biomass, and soil N availability. We hypothesize that seasonal, whole canopy net C assimilation will increase at elevated CO_2, even where other factors limit, or colimit, growth (hypothesis 1, Fig. 1). This increased C assimilation will result in greater belowground and total plant productivity at elevated rather than at current ambient CO_2 (hypothesis 2). Labile soil C availability should increase in response to greater input of root-derived C, and this response will be most evident within the rhizosphere (hypothesis 3). If soil microorganisms are C limited, then greater soil C availability at elevated CO_2 should result in increased soil microbial biomass and greater microarthropod activity (hypothesis 4). Net N mineralization should increase through greater rates of organic matter turnover by a larger or more active microbial population (hypothesis 5). Finally, a larger root system combined with greater N mineralization should result in more plant N acquisition and increased whole plant N content (hypothesis 6). The end result of this series of positive feedbacks should be a sustained, positive effect on ecosystem productivity by rising CO_2.

There are a number of important points to note regarding this conceptual model and our use of it. The predictions of positive responses to CO_2 (that is, positive feedback through the system) represent our working hypotheses, with no CO_2 response being the appropriate null hypothesis. Negative responses to CO_2 are possible as alternate hypotheses and will be explicitly tested for in each case. In addition, the model is most appropriate for the spatial and temporal scale of our CO_2 manipulations. Our time scale has been from weeks to months. Therefore, we have not included C input from leaf, branch, or stem litter in the model. Similarly, the size and number of our open-top chambers limit our examination of the effects of genotypic or landscape heterogeneity on CO_2 responses. Nonetheless, we consider the C and N linkages between above- and belowground components of forest ecosystems, as described in the model, to be of fundamental importance to ecosystem function. As such, tests of these hypotheses should advance our knowledge of forest ecology as well as improve our understanding of the ecological consequences of atmospheric change.

II. Study Site and Experimental Methods

The research summarized here was conducted at the University of Michigan Biological Station, located in northern Lower Michigan near the boundary of two major ecosystems: the boreal coniferous forest/peatland region of Canada and the temperate hardwood forest region of the central and eastern United States. The Biological Station has dedicated approximately 5 ha of open field to long-term research into the direct effects of elevated CO_2 on forest vegetation. This area has been cleared of vegetation, leveled, and supplied with power. Results have been obtained using 0.5-m^3 open-top chambers (Curtis and Teeri, 1992). Carbon dioxide partial pressure inside chambers was raised by dispensing 100% CO_2 into the inlet port of an input blower and was monitored by continuous sampling of chamber air. Treatments were ambient and ambient + 350 μbar CO_2 (elevated). Each chamber sat on top of an open-bottom root box 1.3 m deep, installed in the ground and fitted with two minirhizotron tubes.

A. 1991 Study: *Populus grandidentata*

Root segments from a single clone of *P. grandidentata* were collected at UMBS and used for the propagation of all plant material. The subsoil obtained from root box excavations (sandy, mixed, frigid entic haplorthod) was subsequently used to fill the boxes, giving a growth medium very low in initial organic C (1977 μg C g^{-1}) and total N (70 μg N g^{-1}). Nine ramets were planted in each root box on May 20, 1991. Nitrogen was added to each root box in two stages. In the first stage, intended to simulate natural rates of N mineralization, a total of 745 mg N was applied as aqueous NH_4Cl to each plot over a 29-day period. In the second stage, intended to simulate natural increases in N availability during mid-summer, a total of 1440 mg N was applied as ^{15}N-enriched NH_4Cl over an 8-day period. Total N added (4.5 g m^{-2}) was similar to that mineralized on an annual basis in dry oak forests (Zak and Pregitzer, 1990). Plants were kept well watered.

B. 1992 Study: *Populus euramericana*

Hardwood cuttings of *P. euramericana* cv. Eugenei were obtained from stock propagated at Michigan State University. Five cuttings per chamber were used. Two contrasting soil fertility treatments were established by filling the root boxes with either 100% locally excavated Kalkaska series topsoil (high fertility) or a homogenized mixture of 20% topsoil, and 80% native Rubicon sand (low fertility). Plants were kept well watered.

C. Aboveground Study Methods

Aboveground growth was followed nondestructively by measurements of the rates of leaf initiation, leaf length, leaf width, and stem height. Leaf gas exchange responses to changes in CO_2, temperature, and light were measured with an ADC LCA3 gas exchange system and a Hansatech oxygen electrode (dark respiration). Leaves were harvested periodically and analyzed for soluble carbohydrates and starch and total N.

D. Root Study Methods

Root growth was followed *in situ* by serial video photography by using a minirhizotron camera system (Upchurch and Ritchie, 1984). The length, diameter, number of internodes, number of root tips, and physiological state class of all roots in the sample areas were measured at each sampling date using ROOTS, an interactive personal computer program (Hendrick and Pregitzer, 1992). Fine root turnover was determined by measuring the life span of cohorts of individual roots appearing at the same time in each treatment. At the termination of each experiment, whole pot harvests were performed in which all root systems were excavated and coarse and fine roots were recovered. Total root length, root length density, and size class distribution were calculated.

E. Soil Carbon Availability, Microbial Dynamics, and Nitrogen Transformations

Microbial respiration and N mineralization and labile C and N pools were assayed by using a modified microlysimeter technique (Zak *et al.*, 1993). Product accumulation curves for microbial respiration (CO_2–C) and mineralized N (inorganic N) were constructed for each microlysimeter and data were fit to a first-order equation, allowing direct assessment of changes in organic matter pool size and chemistry. We used ^{15}N to determine whether elevated CO_2 influenced soil N dynamics. A short-term isotope dilution experiment was used to quantify rates of gross mineralization, gross immobilization, and nitrification. Microbial biomass C and N were measured by using the chloroform fumigation–incubation procedure (Voroney and Paul, 1984).

F. Effects on Soil Biota

Soil samples were used to quantify the mass of vesicular arbuscular mycorrhizae (VAM) external hyphae (membrane filter method; Miller and Jastrow, 1992), the length and mass of VAM and nonmycorrhizal roots (trypan blue, gridline intersect method), protozoa (MPN method), nematodes (wet funnel extraction), and collembola and mites (flotation). Minirhizotron video observations were used to quantify populations of earthworms and larger microarthropods.

G. New Research: *Populus tremuloides*

A multiyear exposure using multiple genotypes of *P. tremuloides* begun in 1994 uses larger 3 m diameter × 2.4 m tall open-top chambers. The design is again a two-way factorial experiment with CO_2 (ambient and elevated) and soil fertility (low and moderate) as the main treatments. Each treatment is replicated 5 times for a total of 20 chambers. Fertility treatments utilize soil mixtures similar to those used in 1992. After excavation to a depth of 20 cm, a 3.5 m × 3.5 m × 60 cm raised bed was filled with either low or moderate fertility soil, and eight minirhizotron tubes were mounted horizontally within it. Open-top chambers rest on top of these beds, and each contains 12 *P. tremuloides* seedlings.

III. Results and Discussion

Results from experiments conducted in 1991 and 1992 linking above- and belowground responses to elevated CO_2 are summarized in Tables I and II, respectively. Overall growth of *P. grandidentata* was comparable to that of *P. euramericana*: final mean above- plus belowground mass across treatments was 84 and 76 g plant^{-1}, respectively. In both years, shoots were unbranched and root systems explored throughout the root boxes in all treatments. While the experimental details and species of *Populus* varied, the conceptual and methodological similarities between the two studies allow broad comparisons to be made. For further details and analysis of the 1991 data, see Zak *et al.* (1993) and Curtis *et al.* (1994).

Results from 1991 provided strong initial support for our conceptual framework linking soil carbon to net C assimilation and belowground growth (Table I). Even at very low soil N, photosynthetic capacity was stimulated by elevated CO_2, and high CO_2 grown plants were much more responsive to changing soil N than were ambient CO_2 grown plants. The

Table I Percent Stimulation Due to CO_2 of Above- and Belowground Characters of *Populus grandidentata* Grown for 142 Days in Field Chambers at Ambient (A) and Elevated (E) CO_2[a]

	$(E - A)/A \times 100$[b]
Hypothesis 1: net assimilation	
Aboveground biomass (g)	4
Leaf area (cm^2)	1
Net photosynthesis (μmol m^{-2} s^{-1})	20 → 115*
Hypothesis 2: root growth	
Belowground biomass (g)	60*
Fine root growth (RER, day^{-1})	113*
Root: shoot ratio	54*
Hypothesis 3: Labile soil C	
Rhizosphere soil (μg g^{-1})	18*
Bulk soil (μg g^{-1})	15
Hypothesis 4: microbial C	
Rhizosphere soil (μg g^{-1})	46*
Bulk soil (μg g^{-1})	43*
Hypothesis 5: N mineralization	
Bulk soil rate (μg N g^{-1} day^{-1})	540*
Hypothesis 6: N acquisition	
Plant N content (g N plant^{-1})	15

[a] Responses are grouped according to the hypotheses of the conceptual model of tree responses to elevated CO_2.
[b] *: significantly different from zero; $P < 0.05$.

Table II Percent Stimulation Due to CO_2 of Above- and Belowground Characters of *Populus euramericana* Grown for 158 Days in Field Chambers at Ambient (A) and Elevated (E) CO_2 and in Soil of Low or High Fertility[a]

	$(E - A)/A \times 100$[b]	
	Low fertility	High fertility
Hypothesis 1: net assimilation		
Aboveground biomass (g)	23*	52*
Leaf area (cm^2)	5	35*
Net photosynthesis (μmol m^{-2} s^{-1})	42*	45*
Dark respiration (μmol kg^{-1} s^{-1})[c]	0	-1
Leaf N content (mg cm^{-2})	-5	20*
Hypothesis 2: root/mycorrhizal growth		
Belowground biomass (g)	29*	48*
Fine root biomass (g)	32*	70*
Root:shoot ratio	5	-3
Mycorrhizal root length (cm)	-62	30
Soil hyphae (cm g^{-1} soil)	8	5
Hypothesis 4: soil biota		
Soil microarthropods (no. cm^{-2})	8	5
Hypothesis 5: N mineralization		
Bulk soil rate (μg N g^{-1} day^{-1})	56	9
Rhizosphere rate (μg N g^{-1} day^{-1})	0	103*

[a] Responses are grouped according to the hypotheses of the conceptual model of tree responses to elevated CO_2.
[b] *: significantly different from zero; $P < 0.05$.
[c] Dry weight less nonstructural carbohydrate basis.

only positive growth responses, however, were belowground, where root relative extension rate and final biomass increased significantly. Labile soil C increased significantly in the rhizosphere of high CO_2 grown plants, and microbial C increased in both bulk and rhizosphere soil. Short-term N mineralization rates were assayed in the bulk soil only and showed a significant increase at elevated CO_2. These data suggest that both the pool size and turnover rate of microbial N increased at elevated CO_2. It is also possible that greater amounts of N were released from native soil organic matter.

These results are consistent with our model and working hypotheses, but can only be considered preliminary. They provide limited information on the N dynamics of our system, and further studies using ^{15}N to characterize the rates of gross N mineralization and immobilization in the presence of plant roots are needed. We are also only able to infer that increased labile soil C resulted from increased root activity and have no direct measure of this process. Although root growth and N mineralization rates increased, we saw no significant increase in plant N acquisition. This somewhat surprising result could be due to the very low nutrient binding capacity of our substrate and the short-term nature of the experiment. We concluded, on the basis of the recovery of ^{15}N in plant tissue and soil, that N not taken up soon after application was quickly leached from the sandy soil. In higher fertility soil, or with further input of N either exogenously or via mineralization, plants grown at elevated CO_2, with more extensive roots, would likely acquire more net N than ambient CO_2 grown plants.

Our experiment in 1992 using *P. euramericana* was a refinement of that in 1991 and focused on the role of soil fertility in hypotheses 1 and 2: the photosynthetic assimilation of C and its pattern of allocation belowground. Of central interest were physiological and allocational processes occurring under conditions of growth limiting N availability and the extent to which they influenced the movement of C into the soil at elevated CO_2. Specifically, we sought to address the question of whether low fertility would reduce net C assimilation to such an extent that little or no additional C moved into the soil.

Our fertility treatment had a dramatic effect on above- and belowground responses to elevated CO_2 (Table II). Low fertility completely eliminated positive leaf area responses to elevated CO_2 and reduced to less than half the increase in aboveground dry weight seen at high fertility. Photosynthetic responses to CO_2 were not nearly as sensitive to soil fertility, with comparable levels of stimulation relative to ambient CO_2 seen at both fertility treatments when considered as a seasonal average. However, late season photosynthesis at high CO_2 and low fertility showed evidence of downward regulation via a reduction in rubisco activity (significant reduction in V_{cmax} relative to ambient CO_2 grown plants, data not shown).

Belowground growth was more responsive to fertility than to CO_2, with root biomass increasing roughly in proportion to shoot biomass in both CO_2 treatments. Fine root biomass appeared more responsive to CO_2 than total root biomass, particularly at high soil fertility. Mycorrhizal responses were variable and showed no significant treatment effects. As in 1991, we found a significant increase in N mineralization at high CO_2, although only in rhizosphere soil from the high fertility treatment. Other data on soil responses were highly variable, with no strong trends apparent. Assays for soil carbon are still in progress, so it is premature to draw firm conclusions at this point. The mineralization data, however, suggest no evidence for negative feedback on the time scale of one growing season.

IV. Conclusions

Results from two years' work with *Populus* in a northern forest ecosystem has provided broad support for our conceptual model of above- and belowground linkages through C and N cycling within the plant–soil system. They have also supported our working hypothesis that rising CO_2 will lead to positive feedback within this system, resulting in increased N mineralization and plant N acquisition. These conclusions must be qualified by the short duration of these experiments and the likelihood that negative feedback may only appear over longer time spans. We also have limited knowledge of the dynamics of C movement from plant to soil and of N transformation within the soil. We hope to address each of these shortcomings within the long- term exposure study set to begin in 1994.

Accepting these qualifications, how might we generalize these results to natural *Populus* stands or to other northern deciduous species? It seems clear that, even under very low nutrient conditions, photosynthesis will be stimulated by exposure to elevated CO_2 and that this will translate, at a minimum, into increased belowground growth. The ecological consequences of increased belowground C flux are potentially far reaching, but at present are largely unexplored (Curtis *et al.*, 1995). Our results demonstrate that relatively short-term experiments, under controlled field conditions, can provide valuable insight into some of these ecosystem-level processes.

With moderate to high soil fertility, aboveground growth in young trees should increase, leading to more rapid stand development and, hence, shorter rotation for harvested species. Many fundamental questions remain, however, concerning elevated CO_2 effects on growth and nutrient dynamics in mature stands following canopy closure. Some of these questions may be addressable using methods described here, while others will require new technological or conceptual approaches.

Acknowledgments

Support for this research was provided by grants from the Hasselblad Foundation, the USDA competitive grants program, the U. S. Department of Energy (National Institute for Global Environmental Change), the Ohio State University, the University of Michigan Project for the Integrated Study of Global Change, and the University of Michigan Biological Station.

References

Clarholm, M. (1985). Interactions of bacteria, protozoa and plants leading to mineralization of soil nitrogen. *Soil Biol. Biochem.* 17, 181–187.

Curtis, P. S., and Teeri, J. A. (1992). Seasonal responses of leaf gas exchange to elevated carbon dioxide in *Populus grandidentata*. *Can. J. Forest Res.* 22,1320–1325.

Curtis, P. S., Zak, D. R., Pregitzer, K. S., and Teeri, J. A. (1994). Above and belowground response of *Populus grandidentata* to elevated atmospheric CO_2 and soil N availability. *Plant Soil* 165, 45–51.

Curtis, P. S., O'Neill, E. G., Teeri, J. A., Zak, D. R., and Pregitzer, K. S. (1995). "Belowground Responses to Rising CO_2: Implications for Plants, Soil Biota, and Ecosystem Processes." Kluwer Academic Publ., Dordrecht, The Netherlands.

Diaz, S., Grime, J. P., Harris, J., and McPherson, E. (1993). Evidence of a feedback mechanism limiting plant response to elevated carbon dioxide. *Nature* 364, 616–617.

Graham, R. L., Turner, M. G. and Dale, V. H. (1990). How increasing CO_2 and climate change affect forests. *BioScience* 40, 575–587.

Gunderson, C. A., and Wullschleger, S. D. (1994). Photosynthetic acclimation in trees to rising atmospheric CO_2: a broader perspective. *Photosynth. Res.* 39, 369–388.

Hendrick, R. L., and Pregitzer, K. S. (1992). The demography of fine roots in a northern hardwood forest. *Ecology* 73, 1094–1104.

Melillo, J. M., Naiman, R. J., Aber, J. D., and Linkins, A. E. (1984). Factors controlling mass loss and nitrogen dynamics of plant litter decaying in northern streams. *Bull. Marine Sci.* 35, 341–356.

Miller, R. M., and Jastrow, J. D. (1992). Extraradical hyphal development of vesicular-arbuscular mycorrhizal fungi in a chronosequence of prairie restorations. *In* "Mycorrhizas in Ecosystems" (D. J. Reed, D. H. Lewis, A. Fitter, and I. Alexander, eds.), pp. 172–176. CAB International, Wallingford, UK.

Norby, R. J., Gunderson, C. A., Wullschleger, S. D., O'Neill, E. G., and McCracken, M. K. (1992). Productivity and compensatory responses of yellow-poplar trees in elevated CO_2. *Nature* 357, 322–324.

Stulen, I., den J. Hertog, (1993). Root growth and function under atmospheric CO_2 enrichment. *Vegetatio* 104/105, 99–116.

Thomas, R. B., Richter, D. D., Ye, H., Heine, P. R., and Strain, B. R. (1991). Nitrogen dynamics and growth of seedlings of an N-fixing tree (*Gliricidia sepium* (Jacq.) Walp.) exposed to elevated atmospheric carbon dioxide. *Oecologia* 88, 415–421.

Upchurch, D. R., Ritchie, and J. T. (1984). Battery-operated color video camera for root observations in minirhizotrons. *Agron. J.* 76, 1015–1017.

Voroney, R. P., and Paul, E. A. (1984). Determination of kc and kn *in situ* for calibration of the chloroform fumigation–incubation method. *Soil Biol. Biochem.* 16, 9–14.

Wong, S. C., Kriedemann, P. E., and Farquhar, G. D. (1992). CO_2 × nitrogen interaction on seedling growth of four species of eucalypt. *Austral. J. Bot.* 40, 457–472.

Zak, D. R., and Pregitzer, K. S. (1990). Spatial and temporal variability of nitrogen cycling in northern lower Michigan. *Forest Sci.* 36, 367–380.

Zak, D. R., Pregitzer, K. S., Curtis, P. S., Teeri, J. A., Fogel, R., and Randlett, D. L. (1993). Elevated atmospheric CO_2 and feedback between carbon and nitrogen cycles. *Plant Soil* 151, 105–117.

4

Effects of Tree Maturity on Some Responses to Elevated CO_2 in Sitka Spruce (*Picea sitchensis* Bong. Carr)

Helen S. J. Lee and Paul G. Jarvis

I. Introduction

A. Carbon Balance

Forests cover one-third of the land area and are responsible for approximately two-thirds of current photosynthesis on a global scale (Kramer, 1981). As trees grow they sequester carbon, storing it as wood, and as long as the wood does not decay or is not burnt, that carbon is prevented from returning to the atmosphere. For example, a large area of forest in the United Kingdom is now young, actively growing and sequestering carbon—about 1.8 tons per ha with a doubling time of ca. 20 years.

It is often stated that the rise in carbon dioxide concentration in the atmosphere will lead to increases in plant biomass because of increases in photosynthetic carbon fixation, but this is an oversimplification. The net carbon balance of a forest depends on both the amount of carbon dioxide being assimilated by stands of trees and the carbon dioxide emissions associated with decomposition and respiratory processes. Carbon dioxide assimilation depends upon the area and phase of growth of the stands of different species, the latter depending upon the age and physiological activity of the trees, the climate, the weather, and the silvicultural system practiced. The biological carbon dioxide emissions depend upon the respiratory costs of both growth and maintenance and the allocation of carbon to components of the trees (roots and branches, vis-à-vis stems), production of litter, and turnover of organic matter in soil. These processes are almost all affected by global change. The increase in atmospheric CO_2 concentration and the deposition of atmospheric nitrogen are fertilizers that will increase the rate of assimilation and alter the allocation of carbon and nitrogen within trees. The anticipated rise in temperature is expected to increase all biochemical and physiological processes, particularly respiratory processes in trees and soil.

B. Impacts of Elevated CO_2

Global change research has highlighted the role of forests in sequestration of atmospheric CO_2 (Jarvis, 1989) and has led to research directed toward the impact of elevated CO_2 on trees. It is clear that CO_2 acts on subcellular and cellular processes in the short term, but the *major* consequences for forests are expressed in the long term (years to centuries) over a regional scale. Trees are unique in that their longevity ensures that many species will experience a doubling of atmospheric CO_2 concentration during their lifetime. For largely historical reasons, predictions of the effects of global change have been based on short-term studies of juvenile trees (Eamus and Jarvis, 1989; Luxmoore *et al.*, 1993). It is evident, however, that trees can grow very large and do not fit into simple experimental enclosures. Furthermore, trees and forests are very well coupled to the atmosphere, and this coupling is often greatly reduced when trees are enclosed in chambers, introducing an additional artifact.

C. Juvenility

The problem of phase change in trees has exercised the minds of foresters, horticulturalists, and physiologists for many years (e.g., Wareing, 1959; Kozlowski, 1971; Borchert, 1976). Trees grow for decades to centuries and have juvenile phases that can last up to 40 years and affect those responses to environmental variables that change with age and degree of maturity. Seedling and juvenile trees may differ from mature, adult trees in phyllotaxy, leaf abscission, habit and rate of growth, ability to flower, and other morphological, anatomical, or physiological characteristics (Robbins, 1961; Kozlowski *et al.*, 1991). In the present context, the rooting ability of cuttings and the intermediate maturity of such plants are particularly relevant. Even such processes and properties as water potential, photosynthesis, stomatal conductance, and water use efficiency (WUE) differ in juvenile and mature woody trees [e.g., Donovan and Ehleringer (1991) for *Acer negundo* and *Salix exigua*.] Juvenile trees have more negative predawn water potential and lower WUE than mature trees. Some trees may exhibit juvenile and adult characteristics at the same time—at the base of the trunk and in the upper crown, respectively, or even on the same branch in some eucalyptus species.

D. Root Responses

Early experiments on the impact of elevated CO_2 on trees were carried out in much the same way as experiments on crops—with potted seedlings with fixed nutrient capital for short periods of days or weeks—and have led to much confusion and misinformation (see Eamus and Jarvis, 1989). One of the most notable results was enhanced growth of the root system. This and other such responses were found to change with time, and a view arose that enhancement of growth by elevated CO_2 was a transient phenomenon that would disappear as trees aged. In hindsight, many of these observations were probably the result of both inadequate rooting volume and low rates of nutrient supply in relation to the growth of trees, exacerbated by the readier availability of carbon as a result of the elevated atmospheric CO_2 in which the trees were grown. In other words, the growth responses observed were primarily the result of a lack of sinks for carbon and dilution of nutrients, particularly nitrogen, within the young trees, leading to the eventual development of nitrogen deficiency. There is still a fundamental lack of understanding of the mechanisms, partly related to the failure to distinguish the effects of small rooting volumes (i.e., sink limitation) or poor nutrition on tree growth and function (McConnaughay et al., 1993). A further complication is that the above- and belowground growth of trees is not synchronous, and this too has led to spurious results from short-term experiments with single harvests. Root growth can continue into the winter months when leaves are absent, while in spring preference is given to leaf and shoot extension.

Even with improved understanding of rooting volume and root growth phenology, a wide range of growth responses to elevated CO_2 is still being observed (Poorter, 1993). Much of the variability in response may relate to the way in which experiments have been done and, in particular, to the length of experiments when resources have been limited. In that circumstance, interactions not formally recognized in the design of experiments may have overriding effects on the results obtained. In particular, failure to maintain the availability of nitrogen so that nitrogen uptake keeps pace with CO_2 uptake is likely to lead to complex responses.

E. Tree Size and Maturity

Because of their size it is difficult to perform CO_2 enrichment experiments *in situ* on mature trees, and the majority of experiments have, therefore, utilized seedlings, often over a single growing season or less. Predictions about the growth of trees and forests made on the basis of such short-term experiments with juvenile plants lack credibility and have little relevance, except for seedling establishment (Eamus and Jarvis, 1989). Like most perennial plants, trees lay down new buds during the year before such buds will burst. These buds contain preformed leaves (in determinate trees such as Sitka spruce), but previous carbon and nutrient acquisition also influences bud size and potential growth (e.g., in indeterminate trees such as birch). A tree that has been exposed to elevated CO_2, with nitrogen fertilization, for 1 or more years has more carbohydrate resources in its tissues than a tree grown in ambient CO_2, and these can be more rapidly mobilized for new growth in the spring.

Some use has been made of clonal cuttings from mature trees to avoid this problem. In a number of species flowering will occur on small clonal plants, but in others the potential to form roots relies upon the juvenility of the tissue. This is the case for Sitka spruce (Mason et al., 1986). An alternative immediately available is to enclose branches of mature trees in large cuvettes or bags and to expose them to elevated CO_2 over several years. This approach also has undoubted limitations (see Sprugel et al., 1991). For example, a branch clearly is not wholly autonomous and may exchange carbohydrates, water, and nutrients with other parts of the tree, but this approach does provide an immediate opportunity to investigate the local responses of fully mature tissue to elevated CO_2, in the absence of sink limitation.

Branches are likely to be autonomous for photosynthate, but they rely on the tree for nutrient and water supply. Older, more mature trees can carry out up to 80% internal recycling of nutrients (Kozlowski et al., 1991). In Sitka spruce, where needles may be retained for 8 years, there is some reallocation of nutrients from older to younger needles, so that, even for nutrients, the branches are semiautonomous. Carbohydrates are exported from all but the youngest branches, and when a branch becomes so shaded that respiration exceeds assimilation, the branch no longer imports carbohydrates and dies.

F. Experimental Protocols

It is now clear that long-term research, over several years, is needed to establish the effects of elevated CO_2 on trees and forests and that such research ideally should start with seeds or very young plants, as it is at that time that the future growth pattern is set, but that the research must also encompass older and mature trees where possible. At Edinburgh we have investigated the responses of both mature and juvenile small trees, but in addition we have enclosed branches of mature trees within large cuvettes or bags and exposed them to air enriched with CO_2 (Barton et al., 1993), and we have grown larger trees up to 4 m tall in individual and multiple-tree open-top chambers. The current studies at Edinburgh embrace a variety of ages of trees that have been (and still are) subjected to continuous elevated CO_2 exposure ($+350$ μmol mol^{-1} above ambient). These measurements provide data on seedling, tree, and stand growth and on C sequestration rates and, by using a model, can be scaled up to forests and regions to predict sequestration rates in the future in a changed climate. The results from the first (a period of acclimation) and subsequent years will be compared to underline the temporal distinction between long-term- and short-term-exposure CO_2 experiments.

G. Upscaling

To scale up to larger areas and to make predictions about likely changes in the future, a model is needed and it is highly desirable that confidence should exist in the model used. The process-based model MAESTRO, which is a mechanistic bottom up model that integrates processes from a small scale up to larger scales (see Jarvis, 1994a,b), has been applied. An advantage of this process-based approach is that processes sensitive to CO_2 concentration are included and explicitly defined, so that predictions of the likely impact of global change on forests can be made with greater confidence than with a purely empirical (or statistical) model. For the purpose of predicting the responses of stands to climate change, it is essential that stand-scale models be parametrized with data appropriate to mature trees. These can be validated with measurements of CO_2 and H_2O exchange above an existing forest canopy using eddy covariance (EC) techniques (Jarvis, 1994a,b). All of the measurements can then be compared with the output of existing tree and forest models.

II. Materials and Methods

Seedling and clonal Sitka spruce and branches of mature trees have been exposed to elevated CO_2 for three to four growing seasons near Edinburgh (55° 57′ N, 3° 13′ W). CO_2 is added to both open-top chambers and branch bags to give a concentration of 350 μmol mol^{-1} above current ambient levels. Table I summarizes the tissue, treatment, and duration for each tree

Table I Summary of the Tissue Types, Experimental Facility, and Duration of Experiments

Tissue type	Sample size	Experiment	Duration (years)
1+1 seedlings	500	OTC pots 1990–1992	5
	48	OTC soil-rooted 1992–1994	
four clones from 5-year-old cuttings	600	OTC pots	4
mature 17-year-old branches	18	Branch bags	4

material. The treatment is continued 24 hr a day throughout the year. Measurements of growth, biomass production, biochemistry, physiology, and gas exchange have been carried out at regular intervals and are the subjects of forthcoming publications.

A. Seedlings

Five hundred 1+1 bare-rooted plants were potted into standard potting compost (83[2015]5, Lot 2, Provenance 20, QCI) in 1990. They were placed in eight (four ambient and four elevated CO_2) open-top chambers (glass-sided, 3 m in diameter, and 2.5 m high) at the Institute of Terrestrial Ecology, Bush, near Edinburgh. Plants were initially fertilized with a slow release fertilizer for the first year, and in subsequent years they were further fertilized with nutrient solution at weekly intervals. The concentration of nutrients was increased during the main period of growth in an attempt to avoid any nutrient limitation or deficiency that might confound the analysis of CO_2 effects (see Introduction). Nondestructive measurements of growth (height, stem diameter, and branch number) and sampling of needles from each age class have been carried out throughout the growing season. Chlorophyll, carbohydrate, and nutrient contents and rubisco activity have been assayed every 2 months. Plants were harvested on 10 occasions during the last 4 years and divided into their component parts (root, main stem, branches, and needles) for subsequent analysis. The remaining plants were repotted into larger containers each year before bud burst and now, having outgrown any container, have been planted directly into forest soil in a further set of eight similar, but taller, open-top chambers belonging to the Forestry Commission (Durrant *et al.*, 1993). Details of the open-top chamber environment are given by Unsworth (1986) and Lee and Barton (1994) and details of cultivation are given by Murray *et al.* (1994).

B. Clones

One hundred and fifty cuttings from each of four clones were taken from 5-year-old trees in 1989, rooted under ambient conditions, and then treated as described earlier in the open-top chambers at the Institute of Terrestrial Ecology [see also Murray *et al.*, 1994]. The four clones originated in the Queen Charlotte Islands. Three harvests, as well as nondestructive growth measurements and sampling of needles, stem, and roots for biochemical analysis, were carried out at intervals throughout each growing season.

C. Branches

Three branches of mature (17-year-old) clonal Sitka spruce were selected in spring 1991 on the third whorl of each of six trees within a small stand outside Edinburgh. Two of the branches on each tree were enclosed in branch bags and fumigated with ambient air or elevated CO_2; the third branch acts as a control and is unbagged. The design of the bags allows free movement of branches and trees with good control of CO_2 concentration (± 20 μmol mol^{-1}) and only small increases in ambient temperature. Further details of the construction and performance of this system are given by Barton et al. (1993). The bags were washed regularly and replaced after 2 years when light transmission had decreased below 85% of daylight. Nondestructive measurements of apical shoot extension, branch diameter, and branch number have been made throughout each growing season. Chlorophyll, carbohydrate, and nutrient contents and rubisco activity have been assayed at regular intervals as outlined earlier.

III. Results and Discussion

The responses of the three tissue types in the CO_2 enrichment experiments differed with both time and age of material. Results obtained in the first year of exposure were on tissue that had grown only under elevated CO_2 but had been laid down under ambient CO_2, whereas results from later years were for tissue both laid down and grown under elevated CO_2. This distinction is made because CO_2 treatment may affect the number of needle initials laid down in the new buds and, hence, the potential growth response of new shoots. The individual tissue types will be considered first, and then a comparison between responses in relation to tissue maturity will be undertaken.

A. Seedlings

In the first year of treatment, few significant effects of CO_2 on growth were seen in seedlings, the exception being an increase in root mass that can be attributed to the size of the pot, the initial nutrient regime, and the well-known rapid root growth of bare-rooted trees. Positive effects of elevated CO_2 on biomass production and allocation were found in later years, with no significant imbalance between root and shoot (Figs. 1a,1b). All elevated CO_2 plants showed an increase in needle mass (Table II) during the growing season that is, in large part, attributable to starch accumulation (Fig. 2). It would appear that the starch was mobilized during the winter months of 1991 and 1992 and, as starch was almost absent in January 1993, that this increase in needle biomass was likely to have been in structural dry mass. At the end of each year, the mass of all component parts (needles, stem, and root) had increased in elevated CO_2, but at other times of the year, the relative increases were determined by the time of harvest (Fig. 1a).

B. Clones

All four clones showed a consistent, large positive response to growth in elevated CO_2 (Table III; Centritto, personal communication). From 1992

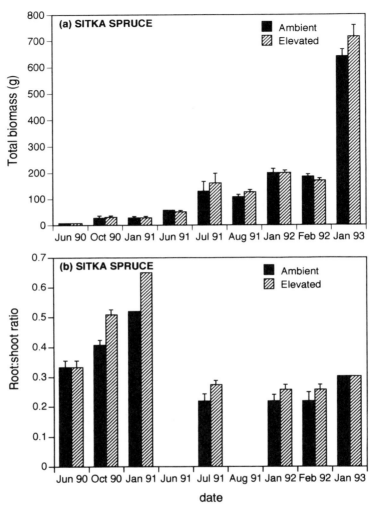

Figure 1 Harvest data from seedling Sitka spruce from initial planting (as bare-rooted 1+1 seedlings in June 1990) grown in ambient or elevated CO_2 ($n = 5-15 \pm$ SE): (a) total biomass; (b) root/shoot ratio.

onward, the trees grown in elevated CO_2 were significantly taller and had larger basal diameters and needle areas and masses than those grown in ambient CO_2. In 1991 the phenology of leaf growth was examined, and the trees in elevated CO_2 were shown to have a shorter growing season because of later bud burst and earlier bud set in elevated CO_2 (Murray et al., 1994). From further experiments on seedling Sitka spruce, Murray et al. concluded that this was related to the nutritional status of the trees.

Table II Percentage Increase or Decrease in Total Biomass of Seedling Sitka Spruce Grown in Elevated CO_2 Relative to Growth in Ambient CO_2 and the Relative Increase (+) or Decrease (−) in Each Component, Needle, Wood, and Root Biomass, at the Times of Harvests[a]

Harvest date	Total biomass	Needle biomass	Woody biomass	Root biomass
June 1990	0	0	0	0
October 1990	+6% ($P = 0.152$)	+ (ns)	+ (ns)	–
January 1991	−0.5% ($P = 0.077$)	–	–	+ ($P = 0.029$)
June 1991	−9%	–	–	–
July 1991	+25% ($P = 0.025$)	+ ($P = 0.013$)	+ ($P = 0.025$)	+ ($P = 0.007$)
August 1991	+15% (ns)	+ (ns)	+ ($P = 0.047$)	+ ($P = 0.044$)
January 1992	0% ($P = 0.987$)			+
February 1992	−9% (ns)	–	–	
January 1993	+12% ($P = 0.05$)	+15% ($P = 0.132$)	+32% ($P = 0.046$)	+53% ($P = 0.036$)

[a] The percentages were calculated from mass of [(elevated − ambient/ambient] × 100. $n = 50$ plants per treatment and per harvest, June 1990 and January 1991. $n = 15$ plants per treatment and per harvest, October 1990 and January, June, July, and August 1991. $n = 5$ plants per treatment and per harvest, January 1993. ns: not significant at $P = 0.05$.

From 1992 onward the trees therefore were given extra liquid fertilizer treatment, which resulted in an increase in the length of the growing season of the elevated CO_2 trees to equal that of the ambient CO_2 trees.

C. Branches

Apical shoot extension of mature branches in the branch bags was not affected by elevated CO_2. However, branch diameter and needle dry/fresh mass of the bagged branches (in both ambient and elevated CO_2) increased (Table IV). In the second and third years this bag effect was seen again, but the unbagged, control branches failed to increase in length to the same extent as the bagged branches or to the extent seen in the first year. The lack of sustained growth in the unbagged control branches may be related to the initial branch biomass and the limited sample size (six branches per treatment) or to the slightly higher temperatures within the bags that are conducive to increased growth and maturation of the needle and stem tissues (see Barton et al., 1993). Doubling of the supply of CO_2 (with no down-regulation of photosynthesis) should lead to an increase in availability of carbohydrates to the tissue. These, however, can be mobilized and exported for use in other parts of the tree (to the stem or roots for storage or to younger whorls for growth). Apparently the supply of carbohydrate was larger than could be exported as the needles were heavier and accumulated starch during the growing season (Fig. 3).

In all three tissue types, photosynthesis was doubled in CO_2-enriched plants, and little or no down-regulation of photosynthesis has been observed (see Barton et al., 1993; Centritto and Murray, personal communications).

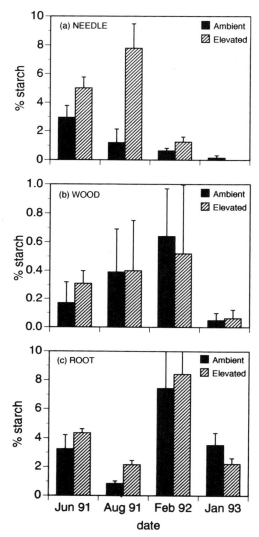

Figure 2 Starch (as % of total dry mass) of selected samples from the seedling Sitka spruce grown in ambient or elevated CO_2. Data relate to some of the harvests in Fig. 1 ($n = 3 \pm$ SE).

High rates of photosynthesis in Sitka spruce under elevated CO_2 parallel the assimilation of CO_2 into biomass (including carbohydrates). Plants that are root or nutrient restricted tend to down-regulate photosynthesis (Thomas and Strain, 1991; McConnaughay *et al.*, 1993), but no such effect

Table III Percentage Increase in Total Biomass of Four Clones of Sitka Spruce (Grown as before) and the Relative Increase (+) in Each Component, Needle, Wood, and Root Biomass, at the Times of Harvest[a]

Harvest date	Total biomass	Needle biomass	Woody biomass	Root biomass
March 1992	+45%	+25%	+41%	+34%
September 1992	+44%	+25%	+33%	+42%
February 1993	+46%	+32%	+34%	+34%
December 1993	+41%	+35%	+35%	+54%

[a] The experiment began in 1991. $n = 5$–10 plants per treatment and per harvest. Total biomass, woody biomass, and root biomass increases were all significant at $P = 0.05$. Needle biomass was only significant at $P = 0.05$ for the December 1993 harvest.

was observed here in the current year needles of mature trees. Similar results have been published for a 3-year study of *Liriodendron tulipifera* and *Quercus alba* (Gunderson *et al.*, 1993).

Losses of CO_2 from respiration by woody tissues, as well as by leaves at night, will influence the carbon balance. No significant differences have been found between rates of dark respiration in the three systems (C. Barton and M. Centritto, personal communications).

D. Comparison of Tissue Types

The repeated harvests of clonal and seedling plants at different times of the year over a number of years have shown the difficulty in interpreting

Table IV Branch Bag Sitka Spruce Change in Dry Mass to Fresh Mass Ratio of Current-Year Needles in the Three CO_2 Treatments ($n = 6$ per Treatment, Mean ± se)[a]

Sampling date	Needle age	Ambient CO_2	Elevated CO_2	Control CO_2
April 1991	1990	0.527 ± 0.005	0.525 ± 0.006	0.530 ± 0.009
June 1991	1991	0.323 ± 0.019	0.328 ± 0.025	0.260 ± 0.026
August 1991	1991	0.453 ± 0.022	0.456 ± 0.019	0.454 ± 0.037
June 1992	1991	0.448 ± 0.034	0.474 ± 0.012	0.455 ± 0.007
	1992	0.412 ± 0.037	0.426 ± 0.058	0.322 ± 0.040
June 1993	1991	0.459 ± 0.017	0.510 ± 0.023	0.467 ± 0.006
	1992	0.451 ± 0.011	0.477 ± 0.039	0.451 ± 0.013
	1993	0.214 ± 0.022	0.232 ± 0.031	0.173 ± 0.007
August 1993	1991	0.465 ± 0.005	0.478 ± 0.005	0.457 ± 0.004
	1992	0.448 ± 0.011	0.456 ± 0.007	0.446 ± 0.006
	1993	0.371 ± 0.009	0.362 ± 0.005	0.388 ± 0.006

[a] April 1991 represents baseline values before installation of the branch bags. Needles in 1991 developed only under elevated CO_2, while needles in 1992 were both laid down and developed under elevated CO_2.

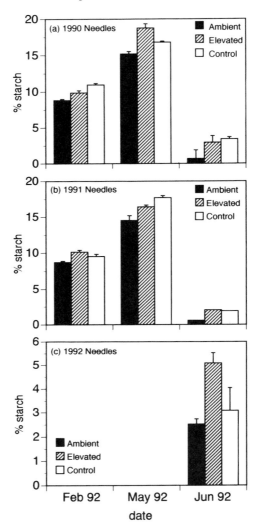

Figure 3 Starch (as % of total dry mass) from three needle age classes on mature Sitka spruce branches in ambient or elevated CO_2 or in control, unbagged branches: (top) 3-year-old; (middle) 2-year-old; (bottom) current-year needles ($n = 3 \pm SE$).

the results of experiments on CO_2 enrichment and the necessity for detailed, long-term experimentation. Had harvests only been made during the winter and had needle samples not been taken during the growing season, the conclusions from the experiment with seedlings would have been that elevated CO_2 does not significantly affect the growth and biomass of Sitka spruce; an increased root to shoot ratio in the first year would have been the only major effect of elevated CO_2 seen (Fig. 1b). By contrast, clonal Sitka spruce responds positively to CO_2 with an increase in biomass of 30–40% from the time that the plants were first exposed (Table III). In the lowland Scottish climate, growth of Sitka spruce seedlings can continue right through the winter months when some photosynthesis takes place (Bradbury and Malcolm, 1978); this, and rapid root growth of bare-rooted seedlings, partially explains the results from the first year harvests of seedlings when root/shoot ratios were found to increase in elevated CO_2 plants (Fig. 1b).

Nutrition also affects plant growth, but this was not limiting in the Sitka spruce experiments described here. Allsopp (1954) related maturation to the nutritional status of the plant and found reversion to juvenility in plants subjected to nutrient stress. However, Steele (1987) found that morphological characteristics of Sitka spruce were not influenced by nutrition, as juvenile tissue grafted into a mature crown remained juvenile even after 2 years of growth. The C to N ratio increased in elevated CO_2 because of increased carbohydrate accumulation (Figs. 2 and 3). Pot size, which has been implicated in a number of experiments as a cause of changed root to shoot mass ratios (Arp, 1991; Thomas and Strain, 1991), is not the underlying cause of the changes in allocation seen in the experiments described here (Fig. 1b). McConnaughay *et al.* (1993) have challenged the view that pots per se limit responses to elevated CO_2 and have clearly demonstrated that nutrition is the determining factor.

Seedling and clonal Sitka spruce responded differently to growth in elevated CO_2 in terms of cumulative increases in height, basal diameter, and total biomass. A major difference between the seedling and clonal trees is the maturity of tissue found in clones (Hackett, 1985). Needle dry mass/fresh mass ratio increased in new needles throughout the growing season in response to elevated CO_2 in both seedling and clonal plants and, in particular, in seedlings at midsummer in 1991 and 1992 (and winter 1993) when biomass was also larger in elevated CO_2 seedlings (Fig. 1a). This is likely to be related to the accumulation of starch, which was then used or broken down and reallocated during the winter (Fig. 2). The larger quantities of such storage carbohydrate accumulated in the elevated CO_2 plants would then be available for continued root and cambial growth in winter.

Lavender (1990) pointed out that the embryos of woody plants contain a very limited number of cells, and this distinguishes their patterns of growth from those of more mature tissue. Such indeterminate growth is found in the seedlings of Sitka spruce, which nearly always have a period of free growth late in the summer (Fig. 4, from day 185), whereas mature trees exhibit a determinate growth pattern. This behavior of seedlings

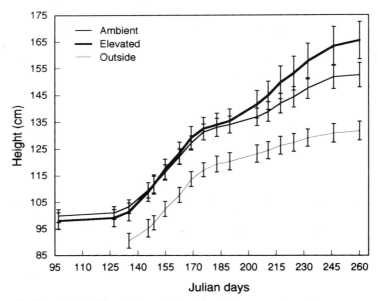

Figure 4 Growth in height of 4-year-old seedling Sitka spruce in ambient or elevated CO_2 or in outside control plots showing two periods of shoot growth: the first, that of the bud laid down the year before, and the second (from day 185), characteristic of juvenility and determined by nutrient status and temperature ($n = 15 \pm SE$).

causes them to be influenced much more by environment than mature, determinate plants, where growth is controlled primarily by the time required to expand initials. Lavender concludes that "studies with seedlings of determinate species may not accurately reflect the growth physiology of mature plants," and our results support this view. Furthermore, cyclophysis [expression in the vegetatively propagated individual of properties associated with the age of the parent plant or the developmental stage of the shoot used as the cutting or scion (Büsgen and Münch, 1929)] can be of major importance in assessing the responses of young trees to elevated CO_2. For example, Fielding (1970) found that the root to shoot ratio of *Pinus radiata* was much larger in cuttings than in seedlings of similar age. However, that was not the case in the current study with Sitka spruce.

The genetic potential, the maturation state of the donor plant, the morphology of the regenerated root system, and the vigor of the propagule influence the growth of clonal cuttings (Foster *et al.*, 1987). A comparison of seedling and clonal plants grown from 1- and 5-year-old source plants of the same parentage of loblolly pine showed that, after 4 years of growth, the plants derived from 1-year-old source plants were taller (and had more branches) than those derived from 5-year-old source plants, but that the latter were the same height as the seedlings. This advantage of the seedlings declined with time as they became more mature. The rooted cuttings continued to mature but at a slow rate, while the seedlings matured at a rapid rate during the first 4 years in the field, effectively catching up with the rooted cuttings in their stage of maturation. In loblolly pine, maturation or phase change seems to progress rapidly during the first 4–5 years for many traits, but then progresses more slowly as age increases. In our experiments with Sitka spruce, the slight reduction in growth rate in the clones in the third year is probably a reflection of this increasing maturity. Similar results were found for Douglas fir by Ritchie *et al.* (1992) and for Sitka spruce by Mason *et al.* (1986), who also found that cuttings were only successful from juvenile parent material. Steele, Yeoman, and Coutts (1990), also working with Sitka spruce, examined juvenility and maturity in relation to rooting ability and found a rapid falloff in rooting ability after 5 years and little or no rooting of trees of 11 years and older (tested up to 39-year-old trees). They concluded that only those trees that are not reproductively mature are able to be used for the production of clonal material and that clonal material does not represent so-called "mature" tissue, but is only an intermediate step.

Generally, seedling trees respond rapidly and positively in terms of total biomass to elevated CO_2 in the first year of growth, and this response gradually decreases with age. One-year-old seedling Sitka spruce in a combined CO_2 and nutrient treatment showed marked increases in height in response to elevated CO_2 when nutrients were nonlimiting (Murray *et al.*, 1994). This is likely to be a direct effect of nutrition, but it may also have been influenced by seed source and it may be reduced or disappear in future years. Only long-term investigation will clarify this. Clonal Sitka spruce consistently outperformed seedlings in terms of biomass increase, even when four unrelated clones were included in the analysis. However, the gap between the two tissue types is narrowing as they mature. The branches on mature trees had a limited potential to respond to elevated CO_2, but still showed small increases in needle and stem biomass. Such interactions of genotype, maturity, and environment are important for future modeling and generally interpreting data.

It is possible that there was an element of chance in the choice of material used in our experiments: other seed provenances and clones may have shown different responses. Other workers have found differing responses between ecotypes and clones of the same species. To test this, three additional unrelated clones and a seed lot (Murray *et al.*, 1994) of Sitka spruce have been subjected to CO_2 treatment. All have shown consistent increases in biomass with elevated CO_2, but the differences were most pronounced in the most nutrient-limited seedlings where allocation of biomass changed in response to elevated CO_2 treatment, favoring root growth and shortening of the growing season. We have now tested two seed batches and eight clones (including those of the branch bag experiment), and all have shown positive responses to long-term growth in elevated CO_2.

IV. Conclusions

1. Sitka spruce (*Picea sitchensis* Bong. Carr) is a major forest crop in the United Kingdom. Our results suggest that the predicted changes in climate are likely to affect the yield of this tree. Long-term data sets are necessary to understand the physiological responses and to model and then predict the likely impact of increasing CO_2 concentrations and temperature. Such modeling exercises will aid decision makers in determining whether to plant different provenances or change forestry practices.

2. Results have been presented for plants of three levels of maturity: seedlings, partially mature clonal plants, and mature branches. In terms of increases in plant mass, clonal Sitka spruce responded to growth in elevated CO_2 more rapidly and more positively than seedlings. A major difference between the two types of plant is the maturity of tissue found in clones. Fully mature tissue (branches) showed the least response to growth in elevated CO_2. The duration of experiments also influences the interpretation of results. Experimental designs for assessing the response of trees to elevated CO_2 must take both the duration and the degree of maturity of tissue into account before attempting to interpret the results or to utilize parameters in upscaling and modeling.

3. Growing plants in chambers at slightly elevated temperatures influences the timing (phenology) and rate of growth. The length of the growing season may also be affected both by a small increase in temperature and by elevated CO_2. The interaction with nutrient availability is a major influence.

4. There is a proliferation of contradictory data in the literature on the effects of elevated CO_2 on plant growth (Eamus and Jarvis, 1989). Continuation of long-term experiments with a variety of plant materials will elucidate physiological effects of elevated CO_2 on both seedlings and mature trees. The use of models, parametrized and tested with our data and against stand eddy covariance measurements, will put us in a better position to predict reliably the likely effects of global change on Sitka spruce forests.

Acknowledgments

This work was supported by the CEC under the EPOCH Programme and is now funded under the Environment Programme. A number of people have been involved in the experiments and in data analysis. We especially thank Craig Barton, Mauro Centritto, and Maureen Murray for allowing us to use their data for this review and Justine Crowe and Ian Leith for technical assistance.

References

Allsopp, A. (1954). Juvenile stages of plants and the nutritional status of the shoot apex. *Nature* 173, 1032–1034.

Arp, W. J. (1991). Effect of source-sink relations on photosynthetic acclimation to elevated CO_2. *Plant Cell and Environment*, 14, 869–876.

Barton, C. V. M., Lee, H. S. J. and Jarvis, P. G. (1993). A branch bag and CO_2 control system for long term CO_2 enrichment of mature Sitka spruce (*Picea sitchensis* (Bong.) Carr.). *Plant Cell Environ.* 16, 1139–1148.

Borchert, R. (1976). Differences in shoot growth patterns between juvenile and adult trees and their interpretation based on systems analysis of trees. *Acta Hort.* 56, 123–130.

Bradbury, I. K. and Malcolm, D. C. (1978). Dry matter accumulation by *Picea sitchensis* seedlings during winter. *Can. J. Forest Res.* 8, 207–213.

Büsgen, M., and Münch, E. (1929). "The Structure and Life of Forest Trees" (English translation by T. Thomson), pp. 436. Chapman & Hall Ltd.

Donovan, L. A., and Ehleringer, J. R. (1991). Ecophysiological differences among juvenile and reproductive plants of several woody species. *Oecologia* 86, 594–597.

Durrant, D., Lee, H. S. J., Barton, C. V. M., and Jarvis, P. G. (1993). A long-term carbon dioxide enrichment experiment examining the interaction with nutrition in Sitka spruce. *Forestry Commission Research Information Note*, No. 238.

Eamus, D., and Jarvis, P. G. (1989).The direct effects of increase in the global atmospheric CO_2 concentration on natural and commercial temperate trees and forests. *Adv. Ecol. Res.* 19, 2–55.

Fielding, J. M. (1970). Trees grown from cuttings compared with trees grown from seed (*Pinus radiata* D. Don). *Silvae Genet.* 19, 54–63.

Foster, G. S., Lambeth, C. C., and Greenwood, M. S. (1987). Growth of loblolly pine cuttings compared with seedlings. *Can. J. Forest Res.* 17, 157–164.

Gunderson, C. A., Norby, R. J., and Wullschleger, S. D. (1993). Foliar gas exchange responses of two deciduous hardwoods during 3 years of growth in elevated CO_2: no loss of photosynthetic enhancement. *Plant, Cell Environ.* 16, 797–807.

Hackett, W. P. (1985). Juvenility, Maturation and Rejuvenation in Woody Plants. *Hort. Rev.* 7, 109–155.

Jarvis, P. G. (1989). Atmospheric carbon dioxide and forests. *In* "Forests, Weather and Climate." *Philos. Trans. R. Soc. Ser. B*, pp. 310–340.

Jarvis, P. G. (1994a). Capture of carbon dioxide by a coniferous forest. *In* "Resource Capture" (K. Scott, M. H. Unsworth, and J. Monteith, eds.), pp. 351–374.

Jarvis, P. G. (1994b). MAESTRO: A model of CO_2 and water vapour exchange by forests in a globally changed environment. *In* "Design and Execution of Experiments on CO_2-Enrichment" (E. D. Schulze and H. A. Mooney, eds.), pp. 107–116.

Kozlowski, T. T. (1971). "The Growth and Development of Trees." Academic Press, New York.

Kozlowski, T. T., Kramer, P. J., and Pallardy, S. G. (1991). "The Physiological Ecology of Woody Plants." Academic Press, New York.

Kramer, P. J. (1981). Carbon dioxide concentration, photosynthesis and dry matter production. *Biosciences* 31, 29–33.

Lavender, D. P. (1990). Measuring Phenology and Dormancy. *In* "Techniques and Approaches in Forest Tree Ecophysiology" (J. P. Lassoie and T. M. Hinckley, eds.), pp. 404–418. CRC Press.

Lee, H. S. J., and Barton, C. V. M. (1994). Comparative Studies on Elevated CO_2 using Open-top Chambers, Tree Chambers and Branch Bags. In "Design and Execution of Experiments on CO_2-Enrichment" (E. D. Schulze and H. A. Mooney, eds.), pp. 239–259.

Luxmoore, R. J., Wullschleger, S. D., and Hanson, P. J. (1993). Forest responses to CO_2 enrichment and climate warming. *Water, Air Soil Pollut.* 70, 309–323.

Mason, W. L., Mannaro, P. M., and White, I. M. S. (1986). Growth and root development in cuttings and transplants of Sitka spruce 3 years after planting. *Scottish Forestry* 46, 276–284.

McConnaughay, K. D. M., Berntson, G. M., and Bazzaz, F. A. (1993). Limitations to CO_2-induced growth enhancement in pot studies. *Oecologia*, 94, 550–557.

Murray, M. B., Smith, R. I., Leith, I. D., Fowler, D., Lee, H. S. J., Friend, A. D., and Jarvis, P. G. (1994). The effect of elevated CO_2, nutrition and climatic warming on bud phenology in Sitka spruce (*Picea sitchensis* (Bong.) Carr) and its impact on frost tolerance. *Tree Physiology* 14, 691–706.

Poorter, H. (1993). Interspecific variation in the growth response of plants to an elevated ambient CO_2 concentration. In "CO_2 and Biosphere" (J. Rozema, H. Lambers, S. C. van de Gejn, and M. L. Cambridge, eds.), pp. 77–98. Kluwer Academic Publishers, Dordrecht (Reprinted from *Vegetatio* 104/105).

Ritchie, G. A., Tanaka, Y., and Duke, S. D. (1992). Physiology and morphology of Douglas-fir rooted cuttings compared to seedlings and transplants. *Tree Physiol.* 10, 179–194.

Robbins, W. J. (1961). Juvenility and induction of flowering. *Recent Adv. Bot.* 2, 1647–52.

Sprugel, D. G., Hinckley, T. M., and Schnap, W. (1991). The theory and practice of branch autonomy. *Annu. Rev. Ecol. Systematics* 22, 309–34.

Steele, M. J. (1987). Morphological and physiological changes with age for Sitka spruce, and their development as indices of physiological age. Ph.D. Thesis, University of Edinburgh.

Steele, M. J., Yeoman, M. M., and Coutts, M. P. (1990). Developmental changes in Sitka spruce as indices of physiological age. II. Rooting of cuttings and callusing of needle explants. *New Phytologist* 114, 111–120.

Thomas, R. B., and Strain, B. R. (1991). Root restriction as a factor in photosynthetic acclimation of cotton seedlings in elevated carbon dioxide. *Plant Physiol.* 96, 627–34.

Unsworth, M. H. (1986). Principles of microclimate and plant growth in Open-Top Chambers. In "Microclimate and Plant Growth in Open-Top Chambers," CEC Air Pollution Research Report 5.

Wareing, P. F. (1959). The problems of juvenility and flowering in trees. *J. Linnean Soc. London (Bot.)* 56, 282–289.

5

Growth Strategy and Tree Response to Elevated CO_2: A Comparison of Beech (*Fagus sylvatica*) and Sweet Chestnut (*Castanea sativa* Mill.)

Marianne Mousseau, Eric Dufrêne, Asmae El Kohen,
Daniel Epron, Denis Godard, Rodolphe Liozon,
Jean Yves Pontailler, and Bernard Saugier

I. Introduction

Studies by a number of research groups indicate considerable interspecific variations in the magnitude and nature of growth responses to elevated CO_2 in temperate deciduous trees [Eamus and Jarvis, 1989; Bazzaz et al., 1990; Mousseau and Saugier, 1992; Bazzaz and Miao, 1993; Ceulemans and Mousseau, 1994; Norby et al., 1996 (this volume)]. These variations may result in substantial changes in the composition and structure of deciduous forest communities in a future, high CO_2 world if the patterns observed in short-term experiments hold over successional time scales in natural situations (Bazzaz, 1990). It is thus of considerable ecological and economic importance to determine and describe the possible behavior of cooccuring major forest species in elevated CO_2.

As yet there is no clear basis for the classification of the CO_2 response of different deciduous tree species. There are indications, however, that the growth and developmental strategy of a given species seem to drive the repsonse to elevated CO_2 through effects on source–sink balance (Kaushal et al., 1989; El Kohen et al., 1993). Here we synthesize our previous research on the performance in elevated CO_2 of two important deciduous tree species of the western European forests, namely, beech (*Fagus sylvatica* L.) and sweet chestnut (*Castanea sativa* Mill.). These species differ in a number of growth and life history characteristics that are expected to influence source–sink balance and, thus, response to CO_2. Sweet chestnut is a relatively fast growing species; its growth rate is comparable to those of some poplar cultivars. It has a high resprouting capacity, which allows its cropping as a coppiced species. In contrast to sweet chestnut, beech is well-known as a shade tolerant and relatively slow growing species. The two species also differ in their growth patterns: sweet chestnut is characterized by indeterminate growth, whereas beech has a determinate growth pattern, sometimes showing two growth flushes per year. Our studies have focused primarily on the seedling stage of individuals raised in whole tree exposure chambers. Results of studies of mature beech trees treated with a new branch bag technique (Dufrêne *et al.*, 1993) are also presented for comparison to responses observed in seedlings.

II. Materials and Methods

A. Study Site and Plant Material

The studies have been conducted on the University campus of Orsay, using a modification of the open-top chamber design. The chambers are covered on top with transparent polypropylene and opened at the opposite ends for ventilation; air flow was horizontal and well mixed by fans (Mousseau and Enoch, 1989), providing either an ambient or a twice-ambient CO_2 atmospheric concentration. Four open-sided chambers (2 m^2 each) accommodated potted trees or young seedlings directly planted into the soil at a density of 100 m^{-2} for beech and 28 m^{-2} for sweet chestnut.

Similar experiments have been performed on sweet chestnut and beech seedlings introduced to the CO_2 facilities when 2 years old. Sweet chestnut, *Castanea sativa* Mill., has been grown both in 12-l pots containing fertilized or unfertilized sandy forest soil (El Kohen *et al.*, 1992) and in miniecosystems consisting of seedlings directly planted into a similar unamended forest soil at the same density (28 m^{-2}). Beech seedlings have also been potted and soil planted (17 m^{-2}). Two types of soils were used for this species: the same forest soil as used for chestnuts in the case of potted plants and a moderately fertile, sandy soil when seedlings were directly planted. In this last case, a high density (100 m^{-2}) was chosen to mimic a natural regeneration stand.

B. Analysis of Growth and Physiological Responses

1. Growth Destructive and nondestructive measurements were performed on all plant parts throughout 1 or 2 years for potted and soil-planted seedlings. Measurements included shoot length, leaf number and area, and dry matter per unit leaf area. Biomass was determined at the end of experiments or inferred from volume indices throughout the experiment.

2. Gas Exchange Measurements of shoot photosynthesis and dark respiration have been conducted by using a whole-plant closed system in the field, without climate control [described in El Kohen *et al.* (1993)], and an open gas exchange system in the laboratory for determination of the response of leaf assimilation to internal CO_2 (A–C_i curves).

3. Chemical Characteristics Nitrogen content of leaf, stem, and root samples was determined by elemental analysis of dried and ground material. Chlorophyll content was determined with the classical method of McKinney (1941) in 80% acetone extracts. Soluble sugar concentration was determined in ethanol extracts by the anthrone colorimetric method (Ashwell, 1957).

C. Statistical Analysis

Student *t*-tests were used to compare the significance of means. A level of probability less than 0.05 was considered significant and is quoted in figures and tables as *.

III. Results

A. Biomass and Partitioning

In sweet chestnut, total biomass production increased by 20 (low fertility) or 26% (high fertility) in elevated CO_2 (El Kohen *et al.*, 1992). The allocation pattern of this increase was dependent on nutrient conditions: in infertile soils, it was allocated mainly to the roots, whereas it was allocated mainly to the aboveground parts in fertilized soil (El Kohen *et al.*, 1992). Leaf area and number (Figs. 1A, 1B), as well as stem diameter and stem volume (Fig. 1D), were enhanced by elevated CO_2 only when nutrition was adequately supplied. The results were very different when sweet chestnut was planted directly in an unfertile soil (Fig. 2A). Here the total biomass increase due to CO_2 enrichment amounted to 45% and was equally allocated to different plant parts.

In potted as well as in soil-planted beech seedlings, the positive effect of elevated CO_2 was significant in all plant parts: leaves, stem, and root dry mass increased by 60–65% (Table I); as a consequence, the root/shoot ratio was not affected by the CO_2 treatment. When directly planted in soil, beech biomass per seedling doubled after 2 years of CO_2 treatment in comparison to control plants (Table IC). In potted plants, 1 year of CO_2 treatment resulted in a greater yearly dry mass enhancement for beech than for sweet chestnut (60 and 20% respectively) (Saugier *et al.*, 1993).

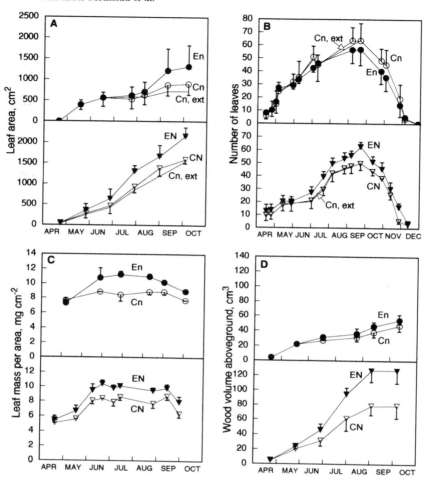

Figure 1 Effects of elevated CO_2 on 2-year-old sweet chestnut potted seedlings in terms of seasonal pattern of total leaf area (A), leaf number (B), leaf dry mass per unit area (C), and total wood volume (cm³) of aboveground parts (D) calculated from volume indices of trunk + branches. Mean values ± sd ($n = 24$). Symbols: E, elevated CO_2; C, control; n, low nutrient forest soil; N, fertilized forest soil; ext, nonchambered plants.

In miniecosystems with soil-planted seedlings, the yearly dry mass enhancement per plant was also greater for beech (94%) than for sweet chestnut (52%).

B. Leaf Area and Leaf Number

In young chestnut seedlings, a doubling in CO_2 concentration generally either did not affect leaf size (Mousseau, 1993) or slightly reduced it,

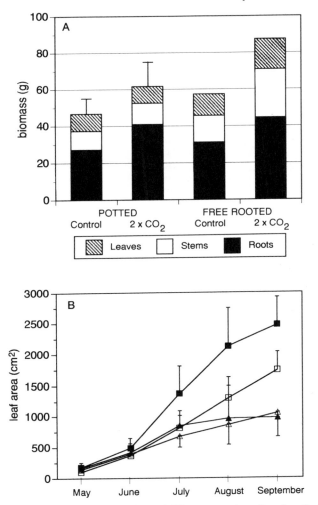

Figure 2 Comparison of the effect of elevated CO_2 on potted or directly soil-planted sweet chestnut seedlings. (A) Effect on the biomass partitioning of potted or free-rooted 2-year-old sweet chestnut seedlings, planted in an unfertile sandy forest soil. Potted plants: mean of 46 individuals ± sd. Free-noted plants: mean of total harvest divided by number of plants (46). (B) Effect on the seasonal pattern of leaf area development of 2-year-old potted sweet chestnut seedlings (triangles) compared to seedlings directly planted in the same soil (squares). Filled symbols, elevated CO_2; open symbols, control. Mean of 46 plants ± sd.

depending on the weather conditions of the year (Mousseau and Enoch, 1989). These results were, however, strongly modified by nutritional status (El Kohen et al., 1992). Figures 1A and 1B show that elevated CO_2 induced a greater number of leaves per plant and thus a greater total leaf area on

Table I Effect of CO_2 Enrichment on the Shoot Growth Characteristics and Biomass Partitioning of 2-Year-Old Beech Seedlings (*Fagus sylvatica* L.)[a]

	Ambient CO_2	Elevated CO_2	% increase
A. Mean Values (±sd) per Potted Seedling after 1-Year CO_2 Treatment (1991), $n = 28$			
Potted seedlings			
Total stem length (10^{-2} m)	83.7 ± 25	115.0* ± 55	+37
Stem diameter (10^{-3} m)	8.5 ± 1.23	9.1 ± 1.45	ns
Leaf area (10^{-2} m^2)	4.0 ± 2.5	5.7* ± 2.7	+42
Leaf dry weight (g)	2.76 ± 1.6	4.04* ± 1.9	+46
Stem dry weight (g)	8.6 ± 4.3	13.8* ± 5.5	+60
Root dry weight (g)	16.8 ± 6.6	27.6* ± 9.3	+64
Plant dry weight (g)	28.2 ± 11.9	45.8* ± 15.8	+62
Root/shoot ratio	1.59 ± 0.40	1.57 ± 0.32	
B. Mean Values (±sd) per Plant at the End of 1-Year CO_2 Treatment (1992) on Soil-Planted Beech Seedlings, $n = 200$			
Soil-planted seedlings			
Total stem length (10^{-2} m)	52.7 ± 1.28	71.2* ± 1.52	+35
Stem diameter (10^{-3} m)	5.1 ± 0.06	5.9* ± 0.12	+16
Volume index (10^{-6} m^3)	8.0 ± 0.27	15.5* ± 0.14	+94
Leaf area (10^{-2} m^2)	1.44	2.42*	+68
C. Mean Values at the End of 2-Years CO_2 Treatment (1992–1993) on Soil-Planted Beech Seedlings[b]			
Soil-planted seedlings			
Total stem length (10^{-2} m)	211.5 ± 7.9	276.3 ± 18.7	+30
Stem diameter (10^{-3} m)	8.1 ± 0.14	10.0 ± 0.27	+ns
Leaf area (10^{-2} m^2)	9.2	14.1	+53
Litter dry weight (g)	3.7	6.2	+67
Stem dry weight (g)	11.7 ± 0.5	22.4* ± 2.2	+91
Root dry weight (g)	11.5	25.8	+124
Plant dry weight (g)	26.9	54.4	+102

[a]* denotes a significant difference between CO_2 treatments ($P < 0.05$).
[b] Values correspond to total harvest divided by plant number (200), except for total stem length, stem diameter, and stem dry weight.

fertilized forest soil, but no significant changes were found when the plants were grown in relatively poor soil.

There was an important increase in total leaf area, unit leaf size, and leaf number in sweet chestnut planted directly into the soil compared to potted plants in both elevated CO_2 and control conditions. Total leaf area of free-rooted sweet chestnut was significantly increased with elevated CO_2 (Fig. 2B), although the bulk soil was the same as in pot experiments (i.e., same low level of nutrition in both cases).

Beech species, known to grow in several flushes especially at the sapling stage, gave rise to different leaf types. During the first year in elevated CO_2, more plants gave rise to a second flush than in ambient air (Table II). This result was found in free-rooted seedlings as well as in potted plants. As the second leaf flush usually presented a larger unit leaf area, elevated CO_2 resulted in a 42% increase in total leaf area per plant in potted seedlings (Table IA) and a 68% increase per plant in the miniecosystem of soil-rooted saplings after 1 year (Table IB). A similar, but less pronouced trend was observed during the second year of growth in high CO_2 (Tables IC and II), yielding an individual total leaf area increase of 53% at the end of the second experimental year.

C. Phenology

In sweet chestnut seedlings, bud burst was delayed 2–3 weeks at the beginning of the second growing season in elevated CO_2 (Fig. 3). Leaves of sweet chestnut also senesced slightly earlier in elevated CO_2 (leaf fall beginning 1 week earlier; Fig. 3).

During the first year of establishment of potted plants or free-rooted beech plants, elevated CO_2 significantly increased both the number of plants presenting a second flush (Table II) and the growth of this second flush (Fig. 4). Stem elongation, mainly due to the second flush growth, was strongly enhanced (+58%) in elevated CO_2 (Fig. 4). The situation was different in the second year of CO_2 treatment, during which the frequency of appearance and the growth of this second flush were not significantly influenced by elevated CO_2 (Table II).

Table II Percentage of Soil-Planted Beech Seedlings Exhibiting a Second Growth Flush and the Relative Importance of This Second Growth Flush during Two Consecutive Growing Seasons in Ambient or Elevated CO_2[a]

	1992		1993	
	Ambient CO_2	Elevated CO_2	Ambient CO_2	Elevated CO_2
Trees with second growth flush	17 a	43 b	87 a	81 a
Second flush length (% of total annual flush) for all trees	8 ± 1 a	26 ± 3 b	62 ± 2 a	48 ± 3 b
Second flush length (% of total annual flush) for trees with second flush)	46 ± 4 a	60 ± 3 b	72 ± 1 a	60 ± 2 b

[a] Within a growing season, significant differences between CO_2 treatments ($P < 0.01$) are designated by different letters.

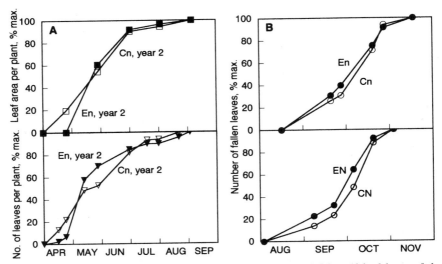

Figure 3 Seasonal pattern of leaf fall after 1 year in elevated CO_2 and bud burst of the following year (2y: second year in elevated CO_2) in potted sweet chestnut seedlings. Same symbols as in Fig. 1.

D. Photosynthesis and Respiration

In potted sweet chestnut, a seasonal change in photosynthetic enhancement showed a net downward acclimation of photosynthesis at low nutrient levels. Slopes and plateaus of $A-C_i$ curves were higher in elevated CO_2 only in May and then were lower from July to the end of the season in CO_2-enriched plants than in ambient (El Kohen and Mousseau, 1994). This acclimation did not occur when fertilizers were applied to the trees (El Kohen and Mousseau, 1994). In contrast, no significant down-regulation of the light-saturated photosynthetic rate was observed throughout the season when plants were directly planted into the soil (Liozon, 1994).

The greater effect of elevated CO_2 on biomass accumulation in beech compared to sweet chestnut was associated with a significantly greater increase in seedling net photosynthesis in beech. In potted beech plants, photosynthesis was higher in elevated than in ambient CO_2 during the whole season, with no evidence of acclimation (El Kohen et al., 1993). Similarly, no downward regulation of photosynthesis was noticeable on branches enclosed in CO_2-enriched bags (Fig. 5) during two consecutive years [for results of the first year experiment, see Saugier et al. (1993), and for the second year, see Fig. 5].

A direct, instantaneous, and reversible effect of elevated CO_2 on the dark respiration rate was found in sweet chestnut (El Kohen et al., 1991). Shoot respiration decreased during the night in response to elevated CO_2 in both

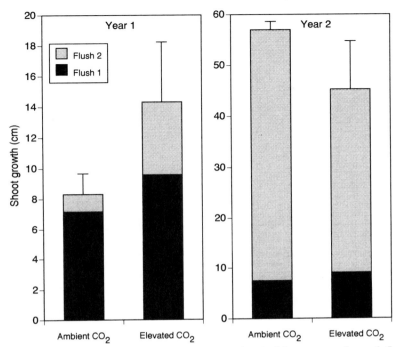

Figure 4 Effect of elevated CO_2 on the mean growth ($\pm$se) of the two annual growth flushes in soil-planted beech seedlings during two consecutive experimental years.

fertilized and unfertilized sweet chestnuts; this decrease was not related to sugar concentration (El Kohen and Mousseau, 1994). The decreasing effect of elevated CO_2 on night respiration was less pronounced after mid-July (Mousseau, 1993), but was shown to occur in the first part of the season in all situations, on a dry mass basis as well as on a nitrogen basis (Fig. 6; El Kohen, 1993).

E. Chemical Characteristics

1. Chlorophyll Content In potted sweet chestnut, chlorophyll concentrations were lower in elevated CO_2 when expressed on a biomass basis, as well as on a leaf area basis (Table III). In soil-planted sweet chestnut and beech, the increase in leaf dry weight per unit area fully compensated for the decrease in chlorophyll per unit dry mass, so that the area–based chlorophyll values were unchanged (Epron, personal comunication).

2. Nitrogen Content In sweet chestnut grown in elevated CO_2, leaf nitrogen was lower on a dry mass basis regardless of nutrient treatment (Fig. 7). However, the decrease in plant nitrogen per unit dry mass in elevated CO_2 was compensated for by an increase in total plant dry mass (Table IV),

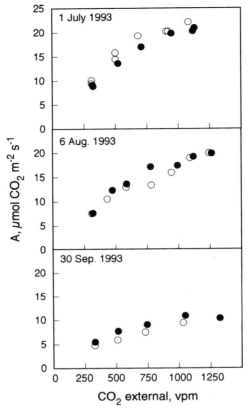

Figure 5 Effect of external CO_2 atmospheric concentration on the net photosynthetic activity of a branch of an adult beech tree submitted to continuous elevated CO_2 treatment. Results were obtained during the second experimental year at three dates during the vegetation cycle: filled circles, branch in elevated CO_2; open circles, branch in ambient air.

leading to a constant N pool per plant (El Kohen *et al.*, 1992). Moreover, after 2 years of CO_2 enrichment, as more fine roots were produced (Rouhier, 1994), the nitrogen pool size of fine roots was higher in elevated CO_2; the plants invested a larger proportion of total nitrogen in fine roots under conditions of decreased plant nitrogen concentration (Rouhier, 1994).

Beech seedlings showed a smaller reduction in nitrogen concentration in elevated CO_2 than chestnut. In terms of the total nitrogen pool, the dry mass increase more than compensated for the reduction in nitrogen concentration, so that beech seedlings in elevated CO_2 contained a larger

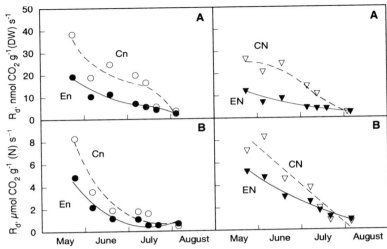

Figure 6 Seasonal pattern of the end-of-night dark respiration rate of potted sweet chestnut seedlings, expressed on a dry weight basis of the total shoot (A) or on a foliar nitrogen basis (B). Symbols: C, control; E, elevated CO_2; n, low nutrient forest soil; N, fertilized forest soil.

total amount of nitrogen per plant at the end of the growing season (Table IV).

3. Carbohydrates In sweet chestnut, elevated CO_2 promoted starch accumulation in both low and high nutrient treatments (Fig. 7), and this accumulation appeared earlier in the low nutrient treatment than in the fertilized one (El Kohen and Mousseau, 1994). The seasonal change in soluble

Table III Chlorophyll Content in Sweet Chestnut Adult Leaves at Two Levels of Atmospheric CO_2 and Two Levels of Mineral Nutrition[a]

	Chl a (mg m^{-2})	Chl b (mg m^{-2})	Chl total (mg m^{-2})	Chl a (mg g^{-1} DW)	Chl b (mg g^{-1} DW)	Chl total (mg g^{-1} DW)
Cn	190 ± 22	60 ± 3	259 ± 25	2.67	0.85	3.52
En	127 ± 11	56 ± 8	183 ± 19	1.29	0.57	1.86
En/Cn	0.66	0.93	0.71	0.48	0.67	0.52
CN	344 ± 34	123 ± 4	467 ± 38	4.41	1.58	5.99
EN	375 ± 6	135 ± 3	510 ± 9	3.83	1.38	5.20
EN/CN	1.08	1.09	1.09	0.86	0.87	0.86

[a] Symbols: C, control; E, elevated CO_2; n, forest soil; N, fertilized forest soil. Means of six pooled samples (±sd) each taken from the fifth leaf. Measurements were made in mid-June.

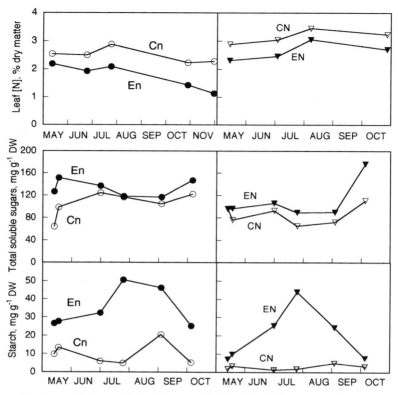

Figure 7 Seasonal changes in leaf nitrogen (top) and soluble sugars and starch (bottom) concentrations of chestnut leaves grown in elevated CO_2 (filled symbols) or ambient CO_2 (open symbols) in two nutritional treatments (N, fertilized; n, unfertilized). No error bars are indicated because analyzed samples (2 g of fresh material) represent the mean values of several leaf disks punched on 5–8 different leaves and plotted together.

sugar content was parallel in elevated CO_2 and in ambient air, with the amount being slightly greater in elevated CO_2. In fertilized plants, the accumulation of soluble sugars appeared later and increased greatly before leaf fall (El Kohen and Mousseau, 1994) (Fig. 7). No carbohydrate measurements were made in beech plants.

IV. Discussion

Comparison of the behavior of these two deciduous species shows that the growth strategy of a species is probably driving its response to elevated CO_2. With respect to the results described here, an essential component of a positive growth response to CO_2 appears to be the ability of a species

Table IV Effects of Elevated CO_2 on Nitrogen Concentration and Content of 2-Year-Old Seedlings of Sweet Chestnut (*Castanea sativa*) and Beech (*Fagus sylvatica*)[a]

	N concentration (%)		N pool size (g organ^{-1})	
	C	E	C	E
Potted Sweet Chestnut (*Castanea sativa*)				
Litter	0.88	0.98	0.10	0.13
Stems	0.75 ± 0.10 a	0.67 ± 0.10 b	0.12 ± 0.03 a	0.12 ± 0.05 a
Coarse roots	1.27 ± 0.23 a	1.01 ± 0.20 b	0.38 ± 0.10 a	0.35 ± 0.04 a
Fine roots	1.92 ± 0.22 a	1.62 ± 0.29 a	0.04 ± 0.01 a	0.06 ± 0.01 b
Total	1.06 ± 0.04 a	0.92 ± 0.05 b	0.64 ± 0.10 a	0.66 ± 0.10 a
Potted beech (*Fagus sylvatica*)				
Litter	1.34 ± 0.22 a	0.95 ± 0.48 b	0.049 ± 0.021 a	0.048 ± 0.012 a
Stems	1.08 ± 0.15 a	0.88 ± 0.09 b	0.113 ± 0.040 a	0.116 ± 0.040 a
Buds	1.27 ± 0.29 a	1.30 ± 0.12 a	0.018 ± 0.009 a	0.022 ± 0.007 b
Roots	1.29 ± 0.80 a	1.25 ± 0.18 a	0.252 ± 0.008 a	0.320 ± 0.080 b
Total	1.22 ± 0.05 a	1.12 ± 0.40 a	0.433 ± 0.146 a	0.512 ± 0.120 b

[a] Nitrogen pool sizes were calculated as the product of mean N concentration and mean dry weight of the considered organ; C = control, E = elevated CO_2. Within a row, means by a different letter differ significantly ($P < 0.05$) between CO_2 treatments; n = 17 for sweet chestnut, n = 8 for beech.

to form new sinks in order to cope with the CO_2-induced surplus of carbohydrates.

In the case of pot-grown sweet chestnuts, a source–sink balance explained much of the variability of the response of the seedlings in the two contrasting nutritional conditions. The same type of source–sink explanation appears to hold for pot-grown beech, with its greater growth response to CO_2 seemingly due to the greater phenological plasticity of its leaf system. In seedlings presenting a second flush, leaves are much larger, thus increasing the total carbon input, as well as constituting new sinks for additional fixed carbon. The absence of down-regulation of photosynthesis in beech in elevated CO_2 may also be related to this capacity to maintain new sinks. Consistent with this view, when branches of mature beech trees are exposed to elevated CO_2 in branch bags, the whole tree may act as a sink for surplus carbohydrates (Dufrêne *et al.*, 1993).

The changes in allocation patterns in elevated CO_2 clearly were nutrition dependent in sweet chestnut. Moreover, in the same unfertile soil there was a great difference between potted and soil-planted plants, the latter behaving like the fertilized potted plants in terms of allocation. The difference in CO_2 responses of potted and free-rooted sweet chestnut was probably related to a "root restriction" factor induced by pot binding (Thomas and Strain, 1991; McConnaughay *et al.*, 1993). It has been shown that, in response to elevated CO_2, root elongation rate and total length increase (Curtis *et al.*, 1990; Rogers *et al.*, 1992, 1993) and belowground spatial deployment changes (Berntson et al., 1993). In bulk soil, the space available for root deployment may favor better access to nutrients, thus mimicking the effect of better nutrition observed in fertilized, potted sweet chestnuts. This emphasizes the fact that studies on the effect of elevated CO_2 on tree seedlings are better conducted with miniecosystems (unpotted plants) than with potted plants, where artifacts due to pot binding may occur.

The total amount of nitrogen acquired per tree did not differ between CO_2 treatments in potted sweet chestnut, whereas it increased in elevated CO_2 in beech, the latter being consistent with a previous study (Overdieck, 1990). From the results described in Table IV, we may assume that potted sweet chestnuts absorbed all of the mineralized nitrogen available in the pot, while the faster growing beech absorbed only part of it. Thus beech, but not sweet chestnut, may be able to increase nitrogen absorption in elevated CO_2. This could lead to decreased soil fertility in beech stands if increased microbial activity in the rhizosphere does not compensate for increased root absorption. as has been described for sweet chestnut (Rouhier et al., 1994).

In the case of beech seedlings planted directly in soil, elevated CO_2 increased the frequency of a second growth flush and resulted in two distinct size classes of plants, which were unchanged after the second growing season. We conclude that elevated CO_2 differentially enhanced the growth of some individuals. These plants displayed a polycyclic pattern of growth and were responsible for accelerating the development of a multilayered height structure in which the lower layer probably will disappear by self-thinning. In a regenerating beech stand, this phenomenon would probably accelerate the phase of seedling competition in elevated CO_2.

Comparison of the behavior of potted chestnut and beech species in elevated CO_2 shows that the developmental pattern was one of the keys to the allocation of carbon: there was increased carbon allocation to the roots in the indeterminate species (*Castanea*) and increased allocation to new stems, branches, and leaves in case of the determinate species (*Fagus*). Thus, for potted plants, the growth pattern seemed to drive the response to elevated CO_2. When these species were directly planted into soil in miniecosystems, however, the difference was less evident. Both responded to elevated CO_2 with similar increases in leaf allocation, the major difference between the species being in the magnitude of the response, which was much larger in beech (63% after 1 year of treatment, Table I) than in sweet chestnut (36.5%, Fig. 2). It must be noted that this comparison does not hold if the results are based on soil area (LAI), because the densities of the seedlings were different in the two miniecosystem experiments. The growth plasticity of a beech seedling thus may be more efficient in terms of increasing sink size than the classical leaf size increase found in sweet chestnut, and this may be related to improved nutrition.

In conclusion, the developmental pattern and plasticity of source and sink structures may determine the specific response of plants to elevated CO_2. In some genotypes, increased allocation of dry matter to roots may be driven by a demand for greater nutrient acquisition, whereas in others the growth strategy may be oriented toward increasing leaf area and light capture. A great number of species reactions will have to be evaluated before we are able to assess functional types within tree genotypes with respect to responses to increasing atmospheric CO_2.

References

Ashwell, G. (1957). Colorimetric analysis of sugars. *In* "Methods in Enzymology" (S. P. Colowick and N. O. Kaplan, eds.), Vol. III, pp. 73. Academic Press, New York.

Bazzaz, F. A. (1990). The response of natural ecosystems to the rising global CO_2 levels. *Annu. Rev. Ecol. Syst.* 21, 167–196.

Bazzaz, F. A., and McConnaughay, K. D. M. (1993). Plant-plant interactions in elevated CO_2 environments. *Aust. J. Plant Phys.* 40, 547–563.

Bazzaz, F. A., and Miao, S. L. (1993). Successional status, seed size, and responses of tree seedlings to CO_2, light and nutrients. *Ecology* 74, 104–112.

Bazzaz, F. A., Coleman, J. S., and Morse, S. R. (1990). Growth respones of seven major co-occuring tree species of the northeastern United States to elevated CO_2. *Can. J. Forest Res.* 20, 1479–1484.

Berntson, G. M., McConnaughay, K. D. M., and Bazzaz, F. N. A. (1993). Elevated CO_2 alters the deployment of roots in "small" growth containers. *Oecologia* 94, 558–564.

Ceulemans, R., and Mousseau, M. (1994). Effects of elevated atmospheric CO_2 on woody plants. Tansley Review No. 71. *New Phytol.* 127, 425–446.

Curtis, P. S., Balduman, L. M., Drake, B. G., and Whigham, D. F. (1990). Elevated atmospheric CO_2 effects on belowground processes in C_3 and C_4 estuarine marsh communities. *Ecology* 71, 2001–2006.

Dufrêne, E., Pontailler, J. Y., and Saugier, B. (1993). A branch bag technique for simultaneous CO_2 enrichment and assimilation measurements on beech (*Fagus sylvatica* L.). *Plant, Cell Environ.* 16, 1131–1138.

Eamus, D., and Jarvis, P. G. (1989). The direct effect of increase in the global atmospheric concentration of CO_2 on natural and commercial temperate trees and forests. *Adv. Ecol. Res.* 19, 1–55.

El Kohen, A. (1993). Effet d'un enrichissement en CO_2 sur la croissance et les échanges gazeux des jeunes plants de châtaignier (*Castanea sativa* Mill.). Theses Doct. Sci., Université Paris-Sud, Orsay, France, 177 pp.

El Kohen, A., and Mousseau, M. (1994). Interactive effects of elevated CO_2 and mineral nutrition on growth and CO_2 exchange of sweet chestnut seedlings (*Castanea sativa*). *Tree Physiol.* 14, 679–690.

El Kohen, A., Pontailler, J. Y., and Mousseau, M. (1991). Effet d'un doublement du CO_2 atmosphérique sur la respiration a l'obscurité des parties aériennes de jeunes châtaigniers (*Castanea sativa* Mill). *C.R. Acad. Sci. Paris* 312, 477–481.

El Kohen, A., Rouhier, H., and Mousseau, M. (1992). Changes in dry weight and nitrogen partitioning induced by elevated CO_2 depend on nutrient availability in Sweet Chestnut (*Castanea sativa* Mill.). *Ann Sci. For.* 49, 83–90.

El Kohen, A., Venet, L., and Mousseau, M. (1993). Growth and photosynthesis of two deciduous forest tree species exposed to elevated carbon dioxide. *Funct. Ecol.* 7, 480–486.

Kaushal, P., Guehl, M., and Aussenac, G. (1989). Differential growth response to atmospheric carbon dioxide enrichment in seedlings of *Cedrus atlantica* and *Pinus nigra* ssp. *Laricio* var Corsicana. *Can. J. Forest. Res.* 19, 1351–1358.

Liozon, R. (1994). Variabilité des parametre foliaires dans des mini-couverts de hêtres et de châtaigniers. Diplome d'étude approfondies spécialité "Ecologie générale et production végétale", Université Paris-Sud Orsay, France, 37 pp.

McConnaughay, K. D. M., Berntson, G. M., and Bazzaz, F. A. (1993). Limitations to CO_2- induced growth enhancement in pot studies. *Oecologia* 94, 550–557.

McKinney, G. (1941). Absorption of light by chlorophyll solutions. *J. Biol. Chem.* 140, 315–322.

Mousseau, M. (1993). Effects of elevated CO_2 on growth, photosynthesis and respiration of sweet Chestnut (*Castanea sativa* Mill.). *Vegetatio* 104–105, 413–419.

Mousseau, M., and Enoch, Z. H. (1989). CO_2 enrichment reduces shoot growth in sweet chestnut seedlings (*Castanea sativa* Mill). *Plant, Cell, Environ.* 12, 927–934.

Mousseau, M., and Saugier, B. (1992). The direct effect of increased CO_2 on photosynthesis and growth of forest tree species. *J. Exp. Bot.* 43, 1121–1130.

Norby, R. J., Wullschlezer, S. D., and Gunderson, C. A. (1996). Tree Responses to Elevated CO_2 and Implications for Forests. *In* "Carbon Dioxide and Terrestrial Ecosystems" (G. W. Koch and H. A. Mooney, eds.). Academic Press, San Diego.

Overdieck, D. (1990). Effects of elevated CO_2 concentration levels on nutrient contents of herbaceous and woody plants. *In* "The Greenhouse Effect and Primary Productivity in European Agro-ecosystems" (J. Goudriaan, H. Van Keulen, and H. H. Van Laar eds.) Wageningen, Pudoc.

Rogers, H. H., Peterson, C. M., McCrimmon, J. N., and Cure, J. D. (1992). Response of plant roots to atmospheric elevated carbon dioxide. *Plant, Cell, Environ.* 15, 749–752.

Rogers, H. H., Runion, G. B., and Krupa, S. V. (1993). Plant responses to atmospheric CO_2 enrichment with emphasis on roots and the rhizosphere. *Environ. Pollut.*, 155–189.

Rouhier, H. (1994). Réponse du Chataignier (*Castanea sativa* Mill.) a l'augmentation du CO_2 atmosphérique: croissance et activité rhizosphérique. Theses Doct. Sci., Université Claude Bernard, Lyon I, France.

Rouhier, H., Billäs, G., El Kohen, A., Mousseau, M., and Bottner, P. (1994). Effect of elevated CO_2 on carbon and nitrogen distribution within a tree (*Castanea sativa* Mill.)-soil system. *Plant Soil* 162, 281–292.

Saugier, B., Dufrêne, E., El Kohen, A., Mousseau, M., and Pontailler, J. Y. (1993). CO_2 enrichment on tree seedlings and branches of mature trees. *In* "Design and Execution of Experiments on CO_2 Enrichment" (E.-D. Schulze and H. A. Mooney, eds.), Ecosystems Research Report No. 6, pp. 221–230. CEC Publications.

Thomas, R. B., and Strain, B. R. (1991). Root restriction as a factor in photosynthetic acclimation of cotton seedlings grown in elevated carbon dioxide. *Plant Physiol.* 96, 627–634.

6
Litter Quality and Decomposition Rates of Foliar Litter Produced under CO_2 Enrichment

Elizabeth G. O'Neill and Richard J. Norby

I. Introduction

Decomposition of senesced plant material is one of two critical processes linking above- and belowground components of nutrient cycles. As such, it is a key area of concern in understanding and predicting ecosystem responses to elevated atmospheric CO_2. Just as root acquisition of nutrients from soils represents the major pathway for nutrient movement from the soil to vegetation, decomposition serves as the major path of return to the soil. For any given ecosystem, a long-term shift in decomposition rates could alter nutrient cycling rates and potentially change the structure, the function, and even the persistence of that ecosystem type within a given region. There is widespread concern that decomposition processes would be altered in an enriched-CO_2 world (Strain and Bazzaz, 1983; Melillo et al., 1990; Schimel, 1993). What is lacking is sufficient experimental data *at the ecosystem level* to determine whether these concerns have merit.

Elevated CO_2 in the atmosphere could have an impact on decomposition rates through several means: through changes in the species composition of ecosystems, through direct effects on decomposer communities, or through changes in the chemical composition of litter ("litter quality"). Changes in the species composition of ecosystems resulting from differential responses to elevated CO_2 may have the greatest impact on decomposition rates, as proportionate litter quality changes might occur (Ågren et al., 1991; Pastor and Post, 1988). Amounts of carbon-based secondary metabolites and recalcitrant materials such as lignin vary among plant species. The relative contributions of differing plant species to an annual cohort of litter will in large part determine ecosystem-level decomposition dynamics. Communities composed of plant species that respond differently to CO_2 enrichment would be the most likely to experience changes in nutrient cycling rates. For example, physiological responses to elevated CO_2 and drought between loblolly pine (*Pinus taeda*) and sweet gum (*Liquidambar styraciflua*) led to a prediction that sweet gum would eventually displace loblolly pine where they co-occur (Tolley and Strain, 1985). Because these two species differ in litter quality, the annual litter fall would presumably mirror changes in dominance within the system, and ecosystem decomposition and nutrient cycling would presumably change through time. Replacement of one species by the other might result not only from the relative responses of the two species to elevated CO_2 but also, in this case, from an increase in nutrient cycling rates as the more recalcitrant litter of pine gives way to the relatively more decomposable sweet gum.

A second hypothesized effect of elevated CO_2 would be direct impact on the decomposer community that results in reduced decomposition. However, although detritivores may be indirectly affected by CO_2-induced changes in litter palatability or nutritional quality, the strong concentration gradient that exists between soils and the atmosphere makes direct effects on soil biota unlikely (O'Neill, 1994).

The greatest area of concern lies in the possibility of changes in the chemical composition of plant litter. Increased C:N ratios of litter combined with increased amounts of soluble phenolics and structural compounds such as cellulose and lignin are hypothesized to be one consequence of growth at elevated CO_2 (Strain and Bazzaz, 1983; Melillo, 1983) and could reduce decomposition rates. Should rates be reduced, either absolute amounts or temporal patterns of availability of some nutrients (particularly N) would be altered, establishing a positive feedback loop that further lowers litter quality and the availability of some nutrients (Pastor and Post, 1988).

This paper will emphasize those changes that result from vegetation responses to atmospheric CO_2 enrichment. However, changes in global climate that result from increased CO_2 in the atmosphere would also affect decomposition and nutrient cycling. Decomposition rates are dependent upon soil moisture and temperature, in addition to initial nutrient concentrations in the fallen litter (Pastor and Post, 1987). Changes in temperature and precipitation patterns could reduce decomposition rates, or even increase them in such a way as to offset decreased litter quality. Decreased stand-level transpiration rates could increase soil water potential (Eamus and Jarvis, 1989), again increasing or decreasing decomposition rates.

Thus, CO_2 effects on decomposition and nutrient cycling could be a complex interaction of plant litter quality, changes in ecosystem structure, and alterations in temperature and moisture through climate change. This paper will consider elevated CO_2 effects on the decomposition of leaf litter. Total decomposition in an ecosystem could also be affected by CO_2 fertilization through changes in the relative input to soil of foliar litter compared to fine root and woody litter or by a long-term change in the species composition of the ecosystem.

Two issues will be discussed: effects of CO_2 enrichment on foliar litter quality and subsequent effects on decomposition rates. The focus will be primarily on N for two reasons: (1) in many terrestrial ecosystems, N is the major nutrient limiting plant growth, and (2) experimental results from diverse ecosystem types have demonstrated that N concentrations are consistently reduced in green foliage produced at elevated CO_2 (Williams *et al.*, 1986; Curtis *et al.*, 1989; Norby *et al.*, 1992). We will also raise two important methodological questions. Much of the available data results from studies on plants grown in pots, often in environmental chambers. To what extent can we extrapolate from litter quality or decomposition data obtained in this manner to real ecosystems? How do we resolve the issue of commonly used "square-wave" experimental exposures that utilize rapid, single step CO_2 elevation to predict changes that will occur in nutrient cycling in ecosystems over decades of slow increases in CO_2 in the atmosphere?

II. Litter Quality and the Decomposition Process

Rates of litter decomposition have been correlated with initial nutrient concentrations in litter, as well as with the ratios of these nutrients to carbon compounds. For example, litter mass loss has been correlated with initial N concentrations (Taylor *et al.*, 1989), initial lignin concentrations (Aber *et al.*, 1990; Berg and McClaugherty, 1987), C:N ratios (Taylor *et al.*, 1989), or lignin:N ratios (Melillo *et al.*, 1982; Aber and Melillo, 1982; Melillo *et al.*, 1984). Predictions of CO_2-induced reductions in decomposition rates have been based primarily upon these observations.

However, in many cases where experimental results have led to the suggestion that decomposition rates will change under elevated CO_2, predictions have been based not upon concentrations in naturally abscised foliage produced under CO_2 enrichment, but on concentrations in green leaves. Strain and Bazzaz (1983) discuss the effects of changing tissue quality on both herbivores and decomposers, without making the distinction between green and senesced foliage concentrations. Woodward *et al.* (1991) state that rates of decomposition should decrease due to the "well established and associated increase in the C/N ratio of individual leaves." However, only two studies on leaf litter (see the following), with conflicting results, are presented to support this contention.

A decrease in green leaf N concentration with elevated CO_2 has been observed in many CO_2-enrichment studies (Curtis et al., 1989; Norby et al., 1992; Koch and Mooney, 1996). If green leaf concentrations were good indicators of litter concentrations, then a prediction of reduced litter quality under elevated CO_2 could be made with more confidence. However, in perennial plants, N concentrations usually differ between green and senesced foliage. In woody deciduous species, approximately one-half of the N content of green foliage is withdrawn from senescing tissue prior to leaf fall and retranslocated to rapidly growing tissues or stored in stem or roots until new growth is initiated (Chapin et al., 1990; Cromack and Monk, 1975; Duvigneaud and Denaeyer-De Smet, 1970). Retranslocation of nutrients in coniferous species occurs before needles are shed, but it may also occur seasonally in species where needles persist over multiple years (Nambiar and Fife, 1987). Nitrogen is one of the primary nutrients remobilized, although other nutrients, as well as soluble carbon compounds, are withdrawn. The proportion of N retranslocated varies by species and may be correlated with the N status of the soil (Chapin and Kedrowski, 1983). If retranslocation results in N withdrawal prior to abscission, why does a CO_2-induced decrease in green leaf N not translate into lowered N concentrations in litter? The explanation may lie in the allocation of leaf N to structural vs soluble forms of N. If reductions in green leaf N reported in the literature are actually reductions in soluble N compounds (such as free amino acids, for example), and soluble N comprises the bulk of leaf N remobilized during abscission, then N concentrations in litter might be independent of the CO_2 concentration under which they were produced.

Recalcitrant C compounds frequently increase in concentration (% of dry mass) during senescence as other compounds are retranslocated. If N alone is translocated, with no corresponding change in nonnitrogenous C compounds, then remobilization could result in an even greater difference in C:N ratios under CO_2 enrichment than is seen in green leaves. Although there may be little retranslocation of carbonaceous materials during senescence (Chapin et al., 1990), there is some evidence to the contrary: that movement of C occurs from leaf to tree and vice versa. Horner et al. (1987) observed that the tannin content of Douglas fir (*Pseudotsuga menziesii*) increased in foliage during senescence in closed-canopy stands and suggested that active sequestration in litter may occur.

Litter decomposition occurs as the result of three interacting processes: (1) leaching of soluble compounds and physical fragmentation by abiotic forces such as wind; (2) fragmentation (or comminution) by soil fauna; and (3) microbial catabolism. These processes may be interdependent, as when leaf fragments are consumed by earthworms and then enriched and inoculated with bacteria and fungi during passage through the gut (Edwards and Fletcher, 1988). The first and third processes are heavily regulated by litter chemical composition; comminution and feeding by soil fauna may also be affected by litter chemistry and "palatability" to soil animals (Anderson, 1973). Because these processes occur simultaneously and/or sequentially, it is often difficult to predict ecosystem decomposition rates simply on the basis of the chemistry of individual litter types (Blair *et al.*, 1990).

The most sensitive process with regard to potential effects of rising CO_2 is the chemical breakdown of litter by microorganisms. Nitrogen transformations are especially important, since in many of the world's ecosystems plant growth is limited by N availability. The decomposition of lignin and production of humic substances are also microbially mediated and result in the immobilization of at least a part of photosynthetically fixed C into long-lived soil C pools.

Several models have been proposed to explain decomposition processes [see Taylor and Parkinson (1988) for a comprehensive listing of commonly used models]. The most widely accepted model, and possibly the most relevant for assessing CO_2 effects on litter decomposition, is the two-phase, negative exponential model. Use of this model allows consideration of the effects of CO_2-induced changes in both N and total C or lignin in litter. According to the two-phase model, the initial N content of litter controls decomposition rates during early decay, while lignin (in substrates with high lignin content) is the major control factor later (Taylor *et al.*, 1989). However, rates of mass loss during early stages may also depend upon the total nonstructural carbon (TNC) content of litter. If TNC concentrations are higher in litter produced at elevated CO_2 (which has been demonstrated in green leaves; Körner and Arnone, 1992), then rates of mass loss initially may be *higher* than litter produced at ambient CO_2 and then lower in the latter stages of decay due to the production of recalcitrant humic materials, as nonstructural C is utilized by microbial decomposers. Reduced mass loss rates due to increased lignin or lignin-to-N ratios would also be manifested in the later stages of decomposition.

Mass loss rates may not be the most important parameter of decomposition with regard to ecosystem-level effects of elevated CO_2. The potential for changes in nutrient cycling rates, especially mineralization of N or P, is equally relevant to ecosystem response. During the early stages of decomposition, two opposing N cycling processes are occurring: immobilization and mineralization. Nitrogen, frequently from exogenous sources, is immobilized by bacteria and fungi; at this stage immobilization greatly exceeds mineralization. Simultaneously, labile C compounds in the litter are utilized by the microbes and C:N ratios decline, approaching those of the decomposer microorganisms. Microbial C:N ratios generally range from 4 to 15 (Brady, 1990; Alexander, 1977), although ratios may vary with substrate nutrient content (Dowding, 1976). Nitrogen mineralization eventually exceeds immobilization and N becomes available for further transformation and plant uptake. The point at which this switch from net immobilization to net mineralization occurs is called the "critical N concentration" (Pastor and Post, 1985) or "critical C:N ratio" (McClaugherty *et al.*, 1985) and is dependent upon several factors: initial N concentration of litter, exogenous N availability, and the composition of the decomposer population, as well as temperature and moisture conditions. Changes in initial litter content of labile C as a result of CO_2 enrichment could also affect the point in time at which the critical N concentration is reached, because soluble nonstructural carbohydrates serve as the primary C source for microorganisms initially colonizing litter.

III. CO_2 Enrichment Effects on Litter Quality and Decomposition

What information do we have that supports the assumption that leaf litter quality or decomposition rates will change under elevated CO_2? Responses to date fall into one of two categories: elevated CO_2 either dramatically increases litter nutrient ratios and decreases N concentrations or has little effect on either (Table I). At the ecosystem level, only herbaceous systems have been examined, although several autecological studies have been conducted on temperate deciduous forest species. Litter quality in tropical forest species has been studied in mixed assemblages in artificial model ecosystems. We are aware of no published information on litter quality responses to CO_2 enrichment in senescent conifer foliage, probably due to the long retention times of conifer needles on the tree. Much of the litter quality data presented here will be presented as unpublished data or as personal communication. This is unfortunate, but necessary because litter quality data frequently is not published until the decomposition study is concluded several years from the initial CO_2 exposure.

In an open-top chamber study conducted on a natural salt marsh, the responses of three communities to CO_2 were compared [Curtis *et al.*, 1989; see also Drake *et al.*, 1996 (this volume)]. Carbon:nitrogen ratios in green leaves of *Scirpus olneyi*, a C_3 sedge, were increased by 20–40% at elevated CO_2 whether grown in a pure or mixed (with *Spartina patens*, a C_4 grass) stand. However, ratios of senescent leaves were not different. It was predicted that, on a seasonal basis, little effect on decomposition or N availability would occur as a result of CO_2 enrichment. Most measures of litter quality also were not different in senesced foliage from a tall grass prairie exposed to twice-ambient elevated CO_2 in open-top chambers [Kemp *et al.*, 1994; Owensby *et al.*, 1993; see also Owensby *et al.*, 1996 (this volume)].

Table I Nutrient Ratios of Litter Produced at Elevated CO_2 for Different Species and with Differing Experimental Approaches[a]

Species	CO_2 enrichment (μmol mol^{-1})				Reference
	C:N		Lignin:N		
	+0	+300	+0	+300	
		Pot Studies			
Liquidambar styraciflua L.			11.8[c]	20.6[c]	Melillo (1983)
Castanea sativa Mill.	40.4	75.0			Coûteaux et al. (1991)
Quercus alba L.			5.7	4.8	Norby et al. (1986)
Quercus alba L. (1989)			12.3	16.3	R. Norby
(1990)			18.8	17.7	(unpublished data)
					R. Norby (unpublished data)
Liriodendron tulipifera L.[b]			1.7	4.4	Boerner and Rebbeck (1993)
Fraxinus excelsior L.	42.0	56.0	5.4	9.1	Cotrufo et al. (1994)
Betula pubescens Ehrh.	35.0	53.0	9.8	17.2	Cotrufo et al. (1994)
Acer pseudoplatanus L.	81.0	106.0	16.2	21.5	Cotrufo et al. (1994)
Picea sitchensis (Bong.) Carr.	18.0	18.0	4.6	6.0	Cotrufo et al. (1994)
		Field Studies			
Quercus alba L. (1989)			13.7	16.8	Norby et al. (1995)
(1990)			14.2	15.7	Norby et al. (1995)
Liriodendron tulipifera L. (1989)			16.3	16.6	Norby et al. (1995)
(1990)			16.3	16.8	Norby et al. (1995)
Scirpus olneyi (pure)	~82.0	~84.0			Curtis et al. (1989)
(mixed)	~46.0	~50.0			Curtis et al. (1989)
Andropogon gerardii	102.4	95.5	37.9	35.9	Kemp et al. (1994)
Sorghastrum nutans	103.8	99.2	32.6	31.2	Kemp et al. (1994)
Poa pratensis	33.1	38.6	9.6	10.6	Kemp et al. (1994)

[a] Values preceded by ~ were either recalculated from published data or estimated visually from the graph.
[b] CO_2 treatments were superimposed on an existing experiment exposing seedlings to twice-ambient levels of O_3.
[c] Structural C (lignin + celluose):N.

Although total N content was significantly reduced in mixed litter produced at elevated CO_2, no differences were seen in decomposition rate nor in final N concentrations in the litter after 2 years of decomposition in the ambient field chambers. Changes in species composition at elevated CO_2 occurred, with C_3 forbs increasing while C_3 grasses declined and C_4 grasses remained constant (C. E. Owensby, personal communication). Differences were seen in species-specific decomposition rates, suggesting that changes in ecosystem species composition under CO_2 enrichment might alter ecosystem-level nutrient cycling (Kemp et al., 1994). In contrast, in a grassland microcosm experiment conducted at the Duke phytotron, C:N was increased in litter by CO_2 enrichment (H. W. Hunt, personal communication). In this study, which also examined interacting effects of CO_2 with changing temperature and moisture, soil cores were taken by driving steel irrigation pipes into the sod and then removing the pipes with cores intact and transporting them to the exposure facility. Although technically a pot study, the use of intact cores permitted the exposure of multiple species at the community level with minimal disturbance.

Estimated litter C:N ratios (estimated litter C = 0.45 × litter biomass) were not different, with a doubling of CO_2 in artificially constructed model tropical ecosystems; C:N for ambient and elevated CO_2 treatments was 42 and 43.5, respectively (Körner and Arnone, 1992). Nitrogen return to the soil (g N m^{-2}) was significantly greater at elevated than at ambient CO_2, due to significantly higher litter production at elevated CO_2.

Litter quality and mass loss rates also were not affected by CO_2 enrichment in yellow poplar (*Liriodendron tulipifera* L.) in an open-top chamber study where saplings were grown in the ground and leaves were allowed to senesce naturally (E. G. O'Neill, unpublished data). Litter was collected as it abscised in the second year of growth at three levels of CO_2 (ambient air + 0, 150, or 300 μmol mol^{-1} CO_2). Green leaf nitrogen concentrations were reduced by CO_2 enrichment (Norby et al., 1992); however, no significant differences in lignin:N or C:N of litter were observed (Table I), although N concentrations were slightly reduced. Litter was placed in mesh bags and laid under the current year's litter in a mesic mixed hardwood stand. After 2 years of decomposition, total mass losses were not significantly different, with 39, 46, and 40% of the original mass remaining in the +0, +150, and +300 μmol mol^{-1} treatments, respectively (Fig. 1). Mass loss rate constants (k) also were not different. Nitrogen concentrations remained lower in the CO_2-enriched litter for the first year of decomposition (Fig. 2). Maximum N concentrations (% original N) during the decomposition period, as well as the point in time where this maximum occurred, were not different. In a subsequent study utilizing yellow poplar litter from the 1990 growing season, again, no differences in litter quality (Table I) or mass loss rates were observed after 2 years of decomposition.

In contrast to work with natural or constructed ecosystems or with litter produced from field-grown saplings, evidence from autecological pot studies supports the hypothesis of reduced litter quality with CO_2 enrichment of the atmosphere. Nitrogen concentrations were decreased and concentrations of soluble phenolics and structural C (cellulose and lignin) were

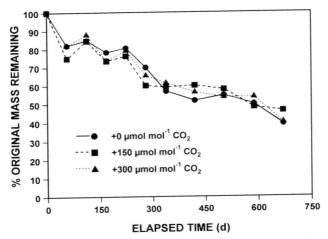

Figure 1 Mass loss of yellow poplar litter produced under three CO_2-enrichment regimes (ambient air + 0, 150, or 300 μmol mol^{-1} CO_2) and decomposed in a mixed hardwood stand using litter bag techniques. Each data point reports a mean of five litter bags.

increased when sweet gum (*Liquidambar syraciflua* L.) was grown in pots in open-top chambers at 935 μmol mol^{-1} CO_2 for one growing season (Melillo, 1983). When elevated CO_2 (300 μmol mol^{-1} CO_2) was added to a twice-ambient O_3 exposure of *Liriodendron tulipifera* L., N concentrations were

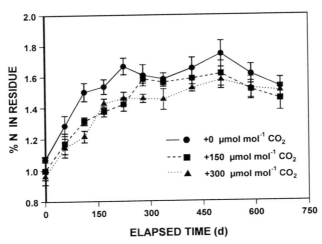

Figure 2 Changes in N concentrations in decomposing CO_2-enriched yellow poplar litter residue through time. Data points represent treatment means ± SE ($n = 5$).

reduced by 31% and lignin:N was increased 2.5 times relative to the O_3-only treatment (Boerner and Rebbeck, 1993). Decomposition and first-year N loss rates were also reduced, although the decrease in N loss rate was correlated to structural changes in foliage rather than to initial N concentrations. This study was also conducted on potted seedlings in continuously stirred tank reactors (CSTRs).

Increases in litter C:N and/or lignin:N with CO_2 enrichment were observed in pot-grown *Fraxinus excelsior* L., *Betula pubescens* Ehrh., and *Acer pseudoplatanus* L. (Cotrufo *et al.*, 1994). In this same study, no changes in ratios occurred in a coniferous species, *Picea sitchensis* (Bong.) Carr., where needles were collected and dried prior to abscission. In all three deciduous species, where foliage abscised naturally, the changes in C:N and lignin:N were due to both reduced foliar N and increased lignin; N was not reduced in *Picea* needles. Both decay and respiration rates were reduced in the CO_2-enriched deciduous litter (although N mineralization rates were not) when the litter was decomposed in soil-free microcosm chambers. A similar study, utilizing microlysimeters, reported temporal differences in N flux from the CO_2-enriched ground litter of *Populus tremuloides* Michx., suggesting increased recalcitrance of residue compounds and reduced decomposition rates (D. W. Johnson and P. Henderson, personal communication). It should be noted that both of these studies examined decomposition in isolation from part or all of the soil decomposer community.

The role of soil fauna community complexity in the breakdown of CO_2-enriched leaves was investigated in a microcosm approach by Coûteaux and co-workers (1991). Sweet chestnut(*Castanea sativa* Mill.) litter, obtained from seedlings grown in 25-l pots, was soaked in demineralized water to equalize soluble phenolic content and then sterilized and reinoculated with increasingly complex mixtures of decomposer fauna. Although initial N concentration was lower and C:N was higher in litter produced at elevated CO_2, reductions in mass loss rates and C mineralization over the 24-week experiment were seen only in the least complex assemblage (microflora plus added protozoans). In the more complex microcosms, total C loss was actually increased for CO_2-enriched litter. Decomposition was divisible into two distinct phases: the initial phase, where rates were dominated by initial litter quality factors and C mineralization was greatest in the control litter, and a later phase, where the primary determinant of decomposition rate was the complexity of the decomposer community. Respiration rates were increased in the CO_2-enriched litter in the more complex communities.

IV. A Few Words to the Wise Decomposer

The paucity of data on decomposition processes under CO_2 enrichment can be attributed to two related factors: (1) the persistence of the assumption that lignin:N and C:N will increase, so that necessary measurements are not made, and (2) the scarcity of realistic exposure regimes to produce foliar litter and measure decomposition.

We must be cautious about forming hard-and-fast assumptions based solely on information gathered from artificial culture conditions. Nutrient ratios, on average, increased by 46 (C:N) and 53% (lignin:N) in litter produced at elevated CO_2 when plants were grown in pots, but were not different due to CO_2 enrichment in any field studies (Table I and Fig. 3). The range of mean values reported was also much greater for pot studies than for studies where unrestricted plants were grown in the ground. To assess how representative reported values of C:N or lignin:N were, we also compared ratios from the ambient treatment of these same studies to ratios reported in the literature for the same species. Kemp *et al.* (1994) included unchambered treatments as part of their experiment; ratios from this treatment were used as the basis for comparison. Results are given in Table II and summarized in Fig. 4. In almost all cases where litter was gathered from plants grown in pots, regardless of pot size, ambient nutrient ratios are lower than those from the literature. In some cases, the difference between ambient and literature values is greater than the increase due to CO_2 enrichment. In contrast, available data for ambient treatments from field studies correspond well with reported values for plants grown in the wild or in chamberless treatments as part of the experiment. The intent here is not to invalidate pot studies, but to urge caution in interpretation of the results and extrapolation to the real world. Results suggest that plants exposed in artificial environments may differ from field-grown plants in either N content or C partitioning, and these differences may confound

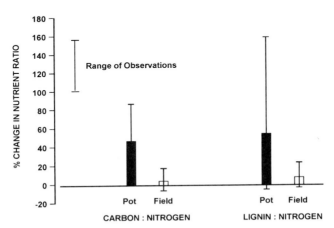

Figure 3 Changes in litter nutrient ratios with CO_2 enrichment in field-grown plants vs pot-grown plants. Data used to generate the graph are presented in Table I.

Table II Comparison of Litter Quality Data from Ambient Treatments of Elevated CO_2 Studies and Reported Values for the Same Species in the Field[a]

Species	C:N AMB	C:N FIELD	Lignin:N AMB	Lignin:N FIELD	References[c]
Pot Studies					
L. styraciflua L.			11.2	52.2	Melillo (1983)
C. sativa Mill.	40.4	63.2			Coûteaux et al. (1991)
Q. alba L.			5.7	17.2	Norby et al. (1986)
Q. alba L (1989)			12.3	17.2	R. Norby (unpublished data)
(1990)			18.8	17.2	R. Norby (unpublished data)
L. tulipifera L.[b]			1.7	14.6	Boerner and Rebbeck (1993)
F. excelsior L.	42.0	30.5			Cotrufo et al. (1994)
B. pubescens Ehrh.			9.8	34.6	Cotrufo et al. (1994)
A. pseudoplatanus L.	81.0	18.1			Cotrufo et al. (1994)
P. sitchensis (Bong.) Carr.	18	80.0			Cotrufo et al. (1994)
Field Studies					
Q. alba L. (1990)			13.7	17.2	Norby et al. (1995)
(1991)			14.2	17.2	Norby et al. (1995)
L. tulipifera L. (1989)			16.6	14.6	Norby et al. (1995)
(1990)			16.3	14.6	Norby et al. (1995)
A. gerardii	37.9	31.8	102.4	72.8	Kemp et al. (1994)
S. nutans	32.6	36.3	103.8	105.0	Kemp et al. (1994)
P. prartensis	9.6	8.7	33.1	28.0	Kemp et al. (1994)

[a] AMB, ambient CO_2 experimental treatment; FIELD, reported values from the field (or from unchambered controls where part of the experimental design).

[b] CO_2 treatments were superimposed on an existing experiment exposing seedlings to twice-ambient levels of O_3.

[c] Field data references: L. styraciflua, E. G. O'Neill and R. J. Norby (unpublished data); C. sativa, Anderson (1973); Q. alba and L. tulipifera, Cromack and Monk (1975); F. excelsior, B. pubescens, and A. pseudoplatanus, Bocock (1964); P. sitchensis, Hayes (1965); A. gerardii, S. nutans, and P. pretensis, Kemp et al. (1994).

responses to CO_2 enrichment. Lignin biosynthesis, for example, may be related to light quality or mineral nutrition (Anderson and Beardall, 1991; Waring et al., 1985). Whether grown in open-top chambers or in environmental cabinets, potted plants may behave differently from unconstrained, field-grown seedlings due to root restriction (Thomas and Strain, 1991), although differences may be due to available nutrients rather than to pot size or shape (McConnaughay et al., 1993). Pot studies can be invaluable

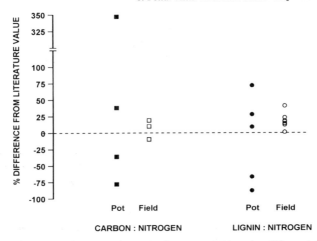

Figure 4 Differences in litter nutrient ratios between ambient (no CO_2 enrichment) treatments in CO_2 effects studies and ratios reported in the literature for the same species under nonexperimental conditions. In cases where unchambered treatments were included in the design, these values were used for comparison against ambient treatments. Data used to generate the figure are presented in Table II.

for the identification and understanding of mechanisms, but care must be taken when culture or exposure conditions have the potential to interact with CO_2 to affect response (Strain and Thomas, 1992).

Likewise, if measurement of decomposition at the ecosystem level is the goal, then researchers must take care not to exclude processes or biota responsible for changes occurring during decomposition. Experiments that do not include exposure to natural and complete decomposer communities, for example, cannot be easily extrapolated to the field (Woodward et al., 1991; Woodward, 1992). Assessment of the potential for changes in decomposition with rising CO_2 only has meaning if it can be extrapolated to the scale of ecosystems. Lignin:N ratios of leaves produced at elevated CO_2 can be measured at the whole plant level. However, to assess the effects of CO_2 enrichment on decomposition, we must produce litter *and* decompose it in a CO_2-enriched ecosystem. Although data exist to suggest either decreased or unaltered litter quality with CO_2 enrichment, no data have been produced where decomposition has occurred under elevated CO_2 to test the hypothesis that decomposition rates will change.

We may also introduce error into predictions of ecosystem response if we rely solely on single-species decomposition studies. Although mass loss rates averaged across several co-occurring plant species and derived from single-species experiments may be comparable to field values, other components of the process may not compare as well. For example, the use of mixed-litter bags for decomposition may give a more accurate estimate of N process rates than averaging across single-species bags. Differences in N flux from litter bags were observed when single-species bags were compared to mixed-species bags (Blair et al., 1990).

Finally, the limitations of square-wave exposure systems must be considered when extrapolating from experimental results to predictions about future ecosystem behavior under elevated CO_2. At the ecosystem level, a rapid, single-step increase in CO_2 (square-wave exposure) is likely to produce different physiological responses than would be observed in ecosystems exposed to slowly increasing CO_2 levels (Eamus and Jarvis, 1989). In fact, what we may be measuring is the response of an ecosystem in the process of adjustment, rather than an ecosystem in dynamic equilibrium. In addition, if litter quality declines slowly as CO_2 increases and available soil N decreases as a result, then trees experiencing twice-ambient CO_2 in the real world will experience it under a different range of soil nutrient conditions than those employed in research (Woodward, 1992). Present experiments are useful for identifying mechanisms, but are limited in their utility as predictors of future ecosystem response.

The questions raised here inevitably lead to the conclusion that ecosystem processes such as decomposition and nutrient cycling must be studied *at the ecosystem level*. In some cases, as with grasslands and other short-stature ecosystems, this has been or is being done. The development of large-scale exposure systems, such as Free-Air CO_2 Exposure (FACE) facilities, offers opportunities for the measurement of ecosystem-level decomposition. However, for some systems, although information can be gained by pot studies, by constructed ecosystems, or by planting multiple individuals of like or unlike species within open-topped chambers, ultimately we will encounter important questions that can only be answered with large-scale, long-term experiments.

V. Summary

Elevated CO_2 may impact ecosystems through its effect on nutrient cycling processes. Decomposition represents a critical process in nutrient cycling and is especially vulnerable to disturbance through changes in nutrient content or nutrient ratios in litter. The evidence is inconclusive, but field results from several ecosystems suggest that rates of decay (and nutrient cycling) will be unaffected by CO_2 enrichment.

Contradictory results appear to be related to experimental approach and scale of observation, e.g., pot vs field studies and single- vs mixed-species decomposition experiments. In order to determine the potential for CO_2-induced changes in decomposition rates to affect ecosystems, research must be conducted at the ecosystem level. This has been accomplished for some ecosystems, but is still lacking for forests, where the large sizes and long life spans of component organisms make large-scale, long-term research as difficult as it is necessary.

Acknowledgments

We thank W. M. Post and C. Trettin for critical review of the manuscript. The authors also applaud the generosity of H. W. Hunt, C. E. Owensby, D. W. Johnson, and P. Henderson in allowing their results to be mentioned and discussed as personal communications. Research was sponsored by the Global Change Research Program of the Environmental Sciences Division, U.S. Department of Energy, under Contract No. DE-AC05-84OR21400 with Martin Marietta Energy Systems, Inc. This is Publication No. 4459, Environmental Sciences Division, Oak Ridge National Laboratory.

References

Aber, J. D., and Melillo., J. M. (1982). Nitrogen immobilization in decaying hardwood leaf litter as a function of initial nitrogen and lignin content. *Can. J. Bot.* 60, 2263–2269.

Aber, J. D., Melillo, J. M., and McClaugherty, C. A. (1990). Predicting long-term patterns of mass loss, nitrogen dynamics, and soil organic matter formation from initial fine litter chemistry in temperate forest ecosystems. *Can. J. Bot.* 68, 2201–2208.

Ågren, G. I., McMurtie, R. E., Parton, W. J., Pastor, J., and Shugart, H. H. (1991). State-of-the-art of models of production-decomposition linkages in conifer and grassland ecosystems. *Ecol. Appl.* 1, 118–138.

Alexander, M. (1977). "Introduction to Soil Microbiology," 467 pp. John Wiley and Sons, New York.

Anderson, J. M. (1973). The breakdown and decomposition of sweet chestnut (*Castanea sativa* Mill.) and beech (*Fagus sylvatica* L.) leaf litter in two deciduous woodland soils. II. Changes in the carbon, hydrogen nitrogen and polyphenol content. *Oecologia* 12, 275–288.

Anderson, J. W., and Beardall, J. (1991). "Molecular Activities of Plant Cells," 384 pp. Blackwell Scientific Publications, Boston.

Berg, B., and McClaugherty, C. (1987). Nitrogen release from litter in relation to the disappearance of lignin. *Biogeochemistry* 4, 219–224.

Blair, J. M., Parmalee, R. W., and Beare, M. H. (1990). Decay rates, nitrogen fluxes and decomposer communities of single- and mixed-species foliar litter. *Ecology* 71, 1976–1985.

Bocock, K. L. (1964). Changes in the amounts of dry matter, nitrogen, carbon and energy in decomposing woodland leaf litter in relation to the activities of soil fauna. *J. Ecol.* 52, 273–284.

Boerner, R. E. J., and Rebbeck, J. (1993). Decomposition of hardwood leaves grown under elevated O_3 and/or CO_2. *Bull. Ecol. Soc. Am. (Suppl.)* 74, 166.

Brady, N. C. (1990). "The Nature and Properties of Soils," 621 pp. Macmillan, New York.

Chapin, F. S., III, and Kedrowski, R. A. (1983). Seasonal changes in nitrogen and phosphorus fractions and autumn retranslocation in evergreen and deciduous taiga trees. *Ecology* 64, 376–391.

Chapin, F. S., III, Schulze, E.-D., and Mooney, H. A. (1990). The ecology and economics of storage in plants. *Annu. Rev. Ecol. Syst.* 21, 423–447.

Cotrufo, M. F., Ineson, P., and Rowland, A. P. (1994). Decomposition of tree leaf litters grown under elevated CO_2: Effect of litter quality. *Plant Soil* 163, 121–130.

Coûteaux, M.-M., Mousseau, M., Celerier, M.-L., and Bottner, P. (1991). Increased atmospheric CO_2 and litter quality: decomposition of sweet chestnut leaf litter with animal food webs of different complexities. *Oikos* 61, 54–64.

Cromack, K., Jr., and Monk, C. D. (1975). Litter production, decomposition and nutrient cycling in a mixed hardwood watershed and a white pine watershed. *In* "Mineral Cycling in Southeastern Ecosystems", (F. G. Howell, J. B. Gentry, and M. H. Smith, eds.), CONF-740513. pp. 609–625. Energy Research and Development Administration.

Curtis, P. S., Drake, B. G., and Whigham, D. F. (1989). Nitrogen and carbon dynamics in C_3 and C_4 estuarine marsh plants grown under elevated CO_2 in situ. *Oecologia* 78, 297–301.

Dowding, P. (1976). Allocation of resources, nutrient uptake and release by decomposer organisms. *In* "The Role of Terrestrial and Aquatic Organisms in Decomposition Processes." (J. M. Anderson and A. Macfadyen, eds.), pp. 169–183. Blackwell Scientific Publications, London.

Drake, B. G., Peresta, G., Beugeling, E., and Matamala, R. (1996) Long-Term Elevated CO_2 Exposure in a Chesapeake Bay Wetland: Ecosystem Gas Exchange, Primary Production, and Tissue Nitrogen. *In* "Carbon Dioxide and Terrestrial Ecosystems" (G. W. Koch and H. A. Mooney, eds.). Academic Press, San Diego.

Duvigneaud, P., and Denaeyer-De Smet, S. (1970). Biological cycling of minerals in temperate deciduous forests. *In* "Analysis of Temperate Forest Ecosystems" (D. E. Reichle, ed.), pp. 199–225. Springer-Verlag, New York.

Eamus, D., and Jarvis, P. G. (1989). The direct effects of increase in the global atmospheric CO_2 concentration on natural and commercial temperate trees and forests. *Adv. Ecol. Res.* 19, 1–55.

Edwards, C. A., and Fletcher, K. E. (1988). Interactions between earthworms and microorganisms in organic matter breakdown. *Agric. Ecosyst. Environ.* 24, 235–247.

Hayes, A. J. (1965). Studies on the decomposition of coniferous leaf litter. I. Physical and chemical changes. *J. Soil Sci.* 16, 121–140.

Horner, J. D., Cates, R. G., and Gosz, J. R. (1987). Tannin, nitrogen and cell wall composition of green vs. senescent Douglas-fir foliage: Within- and between-stand differences in stands of unequal density. *Oecologia* 72, 515–519.

Kemp, P. R., Waldecker, D., Reynolds, J. F., Virginia, R. A., and Owensby, C. E. (1994). Effects of elevated CO_2 and nitrogen fertilization pretreatments on decomposition of tallgrass prairie leaf litter. *Plant Soil* 165, 115–127.

Koch, G. W., and Mooney, H. A. (1996) "Carbon Dioxide and Terrestrial Ecosystems" Academic Press, San Diego.

Körner, C., and Arnone, J. A., III (1992). Responses to elevated carbon dioxide in artificial tropical ecosystems. *Science* 257, 1672–1675.

McClaugherty, C. A., Pastor, J., Aber, J. D., and Melillo, J. M. (1985). Forest litter decomposition in relation to soil nitrogen dynamics and litter quality. *Ecology* 66, 266–275.

McConnaughay, K. D. M., Berntson, G. M., and Bazzaz, F. A. (1993). Limitations to CO_2-induced growth enhancement in pot studies. *Oecologia* 94, 550–557.

Melillo, J. M. (1983). Will increases in atmospheric concentrations affect litter decay rates? *In* "The Ecosystems Center Annual Report", pp. 10–11. Marine Biological Laboratory, Woods Hole, MA.

Melillo, J. M., Aber, J. D., and Muratore, J. M. (1982). Nitrogen and lignin control of hardwood leaf litter decomposition dynamics. *Ecology* 63, 621–626.

Melillo, J. M., Naiman, R. J., Aber, J. D., and Linkins, A. E. (1984). Factors controlling mass loss and nitrogen dynamics of plant litter decaying in northern streams. *Bull. Marine Sci.* 35, 341–356.

Melillo, J. M., Callahan, T. V., Woodward, F. I., Salati, E., and Sinha, S. K. (1990). Effects on Ecosystems. *In* "Climate Change: The IPCC Scientific Assessment." Cambridge University Press, Cambridge, UK.

Nambiar, E. K. S., and Fife, D. N. (1987). Growth and nutrient retranslocation in needles of Radiata pine in relation to nutrient supply. *Ann. Bot.* 60, 147–156.

Norby, R. J., Pastor, J. and Melillo, J. M. (1986). Carbon-nitrogen interactions in CO_2-enriched white oak: physiological and long-term perspectives. *Tree Physiol.* 2, 233–241.

Norby, R. J., O'Neill, E. G., and Wullschleger, S. D. (1995). Belowground responses to atmospheric carbon dioxide in forests. *In* "Carbon Forms and Functions in Forest Soils" (W. W. McFee and J. M. Kelly, eds.). pp. 397–418. Soil Science Society of America, Madison, WI.

Norby, R. J., Gunderson, C. A., Wullschleger, S. D., O'Neill, E. G., and McCraken, M. K. (1992). Productivity and compensatory responses of yellow-poplar trees in elevated CO_2. *Nature* 357, 322–324.

O'Neill, E. G. (1994). Responses of soil biota to elevated atmospheric carbon dioxide. *Plant Soil* 165, 55–65.

Owensby, C. E., Coyne, P. I., and Auen, L. M. (1993). Nitrogen and phosphorus dynamics of a tallgrass prairie ecosystem exposed to elevated carbon dioxide. *Plant Cell Environ.* 16, 843–850.

Owensby, C. E., Ham, J. M., Knapp, A., Rice, C. W., Coyne, P. I., and Auen, L. M. (1996). Ecosystem-level responses of tall grass prairie to elevated CO_2. *In* "Carbon Dioxide and Terrestrial Ecosystems" (G. W. Koch and H. A. Mooney, eds.). Academic Press, San Diego.

Pastor, J., and Post, W. M. (1985). "Development of a linked forest productivity-soil process model," ORNL/TM-5919. Oak Ridge National Laboratory, Oak Ridge, TN.

Pastor, J., and Post, W. M. (1987). Influence of climate, soil moisture and succession on forest carbon and nitrogen cycles. *Biogeochemistry* 2, 3–27.

Pastor, J., and Post, W. M. (1988). Response of northern forests to CO_2-induced climate change. *Nature* 334, 55–58.

Schimel, D. S. (1993). Population and community processes in the response of terrestrial ecosystems to global change. *In* "Biotic Interactions and Global Change" (P. M. Kareiva, J. G. Kingsolver, and R. B. Huey, eds.), pp. 45–54. Sinauer Associates, Inc., Sunderland, MA.

Strain, B. R., and Bazzaz, F. A. (1983). Terrestrial Plant Communities. *In* "CO_2 and Plants" (E. R. Lemon, ed.), AAAS Selected Symposium 84, pp. 177–222. Westview Press, Boulder, CO.

Strain, B. R., and Thomas, R. B. (1992). Field measurements of CO_2 enhancement and climate change in natural vegetation. *Water Air Soil Pollut.* 64, 45–60.

Taylor, B. R., and Parkinson, D. (1988). Aspen and pine leaf litter decomposition in laboratory microcosms. I. Linear vs exponential models of decay. *Can. J. Bot.* 66, 1960–1965.

Taylor, B. R., Parkinson, D., and Parsons, W. F. J. (1989). Nitrogen and lignin content as predictors of litter decay rates: a microcosm test. *Ecology* 70, 97–104.

Thomas, R. B., and Strain, B. R. (1991). Root restriction as a factor in photosynthetic acclimation of cotton seedlings grown in elevated carbon dioxide. *Plant Physiol.* 96, 627–634.

Tolley, L. C., and Strain, B. R. (1985). Effects of CO_2 enrichment and water stress on gas exchange of *Liquidambar straciflua* and *Pinus taeda* seedlings grown under different irradiance levels. *Oecologia* 65, 166–172.

Waring, R. H., McDonald, A. J. S., Larsson, S., Ericsson, T., Arwidsson, E., Ericsson, A., and Lohammar, T. (1985). Differences in chemical composition of plants grown at constant relative growth rates with stable mineral nutrition. *Oecologia* 66, 157–160.

Williams, W. E., Garbutt, K., Bazzaz, F. A., and Vitousek, P. M. (1986). The response of plants to elevated CO_2. IV. Two deciduous-forest tree communities. *Oecologia* 69, 454–459.

Woodward, F. I. (1992). Predicting plant responses to global environmental change. *New Phytol.* 122, 239–251.

Woodward, F. I., Thompson, G. B., and McKee, I. F. (1991). The effects of elevated concentrations of carbon dioxide on individual plants, populations, communities and ecosystems. *Ann. Bot.* 67, 23–38.

7
CO_2-Mediated Changes in Tree Chemistry and Tree–Lepidoptera Interactions

Richard L. Lindroth

I. Introduction

Forests contain 90% of the carbon sequestered in terrestrial biomass and conduct about two-thirds of global photosynthesis (Kramer, 1981; Graham et al., 1990). Moreover, accelerated carbon accumulation is one response predicted for terrestrial ecosystems under CO_2 atmospheres of the future. Improved understanding of the factors affecting forest productivity vis-à-vis enriched CO_2 is essential not only for understanding alterations in the dynamics of some of the world's most expansive biomes but also for anticipating perturbations in the global carbon cycle.

The extent to which atmospheric CO_2 may alter forest productivity is still largely unknown, however, in part because a host of other environmental factors will likely alter tree responses. Prominent among these are *abiotic* factors; evidence is rapidly accruing that the availability of resources such as light, water, and mineral nutrients affects tree responses to enriched CO_2 (Eamus and Jarvis, 1989; Graham et al., 1990; Melillo et al., 1990; Bowes, 1993). To date, however, comparatively little attention has been given to the impact of *biotic* factors on tree responses to atmospheric CO_2.

Of the many biotic factors likely to interact with elevated CO_2 in altering forest ecosystem dynamics, insects are one of the most important. Phytophagous insects are the major primary consumers in temperate forests and can significantly modify species composition, energy flow, and nutrient cycling (Schowalter et al., 1986; Choudhury, 1988; Veblen et al., 1991; Dyer and Shugart, 1992). Shifts in the dynamics of interactions between trees and tree-feeding insects may have pervasive cascade or ripple effects at the levels of populations, communities, and ecosystems. Indeed, the Intergovernmental Panel on Climate Change (Houghton *et al.*, 1990) asserted that "changes in tissue quality could have far-reaching consequences for herbivory, host-pathogen relationships, and soil processes. . . . Much more work is needed in this area before we can make generalizations about the linkages between elevated CO_2, tissue chemistry and ecosystem effects."

This chapter summarizes results from several years of research on the effects of elevated atmospheric CO_2 on the foliar chemistry of deciduous trees and subsequent effects on the performance of lepidopteran herbivores. As will become clear, research on how trophic interactions will be affected by CO_2 enrichment is still in its infancy. Many critical knowledge gaps exist. My intent here is to draw general conclusions as warranted by recent research and to highlight directions for future research.

II. Forest Lepidoptera

To date, all research on the effects of CO_2 on tree–insect interactions has been conducted with folivorous insects. Leaf-feeding insects that cause the greatest impact on forest ecosystems occur primarily in the orders Lepidoptera and Hymenoptera. In North America alone, approximately 85 such species cause periodic and widespread defoliation (Mattson *et al.*, 1991). Of these species, all but a few are lepidopterans. In the United States, an average of 7.6 million ha of forest per year experienced outbreaks of leaf-feeding insects during the period 1957–1987, with four species of Lepidoptera (*Lymantria dispar, Malacosoma disstria, Choristoneura conflictana,* and *Choristoneura fumiferana*) contributing nearly 70% of the total (Mattson *et al.*, 1991).

Chemical composition is probably the single most important determinant of the suitability of foliar tissues for phytophagous insects (Schultz, 1988; Ehrlich and Murphy, 1988). Both primary metabolites (i.e., insect nutrients such as carbohydrates and proteins) and secondary metabolites (allelochemicals) strongly influence insect performance (Slansky and Scriber, 1985; Scriber and Ayres, 1988). Environmental modulation of plant chemistry may play an important role in the onset of insect outbreaks in forests (Mattson and Haack, 1987).

III. Carbon–Nutrient Balance Theory: A Predictive Tool

Carbon–nutrient balance theory (Bryant *et al.*, 1983; Bazzaz *et al.*, 1987; Tuomi *et al.*, 1988) provides a framework for predicting plant chemical responses to shifts in CO_2 availability. This theory proposes that differences in the availability of carbon and mineral nutrients will alter plant carbohydrate stores, which in turn will influence the accumulation of carbon-based secondary compounds (Fig. 1).

The processes of growth, defense, and reproduction compete within individual plants for limited resources of carbon and nutrients (Bazzaz et al., 1987; Fig. 2). Because nutrient limitations generally constrain growth more than they do photosynthesis (Chapin, 1980), environmental conditions that increase carbon availability or decrease nutrient availability alter internal plant reserves in favor of carbon. This in turn can lead to an accumulation of carbon-based allelochemicals or storage compounds (e.g., phenolics and starches).

Carbon–nutrient balance theory predicts that trees grown under conditions of elevated atmospheric CO_2 will increase allocation to carbon-based secondary compounds or carbohydrate storage. Moreover, because fast-growing species exhibit greater plasticity in chemical response to resource availability than do slow-growing species (Bryant et al., 1983, 1987), the former may be expected to respond more strongly to CO_2 than the latter.

IV. Effects of Elevated CO_2 on Tree Chemistry

Here I will describe results compiled from several studies conducted with tree seedlings (1–2 years old) grown for a period of 70 days under ambient

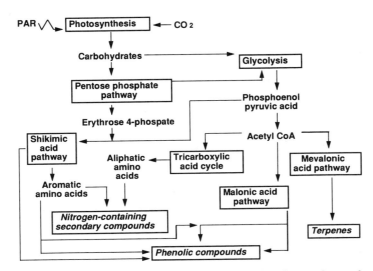

Figure 1 Biosynthetic pathways for the production of major classes of secondary plant compounds [adapted from Taiz and Zeiger (1991)]. PAR: photosynthetically active radiation.

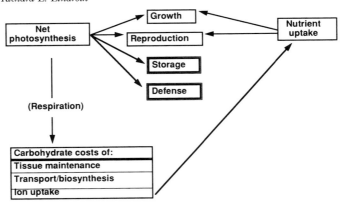

Figure 2 Diagrammatic model of resource allocation in plants [adapted from Ayres (1993)]. Net photosynthesis accounts for carbohydrate losses due to photorespiration. The carbohydrate pool not expended in the process of respiration may be allocated to growth, reproduction, storage, or defense. Because growth and reproduction require mineral nutrients such as nitrogen, these processes are also regulated by nutrient uptake. Many storage and defensive compounds, however, are entirely carbon based, and hence these processes are less affected by nutrient uptake.

and elevated (650–700 ppm) concentrations of CO_2 in the University of Wisconsin biotron. These studies were conducted with tree species typical of the northern hardwoods forest, including quaking aspen (*Populus tremuloides*), paper birch (*Betula papyrifera*), red oak (*Quercus rubra*), and sugar maple (*Acer saccharum*). The species were selected to span a range of inherent growth rates, from rapid (aspen) to slow (maple). The experimental duration encompassed the period during which our selected insect species would be actively feeding in the field. Plant samples for chemical analyses were harvested synchronous with insect feeding bioassays.

To facilitate comparisons among species and experiments, plant chemical responses to CO_2 are presented as relative changes (enriched:ambient CO_2 treatments; Fig. 3). Consistent with numerous other studies (Lincoln *et al.,* 1993), we observed declines in foliar nitrogen of 15–25% in trees grown under enriched CO_2. The only exception to this rule was red oak, which exhibited a small but statistically insignificant increase in foliar nitrogen. Hexose and sucrose concentrations did not change in response to enriched CO_2 for any tree species tested. In contrast, starch concentrations increased dramatically in some species, tripling in aspen and doubling in oak.

A number of studies with herbaceous plants and shrubs failed to detect CO_2 effects on allelochemical concentrations [reviewed in Lincoln *et al.* (1993)], leading to the perception that secondary chemistry may be unresponsive to plant CO_2 environment. Results from our studies contradict

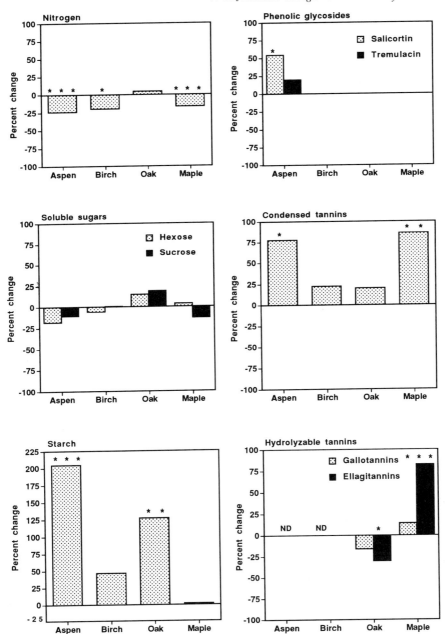

Figure 3 Changes in foliar chemical composition (relative to control plants) for trees grown under enriched atmospheric CO_2 (650 ppm). Phenolic glycosides occur in aspen only; hydrolyzable tannins were not detectable in aspen or birch. Statistical analyses employed a general linear models procedure (GLM; SAS Institute, 1985) followed by comparisons of means for a split-plot experimental design [see Lindroth et al. (1993)]. Asterisks indicate statistically significant changes (*, $P < 0.10$; **, $P < 0.05$; ***, $P < 0.01$). Data were compiled from Lindroth et al. (1993) and Roth and Lindroth (1994).

that notion, at least for particular compounds and tree species (Fig. 3). Concentrations of the phenolic glycoside tremulacin increased in aspen grown under elevated CO_2, although responses for these phenolic compounds vary from year to year (likely due to genotypic variations; R. L. Lindroth, unpublished data). Condensed tannin concentrations tended to increase in all species, especially in aspen and maple. Hydrolyzable tannins, particularly ellagitannins, exhibited a more variable response, declining in oak but increasing in maple. Overall, the strongest allelochemical responses were found in maple, the most slow-growing of the species studied.

Soil nitrogen availability is one of the major limiting factors for forest productivity in eastern North America (Pastor and Post, 1988) and is likely to interact with atmospheric CO_2 concentrations to influence foliar chemistry. We found this to be the case for aspen, oak, and maple seedlings grown as described previously, but with two levels of nitrate availability (1.25 and 7.5 mM; Fig. 4). Foliar nitrogen levels declined in response to both elevated CO_2 and reduced NO_3^- concentrations. Starch levels showed an opposite response, generally increasing in response to both enriched CO_2 and decreased NO_3^-. A significant $CO_2 \times NO_3^-$ interaction term revealed that the accumulation of starch in response to elevated CO_2 was influenced by soil nitrogen availability. Patterns of condensed tannin concentrations roughly paralleled those of starch. In contrast, soil nitrate did not influence hydrolyzable tannin (e.g., ellagitannin) levels. In this study, although plant growth responded more strongly to soil nitrate availability than to CO_2 environment, the reverse was true for foliar chemistry.

Because enriched CO_2 atmospheres may accelerate tree phenology, changes in foliar chemistry attributed to CO_2 may in fact be due to differences in developmental state. Indeed, decreases in foliar nitrogen concentrations and increases in tannin concentrations are common phenological traits of deciduous trees (Slansky and Scriber, 1985; Herms and Mattson, 1992; Waterman and Mole, 1989). Research has employed single rather than multiple leaf sampling periods, synchronized with insect feeding assays rather than with leaf phenology. Consequently, almost nothing is known about how elevated atmospheric CO_2 concentrations may alter plant chemistry independent of phenological effects. Lindroth *et al.* (1995) found that relative increases in the concentrations of starch and condensed tannins in mature leaves of high CO_2 paper birch were 8- and 5-fold greater, respectively, than in similarly treated, young birch leaves in an earlier study (Roth and Lindroth, 1994). These results suggest that the effects of CO_2 on foliar chemistry may be cumulative over time.

In summary, the general phytochemical trends observed in our studies include a decrease in nitrogen with concomitant increases in starch and condensed tannins. Exceptions exist for particular species; work with additional species is required to confirm the generality of these findings. The

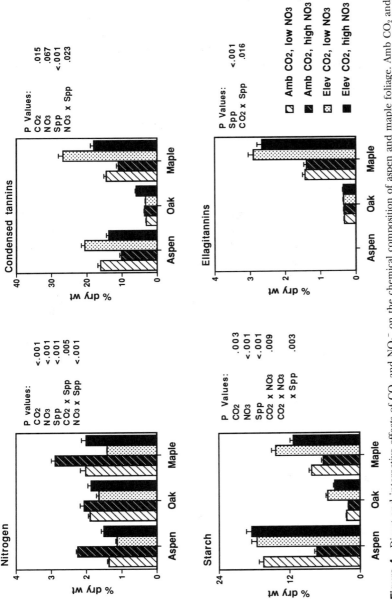

Figure 4 Direct and interactive effects of CO_2 and NO_3^- on the chemical composition of aspen and maple foliage. Amb CO_2 and elev CO_2 indicate ambient and 650 ppm CO_2, respectively; low NO_3 and high NO_3 indicate 1.25 and 7.50 mM NO_3^-, respectively. Vertical lines indicate 95% confidence intervals. General statistical methods used were as described for Fig. 3; only statistically significant ($P < 0.10$) P values are listed. Data are from Kinney et al. (1995).

conclusion that growth under enriched CO_2 increases the carbon:nitrogen ratio of foliage is a robust one. This result occurred for each species in every experiment and was even more pronounced when both CO_2 and soil nitrogen varied (Fig. 5). These findings are corroborated by research by Cipollini et al. (1993) on chemical responses of *Lindera benzoin* to CO_2 and by Julkunen-Tiitto et al. (1993) on chemical responses of *Salix myrsinifolia* to CO_2 and soil nutrient status.

V. Effects on Insect Herbivores

We have conducted performance bioassays with a variety of lepidopteran species common to northern hardwood forests. Most work has involved two spring-feeding outbreak species: the gypsy moth (*Lymantria dispar*, Lymantriidae) and the forest tent caterpillar (*Malacosoma disstria*, Lasiocampidae). To assess the potential for differential impacts of CO_2-mediated changes in chemistry on closely related species, we examined the performance of three species of saturniids that naturally feed on paper birch. These were the cecropia moth (*Hyalophora cecropia*), the luna moth (*Actias luna*), and the polyphemus moth (*Antheraea polyphemus*). Our bioassays focused on short-term growth rates, consumption rates, and food processing efficiencies as performance end points. Lack of sufficient foliage for large-scale feeding experiments precluded long-term trials.

CO_2-mediated changes in foliar chemistry effected changes in insect performance (Fig. 6). As has been observed in other studies (Lincoln et al.,

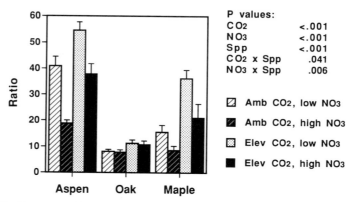

Figure 5 Responses of foliar carbon:nitrogen ratios to atmospheric CO_2 and soil NO_3^- availability. Ratios were calculated from measured concentrations of carbon containing compounds (carbohydrates and phenolics) and nitrogen. See Fig. 4 for a description of figure format.

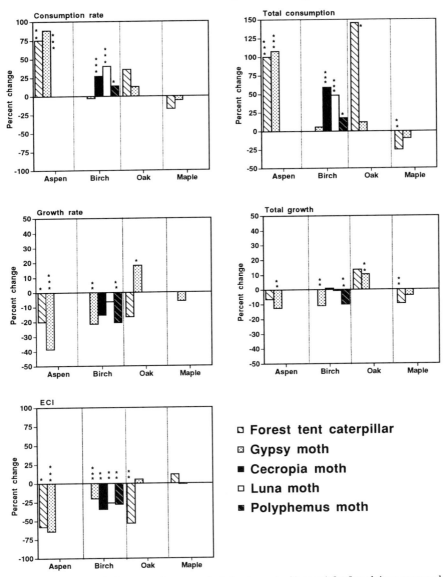

Figure 6 Changes in insect performance (relative to control insects) for fourth instars reared on foliage from trees grown under ambient and enriched (650 ppm) atmospheric CO_2. See Fig. 3 for a description of general statistical methods. Asterisks indicate statistically significant changes (*, $P < 0.10$; **, $P < 0.05$; ***, $P < 0.01$). Data were compiled from Lindroth et al. (1993, 1995) and Roth and Lindroth (1994).

1993), insects generally increased consumption rates and total consumption (over the fourth developmental stadium) when fed high CO_2 foliage. The most dramatic increases were for insects fed aspen, for which total leaf consumption doubled. Increased food consumption may result from compensatory responses elicited by low foliar protein (nitrogen) or from feeding stimulation due to foliar carbohydrates. The only exception to the trend of increased food consumption was for larvae reared on sugar maple.

The growth rates of insects fed high CO_2 leaves generally declined, with statistically significant reductions occurring in about half of our tree–insect combinations. In only one case did the growth rate actually improve: gypsy moths on high CO_2 oak. Insects may compensate for low growth rates by extending their developmental period, a response typically observed in our studies (Lindroth et al., 1993, 1995; Roth and Lindroth, 1994). This compensation was often incomplete, however, as nearly half of the insects in our studies exhibited significant reductions in total growth over the experimental period (fourth stadium).

That insects reared on elevated CO_2 foliage consumed more but grew less than insects reared on control foliage can be attributed to poor food processing efficiencies. The efficiency of conversion of ingested food (ECI) is the product of the efficiency of digestion and the efficiency of conversion of digested food into body mass. Either or both of these parameters typically are reduced in insects reared on high CO_2 foliage (Lindroth et al., 1993, 1995; Roth and Lindroth, 1994). Poor food processing efficiencies likely resulted from the effects of reduced nitrogen and/or increased allelochemicals.

To date, little is known about the impact of CO_2-mediated changes in tree chemistry on insect survival rates. Our research with neonate saturniid larvae showed that high CO_2 birch caused slight reductions in survival through the first larval stadium, particularly in luna (Fig. 7; Lindroth et al., 1995).

Because atmospheric CO_2 and soil NO_3^- interact to affect tree chemistry, it is not surprising that they also affect insect performance. Kinney et al. (1995) found that consumption by gypsy moths was unresponsive to soil nitrate availability at low CO_2 concentrations, but highly responsive to nitrate availability at high CO_2 concentrations (Fig. 8). Such was not the case for insects fed maple, indicating that $CO_2 \times NO_3^-$ interactions will vary depending on the host species. In this study, NO_3^-, but not CO_2 or $CO_2 \times NO_3^-$, affected larval growth (Fig. 8).

Insects can be expected to respond to changes in foliar chemistry under elevated atmospheric CO_2 conditions whether the cause of such changes is CO_2 per se or simply accelerated phenological development. Insect development schedules, although linked to plant development schedules over evolutionary time, are largely driven by thermal conditions in ecological

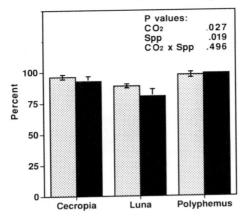

Figure 7 Survival of first instar cecropia, luna, and polyphemus larvae reared on leaves from paper birch grown under ambient (dotted bars) and enriched (700 ppm, solid bars) atmospheric CO_2. Vertical lines indicate $\pm$ 1 SE. Data are from Lindroth *et al.* (1995).

time. Thus, CO_2-mediated acceleration of leaf development (independent of temperature) may pose significant consequences for insects whose fitness is influenced by plant developmental status. Such is the case for many spring-feeding forest insects; for example, the performance of young gypsy moth and forest tent caterpillars is best during a narrow window of opportunity immediately following bud break and markedly declines with leaf maturation (Hunter and Lechowicz, 1992; R. L. Lindroth, unpublished data).

In summary, insect responses to high CO_2 foliage include prolonged development, reduced growth, and increased consumption, although exceptions exist for particular insect and plant species. Decrements in the parameters (e.g., survival, growth) most closely linked to fitness typically have not been large. However, the sum total of multiple relatively minor effects may have a substantial impact on overall insect population dynamics.

VI. Potential Community and Ecosystem Responses

Consideration of community- and ecosystem-level responses to changes in tree chemistry and tree–insect interactions is severely handicapped by the absence of relevant experimental studies. Thus, the following, although based on established patterns and relationships from related work, must be considered conjectural.

Changes in foliar chemical composition are likely to alter interactions between phytophagous insects and their natural enemies. Shifts in diet quality can alter insect susceptibility to pathogens, parasitoids, and preda-

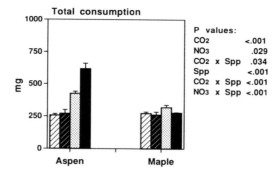

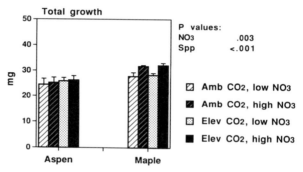

Figure 8 Direct and interactive effects of CO_2 and NO_3^- on the performance of fourth instar gypsy moths fed aspen or maple foliage. See Fig. 4 for a description of figure format. Data are from Kinney et al. (1995).

tors by affecting physiological defense mechanisms, behavior, and periods of exposure or by affecting the performance of the natural enemies themselves (Barbosa and Saunders, 1985; Price et al., 1980). Preliminary work has shown that development is prolonged in larvae of the tachinid parasitoid, *Compsilura concinnata*, when reared on lepidopteran hosts feeding on enriched CO_2 foliage (S. K. Roth and R. L. Lindroth, unpublished data). CO_2-mediated changes in tritrophic interactions may prove to have significant consequences for insect population dynamics. For example, the pathogenicity of a nucleopolyhedrosis virus to gypsy moth larvae is influenced by host tannin concentrations (Hunter and Schultz, 1993). The pivotal role of this virus in gypsy moth population dynamics may be altered under CO_2 conditions of the future.

Altered chemical composition of green and senescent plant tissue will likely change rates of decomposition and nutrient cycling. Tannins and related phenolics resist decomposition and are thought to protract mineralization of detritus (Horner *et al.,* 1988). Thus, decomposition may decelerate under future CO_2 atmospheres, especially in forest systems dominated by species (e.g., maple) whose secondary metabolite profiles are highly responsive to CO_2. The action of insects, however, may serve to accelerate nutrient cycling. On the basis of results from studies of individual insects, one could predict increased energy and nutrient flow through "grazing" (as opposed to "decomposing") food chains under enriched CO_2. Insects eat more and process it less well, indicating that relatively more nutrients will enter the detritus pool as dead animal matter or frass rather than as plant material.

Of primary interest to many, if not most, researchers is not how insects will be affected by enriched atmospheric CO_2 but how altered consumption and population dynamics of insects will feed back to influence forest productivity. Unfortunately, this link cannot yet be established. Although information on individual insect feeding performance is accumulating, we know almost nothing about how insect populations will be affected. Increased plant consumption on an individual insect basis may be offset by reduced insect population densities, resulting from nutritional deficiencies, accumulated secondary metabolites, or natural enemies.

VII. Future Research Directions

Results from these studies suggest that, under the atmospheric CO_2 conditions predicted for the latter half of the next century, the phytochemical composition of deciduous trees will shift (an increase in the C:N ratio) and this in turn will alter interactions with insect herbivores. These changes are also likely to affect other ecosystem processes, such as nutrient cycling and higher level trophic dynamics.

Many avenues of research must be followed to more fully elucidate the consequences of CO_2-mediated changes in tree chemistry on tree–insect interactions and forest dynamics. I close with several suggestions.

1. Expand research to include more plant and herbivore species. To date, conclusions are drawn from a small handful of studies, with restricted taxonomic representation. Identification of robust, general patterns (if they exist) will require greater taxonomic breadth.
2. Assess the impact of altered insect feeding and population dynamics on tree productivity. Population-level studies are required to determine the impact of defoliation under enriched CO_2 conditions on the productivity of trees.
3. Investigate genotypic variation in plant and insect responses. Different plant and insect genotypes are likely to vary in response to elevated CO_2

and changes in host chemistry, respectively. These variations provide the raw material for natural selection and ecosystem "adaptation" to CO_2.
4. Investigate interactions between enriched CO_2 and the availability of other resource factors. Even the small amount of research to date indicates that the impact of CO_2 on tree chemistry and insect performance will be influenced by the availability of other resource factors. Especially important are those factors required for, and often limiting to, plant growth, such as water, mineral nutrients, and light.
5. Address the consequences of CO_2-mediated changes in tree chemistry on higher trophic levels. Because herbivory regulates the flow of energy and nutrients to all higher trophic levels in forest ecosystems, populations of secondary consumers (e.g., insect parasitoids, insectivorous birds) are likely to be impacted.
6. Conduct long-term studies. The degree to which short-term studies with seedlings and saplings accurately represent responses of mature trees over many years is questionable. Long-term studies with larger trees are necessary to understand the effects of CO_2 on forest ecosystem processes and to allow for the assessment of population-level responses of interacting species.

Acknowledgments

Many people contributed to the research described here, especially K. Kinney, S. Roth, C. Platz, S. Jung, and G. Arteel. Two anonymous reviewers provided valuable comments. Research funds were provided by NSF Grant BSR-8918586 and USDA Grant 87-CRCR-2581. This work contributes to the Core Research Programme of the Global Change in Terrestrial Environments (GCTE), Core Project of the International Geosphere Biosphere Programme (IGBP).

References

Ayres, M. P. (1993). Plant defense, herbivory, and climate change. *In* "Biotic Interactions and Global Change" (J. G. Kingsolver, P. M. Kareiva, and R. B. Huey, eds.), pp. 75–94. Sinauer, Sunderland, MA.

Barbosa, P., and Saunders, J. A. (1985). Plant allelochemicals: linkages between herbivores and their natural enemies. *In* "Chemically Mediated Interactions between Plants and Other Organisms" (G. A. Cooper-Driver, T. Swan, and E. E. Conn, eds.), pp. 107–137. Plenum Publishing, New York.

Bazzaz, F. A., Chiariello, N. R., Coley, P. D., and Pitelka, L. F. (1987). Allocating resources to reproduction and defense. *Bioscience* 37, 58–67.

Bowes, G. (1993). Facing the inevitable: plants and increasing atmospheric CO_2. *Annu. Rev. Plant Physiol. Mol. Biol.* 44, 309–332.

Bryant, J. P., Chapin, F. S., III, and Klein, D. R. (1983). Carbon/nutrient balance of boreal plants in relation to vertebrate herbivory. *Oikos* 40, 357–368.

Bryant, J. P., Chapin, F. S., III, Reichardt, P. B., and Clausen, T. P. (1987). Response of winter chemical defense in Alaska paper birch and green alder to manipulation of plant carbon/nutrient balance. *Oecologia* 72, 510–514.

Chapin, F. S., III (1980). The mineral nutrition of wild plants. *Annu. Rev. Ecol. Systematics* 11, 233–260.

Choudhury, D. (1988). Herbivore induced changes in leaf-litter resource quality: a neglected aspect of herbivory in ecosystem nutrient dynamics. *Oikos* 51, 389–393.

Cipollini, M. L., Drake, B. G., and Whigham, D. (1993). Effects of elevated CO_2 on growth and carbon/nutrient balance in the deciduous woody shrub *Lindera benzoin* (L.) Blume (Lauraceae). *Oecologia* 96, 339–346.

Dyer, M. I., and Shugart, H. H. (1992). Multi-level interactions arising from herbivory: a simulation analysis of deciduous forests utilizing FORET. *Ecol. Appl.* 2, 376–386.

Eamus, D., and Jarvis, P. G. (1989). The direct effects of increase in the global atmospheric CO_2 concentration on natural and commercial temperate trees and forests. *Adv. Ecol. Res.* 19, 1–55.

Ehrlich, P. R., and Murphy, D. D. (1988). Plant chemistry and host range in insect herbivores. *Ecology* 69, 908–909.

Graham, R. L., Turner, M. G., and Dale, V. H. (1990). How increasing CO_2 and climate change affect forests. *Bioscience* 40, 575–587.

Herms, D. A., and Mattson, W. J. (1992). The dilemma of plants: to grow or defend. *Q. Rev. Biol.* 67, 283–335.

Horner, J. D., Gosz, J. R., and Cates, R. G. (1988). The role of carbon-based plant secondary metabolites in decomposition in terrestrial ecosystems. *Am. Naturalist* 132, 869–883.

Houghton, J. T., Jenkins, G. J., and Ephraums, J. J. (1990). "Climate Change." Cambridge University Press, New York.

Hunter, A. F., and Lechowicz, M. J. (1992). Foliage quality changes during canopy development of some northern hardwood trees. *Oecologia* 89, 316–323.

Hunter, M. D., and Schultz, J. C. (1993). Induced plant defenses breached? Phytochemical induction protects an herbivore from disease. *Oecologia* 94, 195–203.

Julkunen-Tiitto, R., Tahvanainen, J., and Silvola, J. (1993). Increased CO_2 and nutrient status changes affect phytomass and the production of plant defensive secondary chemicals in *Salix myrsinifolia* (Salisb.). *Oecologia* 95, 495–498.

Kinney, K. K., Lindroth, R. L., Jung, S. M., and Nordheim, E. V. (1995). Effects of atmospheric CO_2 and soil NO_3^- availability on deciduous trees: Phytochemistry and insect performance. *Ecology* (submitted).

Kramer, P. J. (1981). Carbon dioxide concentration, photosynthesis, and dry matter production. *Bioscience* 31, 29–33.

Lincoln, D. E., Fajer, E. D., and Johnson, R. H. (1993). Plant-insect herbivore interactions in elevated CO_2 environments. *Trends Ecol. Evol.* 8, 64–68.

Lindroth, R. L., Kinney, K. K., and Platz, C. L. (1993). Responses of deciduous trees to elevated atmospheric CO_2: productivity, phytochemistry and insect performance. *Ecology* 74, 763–777.

Lindroth, R. L., Arteel, G. E., and Kinney, K. K. (1994). Responses of three saturniid species to paper birch grown under enriched CO_2 atmospheres. *Funct. Ecol.* 9, 306–311.

Mattson, W. J., and Haack, R. A. (1987). The role of drought stress in provoking outbreaks of phytophagous insects. *In* "Insect Outbreaks" (P. Barbosa and J. C. Schultz, eds.), pp. 365–407. Academic Press, Inc., New York.

Mattson, W. J., Herms, D. A., Witter, J. A., and Allen, D. C. (1991). Woody plant grazing systems: North American outbreak folivores and their host plants. *In* "Forest Insect Guilds: Patterns of Interaction with Host Trees" (Y. N. Baranchikov, W. J. Mattson, F. P. Hain, and T. L. Payne, eds.), Gen. Tech. Rep. NE-153, pp. 53–84. USDA Forest Service, Northeastern Forest Experiment Station, Radnor, PA.

Melillo, J. M., Callaghan, T. V., Woodward, R. I., Salati, E., and Sinha, S. K. (1990). Effects on ecosystems. *In* "Climate Change" (J. T. Houghton, G. J. Jenkins, and J. J. Ephraums, eds.), pp. 287–310. Cambridge University Press, New York.

Pastor, J., and Post, W. M. (1988). Response of northern forests to CO_2-induced climate change. *Nature* 334, 55–58.

Price, P. W., Bouton, C. E., Gross, P., McPheron, B. A., Thompson, J. N., and Weis, A. E. (1980). Interactions among three trophic levels: influence of plants on interactions between insect herbivores and natural enemies. *Annu. Rev. Ecol. Systematics* 11, 41–65.

Roth, S. K., and Lindroth, R. L. (1994). Effects of CO_2-mediated changes in paper birch and white pine chemistry on gypsy moth performance. *Oecologia* 98, 133–138.

SAS Institute Inc. (1985). "SAS user's guide: Statistics," version 5 edition. SAS Institute Inc., Cary, NC.

Schowalter, T. D., Hargrove, W. W., and Crossley, D. A., Jr. (1986). Herbivory in forested ecosystems. *Annu. Rev. Entomol.* 31, 177–196.

Schultz, J. C. (1988). Many factors influence the evolution of herbivore diets, but plant chemistry is central. *Ecology* 69, 896–897.

Scriber, J. M., and Ayres, M. P. (1988). Leaf chemistry as a defense against insects. *ISI Atlas Sci.: Animal Plant Sci.* 1, 117–123.

Slansky, F., Jr., and Scriber, J. M. (1985). Food consumption and utilization. *In* "Comprehensive Insect Physiology, Biochemistry, and Pharmacology. Volume 4, Regulation: Digestion, Nutrition, Excretion" (G. A. Kerkut, and L. I. Gilbert, eds.), pp. 87–163. Pergamon Press, New York.

Taiz, L., and Zeiger, E. (1991). "Plant Physiology." Benjamin/Cummings Publishing Co., Redwood City, CA.

Tuomi, J., Niemalä, P., Chapin, F. S. I., Bryant, J. P., and Sirèn, S. (1988). Defensive responses of trees in relation to their carbon/nutrient balance. *In* "Mechanisms of Woody Plant Defenses Against Insects. Search for Pattern" (W. J. Mattson, J. Levieux, and C. Bernard-Dagan, eds.), pp. 57–72. Springer-Verlag, New York.

Veblen, T. T., Hadley, K. S., Reid, M. S., and Rebertus, A. J. (1991). The response of subalpine forests to spruce beetle outbreak in Colorado. *Ecology* 72, 213–231.

Waterman, P. G., and Mole, S. (1989). Extrinsic factors influencing production of secondary metabolites in plants. *In* "Insect-Plant Interactions" (E. A. Bernays, ed.), pp. 107–134. CRC Press, Inc., Boca Raton, FL.

8

The Jasper Ridge CO_2 Experiment: Design and Motivation

Christopher B. Field, F. Stuart Chapin III,
Nona R. Chiariello, Elisabeth A. Holland,
and Harold A. Mooney

I. Introduction

In early 1992, we began CO_2 exposure in an ecosystem-level project that involved experiments in field plots, outdoor microcosms, and growth chambers, as well as modeling. Our basic objective is to quantify the roles of resource availability, species characteristics, and community composition in controlling ecosystem responses to elevated atmospheric CO_2. Our measurements emphasize both integrated ecosystem characteristics (including carbon storage, species composition, soil moisture, and nutrients) and processes (including photosynthesis, evapotranspiration, rhizodeposition, decomposition, trace gas exchange, and many aspects of the dynamics of plants and communities).

The Jasper Ridge CO_2 experiment is a study of a model ecosystem. The study was designed to contribute to the understanding of the principles that regulate ecosystem responses. We chose the Jasper Ridge annual grasslands because they have a number of features that increase the prospects for isolating and quantifying general principles. Some of the features that make these grasslands amenable to useful experiments may also make them unrepresentative of all terrestrial ecosystems or even all grasslands. Our working hypothesis is that the underlying principles are a key to generalization. We hope that increased understanding of the roles of resources, species characteristics, and community composition can be a major element, complementing specific case studies, in the development of improved predictions of the responses of the terrestrial biosphere to naturally increasing atmospheric CO_2. In addition, insight into the principles that regulate ecosystem responses to elevated CO_2 is likely to be useful in understanding other aspects of the regulation of ecosystem processes.

In this paper, we introduce the ideas that motivated the conceptual design of the Jasper Ridge CO_2 experiment, discuss the physical design, and summarize some of the early results.

II. The Challenge

It is very difficult to use a short-term experiment to definitively assess long-term ecosystem responses to elevated atmospheric CO_2 (or to many other factors) because the long-term responses potentially involve many processes that interact on a number of spatial and temporal scales (Field et al., 1992; Mooney et al., 1991a). Even though the direct effects of elevated CO_2 consistently increase photosynthesis, water-use efficiency, and nutrient-use efficiency, these short-term effects may be suppressed or amplified by longer term processes, including biochemical acclimation, structural and chemical changes in plants, changes in plant community composition, and changes in decomposer populations and the decomposability of plant tissue (Bazzaz, 1990; Field et al., 1992).

Some of these processes operate on very long time scales. For example, the replacement of one tree species by another is unlikely on a time scale shorter than one generation, but that can be several centuries for some tree species. CO_2 effects on the decomposability of plant material can be studied in the short term (O'Neill and Norby, 1991), but the time required for all of the carbon pools in an ecosystem to stabilize in response to a CO_2 treatment is at least somewhat sensitive to the turnover time of the slowest pool, which can also be several centuries for old soil carbon (Trumbore, 1993) or large woody material. Models provide a useful approach for estimating the impacts of these slow processes, but none of the models has been tested with a long-term experiment. The fact that most experiments involve step changes in CO_2 concentration, and generally have not explored the relationships among CO_2 and other aspects of global change further complicates the time issue.

The technical difficulty of constructing and operating an elevated CO_2 experiment tends to impose other kinds of limitations. Some of these involve replication. If CO_2 responses are subtle, they are likely to be quantifiable only in well-replicated experiments. But if an experiment is directed at ecosystem-level questions, each replicate needs to cover a reasonable chunk of ecosystem, encompassing enough plants, animals, and microbes to allow some level of population and species dynamics. Ecosystem-level CO_2 experiments on many kinds of forests will require either vast technical infrastructures or novel conceptual and statistical approaches.

Another aspect of the replication issue concerns the number of experiments necessary to address a hypothesis. A single set of treatment and control plots or treatment plots plus chamber and no-chamber controls may be sufficient to quantify the response of one ecosystem type with a particular suite of species on a single soil in a particular climatic regime. With single-treatment ecosystem experiments, attempts to assess the effects of species, soils, or climate must resort to comparisons with experiments on other ecosystems that differ with respect to all of these things. Climate is an important exception, and year-to-year climate variations have already proven to be a valuable resource for assessing climate effects [Owensby *et al.*, 1996 (this volume)]. Yet, in a short-term experiment, separation of the effects of year-to-year climate variations from multiyear CO_2 responses presents a challenge. In general, it would be very useful to have access to a range of comparisons in which some ecosystem characteristics could be held invariant.

The need for a range of comparisons involving the CO_2 responses of a single ecosystem is highlighted by the contrast among papers suggesting principles that might provide a basis for generalizing plant responses to elevated CO_2. Strain and Bazzaz (1983) and Mooney *et al.* (1991a) hypothesized that responses of plant growth should be largest when nutrients are abundant but water is limiting. Idso and Kimball (1993) suggested that the growth of trees is more CO-sensitive than is the growth of other plants, a suggestion that clearly is not supported in all cases (e.g., Norby *et al.*, 1992). Hunt *et al.* (1993) and Poorter (1993) presented data consistent with the conclusion that the CO_2 sensitivity of growth increases with inherent growth potential. A shortage of multifactor experiments, even at the plant level, makes it difficult to eliminate any of these hypotheses.

The community of CO_2 researchers can address the need for multifactor, ecosystem-scale interpretations in a number of ways. These include cross-ecosystem comparisons (e.g., Mooney *et al.*, 1991a), and integration, using models, of plant responses to elevated CO_2 with other results on ecosystem responses to altered carbon, water, or nutrient availability (e.g. Melillo *et al.*, 1993), and multifactor ecosystem experiments. The combination of CO_2 and temperature treatments in the tundra studies of Oechel and colleagues [Oechel and Vourlitis, 1996 (this volume)] and the presence of C_3- and C_4-dominated communities in the salt-marsh studies of Drake and colleagues [Drake *et al.*, 1996 (this volume)] begin to address the need, but with a limited set of factors varied. While these and other studies address critical issues, the fact that they focus on perennials means that they provide little access to potential species changes or to indirect effects driven by species changes.

In the Jasper Ridge CO_2 experiment, we are exploring long-term, multifactor, ecosystem-scale responses to elevated CO_2. The basic approach begins from a focus on an ecosystem type, annual grassland, dominated by short-lived individuals of small stature. The emphasis on ecosystems dominated by annual plants provides two advantages. First, it gives us access to the effects of elevated CO_2 on species composition over several generations and to potential indirect effects of the species changes. Second, the combination of the annual life form plus the small stature makes it feasible to construct artificial ecosystems in which large numbers of individuals complete one or more entire generations. The small stature is also a critical component of the feasibility of examining many combinations of species and resources, with reasonable levels of replication.

A. The Role of Model Ecosystems

Much of the progress in biology in this century has depended on the creative exploration of model systems. The list of the most important model systems includes *Drosophila melanogaster* for population and developmental genetics, *Escherichia coli* for biochemistry and molecular genetics, bacteriophage λ for molecular genetics, and mouse for mammalian genetics. The key characteristics of these experimental models are (1) that they operate with very general mechanisms and (2) that experiments based on them are efficient in terms of space, time, and expense. In many cases, the utility of a model is dramatically increased through the presence of an unusual feature (e.g., the polytene chromosomes in *Drosophila*) or through the absence of a major complication (e.g., the absence of intracellular compartmentation and multiple copies of the genome in *E. coli*).

Model systems have not been widely used in ecology, probably because the objectives of most studies have been to find the unique aspects of a species, community, or ecosystem that let it function in a certain way or to characterize broad patterns of diversity or function. With the emergence of global ecology and global change research, the community of researchers faces new needs. One is to understand the extent to which general features of organism and ecosystem function allow global extrapolation, as well as the limits to useful extrapolation. A second is to design experiments in which it is feasible to subject complete functional units, including complete ecosystems, to a range of novel climates, atmospheres, and disturbance regimes (Mooney *et al.*, 1991b). The Jasper Ridge CO_2 experiment was designed to exploit the Jasper Ridge grasslands as models, paralleling the way that *D. melanogaster* and *E. coli* have been exploited in other fields.

The ultimate utility of an ecological model system will depend on both tractability and generality. Here, we discuss the characteristics that make the Jasper Ridge grasslands tractable model ecosystems. What kinds of features can make the results of a study generalizable? Do ecosystems differ in general relevance?

The earth's ecosystems differ dramatically in the magnitudes of production and nutrient cycling and in the growth forms and identities of the species present. Yet the mechanisms controlling production and nutrient cycling are essentially similar in all ecosystems. Mechanistic controls that operate in all or nearly all ecosystems include the relationships between the solar radiation absorbed by plants and net primary production (Goward *et al.*, 1985), between leaf nitrogen and photosynthesis (Field and Mooney, 1986), and between the C:N or lignin:N ratio of litter and the initial rate of litter decomposition (Melillo *et al.*, 1982). An understanding of the effects of CO_2 on these mechanisms from any ecosystem should have relevance for others.

Plant species characteristics are highly variable, but the suite of major axes of variation is limited. Differences in growth rate, maximum size, tissue composition, biomass allocation, longevity, and phenology can be substantial even within a single ecosystem, allowing study of the consequences of variation in these parameters, even if the range of variation is smaller than the global range.

Ecosystems differ greatly in the nature of the limiting resource or resources and in the intensity of the limitation. Yet the possibility of modifying the availability of light, water, nutrients, or CO_2 makes it possible to explore ecosystem responses to relative excesses or abundance, using any ecosystem as a starting point.

With regard to at least these three issues, the overall uniformity of ecosystem functional mechanisms, the limited suite of plant functional mechanisms, and the potential for manipulating resource levels, appropriate studies on one ecosystem should be relevant to others. Generalization from one ecosystem to the next will not necessarily be easy or direct, but extrapolations based on mechanistic underpinnings should tend to be robust.

III. Jasper Ridge

The Jasper Ridge grassland ecosystems have a number of features that allow us to manage the challenges of detecting and quantifying general principles. To understand the consequences of these features for CO_2 experiments, it is useful to start with a general overview.

The Jasper Ridge Biological Preserve, which is owned and operated by Stanford University, is located at the base of the continental side of the Santa Cruz Mountains (the outer coastal range) in central California. Elevations in the preserve range from 50 to 189 m, with the largest grasslands at the highest elevations. The main arm of the San Andreas Fault passes through the preserve, and its soils reflect a chaotic mix of sedimentary and metamorphic parent materials common along major fault zones (Page and Tabor, 1967). The most abundant soils are thermic mollic Haploxeralfs, mesic typic Argixerolls, thermic lithic Argixerolls, thermic typic Haploxeralfs, and thermic lithic Haploxerolls (Kashiwagi, 1985). Serpentine grassland occurs on Montara clay loam and Obispo clay. Sandstone grassland occurs on the Dibble variant–Millsholm variant complex, the Gilroy–Gilroy variant complex, Botella clay loam, Diablo clay, Francisquito loam, and soils weathered from sandstone, greenstone, and mixed alluvium.

The Jasper Ridge climate is mediterranean, with cool, wet winters and very dry summers (Fig 1). Average annual precipitation over the last 20 years is 582 mm, which is slightly higher than the average for the last 10 years, of 549 mm. Frosts are rare, and midwinter, midday temperatures often reach 20°C.

The Jasper Ridge vegetation is diverse, with well-developed forests, woodlands, shrublands, and grasslands. This diversity of ecosystem types is paralleled by high species diversity, with over 650 species on the list of higher plants (J. H. Thomas, personal communication).

Jasper Ridge has been a site for ecological studies for many years. Areas of emphasis in Jasper Ridge studies include consequences of grazing (e.g., Bartholomew, 1970; Hobbs and Mooney, 1991; Vestal, 1929), population structure in insects (e.g., Brussard and Ehrlich, 1970; Ehrlich *et al.*, 1975), landscape dynamics (e.g., Williams *et al.*, 1987; Wu and Levin, 1994), plant–animal interactions (e.g., Hobbs and Mooney, 1985; Lincoln *et al.*,

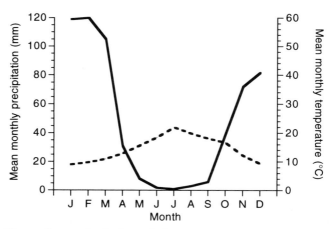

Figure 1 Climate diagram for the Jasper Ridge Biological Preserve, showing mean monthly temperature and precipitation for the period 1974–1975 to 1993–1994.

1982), plant carbon balance (e.g., Merino *et al.*, 1982; Mooney and Chu, 1975), and remote sensing (e.g., Gamon *et al.*, 1993).

Jasper Ridge grasslands fall into two major categories, depending on soil type. Grasslands on shallow, nutrient-poor, serpentine-derived soils are dominated by native forbs and grasses, including both annuals and perennials. On the more fertile sandstone- and greenstone-derived soils, Eurasian annual grasses dominate (Baker, 1989). Prior to European settlement, the more fertile grasslands were probably dominated by native perennial bunchgrasses, and there is some evidence that the bunchgrasses gradually become more important in areas protected from grazing (Murphy and Ehrlich, 1989), although it is possible that climatic factors also play a role in limiting the success of the perennial bunchgrasses (Wagner, 1989)

A. A Model Ecosystem for CO_2 Experiments

The key characteristics of model systems in biology are (1) that they operate with general mechanisms and (2) that experiments based on them are efficient in terms of space, time, and expense. With ecosystem-level experiments, protection of opportunities for the potentially important mechanisms to act is a major challenge. Some of the difficulties relate to the large sizes and long time constants of the players and processes. Others relate to the difficulty of assembling a functional ecosystem without dramatically altering components of the function. One example of the latter problem is the frequent observation that simply removing a soil from nature, mixing it, and adding it to pots result in dramatic increases in nutrient availability [Johnson and Ball, 1996 (this volume)].

Several features of the Jasper Ridge grasslands make them well-suited as experimental models. These include the following.

(1) Dominance by annuals: The annual life form provides three major advantages. The first is the relative freedom from historical effects. With perennial plants exposed to elevated CO_2 after establishment, it is difficult to quantify the role of the preexposure period in modulating the CO_2 response. While some maternal effects are carried across generations (Roach and Wulff, 1987), these are unlikely to be as important as the carryover from the pretreatment growth of an established individual.

The second major advantage of annual plants is that they provide access to all stages of the life cycle. With short-term experiments on perennials, studies are often confined to seedlings. In addition to the obvious effects of life stage on allocation, ecosystem-level nutrient cycling typically is much more closed in mature than in regenerating forests [Schlesinger, 1991; Johnson and Ball, 1996 (this volume)].

Third, a multiyear study of annual plants makes it possible to examine ecosystem effects related to changes in species or genotypic composition. Since a large fraction of the variations in phenology, tissue chemistry, and allocation is interspecific, indirect effects of CO_2 mediated through species changes may be very important relative to effects mediated through changes in the structure or function of established individuals.

(2) All plants are small in stature: Small size is a key element of experimental tractability. The plants of the Jasper Ridge grasslands are small enough that a field chamber 0.65 m in diameter typically contains 1000–3000 individual higher plants on serpentine and 300–1000 plants on sandstone. In the artificial microcosms 0.2 m in diameter, target densities are 205 plants (6500 m^{-2}) for serpentine communities and 110 plants (3500 m^{-2}) for sandstone communities. The small size of the individual plants and of reasonable ecosystems based on them makes it possible to design experiments with large numbers of treatments and/or replicates.

(3) Both grasslands are rich in species and functional types: While both the serpentine and sandstone grasslands are dominated by annuals, the diversity of phenologies and functional types is still quite large. Both grasslands contain some perennials, as well as annuals, that vary in flowering date from early March through November. These variations in flowering date are paralleled by differences in rooting depth, biomass allocation, and tissue chemistry (Mooney et al., 1986). In addition, both grasslands contain legumes with symbiotic N-fixation, as well as mycorrhizal and nonmycorrhizal taxa. While this total range of variation is small relative to the variation across biomes, it provides a foundation for assessing the implications of many kinds of differences against a background of constant climate and soil resources.

(4) The carbon content of the soil is relatively low: With total organic carbon contents in the range of 0.8–1.8%, total soil organic carbon is 3–9 kg m^{-2}, or 10–20 times annual NPP (Table I). This ratio of total soil carbon to NPP is low enough to ensure a reasonable probability of detecting a modest change in carbon storage after a few years. It may or may not be low enough to allow full expression of indirect effects of changes in decomposability within a few years, depending on the decomposition dynamics.

(5) The soil is naturally disturbed with a relatively high frequency: Changes in nutrient availability, hydraulic conductivity, and gas diffusion resulting from soil disturbance generally complicate the interpretation of experiments on plants in pots, unless experimenters use unusual measures to preserve the soil structure and chemistry (e.g., Hunt et al., 1990). On Jasper Ridge, pocket gophers (*Thomomys bottae*) are very abundant and disturb each unit of surface approximately every 5 years (Hobbs and Mooney, 1991). This disturbance by gophers, which prevents the establishment of clear soil profiles, can result in altered nutrient availability (Koide et al., 1987). For the Jasper Ridge CO_2 experiment, the critical aspect of the gopher disturbance is that a thorough churning of the soil is common

and natural, and soil disturbance in the context of establishing microcosms should not introduce conditions outside the natural range.

B. Limitations

While the Jasper Ridge grasslands offer many advantages as a model ecosystem, they do not facilitate access to a number of critical questions and should not be considered an alternative to studies on specific ecosystems. In particular, the absence of woody plants makes it difficult to directly test the hypothesis that increased CO_2 will lead to increased C storage in wood. Other limitations that constrain extrapolations from Jasper Ridge to other ecosystems include the absence of plants with the C_4 photosynthetic pathway, the absence of plants with extremely high growth rates, and the absence of saturated or frozen soils (especially with respect to CH_4 emissions).

IV. The Suite of Experiments

The basic strategy of the Jasper Ridge CO_2 experiment is to combine measurements at a number of levels of organization in order to access mechanisms at different scales, while maintaining a strong connection to ecosystem-scale responses under the most natural possible conditions. The infrastructure for putting the strategy in place consists of open-top chambers in the field and microcosms that provide the opportunity to explore variations in nutrients, water, soils, and plant community properties, including density and species composition. Experiments on plants and communities in more artificial environments [greenhouses, growth chambers, and phytocells (Björkman et al., 1973)] help address mechanisms at the biochemical and physiological scales, while models provide tools for exploring other aspects of the experiments, extending the results to other ecosystems, and extrapolating the results to larger spatial scales.

The field chambers serve both as the focal point for testing some hypotheses and as a reference point for evaluating experiments in the microcosms. The field provides a strong context for assessing changes in the carbon, water, and nutrient budgets as well as trace gas fluxes. The fact that the experiment addresses only two ecosystem types, however, makes it difficult to generalize the responses or to evaluate the role of species or resources. The field provides a reasonably natural setting for following the effects of CO_2 on species composition and population dynamics, although the chambers impose some limitations on pollinator access and seed dispersal. The field provides some information on the consequences of year-to-year climate variations, but this is somewhat confounded with cumulative effects of the CO_2 treatment.

The microcosms or MECCAs (MicroEcosystems for Climate Change Analysis) are the focal point of the experiments designed to quantify components of the ecosystem responses and to test specific hypotheses concerning the roles of species characteristics and resources. In the MECCAs, the experimental units are artificial ecosystems, but the individual units are small enough that we can perform factorial experiments with many combinations of CO_2, nutrients, soils, species, and water. The modest size of the MECCAs makes it possible, for example, to look at enough species to test the hypothesis that responsiveness to CO_2 is related to maximum growth potential (Hunt et al., 1991, 1993; Poorter, 1993) and to look at enough nutrient levels and combinations to assess the relative consequences of limitation by nitrogen and phosphorus. The MECCAs also make it possible to test aspects of the prediction that the most responsive ecosystems should be those with limited water but abundant nutrients (Strain and Bazzaz, 1983) and to test the hypothesis that the abundance of N fixers should increase under elevated CO_2 (e.g., Gifford, 1992). This last hypothesis might also be tested in the field, but chamber-to-chamber differences in initial composition make that a challenging proposition.

The MECCAs potentially provide access to questions at three levels. One-year experiments with single species or communities probe the CO_2 response in the presence of feedback related to physiological acclimation, allocation, and soil moisture, but with little or no effect of decomposition-mediated feedback on nutrient availability. Single-species experiments extended over several years add the potential for decomposition feedback on nutrient availability, while multiyear experiments on self-reproducing communities allow the long-term effects on nutrient availability to influence species composition, which might have other effects on nutrient availability.

The MECCAs are well-suited for measurements of ecosystem-scale gas exchange, trace gas exchange, and soil moisture. They can also be used as a source of material for leaf-level experiments, ranging from fluorescence and daily carbohydrate dynamics to decomposition studies.

Other aspects of the ecosystem response to CO_2 are easier to address with experiments on single plants in pots or with models, and the Jasper Ridge CO_2 experiment is involved in both kinds of activities. Among the experiments to be addressed with plants in pots are studies of maximum growth potential, photosynthetic characteristics, and aspects of plant–herbivore interactions.

Modeling studies are a major component of the Jasper Ridge CO_2 experiment, providing critical tools for interpreting and extending the experimental data. The modeling objectives are staged, with a long-term goal of running ecosystem and population models at least partially on the basis of outputs from plant and canopy models, which are, in turn, driven from the outputs of physiological models. When the models on all of these scales are integrated and validated against the experimental data, we will be able to explore the effects of a broad range of changes in species and resource characteristics. For example, we should be able to estimate the consequences for carbon storage of a change in water-use efficiency or the consequences for nutrient availability of an increase in late season annuals.

The modeling approach in the Jasper Ridge CO_2 project is to build from existing models, emphasizing applicability to the specifics of the Jasper Ridge ecosystems and integration at the interfaces between the models. The three models that form the core of the activity are the CENTURY model of Parton *et al.* (1987, 1993) for decomposition and nutrient cycling, the JASPER model of Moloney *et al.* (1992) and Wu and Levin (1994) for population dynamics, and the GePSi model of Reynolds and colleagues (Reynolds *et al.*, 1993) for plant photosynthesis and growth. To account for the functional diversity of the Jasper Ridge plants, we are also emphasizing models of root allocation (Luo *et al.*, 1994) and phenology.

V. Experimental Facilities

A. Field Chambers

On each substrate type, we selected 30 circular study plots, each 0.65 m in diameter. On the serpentine soil, the plots are bounded by an area approximately 100 m long and 10 m wide. The sandstone plots are in a 30 × 30-m area. Our primary selection criteria were the absence of large rocks (>0.2 m across) and a level or nearly level aspect. In addition, we avoided plots dominated by a large individual of any of the perennial bunchgrasses that occur on both substrates. After identifying the plots, we grouped them into clusters of three nearest neighbors and assigned plots within each cluster to the three treatments (high, chamber plus CO_2; low, chamber without CO_2; and X, no chamber) randomly.

Chambers were installed and the treatments were initiated on January 6, 1992. From that date until June 15, 1992, we maintained the treatment conditions during daylight hours only. Treatments were suspended during the summer drought of 1992. On October 15, 1992, we restarted the treatments, but with continuous, round the clock operation. With the exception of brief shutdowns for maintenance or experimental access, the treatments have been continuous since then.

The individual chambers are circular open-top chambers (OTCs), 0.65 m in diameter and 1 m tall, constructed with an aluminum frame and a skin of 0.1-mm (0.004 in.) polyethylene. The transmittance of the polyethylene film is about 0.9 through the visible and the near-IR ranges, but scattering of the direct beam radiation makes the light more diffuse inside than outside the chambers.

During the experiment's first 2 years, the chambers were single-piece cylinders. During the third year, the chambers were split to facilitate access (Fig. 2). To minimize the impact of the chambers on natural moisture inputs, they are completely open at the top and do not have the frustrum included in many implementations to decrease the penetration of outside turbulence (Baldocchi *et al.*, 1989; Mandl *et al.*, 1989). Instead, the chambers are tall in relation to their width, and they operate at relatively high flow rates. In addition to stabilizing the CO_2 concentration, the additional height also minimizes differences in solar radiation across the chambers. At midday on the equinox, direct solar radiation that has not passed through a chamber wall reaches no lower than 0.14 m above the ground on the chamber's north side.

An independent axial blower, rated at 0.17 m^3 s^{-1} for free-air conditions, draws air from 0.1 m above the ground and injects it into a plenum that forms the lower 0.3 m of each chamber. One hundred 0.01-m holes in the plenum distribute air evenly around the chamber. Beginning in August

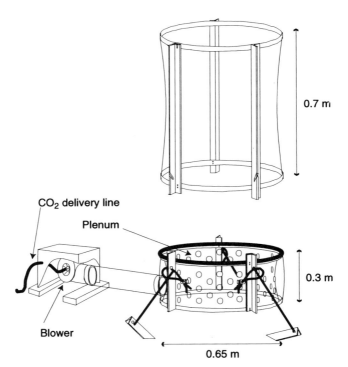

Figure 2 Schematic diagram of one of the field open-top chambers. The top and bottom sections of the chamber normally are connected but separate to allow access to the plants and soil.

1993, the air inlets were raised to 0.4 m above the soil surface to minimize near-surface effects on air temperature and CO_2 concentration. As configured, the blowers deliver 0.08 m^3 s^{-1}, giving a vertical velocity of approximately 0.25 m s^{-1} or 15 air changes per minute. These high flow rates minimize CO_2 variation, but they also tend to make the chambers drier than surrounding sites.

For the chambers receiving elevated CO_2, CO_2 is delivered into the inlet of each blower at such a rate to maintain a daytime average concentration between 67 and 73 Pa (Fig. 3). With low atmospheric turbulence at night, this drifts up by 6–10 Pa. CO_2 levels in the center of each chamber, 0.3 m above the soil, are monitored every 15 min by a data logger (CR–10 Campbell Scientific, Inc., Logan, UT) connected to a CO_2–water vapor analyzer (LI-6262, Li Cor, Inc., Lincoln, NE). CO_2 injection is controlled by a separate needle valve for each chamber. These are adjusted manually, only as necessary.

Although gopher disturbance is an important and common feature of both Jasper Ridge grasslands, we have attempted to eliminate gophers from the experimental area. Our reasoning is that the departure from the natural disturbance dynamics is of less consequence than the extreme spatial and temporal heterogeneity introduced by the gophers.

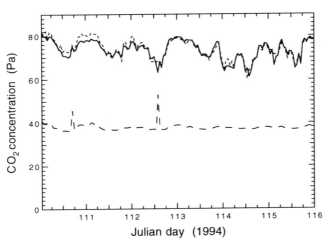

Figure 3 Representative pattern of hourly mean CO_2 concentration for the average of all serpentine chambers at elevated CO_2 (solid line), for a single serpentine chamber at elevated CO_2 (short-dashed line), and for a serpentine chamber at ambient CO_2 (long-dashed line). CO_2 was sampled every 15 min at 30 cm above the ground surface. The data are from April 20–25, 1994. Numbers on the x-axis are at midnight. During this period, the mean CO_2 in the serpentine chambers at elevated CO_2 was 73.4 Pa and the mean ambient CO_2 was 37.9 Pa.

B. MECCAs

The MECCA facility consists of 20 enclosures that contain microcosms with a 0.95-m-deep soil column. Each enclosure contains 2 or 3 microcosms 0.4 m in diameter and either 24 or 28 microcosms 0.2 m in diameter. The enclosures are grouped into two sets of 10, with each set fed with air by a large blower (Fig. 4).

The two sizes of microcosms are used for different types of experiments. The larger (0.4-m diameter) units are used for long-term community experiments, where we are interested in the impacts of elevated CO_2 on plant community composition and on the implications of plant community struc-

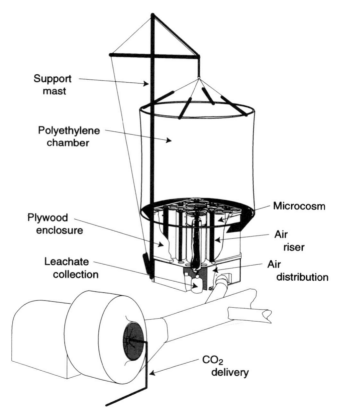

Figure 4 Schematic diagram of one of the MECCA boxes. Air or air enriched in CO_2 is delivered by a large duct into the space below the tubes, and from there it is delivered by nine vertical risers in each box to the level of the tops of the tubes. The box enclosing the microcosm tubes is 1.2 m on a side. The polyethylene chamber is approximately 1.7 m in diameter.

ture for ecosystem processes. The smaller (0.2-m diameter) units are used for two kinds of experiments. One is long-term experiments with single species, where the focus is on the cumulative ecosystem effects of changes in NPP, tissue chemistry, and plant water use. The other is 1-year experiments where the focus is on the initial effects of variation in resources or species characteristics and on the mechanisms driving that variation.

For the first set of experiments, all of the MECCA microcosms contained a synthetic serpentine soil profile. The upper 0.15 m was shredded topsoil from the Kirby Canyon Landfill, a serpentine site 25 km southeast of Jasper Ridge that has been used for a number of experimental studies (e.g., Huenneke et al., 1990; Koide et al., 1988). The subsoil was 0.8 m of a crushed rock mixture of serpentine and nonserpentine minerals. The serpentine in the crushed rock subsoil was excavated at the Acevedo Quarry, 20 km southeast of Jasper Ridge. While the overall nutrient status of the subsoil is very low, the Ca:Mg ratio of about 2 places it near the high end of serpentine soils (Kruckeberg, 1984) and well higher than the 0.1–0.05 in the Jasper Ridge and MECCA topsoils (Table I). The crushed rock subsoil is very coarse, with relatively little water holding capacity. In the field site, for which this material is a surrogate, the serpentine subsoil is fractured bedrock with only occasional patches of topsoil deeper than 0.2 m. But

Table I Chemical Characteristics of the Soils in the Jasper Ridge CO_2 Experiment, Including Field Sandstone and Serpentine, MECCA Sandstone, and MECCA Serpentine Topsoil (Kirby Canyon Landfill) and Subsoil (Acevedo Quarry)[a]

Soil characteristic	Units	Sandstone		Serpentine		
		Field topsoil	MECCA topsoil	Field topsoil	MECCA topsoil	MECCA subsoil
Organic C	%	1.2	0.8	1.8	1.1	0.8
Total N	%	0.12	0.09	0.16	0.11	8
P	ppm	1.8	3.4	3.6	4.6	4.0
Ca:Mg			2.9	0.19	0.10	3.2
K (% total CEC)	%			1.1	0.9	0.4
Mg (% total CEC)	%		27.3	88.7	93.2	33.7
Ca (% total CEC)	%		48.2	9.9	5.6	65.2
Total CEC	meq/100 g		26.2	16.8	27	12.4

[a] The soil tests were performed by ETS (Petaluma, CA) and the Utah State University Soil Testing Lab (Logan, UT).

the abundance of late season annuals is evidence of persistent water at depth, perhaps in cracks in the bedrock.

At the base of the subsoil at the bottom of each microcosm is a 2-cm (uncompressed) polyester wick, which is intended to aid the transfer of moisture out of the soil at the base of the soil column. For the first experiment, a 1-l vessel for leachate was connected to the base of each microcosm, but this volume was increased to 4 l for the second year.

The 0.2-m microcosms (Fig. 5) are constructed from $8 \times 1/8$-in.-thick PVC drain pipe. The bottoms of 1/4-in. PVC rest on rings cemented inside the pipe. Also, 0.2 m from the top of each 0.2-m microcosm, a 13-mm stainless steel tube passes through the pipe, providing a lifting point. Although lifting the 40- to 50-kg microcosms on the ends of this tube works well, we dropped the cross tube from the design for the third experiment because it complicates root harvesting.

The 0.4-m microcosms are constructed from polyethylene trash cans with a slight taper (bottom diameter = 0.33 m). These approximately 200-kg units are lifted with a sling of nylon webbing that passes under the bottom of each.

Each MECCA enclosure consists of an air distribution–leachate space at the bottom, a plywood wall surrounding the soil columns, and an open-top chamber (Fig. 4). Within each enclosure, microcosms are in a tightly packed rectangular arrangement, and the spaces between microcosms generally are not filled. In nine of the spaces between microcosms, air is

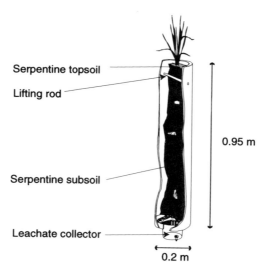

Figure 5 Schematic diagram of a single, small microcosm.

delivered from the bottom air distribution space to the level of the plants through 55-mm (inside diameter) tubes capped with horizontal deflectors.

For the first experiment, each enclosure was topped with a square open-top chamber 1.2 m across and 1.65 m tall. For the second experiment we switched to cylindrical open-top chambers 1.8 m across in order to move the plastic film farther from the plants. The open-top chambers were constructed of 0.15-mm (0.006-in.) UV-resistant polyethylene (3-year greenhouse film, McCalif Grower Supplies, San Jose, CA) for the first experiment, but we switched to 0.1-mm (0.004-in.) UV-resistant polyethylene for increased light transmission for the second experiment. The top of each chamber is covered with a rainfall distributor intended to provide the most uniform possible distribution of rainfall across the microcosms. The rainfall distributors were constructed from polystyrene honeycomb (with 12-mm openings) in the experiment's first year and from black polyethylene hardware cloth (4-mm openings) in the second year. With rain gauges spread across a chamber, gauge-to-gauge variation was less than 10%. Soil moisture in the MECCAs tends to run a little lower than that in the field (data not shown), reflecting the combined effects of the interception loss from the rainfall distributors and effects of soil structure.

The height of the open-top chambers (1.65 m) was selected to ensure that, from just after the autumnal equinox until just before the vernal equinox, all of the direct beam solar radiation striking the microcosm soil surface first passes through a chamber wall. With the smaller zenith angles in the spring and summer, microcosms on the north sides of the enclosure receive some direct beam solar radiation passing through the rainfall distributor. Although the transmittance of the polyethylene film in the visible and near-IR ranges (~0.9) is greater than that of the rainfall distributor (~0.7), the scattering of the direct beam radiation by the polyethylene counteracts the lower transmittance of the rainfall distributor. Measured with a horizontal, cosine-corrected sensor, midday PFD is higher in the areas receiving direct beam radiation through the rainfall distributors, but the light PFD is reasonably uniform when measured with a spherical sensor (LI–193S, Li–Cor, Inc., Lincoln, NE) (data not shown).

The air distribution system is driven by two blowers, each connected with galvanized steel ducting to 10 MECCA enclosures. To ensure uniform flow among enclosures, the ducting is sized so that the minimum cross section is at the final stage, the horizontal deflectors at the tops of the nine distribution tubes in each enclosure. Each blower has a free-air output of 4.0 m^3 s^{-1}, providing a maximum vertical velocity in each chamber of 0.13 m s^{-1}, or 4.7 air changes per minute. Based on CO_2 consumption, the actual blower output is 3.0 m^3 s^{-1}, yielding a vertical velocity in the chambers of 0.083 m s^{-1}.

CO_2 is vaporized at the 5.5-Mg (6-ton) storage tank. The flow rate is regulated by a mass flow controller (Model 840, Sierra Instruments, Monterrey, CA), typically at 0.001 m^3 s^{-1}, to provide a CO_2 increase of approximately 35 Pa at midcanopy height. The CO_2 concentration in each of the 10 high CO_2 chambers, 2 of the low CO_2 chambers, and ambient is measured every 30 min with an IRGA (SBA-1, PP Systems, Stotfield, UK) and stored with a data logger (CR10, Campbell Scientific, Inc., Logan, UT).

At the MECCA site, ambient CO_2 is typically 37–40 Pa during the day and 42–48 Pa at night. High frequency variations in the CO_2 concentration in the elevated chambers are common, with 10–20 short (<30 s) excursions of 100–150 ppm below the target per hour during the daylight when turbulence is greatest.

VI. Results

In the field plots, growth at increased CO_2 led to large increases in leaf-level photosynthesis, decreases in stomatal conductance, and increases in midday leaf water potential in *Avena barbata*, the dominant species in the sandstone grassland (Table II). These patterns at the leaf level persist at the ecosystem scale, although the effects on photosynthesis and conduc-

Table II Leaf- and Ecosystem-Level Midday Gas Exchange for the Field Plots for Mid-April 1993[a]

Grassland	CO_2	A (μmol/m^2 s)	g (mol/m^2 s)	ET (mmol/m^2 s)	WUE (mmol CO_2/mol H_2O)	ψ (MPa)
Sandstone						
Leaf level	Low	8.0*	0.25*	4.7*	1.7*	−1.4*
Leaf level	High	18.0*	0.12*	2.4*	7.5*	−1.1*
Ecosystem level	Low	11.0		1.8	8.8*	
Ecosystem level	High	12.8		1.5	11.5*	
Serpentine						
Ecosystem level	Low	2.3*		1.0	1.9*	
Ecosystem level	High	4.9*		1.1	3.6*	

[a] Values are for net CO_2 exchange (A), stomatal conductance (g), evapotranspiration (ET), water use efficiency (WUE), and leaf water potential (ψ). Leaf and ecosystem values are reported per unit of leaf area and per unit of ground area, respectively. Significant CO_2 effects ($p < 0.05$) are indicated by asterisks. Leaf-level data are from Jackson *et al.* (1994). Ecosystem-level data are from Fredeen *et al.* (1995b).

tance are smaller (Table II). In the serpentine field plots, ecosystem-scale evapotranspiration is generally insensitive to CO_2 treatment, probably because evaporation from the soil surface is a large component of ecosystem evapotranspiration. The greater midday water potential in the high CO_2 sandstone plots suggests that the decreased stomatal conductance leads to increased soil moisture, a result confirmed with measurements based on time domain reflectometry (Fig. 6). Increased late season soil moisture resulting from growth at increased CO_2 could lead to a number of important indirect effects, including altered nitrogen mineralization, species composition, and phenology.

Aboveground production in both Jasper Ridge grasslands is somewhat sensitive to CO_2, but the effect depends on grassland type and year-to-year climate variations. The size (based on a combined index of height and mass) and seed production of *Avena barbata* Link. individuals on sandstone increased significantly in 1993 (Jackson *et al.*, 1994), but not in 1994 (Jackson *et al.*, 1995). On the ecosystem scale, the diversity in developmental schedules makes it difficult to assess trends from harvests at any single date. Aboveground live biomass in the serpentine reaches a maximum in early April. Over the first 2 years of the study, it ranged from 128 to 157 g m^{-2} in the high CO_2 chambers and from 99 to 138 g m^{-2} in the ambient chambers, with no significant CO_2 effects (Table III). Live aboveground biomass in the sandstone chambers peaks in May. For the first 2

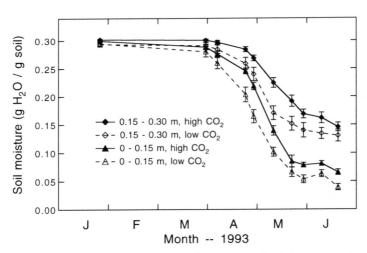

Figure 6 Seasonal trends of soil moisture in the sandstone field site, measured with time domain reflectometry. The effect of CO_2 on the moisture content of the entire profile is significant for two dates in April and one in May. Error bars are standard errors. Redrawn from Fredeen *et al.* (1996).

Table III Aboveground Biomass from the Field Plots for the Date of Peak Biomass and for the Dominant Late Season Annual in each Ecosystem [*Hemizonia congesta* DC. ssp. *luzulaefolia* (DC.) Babc and H. M. Hall in Sandstone and *Calycadenia multiglandulosa* DC. in Serpentine][a]

	Peak biomass harvest					Late annual production		
	High CO_2		Low CO_2			High CO_2	Low CO_2	
	(g m^{-2})	SE	(g m^{-2})	SE	% stimulation	(g m^{-2})	(g m^{-2})	% stimulation
Serpentine 1992	154	9	138	11	12	117	35	234*
Serpentine 1993	128	13	99	17	29	50	6	733*
Sandstone 1992	250	43	264	41	−5			
Sandstone 1993	243	43	279	44	−13	27	7	286*

[a] Means and % stimulation by elevated CO_2 are shown for both the peak biomass harvest and the late annual estimate. Standard errors (SE), shown for the peak biomass harvest, are not appropriate for the estimates of late annual production. Significant CO_2 effects ($p < 0.05$) are indicated with an asterisk.
Unpublished data.

years of the study, it ranged from 243 to 250 g m^{-2} in high CO_2 chambers and from 264 to 279 g m^{-2} at ambient CO_2 (Table III).

Aboveground biomass of late season annuals is more consistently stimulated by increased CO_2, with significant CO_2 effects on the production of the most abundant late season annual in all years measured so far, 1992–1993 in the serpentine and 1993 in the sandstone (Table III). Production by late annuals is a small fraction of the total production in sandstone, but it can be equal to peak season production in serpentine. Total aboveground production is, of course, less than the sum of these two, because the peak season harvest includes a late annual component.

Overall, the tissue composition of the Jasper Ridge annuals is only slightly sensitive to CO_2. For *Avena fatua* and *Plantago erecta*, dominant species in the sandstone and serpentine grasslands, respectively, elevated CO_2 tends to result in small decreases in root lignin and no consistent effects on root or shoot carbon or nitrogen (Table IV). Effects of CO_2 on the schedule of development and the persistence of soil moisture may, in this data set, artificially suppress differences due to CO_2.

Overall, photosynthetic CO_2 fixation increases under ambient CO_2, with down-regulation that is modest, if it occurs at all (Fredeen *et al.*, 1995a; Jackson *et al.*, 1995). Increased CO_2 leads to clear increases in late season soil moisture, at least in sandstone (Fig. 6), which may have important consequences for a range of processes. With ambient nutrient levels, neither biomass nor tissue composition is strongly sensitive to doubled atmospheric CO_2, except in the late season annuals where elevated CO_2 strongly stimulates production. Biomass production and species composition are highly

Table IV Lignin, Carbon, and Nitrogen Contents of Roots and Shoots of *Avena fatua* L. and *Plantago erecta* E. Morris at the Peak Biomass Harvest in 1993, Grown at Elevated and Ambient CO_2 and at Elevated and Ambient Nitrogen, Phosphorus, and Potassium in Serpentine Soil in the MECCA facility[a]

	Treatment		C		N		Lignin		C:N	
	CO_2	Nut	%	SE	%	SE	%	SE		SE
Avena fatua L.										
shoot	Low	Low	42.2	0.5	0.67	0.06	8.1	0.6	66.7	6.7
shoot	High	Low	42.5	0.2	0.68	0.03	7.0	0.3	63.1	2.6
shoot	Low	High	42.8	0.4	0.72	0.04	7.5	0.1	60.4	3.9
shoot	High	High	43.2	0.2	1.00	0.03	7.3	0.4	43.5	1.2
root	Low	Low	41.3	0.5	0.58	0.03	15.9	0.2	73.2	4.7
root	High	Low	41.2	0.6	0.58	0.02	13.8	0.4	71.9	1.9
root	Low	High	41.8	0.8	0.83	0.03	14.0	0.2	51.3	3.2
root	High	High	39.8	0.7	0.78	0.04	12.8	0.3	52.1	2.5
Plantago erecta E. Morris										
shoot	Low	Low	42.8	0.2	1.12	0.04	3.7	0.1	38.7	1.4
shoot	High	Low	42.2	0.6	1.11	0.06	3.7	0.3	38.8	2.5
shoot	Low	High	43.0	0.2	1.97	0.17	4.0	0.2	23.3	3.1
shoot	High	High	42.6	0.2	2.02	0.17	3.1	0.2	22.0	1.9
root	Low	Low	40.2	0.8	0.71	0.06	7.4	0.2	60.0	5.5
root	High	Low	41.9	0.7	0.62	0.03	7.3	0.3	69.4	4.6
root	Low	High	41.4	0.9	1.29	0.14	8.5	0.3	34.8	4.2
root	High	High	40.8	0.9	1.19	0.10	7.6	0.3	35.8	3.3

[a] Means and standard errors are shown. Data are from Chu *et al.* (1996).

variable in the results from the field, emphasizing the importance of large sample sizes in the field and detailed experiments under more controlled conditions.

VII. Concluding Remarks

The Jasper Ridge CO_2 experiment uses a model system approach to address many, but not all, of the variables that may be important in regulating ecosystem responses to elevated CO_2. Results from the Jasper Ridge experiment will be the most useful and most generalizable if they can be interpreted in the context of results from other ecosystem experiments. Some of the other ecosystem experiments should take a model system approach with a focus on a broad array of experiments on tractable ecosystems, but others should focus on ecosystems with unique characteristics or with great potential to impact the global carbon budget. A range of techniques will be required to develop experiments on the entire array of critical ecosystems, and it is important to use less invasive techniques like FACE (free-air CO_2 elevation) to assess the impacts of more invasive techniques like open-top or controlled environment chambers.

The Jasper Ridge CO_2 experiment attempts to characterize CO_2 responsiveness on the basis of species characteristics and environmental resources. Since this experiment can explore only a fraction of the global range of species and environmental characteristics, other ecosystems will provide critical tests of its hypotheses. The community is rapidly approaching the point where the entire body of research on ecosystem effects of elevated CO_2 needs to be viewed as one large experiment, with individual aspects pointing to mechanisms that need to be judged against the larger body of data.

Acknowledgments

The Jasper Ridge CO_2 experiment is supported by NSF grants to the Carnegie Institution of Washington, Stanford University, and the University of California at Berkeley. Many individuals contributed to the design and construction of the experiment, and many others have been involved in data collection and analysis. Special thanks go to Howard Whitted, James Gorham, Brian Welsh, and Sunia Yang for expert assistance with the design and construction of the facilities. Howard Whitted drafted the figures showing the field and MECCA chambers. Celia Chu, Art Fredeen, Bruce Hungate, Geeske Joel, Yiqi Luo, Barbara Mortimer, Jim Randerson, Heather Reynolds, Julie des Rosier, and Julie Whitbeck made major contributions to both the design and the construction of the experiment. Sarice Bassin, Frank Burkholder, Heather Butler, Christa Farmer, Einar Ingebretson, Angela Kalmer, Erik Nelson, and James Wang made significant contributions to the construction.

References

Baker, H. G. (1989). Sources of naturalized herbs and grasses in California. *In* "Grassland Structure and Function: California Annual Grassland" (L. F. Huenneke and H. A. Mooney, eds.), pp. 29–38. Kluwer Academic Publishers, Dordrecht, The Netherlands.
Baldocchi, D. D., White, R., and Johnston, J. W. (1989). A wind tunnel study to design large, open-top chambers for whole-tree pollutant exposure experiments. *Japca* 39, 1549–1556.
Bartholomew, B. (1970). Bare zone between California shrub and grassland communities: the role of animals. *Science* 170, 1210–1212.
Bazzaz, F. A. (1990). The response of natural ecosystems to the rising global CO_2 levels. *Annu. Rev. Ecol. Systematics* 21, 167–196.
Björkman, O., Nobs, M., Berry, J., Mooney, H. A., Nicholson, F., and Catanzaro, B. (1973). Physiological adaptation to diverse environments: approaches and facilities to study plant response in contrasting thermal and water regimes. *Carnegie Inst. Wash. Ybk.* 72, 393–403.
Brussard, P. F., and Ehrlich, P. R. (1970). Contrasting population biology of two species of butterfly. *Nature* 227, 91–92.
Chu, C. C., Field, C. B., and Mooney, H. A. (1996). Effects of CO_2 and nutrient amendment on tissue quality of two California annuals. *Oecologia* (submitted).
Drake, B. G., Peresta, G., Beugeling, E., and Matamala, R. (1996). Long-Term Elevated CO_2 Exposure in a Chesapeake Bay Wetland: Ecosystem Gas Exchange, Primary Production,

and Tissue Nitrogen. *In* "Carbon Dioxide and Terrestrial Ecosystems" (G. W. Koch and H. A. Mooney, eds.). Academic Press, San Diego.

Ehrlich, P. R., White, R. R., Singer, M. C., McKechnie, S. W., and Gilbert, L. E. (1975). Checkerspot butterflies: a historical perspective. *Science* 188, 221–228.

Field, C., and Mooney, H. A. (1986). The photosynthesis-nitrogen relationship in wild plants. *In* "On the Economy of Plant Form and Function" (T. J. Givnish, ed.), pp. 25–55. Cambridge University Press, Cambridge, UK.

Field, C. B., Chapin, F. S., III, Matson, P. A., and Mooney, H. A. (1992). Responses of terrestrial ecosystems to the changing atmosphere: A resource-based approach. *Annu. Rev. Ecol. Systematics* 23, 201–235.

Fredeen, A. L., Koch, G. W., and Field, C. B. (1995a). Effects of atmospheric CO_2 enrichment on ecosystem CO_2 exchange in a nutrient and water limited grassland. *J. Biogeogr.* (in press).

Fredeen, A. L., Randerson, J. T., Holbrook, N. M., and Field, C. B. (1996). Elevated atmospheric CO_2 increases late-season water availability in a water-limited grassland ecosystem. *Oecologia* (submitted).

Gamon, J. A., Field, C. B., Roberts, D. A., Ustin, S. L., and Valentini, R. (1993). Functional patterns in an annual grassland during an AVIRIS overflight. *Remote Sensing Environ.* 44, 239–253.

Gifford, R. M. (1992). Interaction of carbon dioxide with growth-limiting environmental factors in vegetation productivity: implications for the global carbon cycle. *Adv. Bioclimatol.* 1, 24–58.

Goward, S. N., Tucker, C. J., and Dye, D. G. (1985). North American vegetation patterns observed with the NOAA-7 advanced very high resolution radiometer. *Vegetatio* 64, 3–14.

Hobbs, R. J., and Mooney, H. A. (1985). Community and population dynamics of serpentine grassland annuals in relation to gopher disturbance. *Oecologia* 67, 342–351.

Hobbs, R. J., and Mooney, H. A. (1991). Effects of rainfall variability and gopher disturbance on serpentine annual grassland dynamics. *Ecology* 72, 59–68.

Huenneke, L. F., Hamburg, S. P., Koide, R., Mooney, H. A., and Vitousek, P. M. (1990). Effects of soil resources on plant invasion and community structure in Californian serpentine grassland. *Ecology* 71, 478–491.

Hunt, H. W., Detling, J. K., Elliott, E. T., Monz, C. A., and Strain, B. R. (1990). The effects of elevated carbon dioxide and climate change on grasslands i. response of aboveground primary production in intact sods of native shortgrass prairie. *Bull. Ecol. Soc. Am.* 71 (Suppl. 2), 196.

Hunt, R., Hand, D. W., Hannah, M. A., and Neal, A. M. (1991). Response to CO_2 enrichment in 27 herbaceous species. *Funct. Ecol.* 5, 410–421.

Hunt, R., Hand, D. W., Hannah, M. A., and Neal, A. M. (1993). Further responses to CO_2 enrichment in British herbaceous species. *Funct. Ecol.* 7, 661–668.

Idso, S. B., and Kimball, B. A. (1993). Tree growth in carbon dioxide enriched air and its implications for global carbon cycling and maximum levels of atmospheric CO_2. *Global Biogeochem. Cycles* 7, 537–555.

Jackson, R. B., Sala, O. E., Field, C. B., and Mooney, H. A. (1994). CO_2 alters water use carbon gain, and yield in a natural grassland. *Oecologia* 98, 257–262.

Jackson, R. B., Luo, Y., Cardon, Z. G., Sala, O. E., Field, C. B., and Mooney, H. A. (1995). Photosynthesis, growth, and density for the dominant species in a CO_2-enriched grassland. *J. Biogeogr.* (in press).

Johnson, D. W., and Ball, J. T. (1996). Interactions between CO_2 and Nitrogen in Forests: Can We Extrapolate from the Seedling to the Stand Level? *In* "Carbon Dioxide and Terrestrial Ecosystems" (G. W. Koch and H. A. Mooney, eds.). Academic Press, San Diego.

Kashiwagi, J. (ed.) (1985). "Soils map of the Jasper Ridge Biological Preserve." Soil Conservation Service Map, Jasper Ridge Biological Preserve Publication, Stanford, CA.

Koide, R. T., Huenneke, L. F., and Mooney, H. A. (1987). Gopher mound soil reduces growth and affects ion uptake of two annual grassland species. *Oecologia* 72, 284–290.

Koide, R. T., Huenneke, L. F., Hamburg, S. P., and Mooney, H. A. (1988). Effects of applications of fungicide, phosphorus and nitrogen on the structure and productivity of an annual serpentine plant community. *Funct. Ecol.* 2, 335–344.

Kruckeberg, A. R. (1984). "California Serpentines: Flora, Vegatation, Geology, Soils, and Management Problems." University of California Press, Berkeley, CA.

Lincoln, D. E., Newton, T. S., Ehrlich, P. R., and Williams, K. S. (1982). Coevolution of the checkerspot butterfly *Euphydryas chalcedona* and its larval food plant *Diplacus aurantiacus*: larval response to protein and leaf resin. *Oecologia* 52, 216–223.

Luo, Y., Field, C. B., and Mooney, H. A. (1994). Predicting responses of photosynthesis and root fraction to elevated CO_2: Interactions among carbon, nitrogen, and growth. *Plant, Cell Environ.* 17, 1195–1204.

Mandl, R. H., Laurence, J. A., and Kohut, R. J. (1989). Development and testing of open-top chambers for exposing large, perennial plants to air pollutants. *J. Environ. Qual.* 18, 534–540.

Melillo, J. M., Aber, J. D., and Muratore, J. F. (1982). Nitrogen and lignin control of hardwood leaf litter decomposition dynamics. *Ecology* 63, 621–626.

Melillo, J. M., Kicklighter, D. W., McGuire, A. D., Moore, B., III, Vorosmarty, C. J., and Grace, A. L. (1993). Global climate change and terrestrial net primary production. *Nature* 363, 234–240.

Merino, J., Field, C., and Mooney, H. A. (1982). Construction and maintenance costs of mediterranean-climate evergreen and deciduous leaves. I. Growth and CO_2 exchange. *Oecologia* 53, 208–213.

Moloney, K. A., Levin, S. A., Chiariello, N. R., and Buttel, L. (1992). Pattern and scale in a serpentine grassland. *Theor. Pop. Biol.* 41, 257–276.

Mooney, H. A., and Chu, C. C. (1975). Seasonal changes in carbon allocation in *Heteromeles arbutifolia*, a California evergreen shrub. *Oecologia* 14, 295–306.

Mooney, H. A., Hobbs, R. J., Gorham, J., and Williams, K. (1986). Biomass accumulation and resource utilization in co-occuring grassland annuals. *Oecologia* 70, 555–558.

Mooney, H. A., Drake, B. G., Luxmoore, R. J., Oechel, W. C., and Pitelka, L. F. (1991a). Predicting ecosystem responses to elevated CO_2 concentrations. *BioScience* 41, 96–104.

Mooney, H. A., Medina, E., Schindler, D. W., Schulze, E.-D., and Walker, B. H. (eds.) (1991b). "Ecosystem Experiments." John Wiley and Sons, Chichester, UK.

Murphy, D. D., and Ehrlich, P. R. (1989). Conservation biology of California's remnant native grasslands. *In* "Grassland Structure and Function: California Annual Grassland" (L. F. Huenneke and H. A. Mooney, eds.), pp. 201–211. Kluwer Academic Publishers, Dordrecht, The Netherlands.

Norby, R. J., Gunderson, C. A., Wullschleger, S. D., O'Neill, E. G., and McCracken, M. K. (1992). Productivity and compensatory responses of yellow-poplar trees in elevated CO_2. *Nature* 357, 322–324.

Oechel, W. C., and Vourlitis, G. L. (1996). Direct Effects of Elevated CO_2 on Arctic Plant and Ecosystem Function. *In* "Carbon Dioxide and Terrestrial Ecosystems" (G. W. Koch and H. A. Mooney, eds.). Academic Press, San Diego.

O'Neill, E. G., and Norby, R. J. (1991). First-year decomposition dynamics of yellow-poplar leaves produced under carbon dioxide enrichment. *Bull. Ecol. Soc. Am.* 72 (Suppl.), 208.

Owensby, C. E., Ham, J. M., Knapp, A., Rice, C. W., Coyne, P. I., and Auen, L. M. (1996). Ecosystem-Level Responses of Tallgrass Prairie to Elevated CO_2. *In* "Carbon Dioxide and Terrestrial Ecosystems" (G. W. Koch and H. A. Mooney, eds.). Academic Press, San Diego.

Page, B. M., and Tabor, L. L. (1967). Chaotic structure and décollement in Cenozoic rocks near Stanford University, California. *Geol. Soc. Am. Bull.* 7, 1–12.

Parton, W. J., Schimel, D. S., Cole, C. V., and Ojima, D. S. (1987). Analysis of factors controlling soil organic matter levels in Great Plains grasslands. *Soil Sci. Soc. Am. J.* 51, 1173–1179.

Parton, W. J., Scurlock, J. M. O., Ojima, D. S., Gilmanov, T. G., Scholes, R. S., Schimel, D. S., Kirchner, T., Menaut, J.-C., Seastedt, T., Garcia Moya, E., Kamnalrut, A., and Kinyamario, J. L. (1993). Observations and modeling of biomass and soil organic matter for the grassland biome worldwide. *Global Biogeochem. Cycles* 7, 785–809.

Poorter, H. (1993). Interspecific variation in the growth response of plants to an elevated ambient CO_2 concentration. *Vegetatio* 104/105, 77–97.

Reynolds, J. F., Hilbert, D. W., and Kemp, P. R. (1993). Scaling ecophysiology from the plant to the ecosystem: A conceptual framework. *In* "Scaling Physiological Processes: Leaf to Globe" (J. R. Ehleringer and C. B. Field, eds.), pp. 127–140. Academic Press, San Diego.

Roach, D. A., and Wulff, R. D. (1987). Maternal effects in plants. *Annu. Rev. Ecol. Systematics* 18, 209–236.

Schlesinger, W. H. (1991). "Biogeochemistry: An Analysis of Global Change." Academic Press, San Diego.

Strain, B. R., and Bazzaz, F. A. (1983). Terrestrial plant communities. *In* "CO_2 and Plants: The Response of Plants to Rising Levels of Carbon Dioxide" (E. Lemon, ed.), pp. 177–222. American Association for the Advancement of Science, Washington, DC.

Trumbore, S. E. (1993). Comparison of carbon dynamics in tropical and temperate soils using radiocarbon measurements. *Global Biogeochem. Cycles* 7, 275–290.

Vestal, A. G. (1929). Pacific and Palouse Prairies. *Carnegie Inst. Wash. Ybk.* 28, 201.

Wagner, F. H. (1989). Grazers, past and present. *In* "Grassland Structure and Function: California Annual Grassland" (L. F. Huenneke and H. A. Mooney, eds.), pp. 151–162. Kluwer Academic Publishers, Dordrecht, The Netherlands.

Williams, K., Hobbs, R. J., and Mooney, H. A. (1987). Invasion of an annual grassland in Northern California by *Baccharis pilularis* ssp. *consanguinea*. *Oecologia* 72, 461–465.

Wu, J., and Levin, S. A. (1994). A spatial patch dynamic modeling approach to pattern and process in an annual grassland. *Ecol. Monogr.* 64, 447–464.

9

Ecosystem-Level Responses of Tallgrass Prairie to Elevated CO_2

Clenton E. Owensby, Jay M. Ham, Alan Knapp,
Charles W. Rice, Patrick I. Coyne, and Lisa M. Auen

I. Introduction

The increase in atmospheric CO_2 concentration has led to predictions about the responses of natural ecosystems that are based primarily on research on plants grown in pots in growth chambers or greenhouses. Ecosystem-level studies of CO_2 enrichment have been reported for only three natural systems: (1) Arctic tundra [Oechel and Strain, 1985; Oechel and Vourlitis, 1996 (this volume)], (2) estuarine salt marsh [Curtis *et al.*, 1989a; Drake *et al.*, 1996 (this volume)], and (3) tallgrass prairie (Owensby *et al.*, 1993a). The sample area of the plant community exposed to elevated CO_2 has been restricted in all ecosystem-level studies to date and severely limits both the extent and number of parameters that can be measured. Because the consequences of any change in ecosystem resources likely will not be evident in the near term, destructive sampling on small plots usually is not possible. The primary limitation comes with sampling of belowground processes. Also, most ecosystem-level studies have not been conducted for a decade or more on the same area as Mooney *et al.* (1991) suggested. While these ecosystem-level studies are few and constrained by size, exposure time, and fumigation technique (Hendrey, 1992; Ham *et al.*, 1993), they offer an opportunity to compare results with predictions based on data derived from controlled-environment pot studies.

Much of the conjecture concerning ecosystem-level responses to elevated CO_2 has centered around plants with the C_3 photosynthetic pathway, because those plants are often carbon-limited by oxygen competition with CO_2 for the active site on the primary carboxylation enzyme, ribulose-1,5-bisphosphate carboxylase (rubisco). The focus on C_3 plants is not surprising, in that greenhouse and growth chamber trials and the Maryland salt marsh study have shown increased biomass production under elevated CO_2 (Bazzaz, 1990). Conversely, in similar studies, plants with the C_4 pathway have exhibited little growth response to elevated CO_2. The consequence of that research has been to predict changes in competitive relationships in natural ecosystems, with C_3 species populations increasing at the expense of C_4 species. Further impetus to support speculation that C_3 species will gain a competitive advantage has come from competition studies with assemblages of plants in controlled-environment studies (Zangerl and Bazzaz, 1984; Wray and Strain, 1986; Bazzaz and Garbutt, 1988; Williams et al., 1988). Only in the tallgrass prairie ecosystem-level study have plant population changes been measured (Owensby et al., 1993a). The tallgrass prairie research was in ungrazed prairie, but the site was mowed in late February each year. Since grazing imparts different canopy structure and affects carbon acquisition, the results may not reflect the actual changes that will occur in a high-CO_2 world.

C_3 and C_4 plant production responses to elevated CO_2 have been summarized by numerous authors (Kimball, 1983; Bazzaz, 1990; Newton, 1991; Woodward et al., 1991). Net photosynthesis in plants with the C_3 photosynthetic pathway generally increases in response to elevated CO_2 compared to C_4 species, which do not respond at all. That should translate into higher primary production for C_3 vs C_4 species if other resources are not more limiting than carbon dioxide. Scaling estimates derived from single-plant studies up to population- or ecosystem-level primary production generally are not possible. Indeed, Arp (1991) and Thomas and Strain (1991) have cast some doubt on the quantification of photosynthetic responses where the rooting environment was restricted.

Increased atmospheric CO_2 concentration has been shown to reduce plant water loss through changes in stomatal conductance (g_s) (van Bavel, 1974). Whether or not assimilation (A) is increased by elevated CO_2, reduced g_s results in increased water use efficiency (WUE) (Eamus, 1991; Newton, 1991). Even greater water use efficiency is attained when there is a concomitant increase in carbon assimilation. Therefore, elevated CO_2 has the potential to moderate water loss without restricting carbon fixation. If an ecosystem experiences periodic moisture stress, its productivity may be more sensitive to CO_2-induced changes in transpiration than to changes in photosynthesis due to the photosynthetic pathway (Owensby, 1993).

Common speculation has been that increased carbon fixation under elevated CO_2 will wane in the face of reduced nutrient supplies resulting from reduced decomposition rates (Hunt et al., 1991) or that the response will be limited because carbon is not the primary limiting resource (Field et al., 1992). Essentially all elevated CO_2 studies have reported reduced tissue nitrogen concentration (Bazzaz, 1990; Newton, 1991). Dilution of a fixed N supply by increased carbon assimilation has been reported as the primary cause (Field et al., 1992), but reduced tissue N concentration has also been reported in natural ecosystems for species where elevated CO_2 did not increase biomass production (Curtis et al., 1989b; Owensby et al., 1993b). Reduced tissue N concentration developed very early in vegetative growth and persisted to maturity, particularly with C_3 species. That may result from reductions in enzymes associated with photosynthesis, while maintaining assimilation rates comparable to those measured at current atmospheric CO_2 concentrations (Wong, 1979; Bowes, 1991; van Oosten et al., 1992). Reductions in the chlorophyll content of leaves exposed to elevated CO_2 could also contribute to reduced tissue N concentration (Cave et al., 1981). The net impact of elevated CO_2 will likely be a reduced nitrogen requirement for carbon assimilation, which may partially offset the slower nutrient cycling.

Herbivory is an integral part of natural ecosystems, and the effect of elevated CO_2 on diet quality and the associated feedback to carbon acquisition and nutrient cycling may play an important role in ecosystem function in a high-CO_2 world. Insects may increase herbivory as diet quality declines under elevated CO_2, primarily mediated through reduced N concentration in leaves (Field et al., 1992). Fajer et al. (1989) and Lincoln et al. (1986) have concluded that herbivorous insects in a CO_2-enriched atmosphere will increase consumption to offset lower quality forage and that there will likely be longer larval periods and greater mortality. Ruminants and functional caecum animals will likely have reduced consumption and productivity because intake is controlled largely by the rate of passage of ingesta, which is reduced as diet quality decreases (Owensby, 1993). Herbivory affects nutrient cycling by changing the rate at which the plant tissues are broken down and the nutrients released. Ruminants in grassland systems cycle as much as 25% of the aboveground biomass through rumen fermentation, which has a degradation time measured in hours compared to years for surface litter decomposition. It is imperative that herbivory both above- and belowground be addressed in ecosystem-level studies with elevated CO_2.

It is obvious from the preceding discussion that research with elevated CO_2 has generated an enormous amount of speculation concerning ecosystem-level responses. The focus of this paper is to compare those common speculations with the outcome from ecosystem-level research in the tallgrass prairie and summarize and relate the research results to date.

II. Study Site and Experimental Design

The experimental site for the tallgrass prairie research was located in pristine tallgrass prairie north of and adjacent to the Kansas State University campus. Vegetation on the site was a mixture of C_3 and C_4 perennial species and was dominated by big bluestem (*Andropogon gerardii* Vitman; C_4) and indiangrass [*Sorghastrum nutans* (L.) Nash; C_4]. Subdominants included Kentucky bluegrass (*Poa pratensis* L.; C_3), sideoats grama [*Bouteloua curtipendula* (Michx.) Torr.; C_4], and tall dropseed [*Sporobolus asper* var. *asper* (Michx.) Kunth; C_4]. Members of the sedge family (C_3) made up 5–10% of the composition. Principal forbs included ironweed [*Vernonia baldwinii* subsp. *interior* (Small) Faust; C_3], western ragweed (*Ambrosia psilostachya* DC.), Louisiana sagewort (*Artemesia ludoviciana* Nutt.; C_3), and manyflower scurfpea [*Psoralea tenuiflora* var. *floribunda* (Nutt.) Rydb.; C_3]. Average peak biomass occurs in early August at 425 g m^{-2}, of which 35 g m^{-2} is from forbs. Soils in the area are transitional from Ustolls to Udolls (Tully series: fine, mixed, mesic, montmorillonitic, Pachic Argiustolls). Slope on the area is 5%. Fire has been infrequent, occurring two to three times in 10 years. Past history has included primarily winter grazing by cow–calf pairs. The 30-year average annual precipitation is 84 cm, with 52 cm occurring during the growing season.

Nine circular plots were established in early May 1989 on native tallgrass prairie. Each year in late February the area is mowed to a 5-cm height and the cut material is removed. Treatments, replicated three times, were ambient CO_2–no chamber (A), ambient CO_2 plus chamber (CA), and double-ambient (enriched) CO_2 plus chamber (CE) and continued on the same plots through 1993. A nitrogen fertilizer study on the same area had twice-replicated treatments of ambient CO_2 plus nitrogen (AN), chamber plus ambient CO_2 plus nitrogen (CAN), and chamber plus CO_2-enriched plus nitrogen (CEN). Nitrogen was applied as ammonium nitrate at 56 kg N ha^{-1} in late March of both years. Fumigation chambers were placed over the natural vegetation in late March 1990 and retained on the same area for a 2-year period. On the unfertilized and fertilized studies, one-half of each chamber was used to determine biomass accumulation, plant population dynamics, and nutrient status, and the remaining half was grazed by esophageally fistulated sheep to collect samples to determine forage quality differences among treatments.

III. Results and Discussion

The following discussion relates to the preceding study site and experiments. The statement preceding each discussion section represents a synthesis of current speculation concerning elevated CO_2 effects on natural ecosystems.

Speculation: Biomass production of C_3 species will increase more under elevated CO_2 than C_4 species and interspecific competition will favor C_3 species.

Owensby *et al.* (1993a) determined above- and belowground biomass production, leaf area, and plant community species composition and measured and modeled the water status of a tallgrass prairie ecosystem exposed to ambient and twice-ambient CO_2 concentrations. Exposure was in open-top chambers during the entire growing season from 1989 to 1991. Compared to ambient CO_2 levels, elevated CO_2 increased the production of C_4 grass species, but not that of C_3 grass species. Belowground biomass production, estimated in 1990 and 1991 by root ingrowth bags, responded similarly to that of the aboveground, but the relative increase was greater than that above ground. The species composition of C_4 grasses did not change, but *P. pratensis* (C_3) declined and C_3 forbs increased in the stand exposed to elevated CO_2 compared to ambient. The reduction in C_3 grasses was partly due to the lack of grazing, which allowed the taller C_4 grasses to quickly overtop the shorter C_3 species, but the increased biomass and leaf area in CO_2-enriched plots was probably the primary forcing factor in the decline of the C_3 grasses. There was also drought in 2 of the initial 3 years of the study, which also favored the C_4 species. The taller C_3 forbs increased under elevated CO_2, supporting the hypothesis that canopy response (i.e., competition for light) associated with CO_2 enrichment affected interspecific competition.

Tallgrass prairie productivity is commonly limited by N and water availability (Owensby *et al.*, 1969). Increased water use efficiency (WUE) under elevated CO_2 apparently had a greater impact on the productivity of C_4 plants than the photosynthetic pathway of C_3 species. We concluded that the lower the growing season precipitation, the greater the production response of *A. gerardii* and total biomass to elevated CO_2 (Fig. 1). Field *et al.* (1992) indicated that the resource most limiting to ecosystem productivity would dominate even under elevated CO_2. Indications are that in the nutrient-poor tallgrass prairie under elevated CO_2 increased primary production resulted from reduced water stress and improved WUE.

The common speculation that, in mixed C_3–C_4 plant communities, C_3 species will gain a competitive advantage was not supported by observations

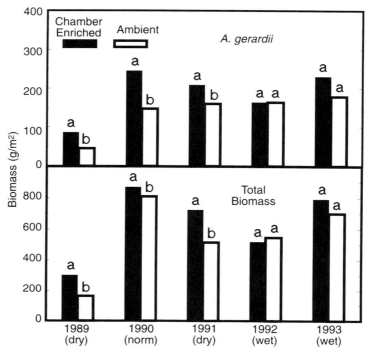

Figure 1 Standing peak biomass (g m^{-2}) of *Andropogon gerardii* and total biomass for the indicated years for native tallgrass prairie exposed to twice-ambient and ambient CO_2 concentrations. Dry, norm, and wet refer to growing season precipitation. Means with a common letter do not differ ($P < 0.05$).

during the 3-year period. The combination of increased biomass and leaf area under elevated CO_2, drought, and lack of grazing may preclude any increase in C_3 grass production or any population increase in tallgrass prairie. Unpublished data from the following 2 years with above-normal precipitation and the same treatments showed an additional decline in C_3 grass production and population. We conclude that the normal limits to C_3 production continued to operate, but that the reduced water stress for the C_4 grasses allowed them to increase production in years with substantial water stress.

Speculation: Ecosystem-level water use efficiency will be enhanced under elevated CO_2.

Knapp *et al.* (1993a) measured responses in the leaf xylem pressure potential (ψ) of *A. gerardii* plants within chambers with ambient and elevated CO_2 and from adjacent unchambered plots throughout the 1991 and 1992 growing seasons. 1991 was a year with below-average rainfall and 1992 was a very wet year. Midday ψ was significantly higher (less negative) in plants grown at elevated CO_2 in both years. When averaged over the growing season, ψ was 0.48–0.70 MPa lower in 1991 than in 1992. In both years, stomatal conductance (g_s) was significantly reduced (21–51%) when measured at 700 vs 350 ml l^{-1} CO_2. These data support the contention that, for C_4 grasses in the tallgrass prairie, positive responses on biomass production to elevated CO_2 may occur only in years with significant water stress, and improved water status will allow for continued growth under elevated CO_2, while growth is abated in ambient CO_2 environments.

In addition to the reduced water stress under elevated CO_2, stomatal response to sunlight may impart additional water savings in *A. gerardii*, the dominant C_4 grass in tallgrass prairie (Knapp *et al.*, 1993b). Within chambers with ambient and twice-ambient atmospheric CO_2, or in adjacent unchambered plots, *A. gerardii* was subjected to fluctuations in sunlight similar to that resulting from intermittent clouds or within canopy shading (full sun > 1500 μmol m^{-2} s^{-1} vs shade = 350 μmol m^{-2} s^{-1}), and stomatal conductance was measured. Time constants describing stomatal responses were significantly reduced (29–33%) at elevated CO_2. As a result, water loss was reduced by 6.5% due to more rapid stomatal responses in twice-ambient CO_2 atmospheres. Leaf xylem pressure potential increased during periods of sunlight variability, indicating that more rapid stomatal responses at elevated CO_2 enhanced plant water status. It is important to note that CO_2-induced alterations in the kinetics of stomatal responses to variable sunlight will likely amplify direct effects of elevated CO_2 and increase WUE in all ecosystems.

In 1993, we measured whole-chamber water vapor fluxes and net carbon exchange (NCE) in CE and CA plots by using the method of Ham *et al.* (1993). Continuous data were collected over a 34-day period when the canopy was near peak biomass (LAI, 4–5 m^2 m^{-2}) and soil water was not limiting. Results showed that twice-ambient CO_2 reduced evapotranspiration by 22% and also increased NCE (Ham *et al.*, 1995). However, the increases in NCE were primarily caused by delayed senescence in the CE plots rather than increased photosynthetic rates during vegetative growth. Reduced evapotranspiration and greater NCE under CO_2 enrichment resulted in higher ecosystem-level water use efficiency (CA, 1.7 g CO_2 kg^{-1} H_2O; CE, 2.6 g kg^{-1}). Additionally, whole-chamber data collected on days with high evaporative demand showed that ecosystem quantum yield (μmol CO_2 μmol PAR^{-1}) in the CE plots remained high in the afternoon period, but decreased in the CA plots (e.g., CA, 0.021 μmol μmol^{-1}, CE, 0.029 μmol μmol^{-1}). These data tend to confirm the leaf-level measurements of Knapp *et al.* (1993a) which showed that more favorable leaf water potentials were maintained under CO_2 enrichment. Measurements of reduced sap flow and increased canopy resistance to water vapor transport in the CE plots provided further evidence that CO_2 strongly influenced the hydrology and plant water relations of the ecosystem (Ham *et al.*, 1994). Data collected at the leaf, whole-plant, and ecosystem scales all suggested that C_4 plant communities exposed to elevated CO_2 will maintain a more favorable water status when subjected to periodic moisture stress.

Our data support the speculation that elevated CO_2 will increase WUE, and we conclude that, in ecosystems that experience moderate to severe water stress, primary production will increase, provided that other resources are adequate. The data from Owensby *et al.* (1993a) and unpublished data from the following 2 years clearly indicate an increase in biomass production in years with water stress and no increase in wetter years. Natural ecosystems, particularly grasslands and shrublands, are often limited by water availability. In the event that global climate change reduces precipitation and increases temperature in a region, the increased WUE in a high-CO_2 world may serve as a buffer against substantial ecosystem change.

Speculation: Photosynthetic capacity of C_3 species will increase, but C_4 species will not have increased photosynthetic capacity under elevated CO_2.

The research presented here is for the photosynthetic capacity of *A. gerardii* under ambient and elevated CO_2 and addresses the speculation that C_4 photosynthetic capacity will be essentially unaffected by elevated CO_2. In 1991 and 1992, *A. gerardii* plants were grown in large pots (as intact soil cores with *A. gerardii* extracted from pristine tallgrass prairie) in the A, CA, and CE plots and periodically were moved to the laboratory and watered to field capacity, and net photosynthetic capacity (A) was measured over a range of CO_2 concentrations, light levels, and air temperatures (Knapp *et al.*, 1993a). In 1992, we also measured maximum photosynthetic capacity (A_{max}), apparent quantum requirement (Q_r), the photosynthetic light compensation point (LCP), and dark respiration (R_d). In 1991, a dry year, A and g_s were significantly reduced (A was 27% lower regardless of measurement CO_2 level) in plants grown at ambient vs. elevated CO_2. Apparently, the more frequent water stress under ambient CO_2 reduced the photosynthetic capacity of *A. gerardii*. Greater water stress also affected A over a broad range of temperatures (17–35°C), with plants grown in A and CA plots having significant reductions in A (as much as 7.1 μmol m^{-2} s^{-1}) compared to those grown in CE plots. In 1992, a wet year, no differences in A, A_{max}, Q_r, LCP, or R_d were detected when ambient and elevated CO_2 grown plants were compared. There was a trend toward lower R_d and LCP in *A. gerardii* grown at elevated CO_2.

Although the photosynthetic capacity of C_4 species may not be directly enhanced under elevated CO_2, we conclude that water stress related reductions in photosynthetic capacity are less likely to occur under elevated CO_2 compared to ambient CO_2 for C_4 grasses. Ultimately, this will lead to higher photosynthetic rates at elevated vs ambient CO_2.

Speculation: Increased biomass production under elevated CO_2 will not be sustained due to nutrient limitations, particularly nitrogen.

While it has often been speculated that increased biomass production under elevated CO_2 will wane as reduced nutrient availability is generated, increased nutrient use efficiency may negate that response. There are both short- and long-term implications regarding nutrient limitation and response to elevated CO_2. The short-term response reflects the inherent low N availability of most natural ecosystems, which affects the ability of the ecosystem to increase productivity with increased carbon availability. The long-term response relates to reduced N availability and results from slower decomposition of litter because of reduced litter quality. We studied the response of nitrogen-fertilized tallgrass prairie to ambient and twice-ambient CO_2 levels over a 2-year period (Owensby et al., 1993c). Comparisons were made with the results of the unfertilized companion experiments of Owensby et al. (1993a,b) at the same research site. Above- and belowground biomass production and leaf area were greater on CEN plots than on CAN or AN plots both years, with a much larger increase measured in 1991, a dry year. Tissue nitrogen concentration was lower for plants in CEN plots than for those in CAN and AN plots, but the total standing crop of N was greater on the CEN plots. The relative differences were the same as those in the unfertilized study, but N concentrations were 15–20% higher with N fertilization. Similar to the unfertilized area, increased root biomass under elevated CO_2 likely increased N uptake. The response of biomass production to elevated CO_2 in N-fertilized plots was much greater than that in the unfertilized plots. Owensby et al. (1993c) concluded that the response to elevated CO_2 was suppressed by N limitation.

These data are for the short term and do not reflect changes in nutrient cycling, but they do indicate the severity of N limitation in tallgrass prairie. If N availability is further reduced under elevated CO_2 by slower nutrient cycling, then the biomass production response to elevated CO_2 may be further limited. The speculation that N limitation will be amplified under elevated CO_2 is supported by our short-term results from the tallgrass prairie.

Speculation: Nitrogen concentration in plant tissues will be reduced and fiber concentration increased under elevated CO_2 and these will slow decomposition and nutrient cycling.

Owensby et al. (1993b) reported on N concentrations in *A. gerardii*, *P. pratensis*, and forb aboveground biomass in tallgrass prairie exposed to ambient and elevated CO_2 over the 3-year period. N concentration of root ingrowth bag biomass was also measured. Total N in above- and belowground biomass was calculated as a product of concentration and peak biomass by species groups. N concentration in *A. gerardii* and forb aboveground biomass was lower and total N was higher in elevated CO_2 plots than in ambient CO_2 plots. N concentration in *P. pratensis* aboveground biomass was lower in elevated CO_2 plots than in ambient, but total N did not differ among treatments in 2 of 3 years. In 1990, N concentration in root ingrowth bag biomass was lower and total N greater in elevated CO_2 than in ambient CO_2 plots. Root ingrowth bag biomass N concentration did not differ among treatments in 1991, but total N in root ingrowth bag biomass was greater in elevated CO_2 plots than in ambient CO_2 plots.

The impact of elevated CO_2 on the decomposition of aboveground biomass and soil microbial processes was also measured. Rice *et al.* (1993) determined the effects of elevated CO_2 in unfertilized and N-fertilized plots on the amounts of carbon and nitrogen stored in soil organic matter and on microbial biomass and microbial activity in 1991 and 1992. The data were taken on the plots that had been exposed to elevated and ambient CO_2 since 1989 for the unfertilized plots and since 1990 for the fertilized. Soil organic C and N increased significantly after three seasons under enriched atmospheric CO_2. In 1991, a dry year, CO_2 enrichment significantly increased microbial biomass C and N compared to ambient, but in 1992, a wet year, microbial biomass C and N did not differ among CO_2 treatments. Added N increased microbial C and N and stimulated microbial activity under CO_2 enrichment. Microbial activity was consistently greater under CO_2 enrichment, probably as a result of better soil water conditions. Ratios of microbial to organic C and N did not differ among treatments. Increased microbial N with CO_2 enrichment was another indication that plant production may become limited by N availability. In order to determine whether the soil system could compensate for the limited N by increasing the labile pool to support increased plant production with elevated atmospheric CO_2, longer term studies are needed to determine how tallgrass prairie will respond to increased C input.

Because translocation of N in late season from the aboveground biomass to belowground storage made the aboveground litter similar in tissue chemistry, the decomposition of surface litter was unaffected by CO_2 treatment (Kemp *et al.*, 1993). They collected standing dead and green foliage litter from *A. gerardii* (C_4), *S. nutans* (C_4), and *P. pratensis* (C_3) plants that had been grown under twice-ambient or ambient CO_2 and with or without nitrogen fertilization. The litter was placed in mesh bags on the soil surface of adjacent prairie and allowed to decay over a 2-year period. Litter bags, retrieved at intervals, showed that the CO_2 treatment had a relatively minor effect on the initial chemical composition of the litter and its subsequent rate of decay. However, there was a large difference among species in litter decomposition. *P. pratensis* leaf litter decayed more rapidly, had higher total N and P, and had different initial carbon chemistry than the C_4 grasses. These differences suggest that there could be an indirect effect of CO_2 on decomposition and nutrient cycling by way of CO_2-induced changes in species composition.

The increased microbial N, higher soil C content of the upper soil layer, and increased microbial respiration with N addition all indicate that there is a reduction in soil organic matter decomposition under elevated CO_2. The addition of N to the ecosystem under elevated CO_2 substantially increased CO_2 evolution from the soil, which lends support to the contention that, in the long term, N limitation may slow the biomass production response to elevated CO_2. Because surface litter chemistry does not appear to differ between ambient and elevated CO_2 treatments, it is not likely that higher atmospheric CO_2 will cause a significant change in surface litter decomposition.

Speculation: Changes in forage quality under elevated CO_2 will affect ruminant herbivory.

Acid detergent fiber (ADF) (primarily cellulose and lignin) of *A. gerardii*, *P. pratensis*, and forb aboveground biomass was estimated by periodic hand-clipping of samples throughout the growing season in 1989 and 1990 (Owensby, unpublished data). In 1991, ADF in peak biomass was estimated by an early August harvest. ADF concentration was higher (indicating lower forage quality) in plots with elevated CO_2 than in those with ambient CO_2 for the C_4 grasses in 1989 and 1990. ADF concentration in *P. pratensis* was similar among treatments in 1989, but in 1990, elevated CO_2 reduced ADF concentration compared to ambient CO_2. Diet quality samples were collected using esophageally fistulated sheep in 1989 and 1990. In both years, *in vitro* dry matter digestibility (IVDMD: disappearance of dry matter following anaerobic incubation in rumen fluid) was lower for sheep-collected samples from CO_2-enriched plots than that for samples from ambient CO_2 plots, with the difference being greater with an up to 32-hr exposure to rumen fluid. We conclude that elevated CO_2 may reduce forage quality and, therefore, may reduce intake and ruminant productivity. Since domestic livestock can be supplemented with harvested food resources, the impact will likely be greater for wild ruminants than for domestic ones.

IV. Summary and Conclusions

A summary of the common speculations about ecosystem-level responses to elevated CO_2 is presented in Table I. The primary impact of elevated CO_2 on the tallgrass prairie ecosystem most likely will be through changes in water use as a result of reduced stomatal conductance and concomitant reduction in water loss from leaves. Whole-chamber water flux measurements and whole-plant sap flow data show a substantial reduction in evapo-

Table I Common Speculations Summarized from the Literature Concerning Responses of Natural Ecosystems to Elevated CO_2 and Results from a CO_2-Enriched, C_4-Dominated Tallgrass Prairie

Common speculation	Results from a CO_2-enriched, C_4-dominated tallgrass prairie
Biomass production of C_3 species will increase more under elevated CO_2 than C_4 species, and interspecific competition will favor C_3 species	C_4 grass species had increased biomass production, and C_3 grass species biomass production was unaffected by elevated CO_2; elevated CO_2 favored C_4 species, with C_3 species declining in the stand
Ecosystem-level water use efficiency will be enhanced under elevated CO_2	Increased biomass production under elevated CO_2 in years with water stress indicated increased water use efficiency; in all years elevated CO_2 reduced water use at the ecosystem level
Photosynthetic capacity of C_3 species will increase, but C_4 species will not have increased photosynthetic capacity under elevated CO_2	Reduced water stress under elevated CO_2 appeared to protect the photosynthetic capacity in dry years compared to ambient, resulting in a greater capacity under those conditions
Increased biomass production under elevated CO_2 will not be sustained due to nutrient limitations, particularly nitrogen	Biomass production response to elevated CO_2 appeared to be nitrogen-limited
Nitrogen concentration in plant tissues will be reduced and fiber concentration will be increased under elevated CO_2 slowing decomposition and nutrient cycling	Under elevated CO_2, the nitrogen concentration was reduced and fiber concentration was increased; belowground organic matter decomposition appeared to be reduced
Changes in forage quality under elevated CO_2 will affect ruminant herbivory	Forage quality of sheep diet samples was reduced under elevated CO_2

transpiration, and less negative diurnal and midday xylem pressure potentials in years with water stress indicate an amelioration of water stress under elevated CO_2. Further, a more rapid stomatal response to sun–shade events also conserves water in plants under elevated CO_2. Soil moisture levels were consistently higher under elevated CO_2, indicating reduced water use by the plant community. Since tallgrass prairie is subject to frequent water stress, biomass production will be enhanced under increased levels of atmospheric CO_2. Our data show that, in years with water stress, biomass production of C_4 species is greater under elevated CO_2 than ambient.

Severe water stress reduces the photosynthetic capacity of *A. gerardii*, the dominant C_4 grass in tallgrass prairie. We found that plants grown under elevated CO_2 maintained greater photosynthetic capacity across a range of air temperatures and CO_2 concentrations after exposure to natural water stress conditions. These data further support the contention that reduced transpiration under elevated CO_2 attenuated the detrimental effects of water stress on carbon fixation.

Interspecific competition in natural ecosystems may not follow the conventional wisdom that C_3 species will gain a competitive advantage [cf. Drake et al., 1996 (this volume)]. In mixed C_3–C_4 communities, those resource limits that currently maintain plant composition may have a greater impact than the CO_2 fertilizer effect on C_3 photosynthesis. In the tallgrass prairie, water stress and time of N availability favor C_4 species, and the partial alleviation of water stress under elevated CO_2 likely gives the C_4 species a greater competitive advantage. Over the 5 years of CO_2 fumigation in tallgrass prairie, the C_3 grasses have declined in the stand. Speculation based on greenhouse studies that C_3 species will have a competitive advantage in natural ecosystems is likely an oversimplification, and care should be taken regarding the prediction of species composition changes.

A reduction in tissue N concentration under elevated CO_2 reported by others was also observed in tallgrass prairie exposed to elevated CO_2 for species with increased biomass production, as well those without increased production. The reduction was likely from dilution of a fixed N supply, as well as from a reduced N requirement. The greater total N in above- and belowground biomass indicates increased availability or increased acquisition capability under elevated CO_2. The apparent reduced N requirement may allow for increased biomass production under elevated CO_2 without increased N availability, or the apparent improved N acquisition may allow for sustained increased biomass production under elevated CO_2. However, the potential reduction in N availability from reduced decomposition rates may result in a N limitation to increased biomass production under elevated CO_2. The addition of N fertilizer to tallgrass prairie shows that biomass production response to elevated CO_2 is already limited by N availability, and further reductions in N availability may preclude substantial responses to elevated CO_2.

Whether natural ecosystems will be sources or sinks for carbon is of great importance to future atmospheric CO_2 concentrations. In tallgrass prairie, preliminary indications are that carbon accumulates under elevated CO_2 in surface soils. It appears that at least a portion of that increased carbon is in the labile pool, since the addition of N to the system results in greater microbial respiration on areas with elevated CO_2. The increased soil microbial N content in CO_2-enriched plots indicates N immobilization due to increased carbon allocation belowground, resulting in increased C:N ratios. In the absence of increased N availability to natural ecosystems, it is likely that the soil carbon quantity will increase in tallgrass prairie and be a substantial sink for carbon as the atmospheric CO_2 concentration rises.

Our research casts some doubt as to whether leaf- or plant-level responses to CO_2 enrichment measured under greenhouse conditions can be scaled to the ecosystem level. Certainly, the mechanisms involved in plant response to elevated CO_2 must be derived at the leaf and plant levels and may be essential to understanding responses at the ecosystem level. However, our data indicate that ecosystem-level responses must be measured under field conditions to avoid erroneous conclusions.

Acknowledgment

This research was supported by the U. S. Department of Energy, Carbon Dioxide Research Division.

References

Arp, W. J. (1991). Effects of source-sink relations on photosynthetic acclimation to elevated CO_2. *Plant, Cell Environ.* 14, 869–875.

Bazzaz, F. A. (1990). The response of natural ecosystems to the rising global CO_2 levels. *Annu. Rev. Ecol. Syst.* 21, 167–196.

Bazzaz, F. A., and Garbutt, K. (1988). The response of annuals in competitive neighborhoods: Effects of elevated CO_2. *Ecology* 69 (4), 937–946.

Bowes, G. (1991). Growth at elevated CO_2: photosynthetic responses mediated through Rubisco. *Plant, Cell Environ.* 14, 795–806.

Cave, G., Tolley, L. C., and Strain, B. R. (1981). Effect of carbon dioxide enrichment on chlorophyll content, starch content and starch grain structure in *Trifolium subterraneum* leaves. *Physiol. Plant.* 51, 171–174.

Curtis, P. S., Drake, B. G., Leadley, P. W., Arp, W. J., and Whigham, D. F. (1989a). Growth and senescence in plant communities exposed to elevated CO_2 concentrations on an estuarine marsh. *Oecologia* 78, 20–29.

Curtis, P. S., Drake, B. G., and Whigham, D. F. (1989b). Nitrogen and carbon dynamics in C_3 and C_4 estuarine marsh plants grown under elevated CO_2 in situ. *Oecologia* 78, 297–301.

Drake, B. G., Peresta, G., Beuyeling, E., and Matamala, R. (1996). Long-term elevated CO_2 exposure in a Chesapeake Bay wetland: Ecosystem gas exchange, primary production, and tissue nitrogen. *In* "Carbon Dioxide and Terrestrial Ecosystems" (G. W. Koch and H. A. Moonney, eds.). Academic Press, San Diego.

Eamus, D. (1991). The interaction of rising CO_2 and temperatures with water use efficiency. *Plant, Cell Environ.* 14, 843–852.

Fajer, E. D., Bowers, M. D., and Bazzaz, F. A. (1989). The effects of enriched carbon dioxide atmospheres on plant-insect herbivore interactions. *Science* 243, 1198–1200.

Field, C. B., Chapin, F. S., III, Matson, P. A., and Mooney, H. A. (1992). Responses of terrestrial ecosystems to the changing atmosphere: A resource-based approach. *Annu. Rev. Ecol. Syst.* 23, 201–235.

Ham, J. M., Owensby, C. E., and Coyne, P. I. (1993). Technique for measuring air flow and CO_2 flux in large open-top chambers. *J. Environ. Qual.* 22, 759–766.

Ham, J. M., Owensby, C. E., Coyne, P. I., and Bremer, D. J. (1995). Fluxes of CO_2 and water vapor from a prairie ecosystem exposed to ambient and elevated CO_2. *Agric. Forest Meteor.* (in press).

Hendrey, G. R. (1992). Global greenhouse studies: Need for a new approach to ecosystem manipulation. *Crit. Rev. Plant Sci.* 11 (2), 61–74.

Hunt, H. W., Trlica, M. J., Redente, E. F., Moore, J. C., Detling, J. K., Kittel, T.G. F., Walter, D. E., Fowler, M. C., Klein, D. A., and Elliott, E. T. (1991). Simulation model for the effects of climate change on temperate grassland ecosystems. *Ecol. Modeling* 53, 205–246.

Kemp, P. R., Waldecker, D., Owensby, C. E., Reynolds, J. F., and Virginia, R. A. (1993). Effects of elevated CO_2 and nitrogen pretreatment on decomposition of tallgrass prairie leaf litter. *Plant Soil* 165, 115–127.

Kimball, B. A. (1983). Carbon dioxide and agricultural yields: An assemblage and analysis of 430 prior observations. *Agron. J.* 75, 779–788.

Knapp, A. K., Hamerlynck, E. P., and Owensby, C. E., (1993a). Photosynthesis and water relations responses to elevated CO_2 in the C4 grass *Andropogon gerardii*. *Int. J. Plant Sci.* 154, 459–466.

Knapp, A. K., Fahnestock, J. T., and Owensby, C. E. (1993b). Elevated atmospheric CO_2 alters stomatal response to sunlight in a C4 grass. *Plant, Cell Environ.* 17, 189–195.

Lincoln, D. E., Couvet, D., and Sionit, N. (1986). Response of an insect herbivore to host plants grown in carbon dioxide enriched atmospheres. *Oecologia* 69, 556–560.

Mooney, H. A., Drake, B. G., Luxmoore, R. J., Oechel, W. C., and Pitelka, L. F. (1991). Predicting ecosystem responses to elevated CO_2 concentrations. *BioScience* 41, 96–104.

Newton, P. C. D. (1991). Direct effects of increasing carbon dioxide on pasture plants and communities. *New Zealand J. Agr. Res.* 34, 1–24.

Oechel, W. C., and Strain, B. R. (1985). Native species responses to increased carbon dioxide concentration. *In* "Direct Effects of Increasing Carbon Dioxide on Vegetation" (B. R. Strain and J. D. Cure, eds.), DOE/ER-0238, pp. 117–154. U.S. Department of Energy, Washington, DC.

Oechel, W. C., and Vourlitis, G. L. (1996). Direct effects of elevated CO_2 on Arctic plant and ecosystem function. *In* "Carbon Dioxide and Terrestrial Ecosystems" (G. W. Koch and H. A. Mooney, eds.). Academic Press, San Diego.

Owensby, C. E. (1993). Potential impacts of elevated CO_2 and above- and belowground litter quality of a tallgrass prairie. *Water, Air Soil Pollut.* 70, 413–424.

Owensby, C. E., Hyde, R. M., and Anderson, K. L. (1969). Effects of clipping and supplemental nitrogen and water on loamy upland bluestem range. *J. Range Manage.* 23, 341–346.

Owensby, C. E., Coyne, P. I., Ham, J. M., Auen, L. M., and Knapp, A. K. (1993a). Biomass production in a tallgrass prairie ecosystem exposed to ambient and elevated levels of CO_2. *Ecol. Appl.* 3, 644–653.

Owensby, C. E., Coyne, P. I., and Auen, L. M. (1993b). Nitrogen and phosphorus dynamics of a tallgrass prairie ecosystem exposed to elevated carbon dioxide. *Plant, Cell Environ.* 16, 843–850.

Owensby, C. E., Auen, L. M., and Coyne, P. I. (1993c). Biomass production in a nitrogen-fertilized tallgrass prairie ecosystem exposed to ambient and elevated levels of CO_2. *Plant Soil* 165, 105–113.

Rice, C. W., Garcia, F. O., Hampton, C. O., and Owensby, C. E. (1993). Microbial response in tallgrass prairie to elevated CO_2. *Plants Environ.* 165, 67–74.

Thomas, R. B., and Strain, B. R. (1991). Root restriction as a factor in photosynthetic acclimation of cotton seedlings grown in elevated carbon dioxide. *Plant Physiol.* 96, 627–634.

van Bavel, C. H. M. (1974). Anti-transpirant action of carbon dioxide on intact sorghum plants. *Crop Sci.* 14, 208–212.

van Oosten, J. J., Afif, D., and Dizengremel, P. (1992). Long-term effects of a CO_2 enriched atmosphere on enzymes of the primary carbon metabolism of spruce tree. *Plant Physiol. Biochem.* 30 (5), 541–547.

Williams, W. E., Garbutt, K., and Bazzaz, F. A. (1988). The response of plants to elevated CO_2-V. Performance of an assemblage of serpentine grassland herbs. *Environ. Exp. Bot.* 28 (2), 123–130.

Wong, S. C. (1979). Elevated atmospheric partial pressure of CO_2 and plant growth. I. Interactions of nitrogen nutrition and photosynthetic capacity in C_3 and C_4 plants. *Oecologia* 44, 68–74.

Woodward, F. I., Thompson, G. B., and McKee, I. F. (1991). The effects of elevated concentrations of carbon dioxide on individual plants, populations, communities and ecosystems. *Ann. Bot.* 67, 23–38.

Wray, S. M., and Strain, B. R. (1986). Response of two old field perennials to interactions of CO_2 enrichment and drought stress. *Am. J. Bot.* 73 (10), 1486–1491.

Zangerl, A. R., and Bazzaz, F. A. (1984). The response of plants to elevated CO_2 II. Competitive interactions among annual plants under varying light and nutrients. *Oecologia* 62, 412–417.

10
Direct Effects of Elevated CO_2 on Arctic Plant and Ecosystem Function

Walter C. Oechel and George L. Vourlitis

I. Introduction

The potential effect of elevated CO_2 on the C exchange dynamics of natural ecosystems is a widely debated issue. Much of the confusion stems from the inability to generalize from leaf-level responses to the whole ecosystem and from the difficulty of integrating the potential CO_2–plant interactions over longer temporal scales (Mooney *et al.*, 1991). In addition, it is difficult to differentiate between direct effects of elevated CO_2 and indirect effects from recent global warming and associated climate changes, as both are interrelated.

Despite these difficulties, substantial progress has been made, and some general patterns have emerged. In some ecosystems, the initial stimulation of photosynthesis is often transient and may disappear after long-term exposure (Körner, 1993). The long-term response of individuals and ecosystems is intimately linked to the interactions between elevated CO_2, resource availability (e.g., nitrogen and water), and temperature (Oechel and Strain, 1985; Bazzaz, 1990; Mooney *et al.*, 1991). In many cases, increased productivity, greater allocation to belowground tissue, and reduced conductance have been observed under elevated CO_2 (Bazzaz, 1990; Mooney *et al.*, 1991; Körner, 1993). There are exceptions, however, and it appears that, in some resource-limited ecosystems, the impact of elevated CO_2 on productivity may primarily result from changes in species composition (Bazzaz, 1990).

Of all natural ecosystems, the Arctic has the potential to be most affected by climate changes associated with elevated CO_2 (Schlesinger and Mitchell,

1987) and the potential to significantly feed back to atmospheric CO_2 and climate changes (Oechel and Billings, 1992). Soil carbon stocks in the Arctic are substantial, and long-term C storage is largely a function of permafrost-mediated soil moisture and temperature conditions (Oechel and Billings, 1992). The Arctic contains roughly 12% of the total soil C pool, but only 5% of the total terrestrial area (Miller *et al.*, 1983; Schlesinger, 1991). Much of this C, roughly 60%, is entombed in the permafrost (Miller *et al.*, 1983), and warmer soil temperatures could lead to thermokarst erosion, eventual loss of permafrost, and an associated loss of substantial soil C (Oechel and Billings, 1992).

Although the growth of Arctic plants is relatively slow (Chapin, 1980), the Arctic has been a sink for atmospheric CO_2 in the historic and recent geologic past because low soil temperatures and the presence of permafrost, which impedes drainage and leads to poor soil aeration, have resulted in low rates of soil decomposition (Oechel and Billings, 1992). Estimates of C accumulation vary widely and are predominately a function of the method or integration time used (Oechel and Billings, 1992). Tussock tundra was estimated by harvest, flux, and simulation models to be accumulating roughly 23 g C m^{-2} yr^{-1} (Miller *et al.*, 1983). By using similar methods, wet sedge tundra was estimated to be accumulating between 27 and 109 g C m^{-2} yr^{-1} (Coyne and Kelley, 1975; Chapin *et al.*, 1980; Miller *et al.*, 1983). Analysis of peat cores using ^{14}C profiling information, however, estimates long-term (last 4000 years) C accumulation rates to be between 2 and 3.5 g C m^{-2} yr^{-1} for tussock tundra and 1.3 g C m^{-2} yr^{-1} for wet sedge tundra (Marion and Oechel, 1993).

Measurements of CO_2 flux in Arctic ecosystems indicate that tussock and wet sedge tundra are now sources of CO_2 to the atmosphere (Oechel *et al.*, 1993; Grulke *et al.*, 1990), possibly due to the reported high latitude warming (Lachenbruch and Marshall, 1986; Hengeveld, 1991; Beltrami and Mareschal, 1991; Chapman and Walsh, 1993) and an associated change in ecosystem soil water content (Oechel and Billings, 1992; Post, 1990; Oechel *et al.*, 1993; Groisman *et al.*, 1994).

Given the huge belowground C stocks in Arctic soils, the release of CO_2 from Arctic ecosystems may exert positive feedback on atmospheric CO_2 and greenhouse warming (Oechel *et al.*, 1993). Conversely, under conditions of elevated CO_2 and global climate change, northern ecosystems could become a sink for carbon, providing negative feedback on atmospheric CO_2 (Oechel and Billings, 1992). The ultimate role of Arctic ecosystems in the global C budget largely depends on both the short- and long-term ecosystem response to elevated CO_2 and concomitant climate changes. The net ecosystem response to elevated CO_2 will, in turn, be a reflection of the whole-plant response, the differential responses of various species and populations, and changes in the competitive relationships within a given community (Bazzaz, 1990).

II. Individual Plant Response to Elevated CO_2

Initial exposure to elevated CO_2 results in significant increases in photosynthetic rates in many Arctic species; however, this stimulation is relatively short-lived (Oberbauer *et al.*, 1986; Tissue and Oechel, 1987). In growth chamber studies, the Arctic sedge *Carex bigelowii*, deciduous shrub *Betula nana*, and evergreen shrub *Ledum palustre* were not significantly affected by double-ambient CO_2 concentrations (Oberbauer *et al.*, 1986). When the long-term photosynthetic response of high-CO_2-grown (675 ppm CO_2) *C. bigelowii*, *B. nana*, and *L. palustre* plants was compared to that of ambient-grown (350 ppm CO_2) plants, there was relatively little difference in photosynthetic rates, and all species exhibited nearly complete homeostatic adjustment to elevated CO_2 in low nutrient conditions (Fig. 1). Similar results were observed *in situ* with the dominant tussock tundra sedge *Eriophorum vaginatum* grown under ambient nutrient levels (Fig. 1; Tissue and Oechel, 1987). Homeostatic adjustment to elevated CO_2 occurred in as little as 3 weeks (Tissue and Oechel, 1987) and within 3 months (Oberbauer *et al.*, 1986).

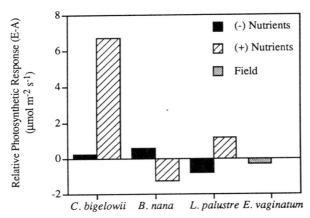

Figure 1 Relative photosynthetic responses of tussock tundra species exposed to double-ambient CO_2 concentration (expressed as the difference between the elevated and ambient responses) and either ambient (solid box), elevated (striped box) nutrients, or *in situ* (stippled box). Data for *C. bigelowii*, *B. nana*, and *L. palustre* are from Oberbauer *et al.* (1986), and data from *E. vaginatum* are from Tissue and Oechel (1987). The interaction of elevated CO_2 and nutrients was not determined for *E. vaginatum*.

Prolonged enhancement of photosynthesis by elevated CO_2 generally requires that nutrients are abundant (Oechel and Strain, 1985; Oechel and Billings, 1992). Plant growth and photosynthesis is strongly limited by nutrient availability in Arctic ecosystems (Shaver and Chapin, 1980; Chapin and Shaver, 1985; Oberbauer et al., 1986). However, fertilization usually acts to stimulate tissue production rather than to increase leaf photosynthesis in Arctic species (Shaver and Chapin, 1980; Chapin and Shaver, 1985; Bigger and Oechel, 1982), and the carbohydrate supply may actually exceed plant need (Tissue and Oechel, 1987). If carbohydrate use is primarily limited by resource availability, then the sinks for additional carbohydrate produced under elevated CO_2 will be limited as well (Tissue and Oechel, 1987; Oechel and Billings, 1992). Without adequate sinks, accumulated carbohydrates may feed back to reduce the amount of RuBP, with a concomitant reduction in photosynthesis (Tissue and Oechel, 1987; Arp, 1991).

This hypothesis is partly supported by the growth chamber study of Oberbauer et al. (1986). With the exception of B. nana, plants exposed to higher nutrient availability had greater CO_2-induced stimulation of photosynthesis than plants exposed to ambient or low nutrient concentrations (Fig. 1); however, the interaction of elevated CO_2 and nutrients was not statistically significant overall (Oberbauer et al., 1986). C. bigelowii was the only species to show a strong positive photosynthetic response to elevated CO_2, and this occurred only at higher nutrient levels (Fig. 1). In contrast, plants either responded weakly to elevated CO_2 and nutrients, as in the case of the evergreen shrub L. palustre, or were inhibited, as with the deciduous shrub B. nana (Fig. 1). This indicates that there are genetic, as well as nutritional, constraints to elevated CO_2 response in these Arctic species.

Although elevated CO_2 was found to have a minimal effect on biomass accumulation in some species, an increase in root:shoot mass ratios has been observed (Oberbauer et al., 1986; Tissue and Oechel, 1987). Root growth, which represents a significant sink for fixed C under ambient CO_2 concentration (Billings et al., 1977), may represent an even larger pool for excess carbohydrates produced under elevated CO_2 (Larigauderie et al., 1988). In addition, changes in the relative growth rate are minimal under elevated CO_2, as are changes in leaf area (Oberbauer et al., 1986; Tissue and Oechel, 1987). Interestingly, tiller production of E. vaginatum increased significantly in high-CO_2-grown plants, so that although elevated CO_2 may not increase the leaf area of mature tillers, it apparently acts to increase the total number of tillers produced (Tissue and Oechel, 1987). This could have profound effects on population dynamics and competitive interactions (Körner, 1993), as increased LAI due to higher tillering rates may lead to greater mutual shading and possible competitive exclusion of intertussock plants.

III. Ecosystem-Level Response to Elevated CO_2

Simulation models indicate that carbon accumulation should increase under elevated CO_2 (Miller *et al.*, 1983; Rastetter *et al.*, 1992; Reynolds and Leadley, 1992). Simulations from the Northern Ecosystems Carbon Simulator (NECS) indicate that C accumulation should increase by about 3–4% in tussock and wet sedge ecosystems over the next 50 years in a double-CO_2 environment coupled with a 4°C temperature increase (Miller *et al.*, 1983). Under elevated CO_2 alone, the relative increase should be much less, but ultimately C accumulation should increase (Miller *et al.*, 1983). The outcome of these simulations is dependent upon initial nutrient availability, the effects of global change on nutrient availability, and the associated increase in temperature (Miller *et al.*, 1983). If nutrient availability remains constant and temperature increases with CO_2 concentration, the systems are predicted to become net C sources (Miller *et al.*, 1983). Measurements indicate that this phenomenon is already occurring (Oechel *et al.*, 1993, 1995; Ciais *et al.*, 1995).

In another simulation, C accumulation under elevated CO_2 alone is expected to increase by 12% over the next 50 years (Rastetter *et al.*, 1992). The cause for the increase in C accumulation is primarily due to differences in the plant and soil C:N ratios. Plant C:N ratios (30:1) are much greater than soil C:N ratios (15:1), and as a result, more C per unit N can be sequestered in plant material than is liberated from the soil through decomposition (Oechel and Billings, 1992; Rastetter *et al.*, 1992; Shaver *et al.*, 1992). Therefore, as soil decomposition increases following soil warming and drying, the nitrogen mineralized should act to stimulate plant productivity, leading to greater C sequestration (Rastetter *et al.*, 1992).

Growth chamber and *in situ* studies clearly show that ecosystem C accumulation is only briefly stimulated by elevated CO_2 and to a limited extent (Prudhomme *et al.*, 1982; Billings *et al.*, 1984; Hilbert *et al.*, 1987; Grulke *et al.*, 1990). Within a simulated season in a phytotron, doubling of CO_2 had little effect on C accumulation of wet sedge tundra microcosms (Billings *et al.*, 1984). Tundra cores grown under 400 ppm CO_2 were net C sources of approximately 50 g C m^{-2} yr^{-1}, while cores exposed to 800 ppm were net sources of only 2 g C m^{-2} yr^{-1} (Billings *et al.*, 1984). Most of this difference is due to an initial stimulation in C accumulation, but this stimulation declined after about 4 weeks of exposure.

A similar pattern was observed with tussock tundra exposed to elevated CO_2 in temperature-controlled enclosures at Toolik Lake, Alaska (Prudhomme *et al.*, 1982; Hilbert *et al.*, 1987; Grulke *et al.*, 1990; Oechel *et al.*, 1992). Initially, C accumulation under elevated CO_2 was stimulated because whole ecosystem photosynthesis increased while respiration declined (Hilbert et al., 1987), although the mechanism for this decline is poorly understood. The initial stimulation in net ecosystem productivity, however, was found to decrease during the first year of the 3-year study, and by the third year, ambient- and high-CO_2-grown plots had similar net ecosystem fluxes and were comparable C sources (Fig. 2; Oechel et al., 1994).

Reciprocal exposure experiments indicate that homeostatic adjustment of net ecosystem CO_2 flux can be much more rapid than intermediate- to long-term fumigation experiments might indicate and that whole ecosystem CO_2 adjustment begins in the first growing season (Prudhomme et al., 1982; Grulke et al., 1990). One possible reason for this may be due to an interaction between phenology and CO_2. Tundra plants store an estimated 400, 250, and 10% of their annual N, P, and carbohydrate requirements, respectively, through the winter (Chapin et al., 1986). Early season respiratory losses can substantially reduce carbohydrate stores and, thus, create a large sink for additional carbohydrates produced under elevated CO_2 (Grulke et al., 1990). The large initial stimulation in photosynthesis observed during the beginning of the growing season (Billings et al., 1984; Grulke et al., 1990) may be due to the large sink that develops as tundra plants break dormancy. This initial stimulation at the beginning of the growing season, however, diminishes with time, so that the stimulation in consecutive years following exposure to elevated CO_2 becomes smaller each year.

In *in situ* reciprocal exposure experiments, photosynthesis rates of ambient-grown tussock tundra increased by nearly 4 times on the first day

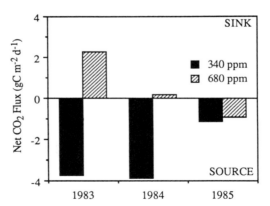

Figure 2 Net CO_2 flux of tussock tundra ecosystems exposed to ambient and double-ambient CO_2 concentrations in *in situ* experiments at Toolik Lake, Alaska. Positive values indicate net ecosystem C accumulation. Data are means of three plots per treatment. Modified from Oechel and Strain (1985); Oechel et al., (1994).

of exposure to elevated CO_2 (Fig. 3). Stimulation peaked on the second day of the experiment and began to decline linearly thereafter (Fig. 3). When the ambient conditions were reestablished 6 days after the reciprocal exposure treatment was initiated, C accumulation rates were depressed relative to the preexposure levels (Fig. 3), indicating a loss in photosynthetic potential due to exposure to elevated CO_2 (Tissue and Oechel, 1987). Similar results were obtained in another reciprocal exposure experiment in tussock tundra (Grulke et al., 1990). During this experiment, significant homeostatic adjustment to elevated CO_2 flux was observed after only 4 days. As during the previous study, a significant loss in photosynthetic potential was observed when plots were reexposed to ambient CO_2 concentrations following short-term CO_2 enrichment.

Exposure to elevated CO_2 apparently acts to delay the onset of ecosystem dormancy over the short term (Grulke et al., 1990). Plots of tussock tundra exposed to high CO_2 showed a significant decline in photosynthetic stimulation during the course of the growing season (Fig. 4A). At the end of the growing season, however, plots exposed to elevated CO_2 were significantly smaller net C sources than ambient-grown plots, and this divergence increased into late September (Fig. 4B). This delay in senescence resulted in approximately 10 additional days of positive flux (Grulke et al., 1990). The increased C gain was attributed, in part, to the prolonged photosynthetic activity of cryptogams and dwarf evergreen shrubs (Grulke et al., 1990). This has important implications for the future C balance of Arctic

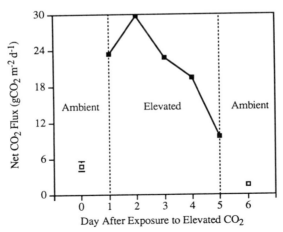

Figure 3 Reciprocal exposure of tussock tundra ecosystems to ambient (340 ppm) and elevated (680 ppm) CO_2 concentrations. Data are expressed as the average net ecosystem CO_2 flux pre- and postexposure to elevated CO_2 (open boxes) and flux during the 5-day period of reciprocal exposure (solid boxes). Modified from Prudhomme et al. (1982).

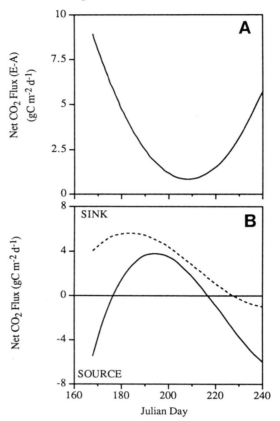

Figure 4 (A) Relative photosynthetic response of tussock tundra ecosystems exposed to elevated CO_2 (elevated photosynthesis − ambient photosynthesis). (B) Seasonal response of tussock tundra ecosystems to 340 ppm CO_2 (solid line) and 680 ppm CO_2 (dashed line). Data are third degrees polynomial curves, with adjusted $R^2 = 0.58$ for ambient and 0.37 for elevated CO_2 treatments. Modified from Grulke et al. (1990).

systems, as one possible outcome of climate change is an increase in growing season length (Waelbroeck, 1993). However, delayed dormancy may also increase the susceptibility of tissue to frost damage (Grulke et al., 1990).

In light of these results, it is unlikely that C accumulation will increase markedly due to the direct effects of CO_2 alone. Rather, an increase in nutrient availability will have to occur to allow the sink required for the utilization of the additional carbohydrates produced under elevated CO_2 (Miller et al., 1983; Tissue and Oechel, 1987; Grulke et al., 1990). In the in situ manipulations at Toolik Lake, tussock tundra exposed to elevated CO_2 and a 4°C increase in ambient temperature was a net sink during the entire 3-year study, with little or no observed acclimation to CO_2 enrichment (Oechel *et al.*, 1994). Increased temperature may directly increase sink strength and activity, as growth and respiration tend to increase at moderately high temperatures in some Arctic species (Limbach *et al.*, 1982; Kummerow and Ellis, 1984; Chapin and Shaver, 1985). More importantly, elevated temperatures may cause increased mineralization and nutrient availability (Nadelhoffer *et al.*, 1992), resulting in greater sink activity and thereby allowing the utilization of additional photosynthate produced under elevated CO_2 (Tissue and Oechel, 1987; Grulke *et al.*, 1990).

IV. Long-Term Ecosystem Response to Elevated CO_2

Our knowledge of the long-term (decade to century) response to elevated CO_2 largely stems from observations of low and high altitude plants (e.g., Körner and Diemer, 1987) and from work near cold, CO_2-emitting springs in Iceland (Cook and Oechel, unpublished data). Low elevation plants are exposed to higher partial pressures of CO_2 than are high elevation plants (Körner *et al.*, 1988), as are plants growing adjacent to CO_2-emitting springs. Therefore, these natural systems should be adequate models for research describing long-term exposure to elevated CO_2.

In general, estimates of the effects of an additional 100 μl l^{-1} atmospheric CO_2 would increase net photosynthesis by approximately 21 and 31% in low and high elevation plants, respectively (Körner and Diemer, 1987). An additional 200 μl l^{-1} would result in an even smaller relative photosynthetic gain (9 and 21% for low and high elevation plants, respectively; Körner and Diemer, 1987). The higher efficiency of CO_2 uptake in high elevation plants is primarily due to greater leaf and palisade layer thicknesses and greater N (protein) content per leaf area (Körner and Diemer, 1987). These results suggest that the reduction in photosynthetic capacity observed under short-term exposure to elevated CO_2 is apparent even after centuries.

Preliminary results from cold CO_2 springs are similar (Cook and Oechel, unpublished data). During the 1994 field season, photosynthesis in *Nardus stricta* plants was measured in ambient and elevated CO_2 habitats in the vicinity of a cold CO_2 spring near Olafsvik, Iceland, over a range of CO_2 concentrations (150, 350, 500, 750, 1000, and 1500 ppm CO_2). Photosynthetic measurements indicate a reduction in the photosynthetic capacity of plants growing in the elevated CO_2 environment (Cook and Oechel, unpublished data). Preliminary measurements show that the reduction in photosynthetic capacity may be due to dramatic reductions in rubisco content, final rubisco activity, and level of chlorophyll in high-CO_2-grown *N. stricta* plants.

In addition, changes in growth and phenology were apparent, presumably in response to long-term exposure to elevated CO_2. Plants growing near the CO_2 spring had a lower leaf area index than plants growing away from the spring and produced fewer flowers and had slower flower development than plants exposed to long-term levels of ambient CO_2 (Cook and Oechel, unpublished data). Furthermore, high-CO_2-grown plants growing near the spring experienced senescence earlier than plants growing in ambient CO_2 environments away from the stream.

At the community level, long-term exposure to elevated CO_2 may be affecting species composition as well. By using the Braun–Blanquet method to compare species abundances in small plots, *Viola palustris*, *Salix herbaceous*, *Carex nigra*, and *Carex bigelowii* were all slightly more abundant adjacent to the spring, while *Thalictrum alpinum*, *Galium uliginosum*, and *Cerastium fontanum* were more abundant away from the spring (Cook, Sveinbjornsson, and Oechel, unpublished data).

V. Summary and Conclusions

The potential response of Arctic plants and ecosystems to elevated CO_2 is complex. Although much progress has been made over the past decade, our predictive ability is hampered by a weak understanding of the interactions between CO_2, resource availability, and climate change and by the difficulty in conducting intermediate- and long-term research.

Even with these difficulties, some general patterns have emerged (Table I). Initial stimulation of plant photosynthetic response and ecosystem C gain is short-lived, primarily due to the lack of sinks for the utilization of excess carbohydrates produced under elevated CO_2 (Table I). The reduction in photosynthetic response appears to hold, even after centuries of exposure to elevated CO_2. Because Arctic plants are limited more by nutrients than by CO_2, prolonged enhancement of photosynthesis and ecosystem C gain will probably require an associated increase in nutrient availability. However, current Arctic genotypes often are unable to respond to elevated CO_2 even when exposed to higher nutrient availability.

Changes in C allocation have been measured in only one experiment (Oberbauer *et al.*, 1986), with root:shoot mass ratios increasing in high-CO_2-grown plants. If this is a general response, then plants may be able to acquire more nutrients in the long term, regardless of whether nutrient availability increases with future climate changes. However, the intermediate- to long-term effects of elevated CO_2 on C allocation are largely unknown. Over the short term, it appears that increased C allocation to belowground tissues either has not created enough of a sink or has increased

Table I Summary of the Observed Response of Arctic Vegetation and Ecosystems to the Direct Effects of Elevated CO_2[a]

	Short term (days to months)	Intermediate term (months to years)	Long term (decades to centuries)
Photosynthesis	+	0	0
Conductance	−/0	?	?
Biomass production	+	+/0	?
Root:shoot ratio	+	?	+
Relative growth rate	+	0	?
Leaf area (per plant)	0	0	−
LAI	+	+/−	−
Tillering	+	+	?
Net ecosystem flux	+	0	?/+[b]

Note. From various sources cited here including Cook and Oechel unpublished data (for long-term responses).

[a] Symbols represent the relative response to elevated CO_2: +, positive response; −, negative response; 0, no response; −/0 or +/0, variable response; and ?, unknown response.
[b] Predicted from simulation models.

the magnitude of nutrient acquisition enough to sustain higher photosynthesis or C accumulation.

Increased tillering of high-CO_2-grown *E. vaginatum* has the potential to decrease growth rates of intertussock species through shading. This could have profound effects on population dynamics and competitive interactions, leading to future changes in plant species composition. Because community-level responses occur over significantly longer time scales than plant-level responses, it may still be too early to tell whether this is occurring. However, differential species responses to elevated CO_2 and results from studies near CO_2-emitting springs indicate the likelihood of compositional changes resulting solely from the direct effects of elevated CO_2.

Whether the Arctic will be a positive or negative feedback to future atmospheric CO_2 concentration and associated climate change is a complex question. Undoubtedly, the effects of climate change associated with the rise of atmospheric CO_2 will have a much greater effect on Arctic ecosystem function than the effects of elevated CO_2 alone. Higher soil temperatures, greater thaw depth, and soil surface layer drying are expected to enhance soil decomposition and nutrient mineralization. Although laboratory and *in situ* experiments do not demonstrate a strong interaction between elevated CO_2 and nutrients, it is likely that ecosystem productivity will ultimately increase. In addition, higher temperatures may increase plant sink strength, which should enhance plant productivity as well.

Unfortunately, the scenarios of enhanced C sequestration do not appear to be occurring, as Arctic ecosystems are losing substantial amounts of CO_2 to the atmosphere (Oechel et al., 1994; Ciais et al., 1995), despite a 25% increase in atmospheric CO_2 concentration in previous decades. Because Arctic plants are inherently slow-growing, it is unclear whether the plant growth and photosynthetic response will be great enough to counteract the CO_2 loss from enhanced decomposition. Much may depend on rates of nutrient loss to streams and lakes following enhanced mineralization and the rate and extent of tree and shrub migration into tundra areas (Smith and Shugart, 1993). Arctic ecosystems are likely to remain sources of CO_2 to the atmosphere for some time before the effects of climate change can stimulate ecosystem-level productivity sufficiently to counteract this loss.

Acknowledgments

This work was supported by the CO_2 Research Program, Environmental Research Division, Office of Health and Environmental Research of the U.S. Department of Energy, and by NSF through the Arctic System Science LAII Trace Gas Flux Study. Long-term support of this project by Boyd Strain, Bill Osburn, Roger Dahlman, and Pat Webber are gratefully acknowledged. We thank J. Stock and K. Turner for help in preparing the manuscript and editing.

References

Arp, W. J. (1991). Effects of source-sink relations on photosynthetic acclimation to elevated CO_2. *Plant, Cell Environ.* 14, 869–875.

Bazzaz, F. A. (1990). The response of natural ecosystems to the rising global CO_2 levels. *Annu. Rev. Ecol. Systematics* 21, 167–196.

Beltrami, H., and Mareschal, J. C. (1991). Recent warming in eastern Canada inferred from geothermal measurements. *Geophys. Res. Lett.* 18, 605–608.

Bigger, C. M., and Oechel, W. C. (1982). Nutrient effect on maximum photosynthesis in Arctic plants. *Holarctic Ecol.* 5, 158–163.

Billings, W. D., Peterson, K. M., Shaver, G. R., and Trent, A. W. (1977). Root growth, respiration, and carbon dioxide evolution in an Arctic tundra soil. *Arctic Alpine Res.* 9, 129–137.

Billings, W. D., Peterson, K. M., Luken, J. D., and Mortensen, D. A. (1984). Interaction of increasing atmospheric carbon dioxide and soil nitrogen in the carbon balance of tundra microcosms. *Oecologia* 65, 26–29.

Chapin, F. S., III (1980). The mineral nutrition of wild plants. *Annu. Rev. Ecol. Systematics* 11, 233–260.

Chapin, F. S., III, and Shaver, G. (1985). Individualistic growth response of tundra plant species to environmental manipulations in the field. *Ecology* 66, 564–576.

Chapin, F. S., III, Miller, P. C., Billings, W. D., and Coyne, P. I. (1980). Carbon and nutrient budgets and their control in coastal tundra. *In* "An Arctic Ecosystem: The Coastal Tundra at Barrow, Alaska" (J. Brown, P. C. Miller, L. L. Tieszen, and F. L. Bunnell, eds.), pp. 458–482. Dowden, Hutchinson, and Ross, Stroudsburg, PA.

Chapin, F. S., III, Shaver, G. R., and Kedrowski, R. (1986). Environmental controls over carbon, nitrogen, and phosphorus chemical fractions in *Eriophorum vaginatum* L., in Alaskan tussock tundra. *J. Ecol.* 74, 167–196.

Chapman, W. L., and Walsh, J. E. (1993). Recent variations of sea ice and air temperature in high latitudes. *Bull. Am. Meteorol. Soc.* 74, 33–47.

Ciais, P., Tans, P. P., Trolier, M., White, J. W. C., and Francey, R. J. (1995). A large northern hemisphere terrestrial CO_2 sink indicated by the $^{13}C/^{12}C$ ratio of atmospheric CO_2. *Science* 269, 1098–1102.

Coyne, P. I., and Kelley, J. J. (1975). CO_2 exchange over the Alaskan arctic tundra: Meteorological assessment by an aerodynamic method. *J. Appl. Ecol.* 12, 587–611.

Groisman, P. Y., Karl, T. R., and Knight, R. W. (1994). Observed impact of snow cover on the heat balance and the rise of continental spring temperatures. *Science* 263, 198–200.

Grulke, N. E., Riechers, G. H., Oechel, W. C., Hjelm, U., and Jaeger, C. (1990). Carbon balance in tussock tundra under ambient and elevated atmospheric CO_2. *Oecologia* 83, 485–494.

Hengeveld, H. (1991). "A State of the Environment Report." Atmospheric Environment Service, Environment Canada.

Hilbert, D. W., Prudhomme, T. I., and Oechel, W. C. (1987). Response of tussock tundra to elevated carbon dioxide regimes: Analysis of ecosystem flux through nonlinear modelling. *Oecologia* 72, 466–472.

Körner, C. (1993). CO_2 fertilization: The great uncertainty in future vegetation development. *In* "Vegetation Dynamics and Global Change" (A. M. Soloman and H. H. Shugart, eds.), pp. 53–70. Chapman and Hall, New York.

Körner, C., and Diemer, M. (1987). *In situ* photosynthetic responses to light, temperature, and carbon dioxide in herbaceous plants from low and high altitude. *Funct. Ecol.* 1, 179–194.

Körner, C., Farquhar, G. D., and Roksandic, Z. (1988). A global survey of carbon isotope discrimination in plants from high altitude. *Oecologia* 74, 623–632.

Kummerow, J., and Ellis, B. A. (1984). Temperature effect on biomass production and root/shoot biomass ratios in two Arctic sedges under controlled environmental conditions. *Can. J. Bot.* 62, 2150–2153.

Lachenbruch, A. H., and Marshall, B. V. (1986). Changing climate: geothermal evidence from permafrost in the Alaskan Arctic. *Science* 234, 689–696.

Larigauderie, A., Hilbert, D. W., and Oechel, W. C. (1988). Effect of CO_2 enrichment and nitrogen availability on resource acquisition and resource allocation processes in a grass, *Bromus mollis*. *Oecologia* 77, 544–549.

Limbach, W. E., Oechel, W. C., and Lowell, W. (1982). Photosynthetic and respiratory responses to temperature and light of three Alaskan tundra growth forms. *Holarctic Ecol.* 5, 150–157.

Marion, G. M., and Oechel, W. C. (1993). Mid- to late-Holocene carbon balance in Arctic Alaska and its implications for future global warming. *Holocene* 3, 193–200.

Miller, P. C., Kendall, R., and Oechel, W. C. (1983). Simulating carbon accumulation in northern ecosystems. *Simulation* 40, 119–131.

Mooney, H. A., Drake, B. G., Luxmoore, R. J., Oechel, W. C., and Pitelka, L. F. (1991). Predicting ecosystem responses to elevated CO_2 concentrations. *BioScience* 41, 96–104.

Nadelhoffer, K. J., Giblin, A. E., Shaver, G. R., and Linkins, A. E. (1992). Microbial processes and plant nutrient availability in arctic soils. *In* "Arctic Physiological Processes in a Changing Climate" (F. S. Chapin, III, R. Jefferies, J. Reynolds, G. Shaver, and J. Svoboda, eds.), pp. 281–300. Academic Press, San Diego.

Oberbauer, S. F., Sionit, N., Hastings, S. J., and Oechel, W. C. (1986). Effects of CO_2 enrichment and nutrition on growth photosynthesis, and nutrient concentration of Alaskan tundra plant species. *Can. J. Bot.* 64, 2993–2998.

Oechel, W. C., and Strain, B. (1985). Response of native species and ecosystems. "Direct Effects of Increasing Carbon Dioxide on Vegetation." Dept. of Energy DOE/ER-0238.

Oechel, W. C., and Billings, W. D. (1992). Anticipated effects of global change on carbon balance of Arctic plants and ecosystems. *In* "Arctic Physiological Processes in a Changing Climate" (F. S. Chapin, III, R. Jefferies, J. Reynolds, G. Shaver, and J. Svoboda, eds.), pp. 139–168. Academic Press, San Diego.

Oechel, W. C., Riechers, G., Lawrence, W. T., Prudhomme, T. I., Grulke, N., and Hastings, S. J. (1992). "CO_2LT" and automated, null-balance system for studying the effects of elevated CO_2 and global climate change on unmanaged ecosystems. *Funct. Ecol.* 6, 86–100.

Oechel, W. C., Hastings, S. J., Vourlitis, G., Jenkins, M., Riechers, G., and Grulke, N. (1993). Recent change of Arctic tundra ecosystems from a net carbon dioxide sink to a source. *Nature* 361, 520–523.

Oechel, W. C., Cowles, S., Grulke, N., Hastings, S. J., Lawrence, B., Prudhomme, T., Riechers, G., Strain, B., Tissue, D., and Vourlitis, G. (1994). Transient nature of CO_2 fertilization in Arctic tundra. *Nature* 371, 500–503.

Oechel, W. C., Vourlitis, G. L., Hastings, S. J., and Bochkarev, S. A. (1995). Change in Arctic CO_2 flux over two decades: Effects of climate change at Barrow, Alaska. *Ecol. Appl.* 5, 846–855.

Post, W. M. (1990). Report of a workshop on climate feedbacks and the role of peatlands, tundra, and boreal ecosystems in the global carbon cycle. Oak Ridge National Laboratory. 32 pp.

Prudhomme, T. I., Oechel, W. C., Hastings, S. J., and Lawrence, W. T. (1982). Net ecosystem gas exchange at ambient and elevated carbon dioxide concentrations in tussock tundra at Toolik Lake, Alaska: An evaluation of methods and initial results. The Potential Effects of Carbon Dioxide-Induced Climate Change in Alaska: Proceedings of a Conference, April 7–8, 1982, School of Agriculture and Land Resources Management, University of Alaska, Fairbanks, Misc. Publication.

Rastetter, E. B., McKane, R. B., Shaver, G. R., and Melillo, J. M. (1992). Changes in C storage by terrestrial ecosystems: How C-N interactions restrict responses to CO_2 and temperature. *Water, Air, Soil Pollut.* 64, 327–344.

Reynolds, J. F., and Leadley, P. W. (1992). Modeling the response of arctic plants to climate change. *In* "Arctic Physiological Processes in a Changing Climate" (F. S. Chapin, III, R. Jefferies, J. Reynolds, G. Shaver, and J. Svoboda, eds.), pp. 413–438. Academic Press, San Diego.

Schlesinger, W. H. (1991). "Biogeochemistry: An Analysis of Global Change." Academic Press, Inc., San Diego, 443 pp.

Schlesinger, M. E., and Mitchell, J. F. B. (1987). Climate model simulations of the equilibrium climatic response to increased carbon dioxide. *Rev. Geophys.* 25, 760–798.

Shaver, G. R., and Chapin, F. R. (1980). Response to fertilization by various plant growth forms in an Alaskan tundra: Nutrient accumulation and growth. *Ecology* 61, 662–675.

Shaver, G. R., Billings, W. D., Chapin, F. S., III, Giblin, A. E., Nadelhoffer, K. J., Oechel, W. C., and Rastetter, E. B. (1992). Global change and the carbon balance of arctic ecosystems. *BioScience* 42, 433–441.

Smith, T. M., and Shugart, H. H. (1993). The transient response of terrestrial carbon storage to a perturbed climate. *Nature* 361, 523–526.

Tissue, D. T., and Oechel, W. C. (1987). Response of *Eriophorum vaginatum* to elevated CO_2 and temperature in the Alaskan tossock tundra. *Ecology* 68, 401–410.

Waelbroeck, C. (1993). Climate-soil processes in the presence of permafrost: a systems modelling approach. *Ecol. Modelling* 69, 185–225.

11

Response of Alpine Vegetation to Elevated CO_2

Christian Körner, Matthias Diemer, Bernd Schäppi, and Lukas Zimmermann

I. Introduction

A. Why Study CO_2 Response in Alpine Vegetation?

There are three reasons why alpine ecosystems are of particular interest in CO_2 enrichment research: (1) The distribution of tundra-like alpine vegetation is unique since it is the only type of plant cover that occurs at all latitudes, although at various altitudes. This worldwide distribution of alpine type vegetation makes it ideal for global monitoring of environmental change. (2) High mountain vegetation evolved under low atmospheric pressure and, thus, low partial pressure of CO_2. If CO_2 is a limiting resource for photosynthesis and plant growth, one might expect it to be more limiting at high elevations. A relief from CO_2 limitation thus might be more effective in plants grown at high altitudes. (3) Mountain vegetation is particularly fragile, and in many parts of the world the functional integrity of mountain ecosystems is an essential prerequisite for the safety of settlements and landscapes at lower altitudes. Environmental changes that include transitory destabilization (change of species composition, migration of ecotones) thus may pose a particular risk.

B. Evidence That Alpine Plants May Be Particularly Responsive to CO_2 Enrichment

It has been shown that plants from high altitudes—at any given partial pressure of CO_2—fix more carbon dioxide per unit leaf area than comparable plants from low altitudes (Körner and Diemer, 1987). A global comparison of carbon isotope discrimination in mountain and lowland plants indicates that this higher efficiency of CO_2 utilization in mountain plants is a general phenomenon (Körner et al., 1988). The increased efficiency of CO_2 use seems to be correlated with higher nitrogen content per unit leaf area and greater leaf thickness. Both of these features are associated with leaf formation under low temperatures and possibly are not directly linked to atmospheric pressure. The reduction in oxygen partial pressure with altitude further adds to better utilization of CO_2 (Körner et al., 1991). Experiments with alpine plants grown for two full seasons in an alpine climate simulator under various CO_2 levels indicate that this greater efficiency of CO_2 utilization persists over long periods and thus may translate into increased carbon gain in the future (Körner and Diemer, 1994).

It is well-known that CO_2 responses of plants strongly depend on temperature and nitrogen supply. Low temperatures are expected to preclude pronounced photosynthetic CO_2 responses, while high temperatures should facilitate strong responses (Long, 1991). Contrary to popular belief, mountain plants often photosynthesize under quite warm conditions, despite low air temperatures. Under strong radiation, canopy temperatures between 20 and 30°C are not unusual, and thus, temperature is not particularly limiting for CO_2 assimilation in mountain plants (Körner and Larcher, 1988). These microclimatic effects may further support the view that CO_2 enrichment will be particularly stimulating to photosynthesis in mountain plants. Finally, leaf nitrogen contents in mountain plants have been found to be rather high (Körner, 1989a), a feature also thought to favor CO_2 responsiveness.

C. Possible Constraints on Alpine Plant Response to Elevated CO_2

The majority of CO_2 enrichment studies have been conducted with fast growing crop plants and seedlings in their exponential growth phase, mostly without any restrictions on nutrient and water supply (Körner, 1993). In contrast, there is little data on how low productivity, slow growing vegetation, which represents a large part of the biosphere, may respond to elevated CO_2. Studies in the Arctic tundra by Oechel and co-workers have shown little stimulation of biomass production by CO_2 enrichment [Tissue and Oechel, 1987; Oechel and Vourlitis, 1996 (this volume)]. Low productivity annual Mediterranean climate grassland communities have shown little stimulation of productivity under elevated CO_2 [Field et al., 1996 (this volume)], a result substantiated by observations of Mediterranean grassland growing in a naturally CO_2-enriched environment (Körner and Miglietta, 1994 and unpublished data). A moderate stimulation of growth (of C_4 species) is reported for tallgrass prairie [Owensby et al., 1996 (this volume)].

Most alpine species have low potential growth rates. It has been amply demonstrated that the intrinsic growth rate of plants is an important criterium for CO_2 responsiveness (e.g., Poorter, 1993). Apart from this genetic component, there is also a direct influence of low night temperatures. In contrast to the favorable microclimatic conditions during periods of sunshine, low night temperatures possibly constrain cell division and cell differentiation (Körner and Pelaez Menendez-Riedl, 1989). This discrepancy between high photosynthetic efficiency and slow growth potential may explain the well-known high levels of nonstructural carbohydrates in the tissues of mountain plants (e.g., Mooney and Billings, 1965). A final constraint, at least in temperate and subpolar mountain areas, is the short growing season. As in tundra, it may take several years for CO_2 effects to become manifested in CO_2 enrichment experiments. From this, it may be concluded that significant CO_2 responses are more likely to occur if atmospheric warming will accompany CO_2 rise, particularly when night temperatures increase appreciably.

In summary, there are a number of features of alpine vegetation that suggest high sensitivity to CO_2 enrichment, while a number of constraints may prevent a strong growth response. Will alpine ecosystems respond in a fashion similar to the low productivity systems mentioned earlier (little or no response), or will the specific growth conditions in the alpine zone still permit a significant CO_2 response? In order to clarify this situation, we initiated a CO_2 enrichment study in the Swiss Central Alps in 1992. Here, we describe the experimental design and report the results of the first 2 years of CO_2 treatment in the field.

II. Methods

A. Vegetation, Climate, and Soils

We selected the most abundant type of truly alpine, closed vegetation in the European Central Alps: a sedge-dominated plant community with *Carex curvula* as the main species, with a maximum green canopy height of 6 cm (Körner, 1989b). Communities of this type are found at altitudes between 2300 and 3000 m (tree line is around 2000 m). The mean length of the snow-free period is 15 weeks, with snow melt usually occurring during the last 10 days of June and the first heavy snow falls by mid-October. Growth of *C. curvula* starts a few days after snow melt but ceases (possibly photoperiodically controlled) by mid-September, substantially before the winter snowcover develops. Thus, the active growth period lasts about 10 weeks per year. Our experimental site is situated at 2470-m altitude near the Furka Pass on the Bidmer Plateau at the upper end of the Rhone Valley (46° 35' N, 8° 23' E). Long-term climate records are available from a station located at 1980-m altitude, i.e., 500 m lower and roughly 5 km northwest of our field site. Annual means from the preceding 5 years are summarized in Table I. Precipitation is distributed uniformly over the year, but monthly variations between years are considerable. Day by day comparisons between temperatures at this station and our research site for July and August 1993 indicate that air temperatures at our site are, on average, 2.7 K lower (i.e., −0.55 K per 100 m of altitude). Day by day comparisons of summer precipitation between the stations indicate 2–3 mm less rainfall per rainy day at our research site.

The soil is a podzolized alpine brown earth on siliceous schists, with high organic content and a pH around 4 (Table II). The total amount of fine substrate (<2 mm) per unit land area amounts to a column of ca. 430 mm down to a depth of 600 mm, but may be smaller in particularly rocky areas. The total soil carbon pool in this volume is on the order of 12 kg C m^{-2}, which is approximately 15 times the amount of C stored in biomass. Although most roots occur in the top 10 cm, some roots are found at 1-m soil depth. Table II illustrates midsummer profiles of soil temperature and soil moisture. Given these precipitation and soil moisture data and daily midseason means of evapotranspiration of around 3 mm per day (Körner *et al.*, 1989; the mean for the whole snow-free period is around 2 mm per day), physiological water stress of plants is impossible at this site. However, dehydration of the top 10 cm of the soil profile during dry summer weather may impair microbial activity and thereby limit plant access to mineral nutrients.

Figure 1 shows the biomass distribution among major plant species or groups of plant species in this plant community (1991 harvest). A total of 20 different species of higher plants is found at the experimental site, with *C. curvula* representing two-thirds of the total biomass. Lichens (not

Table I Climate Data for the Grimsel Meteorological Station[a] Located at 1980-m Altitude for 1988–1992 (Last 5 Years) and the Bidmer Field Site for 1992 and 1993

	Grimsel 1980 m 1988–1992	Bidmer 2470 m	
		1992	1993
Annual mean temperature (°C)	2.3	−1 to 0	
Mean temperature July–August (°C)	10.6	10.3	6.7[b]
−15 cm soil temperature July–August (°C)			11.6
Mean annual precipitation (mm)	1890	>1500	
Mean precipitation July–August (mm)	241	>180[c]	
Mean air pressure July–August (hPa)	808	765	

[a] Records kindly provided by the Swiss Meteorological Institute.
[b] An estimate for the last 5 years using an adiabatic laps rate of −0.55 K per 100 m yields a mean of 7.9°C.
[c] Some precipitation events are missing from our own records.

Table II Soil pH, Fine Soil Volume, and Midseason Moisture and Temperature Profiles

Soil depth (cm)	pH[a]	Volume of <2-mm fraction (%)	Water content[b] (mm)	Soil temperature[c] (°C)
0–2.5[d]	3.62	97	5.0	18.7
2.5–10[e]	3.58	86	22.6	16.0
10–20	4.17	60	24.0	14.2
20–30	4.48	75	25.0	13.0
30–40	4.64	82	21.0	12.3
40–50	4.72	75	20.5	11.9
50–60	4.84	50	17.0	11.8
>60		10	5.0	
Means (2.5–60) or totals	4.42	71	140[f]	13.2

[a] 10 ml <2-mm fraction of soil in 20 ml 1 M KCl.
[b] Midseason (July 27, 1994) values after 5 weeks of good weather and subnormal precipitation. TDR readings in soil volumes free of stones >10 mm converted to mm water column after subtraction of >10-mm stone volume in the profile.
[c] 15.00 hr true local time on a bright midsummer day (July 27, 1994).
[d] Raw humus and root felt.
[e] Gray bleaching horizon.
[f] The freely available moisture by July 27, 1994, was approximately 70 mm, which is sufficient for evapotranspiration of 23 bright days without precipitation (see text).

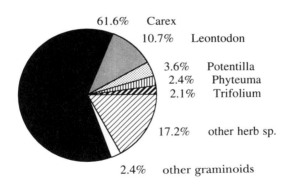

Total aboveground biomass 125 g/m^2

Figure 1 Aboveground biomass distribution in the alpine grassland on the Bidmer Plateau at 2470 m in 1991. The aboveground biomass of the main species is presented as percentage of the total biomass per land area, excluding lichens (ca. 124 g m^{-2}; lichen mass is ca. 46 g m^{-2}). Species with low contributions to total biomass, such as *Homogyne alpina*, *Soldanella alpina*, *Geum montanum*, *Ligusticum mutellina*, are included in the "other herb sp." fraction. *Poa alpina*, a species possibly being stimulated by elevated CO_2, represents less than 1% of aboveground biomass.

included in Fig. 1) are an important component, contributing 27% of the aboveground biomass total of approximately 170 g m^{-2}. Besides its great abundance in the Central Alps, this plant community was also selected because many years of previous research, mainly in the Austrian Alps, has provided an excellent database on the structure and function of this type of vegetation (Cernusca, 1977, 1989; Franz, 1980; Klug-Pümpel, 1981, 1982; Grabherr *et al.*, 1980; Körner, 1982).

B. Experimental Design and Field Installations

Our intention was to expose intact plant communities to ambient and elevated CO_2 (i.e., 350 vs 680 ppm CO_2), while affecting the microclimate as little as possible. The only practical method meeting these requirements in alpine terrain is the open-top chamber. An initial biomass harvest indicated that a random sample of 12 plots of 25 × 25 cm is sufficient to achieve a biomass estimate that lies within the 95% confidence interval of the community mean. Consequently, we installed 12 open-top chambers supplied with ambient CO_2 and 12 supplied with CO_2-enriched air. Twelve additional plots were assigned as controls without chambers. Four additional replicates of each plot type received nutrient amendments (see below). Thus, we installed 32 open-top chambers, plus controls, for a total of 48 plots, which were grouped into four blocks. Cylindrical chambers of diameter and height of 38 cm were constructed from 2-mm-thick, UV-transmissive plexiglass. The ground area within a chamber contained ca. 500 tillers of *C. curvula* or roughly 2000 leaves. The chamber height exceeded canopy height by a factor of 6–7. The northern interior half of the plexiglass cylinders was covered with a 1-mm nylon mesh in order to reduce reflection and light focusing (see microclimate results).

In order to evaluate the role of mineral nutrients in influencing CO_2 effects, four plots of each type (ambient, elevated CO_2, and no chamber) received a complete fertilizer solution (Typ Δ, Hauert, Grossaffoltern, Switzerland) together with rainwater during the first part of the season (N provided as ammonium nitrate). The total amount of nutrients supplied was adjusted to fit a soluble nitrogen deposition that represents approximately twice the current wet nitrogen deposition at low altitudes in the northern part of Switzerland (40 kg N ha^{-1} yr^{-1}). The background wet deposition of nitrogen is not yet known for the area of this study, but it most likely will be found to be close to 10 kg N ha^{-1} yr^{-1}; hence, plants received a 4-fold "deposition" of nitrogen on these plots. Nonfertilized treatments received an equivalent amount of rainwater. The nutrient addition clearly has second priority in this experiment, since we believe that the ecological significance of fertilizer experiments with natural ecosystems is rather poor. In the long term, nutrient addition will always lead to a new community structure at the loss of the former; hence, it is an inadequate means to illustrate "limitation" at the community level (Körner, 1989a). In our case, we simply wanted to illustrate whether and to what extent interactions exist between CO_2 supply and soil resources, taking the current communities as a reference.

Besides monitoring ambient climate, we made microclimate measurements in three plots per treatment. Measurements included quantum flux density (QFD, cross-calibrated photodiodes), soil temperature (5-cm depth), canopy air temperature (thermistor), canopy air humidity (ventilated Vaisala sensors), and three leaf temperatures (thermistors). Soil moisture was measured at regular intervals by time domain reflectometry (TDR; TRIME-FM, IMKO, Ettingen, Germany). TDR forks were installed permanently to 4-cm soil depth, where most fine roots occur. Readings were calibrated against *in situ* volumetric soil water content.

Two large blowers produced a constant flow of fresh air to the 16 open tops per treatment. The air was distributed to the chambers by manifolds. One fan received computer-controlled CO_2 additions to the airstream to achieve 680 ppm CO_2. In order to minimize possible confounding effects contributed by the air delivery system, blowers and hoses are switched between treatment groups every fortnight. The volume of air in each chamber is exchanged approximately 11 times per minute.

Field equipment was installed during 1991, and CO_2 enrichment began after snowmelt in early July 1992. The data presented here cover the first two growing seasons under CO_2 enrichment, namely, the summers of 1992 and 1993.

C. Methods of Plant and Ecosystem Studies

1. Ecosystem Gas Exchange   Ecosystem CO_2 exchange is measured by fitting open tops with a top and using them as gas exchange cuvettes in an open-differential system for a short period of time (several minutes). CO_2 concentrations were measured with an infrared gas analyzer (LI-6252, LI-COR, Lincoln, NE), and flow rates were determined by a laminar flow element with a differential pressure transducer (LFE 3 and FP 19", Special Instruments, Nördlingen, Germany). Details concerning the gas exchange system are described in Diemer (1994). The separation of soil CO_2 evolution from aboveground plant CO_2 exchange was achieved by inserting steel cups 3 cm in diameter into canopy gaps for approximately 6 min and measuring the rate of CO_2 accumulation (samples were taken through a rubber diaphragm by a syringe at 2-min intervals and analyzed in a pure nitrogen airflow by using a LI 6252 IRGA and a voltage integrator, Hewlett-Packard, Type 3396 II).

2. Biomass   Aboveground biomass is determined by harvesting and by biometric analysis. After 3 years of CO_2 enrichment, the first destructive harvest was made at peak vegetative development in August 1994. All of the aboveground plant biomass in a 2-cm strip of each plot was removed (approximately 40 shoots of *C. curvula*). Dead (necromass) and live (biomass) components of phytomass were separated and further subdivided by species. A final and complete harvest is planned for August 1995. This split harvesting procedure is a compromise between the need for a long-term (4 years) noninvasive treatment and the need to obtain intermediate results after at least 3 years of treatment (and 4 years of funding). Hence, no harvest biomass data can be presented at this time.

The biometric nondestructive biomass determination has been carried out continuously since the beginning of CO_2 enrichment and nutrient additions. In *C. curvula*, the length and width of leaves were measured eight times a season (at 7-day intervals during the rapid phase of growth and biweekly during the slower shoot dieback phase). Overall, 480 labeled shoots (10 per plot) were censused each time, amounting to more than 15,000 leaf measurements per season. Ninety-eight shoots (2 per plot) were harvested every 14 days, analyzed for leaf length, width, and biomass, and subsequently used to calibrate the biometric measurements by linear regression. A less intense and simpler procedure was used for key herbs (no data presented here). All individual plants of herbaceous species were counted each year at peak season.

Belowground biomass has been determined in two ways. Each year, two soil cores of 2-cm diameter and 7-cm depth are taken at the end of the season. The small diameter allows sampling in intertussock gaps. The depth covers the horizon in which 60–70% of overall root mass is found in this alpine grassland. Thus, a total of 96 cores has been sampled each year. Roots were washed, and root length (Root Length Measurement System type RLS, Delta-T Devices, Burwell, England) as well as root dry mass was determined within 2 weeks after harvest (storage at 4°C). The second type of root biomass data was derived from ingrowth cores. The 2-cm-wide holes produced by the preceding harvesting procedure were refilled immediately with sieved soil and the core was labeled. These soil cylinders were then recored by the end of the next season. In this way, the dynamics of reoccupation of newly opened soil space has been studied. The appearance of new roots in these cores represents a rough estimate of annual fine root production, although the disturbance may alter root production. In any case, ingrowth core data provide a relative measure of root dynamics for treatment comparisons.

3. Plant Development The aboveground biometric measurements mentioned earlier also yield phenometric data for the vegetative development of plants. In addition, the timing of flowering and the number of flowers per plot have been determined in order to compare phenologies among treatments.

4. Tissue Quality All biomass samples taken throughout the season have been used for analyses of total nonstructural carbohydrates (TNC), separated into sugars and starch, and nitrogen content (CHN-analyzer, LECO 900, LECO Corp., St. Joseph, MI). The TNC assay follows Wong (1990) and includes boiling of dried, ground, and weighed samples in bidistilled water for 1 hr. Aliquots of the solution are then treated with isomerase and invertase and analyzed for glucose. The decanted material (including starch) is incubated for 15 hr at 37°C with a dialyzed crude enzyme ("Clarase", fungal α-amylase from *Aspergillus oryzae*, Miles Products, Elkhart, IN). In all cases, glucose is determined spectrophotometrically using a glucose test (hexokinase reaction; hexokinase from Sigma, St. Louis, MO).

5. Carbon Flow The fate of carbon within plants and the ecosystem has been studied by using the stable ^{13}C isotope fed to plants via $^{13}CO_2$ (Cambridge Isotope Laboratories, Woburn, MA). Canopy sections of 10-cm diameter were exposed for 6 hr (on an overcast day to prevent cuvette heating) to 600–800 ppm $^{13}CO_2$ atmosphere in both ambient and elevated chambers during July 1993. Concentrations of the label in various compartments of the ecosystem were then determined by mass spectrometry (no data presented here).

III. Results of Two Years of Field Experimentation

The data presented here are preliminary because the duration of the experiment is not yet long enough to permit conclusive statements, and the two experimental seasons have been extremely different in terms of weather. 1992 represented one of the warmest summers on record, while 1993 was very cool and wet. Figure 2 illustrates the frequency distribution of quantum flux density (QFD) during periods with QFD > 30 μmol m^{-2} s^{-1} in July and August of 1992 and 1993. That the quantum supply was clearly much greater in 1992 makes it very difficult to interpret year to year response trends. We will first present data illustrating the CO_2 regimes and the microclimate in the open tops. Next, we will discuss some initial results of ecosystem gas exchange and then summarize observations on growth dynamics and biomass production. Finally, we will present a brief account of changes in plant tissue quality following CO_2 enrichment.

A. CO_2 Concentration and Microclimate

1. CO_2 Regime in Open Tops Figure 3 illustrates the frequency distribution of CO_2 concentrations in high-CO_2 open tops during the 1993 growing season. In order to achieve an average 680 ppm CO_2 level, the set point in the air supply system (top line in Fig. 4) was maintained at 750 ppm.

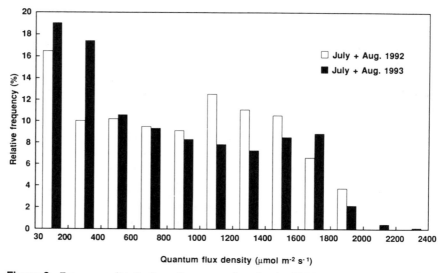

Figure 2 Frequency distribution of quantum flux density (QFD) during July and August 1992 and 1993. Only periods with QFD > 30 μmol m^{-2} s^{-1} were used (half-hourly means of 10-min records).

As a result of gusty atmospheric conditions, 4% of all hours had CO_2 levels below 450 ppm. A total of 77% of all hours was above 550 ppm, with about 15% above 700 ppm. Figure 4 illustrates the short-term variations in CO_2

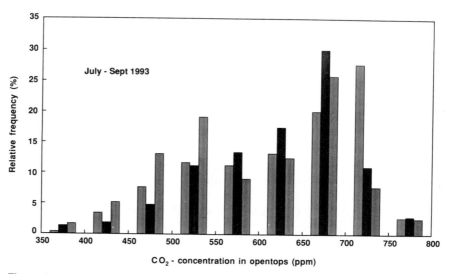

Figure 3 Frequency distribution of CO_2 concentration in three randomly selected high-CO_2 open tops during the 1993 growing season.

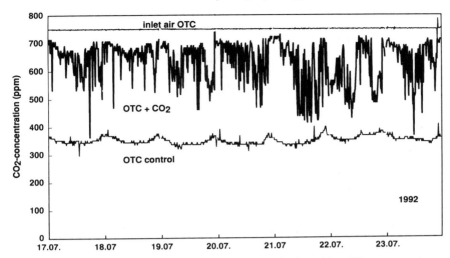

Figure 4 Short-term variations in CO_2 concentration in high- and low-CO_2 open-top chambers at the canopy level and in the high-CO_2 supply system (inlet air OTC) during a typical midsummer week (July 17–23, 1992).

during a midsummer week in 1992. The mean daytime ambient CO_2 level during the measurement campaign was 350 ppm.

2. Microclimate in Open Tops There were virtually no differences in canopy air temperature between chambered and control plots (Fig. 5a). Vapor pressure deficits of canopy air were slightly lower in open tops than in unchambered plots. The maximum difference on clear days was between 2 and 3 Pa early in the afternoon (Fig. 5b). This difference is mainly caused by the difference in ambient air temperature between 1 m above the ground (where the air for chamber supply was sampled) and the canopy level. $\delta^{13}C$ values for plants inside ambient CO_2 chambers (24.58 ± 0.69 ‰) and outside chambers (25.00 ± 0.38‰) are not significantly different ($n = 6$), suggesting no significant effect on stomata due to these small differences in vapor pressure deficit.

On clear days, the quantum flux density at canopy level within the chambers was reduced by 6% at noon and by about 24% in the early morning and evening relative to the unchambered condition. The daylong mean reduction was 13%. Small areas of the canopy received higher than ambient QFD due to reflection (and focusing) phenomena in the chamber. These high-QFD spots changed position from west in the morning to east in the afternoon (+17% at noon and peak values of +26% at midmorning and midafternoon).

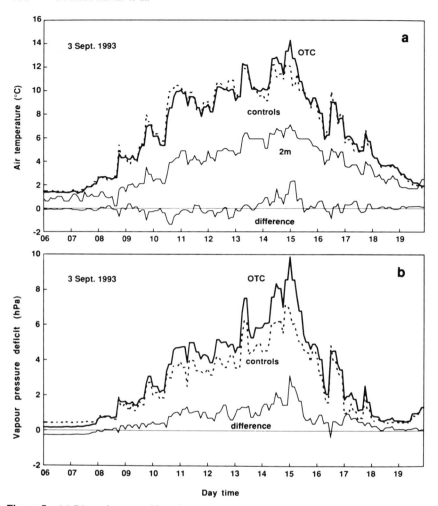

Figure 5 (a) Diurnal course of 2-m air temperature and canopy air temperature (corresponding to leaf temperature of *C. curvula*), 3 cm above the ground, outside (controls, dashed line) and within open tops (high- and low-CO_2 were pooled since these were not different). Bottom: Difference between open-top chambers and controls. (b) Diurnal courses of vapor pressure deficit of the air in the plant canopy within and outside open tops.

Leaf temperatures of *C. curvula* were not significantly different from canopy air temperatures, possibly because of the high turbulence in the chambers and the very small boundary layer resistance of these narrow leaves (0.8- to 1.2-mm leaf width, data not shown).

B. Ecosystem Gas Exchange and Carbon Balance

Ecosystem gas exchange is significantly enhanced by both CO_2 and mineral fertilizer applications ($p < 0.01$, two-way ANOVA, Table III). In full sunlight, the enhancement of photosynthesis attributable to the elevated CO_2 treatment amounts to 41 and 81% in unfertilized and fertilized plots, respectively (Diemer, 1994). The mean enhancement due to the mineral fertilizer additions, irrespective of CO_2 and chamber, is 54%. Biomass estimates suggest that additional stimulation of ecosystem CO_2 uptake due to mineral fertilizer is largely the result of increased leaf mass (Schäppi and Körner, 1995). Effects of elevated CO_2 on gas exchange, particularly in unfertilized plots, are directly attributable to CO_2, and realized carbon gains apparently do not cause increases in the biomass of *C. curvula* (see also Fig. 6). Preliminary data on nighttime respiratory CO_2 efflux suggest that ecosystem dark respiration is not significantly altered by elevated CO_2 (Diemer, 1994). Daytime CO_2 evolution from soil measured in canopy gaps does not differ between CO_2 treatments, but is significantly increased in fertilized plots (R. Stocker, unpublished data).

The extent to which elevated CO_2 increases ecosystem CO_2 uptake, namely, by 41% in unfertilized plots, closely corresponds to the 49% value obtained by Leadley and Drake (1993) in a *Scirpus*-dominated salt marsh. Thus, in both graminoid ecosystems the photosynthetic carbon gain is significantly enhanced by elevated CO_2, but this does not appear to stimulate nighttime or daytime release of CO_2 from soils and vegetation. Since growth does not respond appreciably to CO_2 in our alpine grassland, excess carbon

Table III Net CO_2 Uptake per Unit Ground Area (A) in Full Sunlight (QFD > 1400 μmol m^{-2} s^{-1}) at Midseason 1993 (Means ± se)[a]

CO_2 treatment	A (μmol CO_2 m^{-2} s^{-1})	n
	Unfertilized Plots	
Control 355 ppm CO_2[b]	9.5 ± 0.6	12
OTC 355 ppm CO_2	9.6 ± 1.0	12
OTC 670 ppm CO_2	13.5 ± 0.9$_a$	12
	Fertilized Plots	
Control 355 ppm CO_2	12.3 ± 1.1$_b$	4
OTC 355 ppm CO_2	13.7 ± 1.4$_b$	4
OTC 670 ppm CO_2	24.6 ± 5.0$_{ab}$	4

[a] Subscripts denote significant differences at the 1% level (two-way ANOVA).
[b] Note that, due to reduced total atmospheric pressure at this altitude, the mixing ratios of CO_2 and air correspond to CO_2 partial pressures of approximately 26 and 53 Pa.

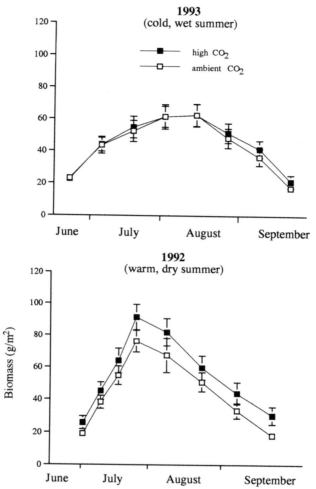

Figure 6 Growth dynamics of *C. curvula* shoots per unit ground area during the 1992 and 1993 seasons. Means and standard errors of biomass are from repeatedly censused shoots. Differences are not significant at the 5% level in both 1992 and 1993 (repeated measures ANOVA with polynomial differences).

must be utilized elsewhere, possibly in soil organic matter, since root biomass did not yet show a significant response (see the following). However, the preliminary data of our ^{13}C labeling experiment do not indicate a more rapid depletion of the label in the biomass under elevated CO_2 (B. Schäppi, unpublished data). Given the enormously large soil carbon pool compared to the biomass carbon pool, changes in soil carbon cannot be detected directly.

C. Biomass and Phenology Responses

After one and two seasons of CO_2 enrichment, there were no statistically significant differences in peak season shoot biomass and LAI of *C. curvula* between CO_2 treatments (Fig. 6; LAI data not shown, since SLA was not different; $p = 0.53$ in 1992 and $p = 0.79$ in 1993, repeated measures analysis of variance). The addition of nutrients alone had a significant effect on aboveground biomass after 2 years of fertilization (+27%, $p < 0.05$), but elevated CO_2 supply did not induce an additional rise in biomass production; thus, there was no significant CO_2 × nutrient interaction ($p = 0.93$ in 1992 and $p = 0.22$ in 1993). The enhancement of growth due to fertilizer application was most pronounced in the nutrient × chamber effect. No changes at the population level (shoot density, number of leaves) for the dominant species *C. curvula* and *Leontodon helveticus* have been observed so far. Controls without open tops showed slightly more biomass production in 1992 and less in 1993 than open-top plots, but differences were not significant.

The increase in biomass in fertilized plots was merely due to the response of herbaceous dicotyledonous species and grasses, whereas the sedge *C. curvula* did not respond—until now—to the additional nutrients provided.

Seasonal dynamics of shoot growth (Fig. 6) in *C. curvula* showed a slight trend of growth stimulation by enhanced CO_2 supply in the warm, dry season of 1992. In essence, this difference built up during the first few days of the season. However, no such trend was observed in the cold, wet season of 1993. Aboveground biomass production in 1993 was 23 (ambient CO_2) to 38% (elevated CO_2) lower than that in 1992. These data confirm that the seasonally integrated quantum flux density (and to a lesser extent temperature) plays a major role in growth and biomass production in alpine ecosystems (Körner, 1982). Carbon dioxide supply might become more effective as temperatures rise (Long, 1991). An alternative hypothesis is that the effect of CO_2 is damped in subsequent years.

Two years after the beginning of CO_2 enrichment, no difference in flowering behavior and tillering could be detected in *C. curvula*. However, there are indications that potentially fast growing grasses such as *Poa alpina*, which were barely present at the beginning of the experiment, may be increasing in prevalence.

The total amount of root biomass (excluding belowground shoots and leaf bases) in 50-cm soil profiles of undisturbed plots is about 700 g m^{-2}. About one-half of this mass is found between the soil surface and a 5-cm depth, one-third occurs between 5- and 15-cm depth, and the rest is distributed over deeper horizons. Analysis of ingrowth cores for 2 years yielded an annual production of approximately 100 g m^{-2} new roots in the top 10 cm of the soil profile. In both years, plots treated with elevated CO_2 produced more roots (+63% in 1992 and +25% in 1993), but the difference between treatments was significant only at the 10% level in 1992 (119 ± 54 vs 73 ± 35 g, $n = 7$, $p < 0.1$) and not significant in 1993 (121 ± 29 vs 97 ± 36 g, $n = 11$, $p > 0.1$). An estimation of the total above- and belowground biomass response is not yet possible, since the responses of undisturbed roots are not known. However, if there is an overall stimulation of biomass production by elevated CO_2, it appears most likely to be belowground.

D. Accumulation of Nonstructural Carbohydrates and Responses of Tissue Nitrogen Concentration

1. Nonstructural Carbohydrates With the exception of the legume *Trifolium alpinum*, all species investigated were found to store large amounts of nonstructural carbohydrates (TNC) even under the present ambient CO_2 conditions (15–30% of leaf dry mass). TNC was primarily present as soluble sugars (glucose, fructose, saccharose; total 12–24% dry mass). The starch fraction seemed to be of minor importance (Fig. 7). These findings are in good agreement with published data (Mooney and Billings, 1965; Sakai and Larcher, 1987; Guy, 1990) supporting the general opinion that plants from cold habitats accumulate substantial amounts of soluble sugars, thereby maintaining a high degree of frost resistance throughout the vegetation period. However, we do not know whether this massive accumulation of sugars is really necessary for adequate cold hardiness, or whether it reflects a certain degree of carbon saturation already at present CO_2 levels, given the low-temperature-limited sink activity (Körner and Pelaez Menendez-Riedl, 1989).

In the first season (1992), CO_2 enrichment effects on TNC in plant leaves varied from weak increases for *C. curvula* (+5%) to significant accumulations for *P. alpina* (+17%, $p < 0.05$). For all species except *T. alpinum*, this increase was solely due to an increase in the sugar fraction. The starch fraction remained unaltered. In contrast, the carbon metabolism of *T. alpinum* in both ambient and high-CO_2 treatments showed a completely different pattern. TNC levels under ambient conditions were much lower than other species (only 8% dry mass), and soluble and nonsoluble TNC fractions were of the same magnitude. In this species exposure to elevated CO_2 mainly resulted in starch accumulation.

Overall, the magnitude of TNC accumulation under enhanced CO_2 in the leaves of these alpine plants is small. There is also no indication of significant TNC enhancement in the belowground plant parts during the first season of CO_2 treatment (1993 data not yet available). These results suggest that part of the additional net assimilation of carbon under elevated

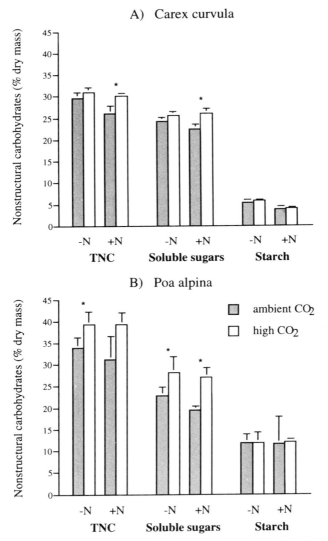

Figure 7 Nonstructural carbohydrates in the leaves of *C. curvula* and *P. alpina* at midseason 1993. Nonstructural carbohydrates are presented as the total (TNC) and as starch and soluble sugar fractions. N: addition of NPK fertilizer (40 kg ha^{-1} per season). * indicates $p < 0.05$.

CO_2 must have been sequestered in the soil, perhaps by faster turnover of carbohydrates in roots or root exudation. The addition of mineral nutrients to high-CO_2 plots did not reduce the TNC pools in any of the species, suggesting that the observed increase in TNC under elevated CO_2 was not due to nutrient-limited sink activity.

2. Leaf Nitrogen Content Nitrogen contents in the leaf biomass under ambient CO_2 conditions differed substantially between species (1.8% in *C. curvula*, and 4.2% in *L. helveticus*). In the high-CO_2 treatment, nitrogen levels based on TNC-free dry matter showed different responses, from weak (*C. curvula*) to significant reductions (in *L. helveticus*, -26%, $p < 0.05$). Thus, nitrogen reduction went beyond what could be explained by dilution by accumulating TNC. Presumably a major component of the nitrogen pool, such as, for example, Rubisco, became reduced under high-CO_2 conditions, at least in *Leontodon*. Since lowered N content indicates reduced leaf quality for herbivores, further implications for nutrient cycling are to be expected. Presently, grasshoppers remove about one-sixth of peak season leaf biomass (P. Blumer, unpublished). A change in their feeding behavior or fecundity is likely to have a substantial influence on ecosystem function.

IV. Conclusions

CO_2-enriched communities did take up more carbon dioxide than controls. However, we can find little if any effect on biomass production. The small warming effect in the open tops alone, or in combination with CO_2 enrichment, had no stimulating effect on plant growth. In fact, during the dry summer of 1992 the chamber effect alone led to slightly reduced plant growth. In contrast, there seems to be a strong mineral fertilizer effect on growth. The small amounts of supplemental mineral nutrients led to significant stimulation of plant growth and further enhancement of CO_2 uptake under elevated CO_2. However, it appears that species other than *C. curvula* are likely to profit from this treatment. For instance, *P. alpina*, which is sparsely distributed in the sedge community but very abundant on more nutrient-rich alpine grazing lands, is increasing in abundance. After two seasons it can be concluded that CO_2 enrichment leads to persistently higher tissue concentrations of nonstructural carbohydrates (mainly sugars).

In summary, our present status of knowledge suggests that these dense mountain sedge mats are, in their present species composition, not particularly responsive in terms of growth to CO_2 enrichment alone and, thus, seem to behave similarly to Arctic tundra [Oechel and Vourlitis, 1996 (this volume)]. However, in contrast to Arctic tundra, the midseason net carbon uptake is significantly increased under elevated CO_2, which necessitates the presence of an active belowground carbon sink. Moreover, the observed changes in tissue composition may affect ecosystem function in the long term via nutrient cycling and influences on the food web. If treatments lead to altered species composition or dominance (some preliminary trends have been noted), more responsive species may cause additional alterations at the ecosystem level. Thus, pronounced effects of CO_2 fertilization, alone or in combination with mineral nutrient addition, may be exaggerated through changes in community structure. However, these processes are very slow and difficult to detect within the time frame of the present research project.

Acknowledgments

We are grateful to F. Ehrsam for his technical help. Deta-T Instruments kindly allowed us to copy their circuitry for humidity sensors in open tops. This project is funded by the Swiss National Science Foundation under Grant Nos. 31-30048.90 and 4031-033431.

References

Cernusca, A. (1977). Alpine Grasheide Hohe Tauern. Ergebnisse der Ökosystemstudie 1976. Veröffentl. Österr. MaB-Hochgebirgsprogramm Hohe Tauern 1, Universitätsverlag Wagner, Innsbruck.

Cernusca, A. (1989) Struktur und Funktion von Graslandökosystemen im Nationalpark Hohe Tauern. Veröffentl. Österr. MaB-Programms, Wagner, Innsbruck.

Diemer, M. (1994). Mid-season gas exchange of an alpine grassland under elevated CO_2. *Oecologia* 98, 429–435.

Field, C. B., Chapin, F. S., III, Chiariello, N. R., Holland, E. A., and Mooney, W. A. (1996). The Jasper Ridge CO_2 Experiment: Design and Motivation. *In* "Carbon Dioxide and Terrestrial Ecosystems" (G. W. Koch and H. A. Mooney, eds.). Academic Press, San Diego.

Franz, H. (1980). Untersuchungen an alpinen Böden in den Hohen Tauern 1974–1978, Stoffdynamik und Wasserhaushalt. Veröffentl. Österr. MaB-Hochgebirgsprogramm Hohe Tauern 3, Universitätsverlag Wagner, Innsbruck.

Grabherr, G., Brzoska, W., Hofer, H. and Reisigl, H. (1980). Energiebindung und Wirkungsgrad der Nettoprimärproduktivität in einem Krummseggenrasen (*Caricetum curvulae*) der Ötztaler Alpen, Tirol. *Oecol. Plant.* 1, 307–316.

Guy, C. L. (1990). Cold acclimation and freezing stress tolerance: role of protein metabolism. *Annu. Rev. Plant Physiol. Plant Mol. Biol.* 41, 187–223.

Klug-Pümpel, B. (1981). Streufall und Streuschwund in einem *Caricetum curvulae*. *Flora* 171, 39–54.

Klug-Pümpel, B. (1982). Effects of microrelief on species distribution and phytomass variations in a *Caricetum curvulae* stand. *Vegetatio* 48, 249–254.

Körner, Ch. (1982). CO_2 exchange in the alpine sedge *Carex curvula* as influenced by canopy structure, light and temperature. *Oecologia* 53, 98–104.

Körner, Ch. (1989a). The nutritional status of plants from high altitudes. A worldwide comparison. *Oecologia* 81, 379–391.

Körner, Ch. (1989b). Der Flächenanteil unterschiedlicher Vegetationseinheiten in den Hohen Tauern: eine quantitative Analyse grossmaßstäblicher Vegetationskartierungen in den Ostalpen. *In* "Struktur und Funktion von Graslandökosystemen im Nationalpark Hohe Tauern" (A. Cernusca, ed.), pp. 33–47. Veröffentl. Österr. MaB-Hochgebirgsprogramm Hohe Tauern 13, Universitätsverlag Wagner, Innsbruck.

Körner, Ch. (1993). CO_2 fertilization: the great uncertainty in future vegetation development. *In* "Vegetation Dynamics & Global Change" (A. M. Solomon, H. H. Shugart, eds.), pp. 53–70. Chapman & Hall, New York, London.

Körner, Ch., and Diemer, M. (1987). *In situ* photosynthetic responses to light, temperature and carbon dioxide in herbaceous plants from low and high altitude. *Funct. Ecol.* 1, 179–194.

Körner, Ch., and Larcher, W. (1988). Plant life in cold climates. *In* "Plants and Temperature" (S. F. Long and F. I. Woodward, eds.), pp. 25–57. The Company of Biol. Ltd., Cambridge, UK.

Körner, Ch., and Pelaez Menendez-Riedl, S. (1989). The significance of developmental aspects in plant growth analysis. *In* "Causes and Consequences of Variation in Growth Rate and Productivity of Higher Plants" (H. Lambers, *et al,*. eds.), pp. 141–157. SPB Academic Publ., The Hague.

Körner, Ch., and Diemer, M. (1994). Evidence that plants from high altitude retain their greater photosynthetic efficiency under elevated CO_2. *Funct. Ecol.* 8, 58–68.

Körner, Ch., and Miglietta, F. (1994). Long-term effects of naturally elevated CO_2 on Mediterranean grassland and forest trees. *Oecologia*, 99, 343–351.

Körner, Ch., Farquhar, G. D. and Roksandic, Z. (1988). A global survey of carbon isotope discrimination in plants from high altitude. *Oecologia* 74, 623–632.

Körner, Ch., Wieser, G., and Cernusca, A. (1989). Der Wasserhaushalt waldfreier Gebiete in den österreichischen Alpen zwischen 600 und 2600 m Höhe. *In* "Struktur und Funktion von Graslandökosystemen im Nationalpark Hohe Tauern" (A. Cernusca, ed.), pp. 119–153. Veröffentl. Österr. MaB-Programms 13, Wagner, Innsbruck.

Körner, Ch., Farquhar, G. D. and Wong, S. C. (1991). Carbon isotope discrimination by plants follows latitudinal and altitudinal trends. *Oecologia* 88, 30–40.

Leadley, P. W. and Drake, B. G. (1993). Open top chambers for exposing plant canopies to elevated CO_2 concentration and for measuring net gas exchange. *In* "CO_2 and Biosphere" (J. Rozema, H. Lambers, S. C. van de Geijn, and M. L. Cambridge, eds.), pp. 3–15. Kluwer Academic Publ., Dordrecht, Boston, London.

Long, S. P. (1991). Modification of the response of photosynthetic productivity to rising temperature by atmospheric CO_2 concentrations: Has its importance been underestimated? *Plant, Cell, Environ.* 14, 729–739.

Mooney, H. A., and Billings, W. D. (1965). Effects of altitude on carbohydrate content of mountain plants. *Ecology* 46, 750–751.

Oechel, W. C., and Vourlitis, G. L. (1996). Direct Effects of Elevated CO_2 on Arctic Plant and Ecosystem Function. *In* "Carbon Dioxide and Terrestrial Ecosystems" (G. W. Koch and H. A. Mooney, eds.). Academic Press, San Diego.

Owensby, C. E., Ham, J. M., Knapp, A., Rice, C. W., Cayne, P. I., and Aven, L. M. (1996). Ecosystem-Level Responses of Tallgrass Prairie to Elevated CO_2. *In* "Carbon Dioxide and Terrestrial Ecosystems" (G. W. Koch and H. A. Mooney, eds.). Academic Press, San Diego.

Poorter, H. (1993). Interspecific variation in the growth response of plants to an elevated ambient CO_2 concentration. *In* "CO_2 and Biosphere" (J. Rozema, H. Lambers, S. C. van de Geijn, and M. L. Cambridge, eds.), pp. 77–97. Kluwer Academic Publ., Dordrecht, Boston, London.

Schäppi, B., and Körner, Ch. (1995). Growth responses of an alpine grassland to elevated CO_2. *Oecologia*, in press.

Sakai, A., and Larcher, W. (1987). "Frost Survival of Plants. Responses and Adaptation to Freezing Stress." Springer, Ecological Studies 62, Berlin, Heidelberg, New York.

Tissue, D. T., and Oechel, W. C. (1987). Response of *Eriophorum vaginatum* to elevated CO_2 and temperature in the Alaskan tussock tundra. *Ecology* 68, 401–410.

Wong, S. C. (1990). Elevated atmospheric partial pressure of CO_2 and plant growth. II. Nonstructural carbohydrate content in cotton plants and its effect on growth parameters. *Photosynth. Res.* 23, 171–180.

12

Long-Term Elevated CO_2 Exposure in a Chesapeake Bay Wetland: Ecosystem Gas Exchange, Primary Production, and Tissue Nitrogen

Bert G. Drake, Gary Peresta, Esther Beugeling, and Roser Matamala

I. Introduction

It is ironic that in this paper we will consider changes that might be brought about in wetlands by rising atmospheric CO_2, the source of which is largely carbon from fossil fuels formed in wetlands during the Cretaceous period. Wetlands occupy approximately 6% of the land surface of the earth, and they have characteristics that are somewhere between those of the terrestrial and the aquatic or marine ecosystems of which they are the interface. Coastal and estuarine wetland soils are composed of very highly compacted, waterlogged silt. These conditions produce anoxic, reduced substrates with high concentrations of toxic compounds, including salt, hydrogen sulfide, and other sulfur containing products of anaerobic respiration.

The dominant effects of rising atmospheric CO_2 concentration and climate changes are likely to be very different for wetlands of different geomorphological and climate zones. Rising sea level and increased storm frequency may increase erosion and disturb animal habitats of coastal wetlands, and rising temperatures may dry northern wetlands releasing additional carbon to the atmosphere as has been shown to occur in Arctic tundra [Oechel *et al.*, 1994; Oechel and Vourlitis, 1996 (this volume)]. Rising atmospheric CO_2 may lead to increased photosynthesis and carbon sequestration in wetlands in temperate, tropical, and subtropical regions, with very different effects from those recorded for cool wetlands, where the response of photosynthesis to rising atmospheric CO_2 may be relatively small. Based on the results of the study reported here, the list of potential effects is very long and includes such important functions as increased nitrogen fixation and methane production, as well as reduced water stress, a consequence of the effect of CO_2 on stomatal conductance and transpiration. Whether or not wetlands are a net source or sink for carbon depends on the degree to which plant production is stimulated by the CO_2 fertilization effect when this is balanced against the potential effects of temperature on soil respiration and decomposition of litter, which largely determine the fraction of primary production accumulated in wetland soils.

A field project to determine the effects of elevated CO_2 on a Chesapeake Bay wetland has been in progress since 1987. Open-top chambers have been used to create test atmospheres of normal ambient and elevated (=normal ambient + 340 ppm) CO_2 concentrations in monospecific stands of the C_4 grass *Spartina patens*, the C_3 sedge, *Scirpus olneyi*, and a mixed community consisting of these two species along with a third species, the C_4 grass *Distichlis spicata*. Details of the chamber design and experimental procedures are given elsewhere (Drake, *et al.*, 1989; Curtis *et al.*, 1989a,b; Leadley and Drake, 1993).

The goals of the project are to determine the responses of key processes that regulate ecosystem carbon metabolism in communities dominated by the three perennial species and to measure the impact of these effects on ecosystem carbon accumulation. A number of research reports on this project have been published discussing the effects of elevated CO_2 on biomass production (Curtis *et al.*, 1989a,b), photosynthesis and respiration (Arp and Drake, 1991; Arp *et al.*, 1993; Azcon-Bieto *et al.*, 1994; Drake and Leadley, 1991; Jacob *et al.*, 1995; Long and Drake, 1991; Ziska *et al.*, 1990), and ecosystem processes, including nitrogen dynamics (Curtis *et al.*, 1990), methane production (Dacey *et al.*, 1994) plant–fungi–insect interactions (Thompson and Drake, 1994), and plant competition (Arp *et al.*, 1993). A summary of initial findings has also been reported (Drake, 1992). The study is scheduled to continue through 1997, and other projects now being completed will report on ecosystem respiration and carbon balance, soil respiration, mineralization, fixation and cycling of nitrogen, and soil carbon content.

Papers by Curtis *et al.* (1989a,b, 1990), Drake and Leadley (1991), and Arp and Drake (1991) have shown that elevated CO_2 increased the photosynthesis of single shoots and of the ecosystem, and this resulted in greater production of biomass, particularly of roots during the first years of the project. Here we ask whether the initial effects of elevated CO_2 have been sustained throughout the course of the study. We compared the effects of the CO_2 treatment on three key factors for the period 1987–1993: (1) net ecosystem CO_2 exchange, (2) root and shoot production, and (3) nitrogen concentration of above- and belowground tissues. We measured ecosystem gas exchange in order to determine the effect of elevated CO_2 on carbon uptake by the system, and we report the integrated values from a 3-hr period (11:30–14:30) that was determined periodically throughout July and August of each year by using the approach of Leadley and Drake (1993). During the first 2 weeks of August, we determined shoot density in pure stands of the C_3 sedge and the mixed community, as well as the biomass of shoots of the C_4 grasses, by using the methods outlined in Curtis et al. (1989a). Shoot density production of C_3 sedge was determined by census as was done by Curtis et al. (1989a). Biomass production in the C_4 community was determined by harvesting aboveground biomass from 5–10 randomly selected quadrants (25 cm^2 each) in each chambered and unchambered control plot of the C_4 and mixed C_3 and C_4 communities. We determined root biomass by the method of regrowth cores harvested at the end of the growing season (Curtis et al., 1990). We also report nitrogen concentrations determined on shoots collected during the midseason census and on roots from regrowth cores harvested at the end of the growing season [methods outlined in Curtis et al. (1990)].

II. Results

A. Net ecosystem CO_2 Exchange

Elevated CO_2 increased seasonal net ecosystem CO_2 exchange (NEC) in all three communities, but it increased the most in the pure C_3 community dominated by *S. olneyi*. The relative stimulation of ecosystem carbon accumulation *per unit ground area* increased throughout the season in all communities (Drake and Leadley, 1991). Stimulation of ecosystem CO_2 assimilation *per unit of green biomass* in the C_3-dominated *S. olneyi* community for 1988 was nearly constant at about 55% (Drake and Leadley, 1991).

The effect of long-term exposure to elevated CO_2 on NEC was estimated by comparing the midday rate of NEC for the middle part of the growing season (Fig. 1). We chose the period 11:30–14:30 hr because the conditions for measurement of ecosystem gas exchange are most favorable at this time. For this analysis, we used only data collected on days when the photosynthetically active solar radiation was above 1000 μmol m^{-2} s^{-1}, when NEC would approximate maximal rates. We chose daily averages from the period of late June through late August, corresponding to the period of maximum biomass production (Curtis et al., 1989a) and maximum photosynthesis (Leadley and Drake, 1993). While this measure of NEC differs from total daily carbon accumulation, it does encompass the time of day when plants

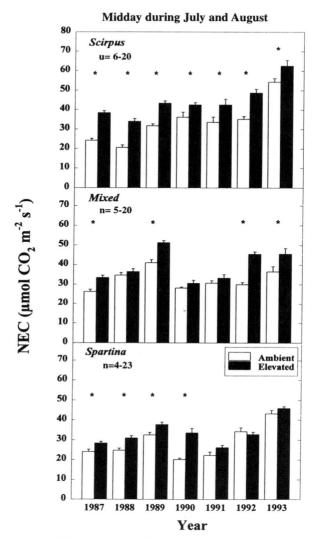

Figure 1. Maximum daily net ecosystem CO_2 exchange (NEC) measured by gas exchange in three wetland communities on Chesapeake Bay, Maryland. Bars are means with standard errors for five chambers in each treatment. Open bars represent chambers in normal ambient CO_2, and solid bars represent chambers in elevated CO_2. Asterisks indicate years of significant differences ($p < 0.05$).

accumulate a large fraction of their daily carbon for growth and when the water stress and heat stress are likely to be most severe.

NEC increased in all treatments from the beginning of the study, with highest rates during 1992 and 1993 and the lowest rates during the first 2 years of the study, 1987 and 1988 (Fig. 1). Elevated CO_2 had the largest effect on the C_3 community (60 and 65% stimulation) during 1987 and 1988 when the absolute rates of NEC were the lowest. The effect of elevated CO_2 on NEC in the C_3 sedge community was significant in all years, but in the C_4 grass community, elevated CO_2 had a statistically significant effect only during the first 4 years and in the mixed community only during the first, third, sixth, and seventh years (Fig. 1).

Year to year variations in the relative effects of elevated CO_2 on NEC in the C_3 sedge community were linearly dependent on air temperature during the time of NEC measurements (Fig. 2). The stimulation of NEC by elevated CO_2 was computed as $[(NEC_e - NEC_a)/(NEC_a)]100$, where the subscripts e and a indicate elevated and ambient treatments, respectively. Values computed from the yearly means in Fig. 1 for each year were plotted against the mean air temperature recorded at 2-m height by a shaded thermocouple during the period of gas exchange measurement (11:30–14:30) on those days when gas exchange measurements were made. The mean air temperature varied between approximately 28 and 33.5°C, with the highest temperatures occurring during the first 2 years of the study. The coefficient of determination for the relationship between stimulation of NEC and air

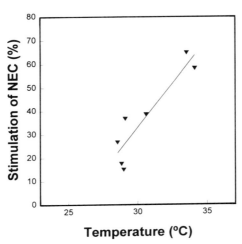

Figure 2. Dependence of CO_2 stimulation of NEC on temperature in the C_3 sedge community. Temperature was recorded at a 2-m height for all years except 1987, when the temperature at the top of the C_3 sedge community in one of the unchambered control plots was used.

temperature was very high ($r^2 = 0.92$; Fig. 2). The slope of the dependence of NEC on temperature suggests that the effect of elevated CO_2 increased by approximately 8%/°C in the range 28–34°C.

B. Shoot and Root Production

The long-term effects of elevated CO_2 on average shoot density in the C_3 sedge and on biomass in the C_4 grass communities are shown in Fig. 3. Stimulation by elevated CO_2 of peak season shoot density of the C_3 sedge in the pure community ranged from 10% in 1989 to 25% in 1992; the mean for the period was 16%. The CO_2 stimulation was statistically significant only during the first year, 1987. Curtis et al. (1989a) reported no effect of elevated CO_2 on shoot production before midseason, but stimulation increased thereafter. We chose to measure shoot density at the peak of the season because it was greatest at this time. It was also shown by Curtis et al. (1989a) and subsequently confirmed through 7 years of CO_2 exposure that there was no treatment effect of the relationship of weight to surface area or shoot height. Because estimates of primary production based on these allometric relationships would be dependent only on variations in length and shoot density, we report only shoot density. The increase in primary production of aboveground foliage was relatively small in comparison with the effects of elevated CO_2 on biomass production in crop species, which averaged about 34% for a doubling of the CO_2 concentration (Kimball, 1983). Elevated CO_2 had no measurable effect on aboveground biomass production in the pure C_4 community nor in the C_4 component of the mixed community (Fig. 3).

In the mixed community, elevated CO_2 has had a relatively large effect on the number of shoots of C_3 sedge *S. olneyi* [Fig. 3; see also Curtis et al. (1989a)]. Throughout the course of the study, elevated CO_2 stimulated the production of C_3 *S. olneyi* stems in the mixed community: the range was from a low of approximately 60% in 1987 to a high of 225% (Fig. 4). By repeated measures analysis of variance, it was determined that elevated CO_2 had a highly significant effect on shoot density. The greatest stimulation of shoot density in the C_3 component of the mixed community occurred when the shoot densities of the controls and ambient treatments were lowest, during 1988 and 1989, and as the absolute values of shoot density increased, the relative stimulation by elevated CO_2 decreased from 1988 through 1993. Whether this is an indication of a transient effect of elevated CO_2 on the production of aboveground biomass in *S. olneyi* is not clear because the years when shoot density was lowest were also years of high temperatures and drought (Arp et al., 1993). It may be that the C_3 *S. olneyi* does well during times of greatest stress in which the C_4 species are least productive.

We have studied the effects of elevated CO_2 on the production of roots by the method of regrowth cores (Curtis et al., 1990), which gives a measure of the relative effect of elevated CO_2 on root and rhizome production, but does not provide an estimate of actual ecosystem production. Although there were large variations in both the absolute production of biomass and the relative stimulation of production by elevated CO_2, the greatest CO_2 effect on root production was in the pure C_3 community, which showed an average stimulation of belowground biomass of 44% (ranging from 83% in the first year to no effect at all during the last year, 1993). Although means for fine root biomass in the mixed community were higher in elevated CO_2, they did not significantly differ from means for the normal ambient or control treatments. There is no clear trend in these data on roots and rhizomes, apart from the fact that the absolute amount of recoverable biomass in the regrowth cores increased throughout the course of the study in the pure C_3 and mixed communities approximately in parallel with the increase in net ecosystem CO_2 exchange.

C. Tissue Nitrogen Concentration

We determined the nitrogen concentration ([N]) in the green tissues of aboveground biomass during August for each year between 1987 and 1992 (Figs. 5A and 5B). [N] of shoots and roots decreased in the C_3 sedge, but not in the two C_4 species, consistent with early reports from this project (Curtis, et al., 1989b, 1990). [N] of plants grown in normal ambient CO_2 was between 1.1 and 1.4% of dry weight and, with the exception of 1989, the lowest values occurred in the earliest years of the study. In elevated CO_2, [N] followed this same pattern, but was reduced by an average of approximately 18%. We have shown that the reduction in [N] is due to increased carbohydrate and reduced soluble protein concentrations, particularly of rubisco, which is reduced by 30–58% on a dry weight basis (Jacob et al., 1995).

III. Discussion

Kramer (1981) and others have suggested that the CO_2 fertilization effect could not be sustained because it would require additional nitrogen and other resources generally thought to be limited in mature ecosystems. Plant production is believed to be limited by resource availability, particularly nitrogen and phosphorus, and not by carbon, a view that is commonly held by many reseachers working on this problem (Melillo et al., 1990). Support for this hypothesis has been found in the elevated CO_2 studies on the Arctic tundra ecosystem by Oechel and his co-workers [Oechel and Vourlitis, 1996 (this volume)]. A related, but not necessarily alternative, hypothesis is

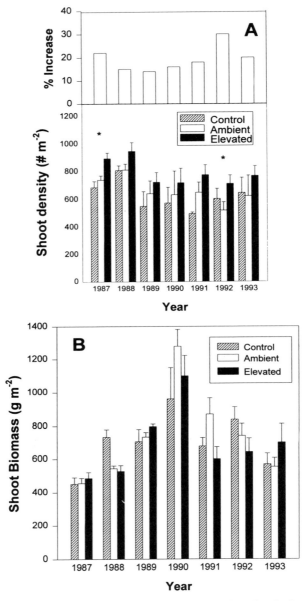

Figure 3. Shoot density in a community dominated by the C_3 sedge *S. olneyi* (A) and in a mixed C_3 and C_4 community (B) and shoot biomass of a monoculture of the C_4 grass *Sp. patens* (C) and of the two C_4 species *Sp. patens* and *Distichlis spicata* in stands of a mixed community. Bars indicate means and standard errors for five plots.

12. Elevated CO₂ Exposure in a Chesapeake Bay Wetland   205

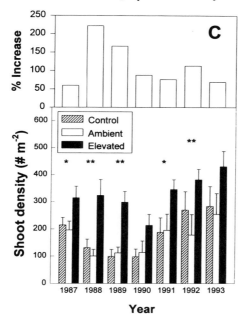

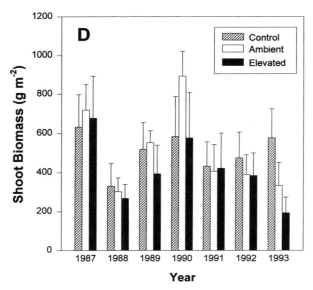

Figure 3—*Continued*

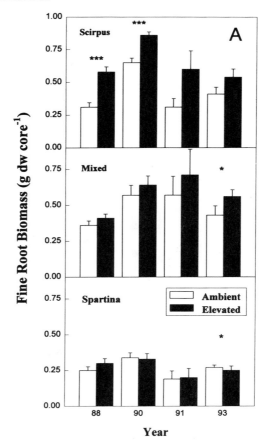

Figure 4. Biomass per core (0–20cm deep) of fine roots (A) and fine roots plus rhizomes (B) in three communities. Asterisks indicate the level of significance (*, $p < 0.05$; **, $p < 0.01$; ***, $p < 0.001$). Cores were harvested between December and February.

that increased resource use efficiency, a consequence of effects of rising atmospheric CO_2 on physiological processes, may provide additional energy to ecosystems, which will, in turn, stimulate soil microbial processes, increasing nutrient turnover and ultimately leading to the accumulation of soil carbon [Luxmore, 1981; Gifford, 1994; Curtis *et al.*, 1996 (this volume)]. Experimental support for this view has been found in the literature on the well-known responses of plants to elevated CO_2 (e.g., Kimball, 1983; Long and Drake, 1992; Gifford, 1994) and in model ecosystems such as that of Zak *et al.* (1993).

The hypothesis that responses to elevated CO_2 would not be sustained implies that photosynthesis must be down-regulated. Photosynthetic capac-

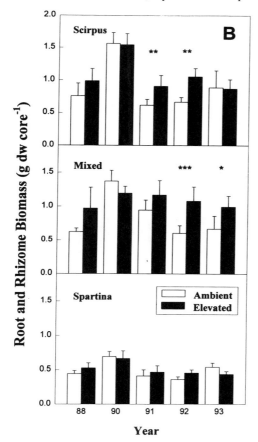

Figure 4—*Continued*

ity is regulated by sink demand for photosynthate (Herold, 1980). Wardlaw (1990) suggested that the effects of long-term exposure to elevated CO_2 on photosynthesis are caused by changes in the sink demand for photosynthate, over which environmental factors exert a significant regulating effect. Jacob *et al.* (1995) show that, although photosynthesis of the dominant C_3 plant in the communities studied here quickly acclimates to the elevated CO_2 treatment through reduced rubisco levels, rates of photosynthesis in elevated CO_2 remain 30–60% higher than those in normal ambient CO_2. [In a review, Long and Drake (1992) found that the rate of photosynthesis is nearly always greater in plants grown and tested in elevated CO_2 than in normal ambient CO_2.] Jacob *et al.* (1995) also show that there is sufficient sink capacity in the system to absorb the additional carbohydrate produced in elevated CO_2. Consistent with the sustained stimulation of shoot photo-

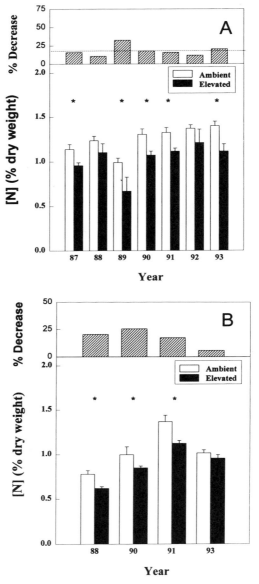

Figure 5. Nitrogen concentration (% of dry weight) in shoots (A) and roots (B) of the C_3 sedge *S. olneyi*. Data as in Fig. 1.

synthesis by the C_3 *Scirpus* in elevated CO_2 mentioned earlier, we showed here that midday NEC in the C_3 community averaged 44% higher in elevated CO_2 over the course of 7 years, varying from a low of 16% in 1993 to a high of 65% in 1988. These findings suggest that the speculated limitation on primary production by nutrient supply in the Arctic ecosystem [Oechel and Vourlitis, 1996 (this volume)] does not occur in the temperate wetland ecosystem.

It seems likely that CO_2 stimulation of NEC in the C_3 community involved effects on canopy structure, as well as photosynthesis. Elevated CO_2 stimulated the number of shoots by about 18% and canopy photosynthesis by 37%. Thus, about 50% of the stimulation of NEC could be ascribed to direct effects of elevated CO_2 on photosynthesis at the leaf level and the remainder to increased leaf area.

When growth is limited by environmental factors such as drought or high temperatures, ecosystem carbon assimilation will be relatively small and ecosystem production will be reduced. In this study, NEC was lowest in the years of highest temperature, 1987 and 1988. The effect of elevated CO_2 on NEC was a function of temperature (Fig. 2), such that the relative increase in NEC in elevated CO_2 was greatest when the absolute value of NEC was lowest and when shoot production, at least for the C_4 plants, was also low. In this case, the effects of elevated CO_2 seem to be generated by the interaction of the physiological effects of elevated CO_2 and temperature on photosynthesis, as discussed in Long and Drake (1992).

In addition to its effects on photosynthesis, however, it appears that elevated CO_2 also moderated the impact of stress on productivity. This would explain the effects of elevated CO_2 on NEC of the C_4 community, in which elevated CO_2 did not stimulate leaf-level photosynthesis (Ziska *et al.,* 1990), but did produce a small but statistically significant increase in NEC during the first 4 years of the study when the highest temperatures and low rainfall occurred (Arp *et al.,* 1993). This may have involved changes in water balance, concentration of CO_2 within the plant canopy, or dark respiration. Arp *et al.* (1993) show that elevated CO_2 increased midday water potential in both the C_3 and C_4 species. The density of the canopy of the *Spartina patens* community increases in midsummer as a result of lodging. At this time, the canopy may be so dense that boundary layer resistance may impede CO_2 supply, resulting in a CO_2 deficit that could be mitigated by increased atmospheric CO_2 (Turitzin and Drake, 1981). Finally, we have shown that elevated CO_2 reduces dark respiration in the C_4 grass (unpublished data), and this would have the effect of increasing the duration and the rate of carbon uptake. A combination of these indirect effects may be responsible for the effect of elevated CO_2 on NEC in the C_4 community. Thus, the stimulation of the rate of ecosystem carbon uptake was sustained over the course of the study, but modified by the environmental conditions, notably temperature. These findings place relatively greater emphasis on the interaction of climate variables with the effects of rising CO_2 and less emphasis upon nutrient cycling as regulating factors in the sustained responses of ecosystem carbon balance to elevated CO_2.

Belowground production increased much more, on average, than did shoot production. The most consistent result was that elevated CO_2 increased fine root production in the C_3 and mixed communities. Unfortunately, root and rhizome biomasses were determined together in 1992, and the regrowth cores were left in the ground in 1989 and harvested in 1990. The high variability of rhizome biomass production made the determination of the elevated CO_2 stimulation of root production very difficult. The use of simple measures of biomass production to determine the effects of elevated CO_2 on belowground processes is in serious need of attention.

The major part of the additional carbon from the large stimulation of ecosystem carbon dioxide assimilation in the C_3 community (37% averaged over 7 years) is allocated belowground, ultimately as soil carbon. It was shown that soil carbon increased by 5% in the upper 5 cm of the C_3 sedge community (Ruiz, unpublished data), and in a number of ongoing studies, we have shown that the additional soil carbon is accompanied by increases in the rates of energy dependent processes in the soil and rhizosphere. These include greater numbers of soil invertebrates (Ruiz, unpublished data), increased soil respiration (Ball, unpublished data), increased nitrogen fixation (Dakora, unpublished data) and increased methane production (Dacey et al., 1994).

The most important observations to come out of this analysis of 7 years of data are that there is a sustained effect of elevated CO_2 on net ecosystem carbon accumulation in the wetland community dominated by the C_3 sedge, *S. olneyi*, and that the additional carbon is primarily allocated to fine roots. This finding is consistent with an increasing mass of data that shows that elevated atmospheric CO_2 will alter various processes of carbon metabolism in plants and ecosystems. In the present study, the most important of these was photosynthesis.

Acclimation of photosynthesis was associated with reduced protein and increased carbohydrate concentrations and, as a result, increased tissue C:N ratio (Curtis et al., 1989b). Accompanying these chemical changes were reductions in rates of dark respiration (Azcon-Bieto et al., 1994), insect grazing and fungal damage (Thompson and Drake, 1994), decomposition (Whigham, unpublished results).

IV. Conclusions

The question this paper set out to address is whether the data support one or the other of two hypotheses, namely, that (1) the effects of elevated

CO_2 would be transient (Kramer, 1981) or (2) the increased capacity of the system to assimilate CO_2 would increase the efficiency of key processes, thereby sustaining carbon accumulation (Luxmore, 1981). Stated this way as an either/or proposition, the data are inconclusive according to the logic of either argument. Individual lines of evidence taken out of the context of the whole study could support either hypothesis. In favor of the hypothesis that, as a consequence of the acclimation of photosynthesis and the depletion of resources, the effects of elevated CO_2 would be transitory, we have shown that photosynthesis does acclimate, that the greatest stimulation of ecosystem gas exchange and primary production occurred in the early years of the study, and that the tissue nitrogen concentration was reduced as would be expected if soil nitrogen supply is progressively depleted. On the other hand, although photosynthesis did acclimate, rates of photosynthesis in shoots of the C_3 sedge, *S. olneyi* were much higher in elevated CO_2 than in normal ambient CO_2, and the interannual pattern in the response of ecosystem gas exchange to elevated CO_2 could be explained by variations in temperature (Fig. 2).

In support of the hypothesis that increased efficiency of carbon use would stimulate carbon accumulation, we have shown that light and nitrogen use efficiencies increased in elevated CO_2 (Long and Drake, 1991; Jacob *et al.*, 1995), that the additional carbon in the system was in the fine roots, and that increased allocation of carbon belowground stimulated biogeochemical processes, including methane production (Dacey *et al.*, 1994) and nitrogen fixation (Dakora and Drake, unpublished data).

In the end, these two paradigms for the interpretation of rising atmospheric CO_2 and climate changes may not be mutually exclusive, and both find support in the field studies, as we report here. This would suggest that the effects of climate change are likely to be pervasive and cannot be fully accounted for by current paradigms as represented by Kramer (1981) and Luxmore (1981). The number of processes investigated has been a major strength of this study and is consistent in showing that elevated CO_2 has increased carbon in the stands dominated by the C_3 sedge, *S. olneyi*. An important conclusion is that elevated CO_2 has had a powerful impact on the wetland ecosystem at all levels of organization.

V. Summary

Field experiments using open-top chambers to create test atmospheres of normal ambient or elevated CO_2 (= normal ambient + 340 ppm) were carried out in three communities (pure C_3 sedge, pure C_4 grass, and a mixed C_3 and C_4 community) on a brackish wetland on Chesapeake Bay, Maryland. Data for net ecosystem CO_2 exchange (NEC, μmol m^{-2} s^{-1}

ground area basis), shoot density, root and rhizome production and nitrogen concentration of above- and belowground tissues are reported for each year of the study, 1987–1993. NEC averaged 37% higher (significant in all years) in stands of the C_3 sedge (*S. olneyi*) growing in the elevated CO_2 treatment than in their counterparts growing in normal ambient CO_2 concentration. The stimulation of NEC by elevated CO_2 in the C_3 sedge community was proportional to air temperature in the range 27–35°C. NEC increased by an average of 21% in both the C_4 *Sp. patens* monospecific stands and the mixed C_3 and C_4 community exposed to elevated CO_2 (difference was significant in 4 out of 7 years). The shoot density of the C_3 sedge exposed to elevated CO_2 was not significantly greater than that in the normal ambient CO_2 treatment although the mean value for CO_2 stimulation of shoot density for 7 years was 19% with a range of 13–30%. In the mixed C_3 and C_4 community, the elevated CO_2 treatment increased C_3 shoot density from 50 to 225%, and by repeated measures analysis of variance, the treatment effect was highly significant. In the mixed community, the shoot density of the C_3 sedge was inversely related to the biomass production of the C_4 component of the community. There was no measurable effect of elevated CO_2 on aboveground biomass in the pure C_4 community nor on the C_4 component in the mixed community. Elevated CO_2 stimulated the production of fine roots and rhizomes in the C_3, but not in the C_4 community. Nitrogen concentration ([N]) was reduced by about 20% in the green tissues and roots of the C_3 sedge, but not in the C_4 grasses. A reduction in [N] of approximately 20% in roots and rhizomes of the C_3 sedge community occurred in 3 out of 4 years. The overall effect of elevated CO_2 on the C_3 plant community was to increase carbon accumulation in fine roots. The major pattern in the long-term response to elevated CO_2 was that the effect on carbon accumulation was greatest at times when the NEC of the ambient community was lowest, and this occurred during the first 3 years, which were also the hottest, driest years of the study.

Acknowledgments

This work was supported by the Smithsonian Institution and by the U.S. Department of Energy.

References

Arp, W. J., and Drake, B. G. (1991). Increased photosynthetic capacity of *Scirpus olneyi* after four years of exposure to elevated CO_2. *Plant, Cell Environ.* 14 (9), 1003–1006.

Arp, W. J., Drake, B. G., Pockman, W. T., Curtis, P. S., and Whigham, D. F. (1993). Effects of fouryears exposure to elevated atmospheric CO_2 on competition between C_3 and C_4 salt marsh plant species. *Vegetatio* 104/105, 133–143.

Azcon-Bieto, J. W., Dougherty, W., Gonzales-Meler, M. A., and Drake, B. G. (1994). Acclimation of respiratory O_2 uptake in green tissues from field grown plants after long-term exposure to elevated atmospheric CO_2. *Plant Physiol.* 106, 1163–1168.

Curtis, P. S., Drake, B. G., Leadley, P. W., Arp, W. J., and Whigham, D.F. (1989a). Growth and senescence in plant communities exposed to elevated CO_2 concentrations on an estuarine marsh. *Oecologia* 78, 20–26.

Curtis, P. S., Drake, B. G., and Whigham, D. F. (1989b). Nitrogen and carbon dynamics in C_3 and C_4 estuarine marsh plants grown under elevated CO_2 in situ. *Oecologia* 78, 297–301.

Curtis, P. S., Balduman, L. M., Drake, B. G., and Whigham, D. F. (1990). The effect of elevated atmospheric CO_2 on belowground processes in C_3 and C_4 estuarine marsh communities. *Ecology* 71, 2001–2006.

Curis, P. S., Zak, D. R., Pregitzer, K. S., Lussenhop, J., and Teeri, J. A. (1996). Linking Above- and Belowground Responses to Rising CO_2 in Northern Deciduous Forest Species. *In* "Carbon Dioxide and Terrestrial Ecosystems" (G. W. Koch and H. A. Mooney, eds.). Academic Press, San Diego.

Dacey, J. W. H., Drake, B. G., and Klug, M. J. (1994). Stimulation of methane emission by carbon dioxide enrichment of marsh vegetation. *Nature* 370, 47–49.

Drake, B. G. (1992). A field study of the effects of elevated CO_2 on ecosystem processes in a Chesapeake Bay Wetland. *Austral. J. Biol.* 40, 579–595.

Drake, B. G., and Leadley, P. W. (1991). Canopy photosynthesis of crops and native plant communities exposed to long-term elevated CO_2 treatment. *Plant, Cell Environ.* 14 (8), 853–860.

Drake, B. G., Leadley, P. W., Arp, W. J., Nassiry, D., and Curtis, P. S. (1989). An open top chamber for field studies of elevated atmospheric CO_2 concentration on saltmarsh vegetation. *Funct. Ecol.* 3, 363–371.

Gifford R. M. (1994). The global carbon cycle: A viewpoint on the missing sink. *Aust. J. Plant Physiol.* 21, 1–15.

Herold, A. (1980). Regulation of photosynthesis by sink activity—the missing link. *New Phytologist* 86, 131–144.

Jacob, J., Greitner, C., and Drake, B. G. (1995). In press. The effect of long-term elevated CO_2 exposure on the content and activation state of Rubisco in the sedge, *Scirpus olneyi* and the shrub, *Lindera benzoin*. *Plant, Cell Environ.* 18, 875–884.

Kimball, B. A. (1983). Carbon dioxide and agricultural yield: an assemblage and analysis of 430 prior observations. *Agron. J.* 75, 779–788.

Kramer, P. J. (1981). Carbon dioxide concentration, photosynthesis, and dry matter production. *BioScience* 31, 29–33.

Leadley, P.W, and Drake, B. G. (1993). Open top chambers for exposing plant canopies to elevated atmospheric CO_2 and for measuring net gas exchange. *Vegetatio* 104/105, 3–15.

Long, S. P., and Drake, B. G. (1991). The effect of the long-term CO_2 fertilization in the field on the quantum yield of photosynthesis in the C_3 sedge, *Scirpus olneyi*. *Plant Physiol.* 96, 221–226.

Long, S. P., and Drake, B. G. (1992). Photosynthetic CO_2 assimilation and Rising Atmospheric CO_2 Concentrations. Commissioned Review. "Topics in Photosynthesis," Volume 11. (N. R. Baker and H. Thomas, eds.). Elsevier Science Publishers, B. V., Amsterdam.

Luxmore, R. J. (1981). CO_2 and phytomass. *BioScience* 31, 626.

Melillo, J. M., Callaghan, T. V., Woodward, F. I., Salati, E., and Sinha, S. K. (1990). Effects on Ecosystems. *In* "Climate Change: the IPCC Scientific Assessment" (J. T.Houghton, G. J. Jenkins, and J. J. Ephraums, eds.), pp. 282–310. Cambridge University Press, Cambridge, UK.

Wechel, W. J., and Vourlitis, G. L. (1990). Direct Effects of Elevated CO_2 on Arctic Plant and Ecosystem Function. *In* "Carbon Dioxide and Terrestrial Ecosystems" (G. W. Koch and H. A. Mooney, eds.). Academic Press, San Diego.

Oechel, W. J., Cowles, S., Grulke, N., Hastings, S. J., Lawrence, B., Prudhomme, T., Riechers, G., Strain, B., Tissue, D., and Vourlitis, G. (1994). Transient nature of CO_2 fertilization in arctic tundra. *Nature* (in press).

Thompson, G. B., and Drake, B. G. (19940. Insects and fungi on a C_3 sedge and a C_4 grass exposed to elevated atmospheric CO_2 concentrations in open top chambers in the field. *Plant, Cell Environ.* 17, 1161–1167.

Turitzin, S. N., and Drake, B. G. (1981). Canopy structure and the photosynthetic efficiency of a C_4 grass community. In "Photosynthesis VI. Photosynthesis and Productivity, Photosynthesis and Environment" (G. Akoyunoglou, ed.), pp. 73–80. Balaban International Science Services, Philadelphia.

Wardlaw, I. F. (1990). The control of partitioning in plants. *New Phytologist* 116, 341–381.

Zak, D. R., Pregistzer, K. S., Curtis, P. S., Teeri, J. A., Fogel, R., and Randlett, D. L. Randlett. 1993. Elevated atmospheric CO_2 and feedback between carbon and nitrogen cycles Plant Soil 151, 105–117.

Ziska, L., Drake, B. G. and S. Chamberlain. 1990. Long-term photosynthetic response in single leaves of a C_3 and C_4 salt marsh species grown in elevated atmospheric CO_2 in situ. Oecologia 83, 469–472.

13

Free-Air CO_2 Enrichment: Responses of Cotton and Wheat Crops

Paul J. Pinter, Jr., Bruce A. Kimball,
Richard L. Garcia, Gerard W. Wall,
Douglas J. Hunsaker, and Robert L. LaMorte

I. Introduction

Net photosynthetic rates for many C_3 food and fiber crops are limited by present concentrations of atmospheric carbon dioxide (CO_2) (Pearcy and Björkman, 1983). As a result, anticipated increases in CO_2 during the next century are likely to have important consequences for agricultural systems. In fact, a projected doubling of CO_2 could boost biomass production and marketable yields of agricultural crops by one-third (Kimball, 1983) or even more if other constraints to productivity are limiting (Idso and Idso, 1994). Until only recently, these projections have been based solely on results obtained from controlled-environment enclosures, glasshouses, and open-top field chambers, where restricted soil rooting volumes and/or microclimatic changes caused by the chambers themselves may have influenced the outcome (Lawlor and Mitchell, 1991; Arp, 1991; Kimball *et al.*, in press). Despite a preponderance of evidence that shows a positive effect of CO_2 on plant growth under enclosed conditions, justifiable concern has focused on possible artifacts caused by the test environments themselves (Allen *et al.*, 1992). Would the results have been the same if an entire community of plants was exposed to high CO_2 levels under more natural field conditions?

Responding to the need for appropriate experimental methodology to answer this question, scientists and engineers from Brookhaven National Laboratory (BNL) developed an innovative system in the mid- to late-1980s to fumigate plants with CO_2 in a open-field setting (Allen, 1992). The experimental facility they created was given the acronym FACE, which stands for free air carbon dioxide enrichment. FACE was found to be free of most of the microclimatic artifacts that had plagued earlier studies in chambers and, thus, was expected to provide more realistic information on plant responses to CO_2 conditions that might be encountered in the future. Furthermore, since FACE enabled large areas ($\sim$500 m^{-2}) to be fumigated as a single unit, plants were also expected to function as a complex community, with natural competitive interactions for environmental resources. The large number of plants treated by FACE also implied that many different types of studies could be carried out on the same community and that destructive sampling could be accommodated without affecting the response of remaining plants.

The first FACE field trials took place in 1987 and 1988 in a cotton field near Yazoo City, MS, where the enrichment system was operated for varying numbers of weeks during the season (Allen, 1992). In 1989, the FACE facility was moved to The Maricopa Agricultural Center (MAC), about 40 km south of Phoenix, AZ. Since then, the system has been used to evaluate the effects of elevated CO_2 on cotton (three seasons from 1989 until 1991) and wheat (two seasons, 1992–1993 and 1993–1994) within an agricultural production environment. Many of the biological investigations have been conducted by USDA Agricultural Research Service (ARS) personnel. However, more than 100 scientists and graduate students from 43 different research locations and 8 countries have participated in FACE experiments, which examine the effects of increased CO_2 on the dynamics of carbon and water cycling of plants growing in realistic, open-field conditions (see Appendix).

Results from our cotton FACE experiments have been summarized and published in special dedicated issues of *Critical Reviews in Plant Sciences* (Vol. 11; Hendrey, 1992) and *Agricultural and Forest Meteorology* (Vol. 70; Dugas and Pinter, 1994). Discussion in this chapter will review the effects of CO_2 on photosynthesis, conductance, growth, and yield of cotton from those FACE experiments in Arizona. Analyses of data from wheat grown in the FACE facility are underway. Nevertheless, enough of the preliminary findings is presented here to provide interesting comparisons and contrasts with cotton.

In addition to the studies we will present here, open-air field systems have been used for a number of years to test the effects of gaseous pollutants on vegetation (McLeod *et al.*, 1985). FACE facilities have been operational for several growing seasons in Switzerland, where investigations include growth responses of ryegrass (*Lolium perenne* L.) and white clover (*Trifolium repens* L.) (Blum, 1993). A prototype FACE system for forests has been tested in an experimental plantation of loblolly pine (*Pinus taeda*) at Duke

University in Durham, NC (Hendrey and Kimball, 1994). Plans are also underway to use FACE and FACE-like systems for evaluating the effect of CO_2 on production and animal utilization of pasture forage in New Zealand, of forests and crops in Europe, and of chaparral and desert shrubs in the western United States.

A. The Maricopa Agricultural Center (MAC)

MAC is a research and demonstration farm administered by The University of Arizona. It is located in the midst of an extensive agricultural region in central Arizona (33.07° N latitude, 111.98° W longitude, 358-m elevation). Primary crops in the area are cotton, alfalfa, barley, spring wheat, and pecans; all must be irrigated by surface, subsurface (drip), or sprinkler irrigation to realize an economic yield. The experimental FACE site was located in two adjacent 4.5-ha fields at MAC, where soil was classified as a reclaimed Trix clay loam [fine loamy, mixed (calcareous), hyperthermic Typic Torrifluvents].

B. FACE Facility at MAC

The FACE system used in the Arizona experiments was designed by BNL and operated jointly by BNL and ARS personnel. The apparatus consisted of four circular plenums, each connected to an array of 32 vertical vent pipes, encompassing an area about 25 m in diameter. Pure gaseous CO_2 (from a 48-Mg liquid storage vessel) was metered into the intake of a high volume fan attached to the plenum, mixed with ambient air, and injected from emitter ports of vent pipes on the upwind side of the array. An algorithm based on wind speed, wind direction, and CO_2 concentration in the center of each array was used to maintain CO_2 treatments at desired levels. Each FACE plot was paired with a similarly sized control plot located 90–100 m away in the same field (Wall and Kimball, 1993). During the 1990 and 1991 cotton experiments and both wheat experiments, controls were equipped with a plenum and vent pipe construction similar to that of the FACE arrays, but without fans for air injection. Additional information on FACE design, construction, and control algorithms can be found in Lewin *et al.* (1994) and Nagy *et al.* (1994).

C. FACE Performance and Reliability

FACE technology depends on atmospheric dispersion for the uniform distribution of injected CO_2 across the experimental area (Lipfert *et al.*, 1991, and 1992). During our experiments, it was possible to control CO_2 at the center of FACE arrays at the desired set point (550 ± 20% μmol mol^{-1}) for 99% of the time, thus meeting or exceeding original system design criteria (Hendrey *et al.*, 1993a; Nagy *et al.*, 1994). Three-dimensional analyses of gas concentrations showed that CO_2 had a tendency to build up beneath dense cotton canopies on calm days late in the season (Hileman et al., 1992), and concentrations were slightly higher on the upwind side of the array (Hendrey et al., 1993b). Nevertheless, on a spatial basis, most CO_2 concentrations also fell within the original design criteria. Analyses of the starch content of leaves (Hendrix, 1992), yield parameters (Mauney et al., 1994), and stable carbon isotopes of lint ($\delta^{13}C$; Leavitt et al., 1994) were sufficiently uniform across the FACE array to indicate that the cotton plant integrated most of the shorter term fluctuations in CO_2. When comparing FACE performance characteristics to other CO_2-controlled apparati, it is relevant to note that temporal and spatial concentrations of CO_2 can be quite variable under natural field conditions. Standard deviations of CO_2 ranging from 30 to 100 μmol mol^{-1} have been reported in open-top chambers (Leadley and Drake, 1993).

II. Materials and Methods

A. Experimental Crops and Treatments

1. Cotton and Wheat Test Crops An early goal of the FACE program was to examine the effects of supraambient CO_2 on the response of simple agricultural monocultures before applying FACE technology to more complex natural vegetation communities. Accordingly, we selected two C_3 crop species that are important to world agriculture and that also represent contrasts in patterns of growth and development. A short-staple cotton (*Gossypium hirsutum*, L., cv. Deltapine 77) was chosen because it was a warm-season, perennial, woody plant that had an indeterminate fruiting pattern. A hard, red, spring wheat (*Triticum aestivum*, L., cv. Yecora Rojo) was selected as being representative of cool-season annual grasses with determinate grain-filling characteristics. The cultivars chosen for our experiments were well-adapted to central Arizona climatic conditions and grown commercially by local farmers.

Culture practices (cultivation, insect and weed control, soil nutrient levels, etc.) were managed according to the recommendations of Arizona State Extension Service and University of Arizona research and farm support staff. Cotton was seeded during mid- to late-April on raised soil beds spaced 1.02 m apart. After emergence, the population was thinned to about 10 plants m^{-2}. Cotton was supplied with an average of 142 kg ha^{-1} N during each year. Irrigation was terminated and seed cotton harvested in mid-September.

Wheat has been studied for two complete growing seasons in the FACE facility. It was sown in December (1992 and 1993) and harvested during late May in 1993 and June in 1994. Plants were grown on a flat, unbedded soil surface in rows spaced 0.25 m apart; density at the time of harvest was 109 and 153 plants m^{-2} in 1993 and 1994, respectively. Wheat received 277 kg N ha^{-1} and 44 kg P ha^{-1} in 1992–1993 and 261 kg N ha^{-1} and 29 kg P ha^{-1} in 1993–1994.

2. Carbon Dioxide Treatments Wheat and cotton crops were exposed to ambient (control, ~370 μmol mol^{-1}) and enriched (FACE, ~550 μmol mol^{-1}) levels of CO_2; paired treatment plots were replicated four times. In the FACE plots, CO_2 enrichment began shortly after emergence and continued until shortly before harvest. During the cotton experiments, CO_2 was injected into FACE plots during daylight hours. At night, FACE and control plots were exposed to similar ambient CO_2 concentrations, which often exceeded 400 μmol mol^{-1} (Nagy et al., 1994). In the wheat experiments, treatments continued 24 hr a day except for the last 2 weeks of January 1993, when heavy rains prevented regular CO_2 delivery and enrichment was shortened to daylight hours to conserve supplies.

Irrigation Treatments Cotton plants were irrigated using microirrigation (drip) tubing that was buried at a 0.18- to 0.25-m depth and spaced 1.02 m apart. In the first year of the experiment (1989), all plots were well-watered (WET treatment). Irrigation timing and amounts were based on full-season consumptive use requirements of cotton as determined from estimates of potential evapotranspiration (PET) obtained from an on-farm meteorological station. In 1990 and 1991, main CO_2 treatment plots were split to test the interactive effects of deficit irrigation (DRY treatment) and elevated CO_2 on cotton. WET treatments received irrigation amounts equivalent to that evaporated from a class A pan (1990) and PET (1991). DRY plots were irrigated on the same days as WET plots, but received only 75 and 67% of that supplied to the WET plots during 1990 and 1991, respectively. The differential irrigations were initiated on July 3, 1990, and on May 20, 1991. The combined amounts of irrigation and rainfall from planting to harvest averaged 1232 mm across all years for the WET treatment, while DRY treatments received 1185 mm in 1990 and 833 mm in 1991 (Mauney et al., 1994; Hunsaker et al., 1994a).

In the FACE wheat project, similar to the preceding cotton experiments, each of the main plots was split in half to provide two irrigation levels as subplots in strips within replications. Microirrigation tubing was spaced 0.51 m apart (parallel to plant rows) and buried 0.18–0.25 m deep. An irrigation scheduling program based on a soil water balance approach was used to determine irrigation of the WET treatment when ~30% of the available soil moisture in the root zone had been depleted. Amounts given to the WET subplots were based on estimates of daily PET multiplied by an appropriate crop coefficient. In 1992–1993, plants in the DRY treatment were irrigated on the same day as those in the WET treatment, but received only 50% of the amount. During 1993–1994, plants in the DRY subplots received the same amount as the WET subplots, but only on every other irrigation. Irrigation totals from emergence to harvest averaged 600 mm for the WET treatment during both years. Dry treatments received 275 mm in 1992–1993 and 257 mm the following season. Rainfall over the same periods was 76 and 61 mm, respectively, for the 2 years.

B. Experimental Measurements

1. Cotton Plant Sampling Protocol Cotton biomass was measured by destructive plant sampling at 1- to 2-week intervals during the growing season (Mauney *et al.*, 1994). Sampling consisted of pulling every third cotton plant (with as much of the lateral and tap root structures as possible) from three separate, 1-m-long segments of plant row in all replicates of each CO_2–irrigation treatment combination (subplot). This was equivalent to sampling all the plants in a contiguous 1.02-m^2 area, but did not leave large gaps in the canopy that might have adversely affected the response of remaining plants or the aerodynamics of CO_2 flow. Leaf area was obtained by using an optical planimeter. Dried biomass was measured separately for root, stem, leaf, and fruiting structures.

2. Wheat Plant Sampling Protocol Wheat was sampled at 7- to 10-day intervals during the 1992–1993 and 1993–1994 seasons (18 and 22 sampling periods, respectively). A minimum of six plants was obtained from four sampling zones in each subplot (24 plants total). Plant phenology (according to both the Zadoks and Haun development scales; Bauer *et al.*, 1983), number of stems (tillers), green stem area, and green leaf area were determined on a subsample of 12 median-sized plants per subplot. Heads were counted when awns first became visible above the ligule. Dried biomass was determined for crown, stem, green leaf, non-green leaf, and head components of all 24 plants after oven-drying at 65–70°C. Leaf area index was computed from specific leaf weight of the subsample, green leaf biomass of all plants, and plant density. Beginning 1 week after anthesis, developing grains were separated from the chaff by a combination of hand and machine threshing of the heads and oven-dried for a total of 14 days at 65–70°C. Final grain yields were determined by machine harvest of ~20 m^{-2} of each subplot on May 25–27, 1993 and on June 1, 1994.

3. Measurements of Photosynthesis and Transpiration Carbon assimilation was measured by using several different techniques during the FACE experiments. At the individual leaf level, we used portable LI-COR Model 6200[1]

---

[1] Names are necessary to report factually on available data; however, the USDA neither guarantees nor warrants the standard of the product, and the use of the name by the USDA implies no approval of the product to the exclusion of others that may also be suitable.

photosynthesis systems (LI-COR, Inc., Lincoln, NE) that were equipped with either 0.25 or 1.0-l cuvettes and moved between subplots from dawn until dusk (Hileman et al., 1992, 1994; Idso et al., 1994). At the canopy level, two types of chamber systems were used. The closed chamber systems used during the first 2 years of cotton experimentation have been described by Hileman et al. (1994).

During the third year of cotton and both FACE wheat experiments, continuous-flow, gas exchange chambers were used to measure canopy gas exchange rates within the perimeter of FACE and control arrays. These systems were engineered similar to those described by Garcia et al. (1994). They consisted of an aluminum frame covered with Propafilm C[1] (ICI Americas, Inc., Wilmington, DE), a material having a high transmittance of thermal infrared radiation that minimized temperature increases when chambers were exposed to direct sunlight. Differences in CO_2 and H_2O vapor concentrations between inlets and outlets of the chamber were measured with a LI-COR Model LI-6262[1] infrared gas analyzer (IRGA). In FACE arrays, absolute CO_2 was controlled at 550 μmol mol^{-1} by injecting CO_2 into the inlet stream and monitoring it via a second IRGA. A calibrated infrared thermometer (IRT) and quantum sensor were positioned inside the chamber to measure foliage temperatures and incident photosynthetic photon flux density (PPFD), respectively. Chamber ventilation afforded four complete air exchanges per minute and a slight positive pressure, which reduced the potential contribution of soil CO_2 flux to the overall net measurement. Gas exchange (CO_2 and H_2O) was measured continuously (10- to 15-min averages) for 7–14 days, after which systems were moved to a different treatment replicate and plant material was harvested for the determination of biomass and leaf area index. Chambers used in the cotton experiments measured 1.15 × 1.0 m wide and stood 2 m tall, enclosing one row of cotton plants that was 1 m in length; in wheat, the chambers were 0.75 × 0.75 m wide by 1.3 m high, enclosing three adjacent rows of plants, each 0.75 m in length.

III. Results and Discussion

A. Net Photosynthesis

1. Leaf Photosynthesis Effects of elevated CO_2 on net photosynthesis (P_n) at the individual leaf level provided comparative data for cotton and wheat. In well-watered cotton, for example, there was about a 30% increase in midday P_n associated with a change in CO_2 from control to FACE conditions (Hileman et al., 1992, 1994; Idso et al., 1994). Significant differences related to CO_2 treatments were also found in light response functions (Idso et al.,

1994). Elevated levels of CO_2 increased the asymptotic limit of P_n by 22%, increased initial light conversion efficiency by 28%, and decreased the light compensation point from 63 to 10 μmol CO_2 m^{-2} s^{-1} compared with controls. The latter factors implied greater stimulation of P_n by CO_2 at lower light intensities found at levels deep within the canopy, although this was not explicitly tested with leaves from the lower levels.

Analysis of P_n data from upper canopy leaves during the 1992–1993 wheat study showed similar results. Under well-watered conditions, FACE increased midday P_n values by 28% compared with control values (Garcia et al., 1995), without any evidence of acclimation in the photosynthetic process (Nie et al., 1995 and the FACE Team, unpublished data in press). The CO_2 effect was even larger (>50%) for the DRY irrigation treatments (G. Wall et al., personal communication). The stimulatory effect of CO_2 was also much greater for leaves in the middle and lower levels of the canopy, where they were exposed to lesser intensities of photosynthetically active radiation (PAR) (Osborne et al., 1995). Unlike cotton, however, CO_2 stimulation of single leaf P_n disappeared toward the end of the grain filling period in wheat (Garcia et al., 1995). These observations were consistent with accelerated developmental rates and senescence of leaves in FACE and decreased synthesis of major proteins involved in the photosynthetic process (Nie et al., in press).

2. Canopy Photosynthesis Canopy-level measurements revealed increases in the rates of carbon assimilation that mirrored CO_2 enhancement effects observed on a single leaf basis and were similar to studies reported in the literature for other species (Drake and Leadley, 1991). Hileman et al. (1994), using closed chambers in the 1989 and 1990 experiments, reported a midday P_n for well-watered cotton canopies in the FACE treatment that was approximately 30% higher than control values. The P_n enhancement caused by CO_2 was greatest during the middle part of the season when plants in the FACE treatments were larger and intercepted more light (Pinter et al., 1994).

Representative data from flow-through chamber systems operated during the 1991 cotton and 1992–1993 wheat FACE experiments showed CO_2 exchange rates that varied with treatment and time of year (Figs. 1 and 2). During early August 1991, when cotton canopy cover was 100% and plants had a moderately heavy boll load, P_n in the control WET canopy rose to 60–70 μmol CO_2 m^{-2} s^{-1} during midmorning and remained constant for about 5 hr (Fig. 1a). Assimilation rates in the FACE WET treatment on the same day were 35–40% higher, a difference that remained fairly consistent each day during this period of canopy development. As the season progressed, however, absolute P_n values declined. By early September (Fig. 1b), the P_n of control WET was only 30 μmol CO_2 m^{-2} s^{-1}, a reduction

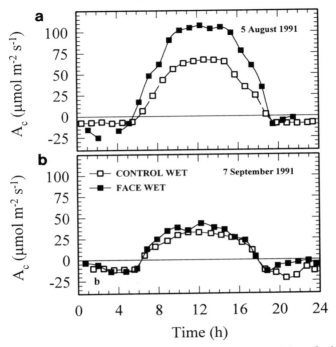

Figure 1 Diurnal trajectories of net canopy assimilation (A_c, μmol CO_2 m^{-2} s^{-1}) of cotton exposed to FACE ($\sim$550 μmol mol^{-1}) and control CO_2 ($\sim$370 μmol mol^{-1}) treatments. Data were measured under clear sky conditions using continuous-flow chambers on representative days during early-August (a) and early-September (b) 1991. Measurements were made only in the WET irrigation treatment.

of 50% from values measured a month earlier. Despite this decline, the enhancement due to extra CO_2 was still evident during midday, with FACE showing an approximate 20% increase over control. Nighttime gas exchange data indicated that dark respiration of cotton was reduced by elevated CO_2, but the results were inconclusive.

Daily values of net canopy photosynthesis were measured with similar continuous-flow chambers during both years of our wheat experiments. Typical trends in these data are shown for mid- and late-season crops in 1993 (Fig. 2). During mid-March, wheat plants in all of the treatments were just beginning to flower, canopy cover was nearly 100%, and the green leaf area index (GLAI) was approaching seasonal maximum values of 5 or 6. The FACE WET canopy had P_n rates on that day that remained above 50 μmol CO_2 m^{-2} s^{-1} for more than 4 hr (Fig. 2a). Compared with the control WET canopy, this represented a P_n enhancement of about 20%. Above-average winter precipitation prevented the development of measurable

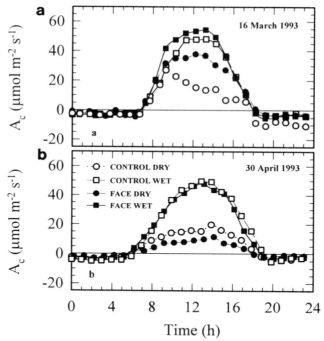

Figure 2 Diurnal trajectories of net canopy assimilation (A_c, μmol CO_2 m^{-2} s^{-1}) of wheat exposed to FACE (~550 μmol mol^{-1}) and control CO_2 (~370 μmol mol^{-1}) treatments. Data were measured under clear sky conditions using continuous-flow chambers during mid-March (a) and late-April (b) 1993. Measurements were made in the WET and DRY irrigation treatments.

differences in biomass or GLAI between our WET and DRY treatments until about 1 week later in the season, but the effects of water stress were expressed clearly in reduced canopy P_n (Fig. 2a). The interaction between water stress and CO_2 was also evident in these data; the CO_2 stimulation of P_n in the DRY treatment was considerably larger than the effect measured in the WET treatment.

Stimulation of canopy P_n by elevated CO_2 continued from anthesis through the milk and early dough stages of grain development (from March through the middle of April). During this period, plants also continued to respond more favorably to elevated CO_2 under deficit irrigation regimes than they did in the WET treatments. This presumably was because the larger root systems of FACE DRY (Wechsung and Wechsung, personal communication) enabled plants to extract soil moisture more efficiently. By April 30, however, CO_2 enhancement of P_n had disappeared from the

WET treatment and reversed itself in the DRY treatment (Fig. 2b). This phenomenon was similar to the response seen in individual leaf P_n data and was caused by different rates of canopy and leaf senescence that will be presented later. The gas exchange methodology was not sufficiently sensitive to resolve differences in respiration between CO_2 treatments.

B. Water Relations

Effects of CO_2 on water relations and latent energy exchange processes of cotton and wheat were examined at the leaf, whole-plant, and canopy levels by using a number of diverse experimental techniques.

1. Leaf Water Potentials Total leaf water potentials (ψ_l) were measured on a diurnal basis on numerous occasions during our FACE experiments by using a pressure chamber. At midday there was a tendency for both species in the WET treatment to have a slightly less negative total leaf water potential in the FACE than in the control. However, these differences generally were small and disappeared altogether earlier and later in the day [for cotton, see Bhattacharya *et al.* (1994); for wheat, see Wall *et al.* (1995)]. Toward the end of the season in cotton and for much of the season in wheat, ψ_l in the FACE DRY irrigation treatments was often significantly less negative than that in control DRY. We suspect that this interaction between CO_2 and irrigation treatment was a compounded result of two phenomena in the FACE treatment: (1) improvement in internal leaf water status caused by partial stomatal closure and (2) the larger volume of soil being exploited by roots (Prior *et al.*, 1994; Wechsung and Wechsung, personal communication).

2. Stomatal Numbers and Conductance No direct effects of CO_2 exposure were observed in the stomatal density (number of stomata per unit area) or the stomatal index (number per epidermal cell) in either the cotton (S. Malone, personal communication) or wheat plants grown in the FACE facility (Estiarte *et al.*, 1994). Single leaf conductances (g_s) obtained during the cotton experiments by using a LI-COR Model LI-1600[1] steady state porometer and during the wheat experiments by using a LI-COR Model LI-6200[1] closed, gas exchange system revealed a consistent closure response of leaf stomata to elevated CO_2. During the 1990 cotton experiment, Hileman *et al.* (1994) reported that average midday g_s values of leaves exposed to 550 μmol mol^{-1} in the WET treatment were 22% lower than those in controls. Bhattacharya *et al.* (1994) showed similar g_s differences in cotton toward the end of the 1991 FACE season and found an even larger g_s closure in response to CO_2 in the DRY irrigation regime. Midday g_s values of wheat leaves in the FACE WET treatment were 38% lower than in control WET throughout the 1992–1993 wheat experiment (Garcia *et al.*, 1995). Reductions in g_s were also accompanied by small, but consistent increases in canopy temperatures (T_c) measured with infrared thermometers. Compared with the control WET treatment, midday T_c values of plants in the FACE WET treatment were about 0.8°C warmer for cotton (Kimball et al., 1992) and 0.6°C warmer for wheat (Kimball et al., 1994b).

3. Sap Flow Transpiration Estimates Constant power sap flow gauges were used to quantify the effect of elevated CO_2 on the transpiration of whole cotton plants and individual wheat tillers. On a ground area basis, mean transpiration rates of cotton in the well-watered treatments varied between 4.4 and 9.5 mm day^{-1} (Dugas et al., 1994). No consistent differences in sap flow were found between FACE and control wheat on either a diurnal or a seasonal basis. When calculated over 2-week intervals, integrated cotton sap flows were similar to cumulative evapotranspiration obtained via soil water depletion data (Hunsaker et al., 1994a).

The effects of elevated CO_2 on sap flow transpiration estimates of individual wheat tillers were variable and confounded by spatial variation of soil water distribution from the buried microirrigation line (Senock et al., in press). Daily differences in transpiration between CO_2 treatments were small and often nonsignificant. However, the cumulative water use (on a ground area basis) by well-watered FACE plants was 7–23% less than that by control plants during heading and early grain fill.

4. Evapotranspiration Estimates from Soil Moisture Depletion Cumulative evapotranspiration (ET_c) was estimated from temporal changes in soil water contents within the rooting zone of the plants (Hunsaker et al., 1994a,b). The effect of CO_2 on ET_c was not statistically significant for any year of our study for either cotton or wheat. However, when ET_c of the WET treatment was averaged over both years of the wheat experiments, it was 4.5% less in FACE than in controls. In the DRY treatment the CO_2 effect was reversed, with the ET_c of FACE averaging 3% higher than that of control (most likely because the larger plants in the FACE DRY treatment were able to exploit a larger volume of soil than control DRY plants).

5. Evapotranspiration Estimates from the Canopy Energy Balance The effects of CO_2 on ET_c of crops under well-watered conditions were also investigated by calculating the latent energy exchange of the canopy as a residual in the surface energy balance. Kimball et al. (1994a) concluded that ET_c differences between CO_2 treatments in the FACE cotton study were within the uncertainties associated with measuring net radiation (R_n). By using improved R_n measurement techniques during the wheat experiments, daily totals of latent heat flux and ET_c of plants in the elevated CO_2 treatment were shown to be 8% lower than in controls in 1992–1993 and 11% lower in 1993–1994 (Kimball et al., 1994b).

6. Ecosys-Modeled Evapotranspiration in Wheat Energy exchange between the atmosphere and wheat canopy in the FACE experiments was also examined by using a mathematical model, ecosys, which predicts the hourly growth and consumption of water by a wheat crop (Grant et al., 1995). The overall net effect of increasing CO_2 to 550 from 370 μmol mol^{-1} was to reduce simulated, season-long transpiration under well-watered conditions by 7%. This value is comparable to the 5% reduction in seasonal evapotranspiration using soil water balance (Hunsaker et al., 1994b), the 10% average reduction estimated from daily measurements of latent energy exchange (Kimball et al.,,1994b), and the 7–23% average reduction in transpiration measured during peak leaf area with sap flow gauges (Senock et al., in press). However, these observations and predictions for the canopy contrast sharply with the 38% reduction in conductance measured using single leaf cuvette systems (Garcia et al., 1995). In the ecosys simulation and processes measured on the community scale, CO_2-mediated rises in canopy temperature and early season increases in GLAI with elevated CO_2 appear to eliminate or moderate the transpiration effects observed at the leaf surface.

C. Growth and Yield Responses

1. Cotton Biomass, Leaf Area, and Fruit Production Deficit irrigation and elevated CO_2 had significant effects on the growth of cotton during each year of our study (Mauney et al., 1992, 1994). Destructive plant samples obtained at intervals throughout the season showed that water stress significantly reduced total biomass and GLAI in the DRY treatments compared with the WET treatments (Fig. 3). Although cotton fruit production was also lower in the DRY treatments, harvest indices (defined here as the ratio of flower plus boll biomass to total biomass) were larger.

Exposure of cotton plants to higher CO_2 levels increased total biomass significantly compared to the ambient CO_2 treatment. It also increased the dry weight density and weight per unit length of cotton roots over the whole soil profile during the early vegetative and midreproductive developmental stages of growth (Rogers et al., 1992; Prior et al., 1994). The overall effects of elevated CO_2 on photosynthetically active components of the canopy were more varied and depended upon the degree of water stress to which the crops were subjected. In the DRY treatments, elevated CO_2 conferred a modest advantage on GLAI and light interception during most of the season (Pinter et al., 1994). An advantage was also present during the first part of the growing season in the WET treatments. We believe that these early differences in GLAI had cumulative effects on productivity that were important in developing a bigger plant having a larger photosynthetic apparatus and higher fruiting potential under the elevated CO_2 conditions.

Resource partitioning by the cotton plants during 1991 is shown in Fig. 4. In early July, when the plants were just starting to flower profusely, stem tissue composed about 50% of the total plant biomass in the FACE treatment compared with about 45% for controls. At the same time, the

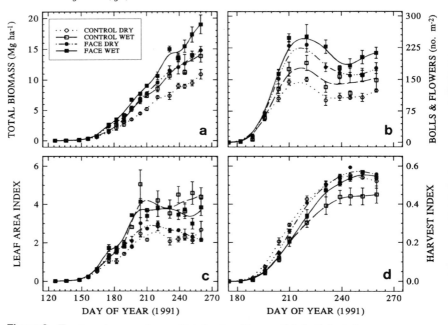

Figure 3 Treatment comparisons of total cotton biomass (Mg ha^{-1}, includes most tap roots but not small lateral and fine roots), boll and flower density (no. m^{-2}), green leaf area index (m^2 m^{-2}), and harvest index (boll biomass divided by total biomass) obtained from 17 weekly samples of cotton in the 1991 FACE experiment. Data are the means of four replicates per treatment combination. Bars show ±1 standard error.

green leaf component in FACE was ca. 37% of the total versus 40% in controls. From that point until the end of August, plants in all treatments translocated a proportionately larger share of photosynthate into maturing fruit, and the relative allocation of resources to stem and leaf tissues declined markedly. Stem and leaf biomass (on an absolute basis) continued to increase slightly during July and then leveled off for the remainder of the season. Despite similar sizes in the photosynthetically active biomass of their canopies, plants exposed to elevated CO_2 had a 25% higher light utilization efficiency than controls (Pinter *et al.*, 1994). We believe that this enabled FACE plants to initiate and sustain significantly more fruit (Fig. 3b) prior to cutout (a source-limited hiatus in flowering that normally occurs during August). Large differences between the harvest indices of FACE and control WET (Fig. 3d) may likewise be explained by a larger plant with increased sinks (fruit) that could benefit more from the fertilization effect of extra CO_2 late in the season. At the end of the season, the ratio between green and mature bolls rapidly declined in the DRY treatments because the recruitment of new fruit was limited by water stress.

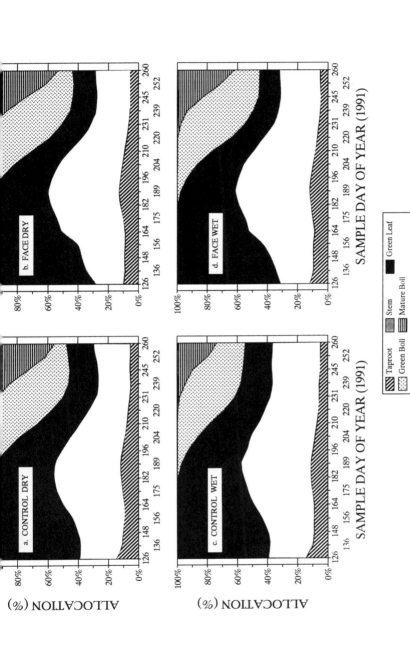

Figure 4 Allocation of biomass (expressed as a percentage of the total dry weight) to different cotton plant compartments in the 1991 FACE experiment. Data are shown for tap roots obtained when plants were pulled from soil, stems, green leaves, green fruit (bolls), and mature bolls (burr, seed, plus lint). Data are the means of four replicates per treatment combination that have been smoothed with three-term running average.

Although treatment-related differences in biomass allocation were not particularly evident in the partial tap root component of plants sampled at frequent intervals during the season (Fig. 4), Mauney *et al.* (1994) show a trend toward higher root to shoot ratios during boll development that was associated with FACE. The more precise root core studies of Rogers *et al.* (1992) and Prior *et al.* (1994) revealed that the cotton plant's belowground investment in total roots was increased significantly by elevated CO_2.

2. Wheat Biomass, Leaf Area, and Head Production Seasonal trajectories of total biomass accumulation revealed relatively small CO_2 and irrigation effects and slow growth rates during the winter when temperatures were cool (Fig. 5). As temperatures warmed and PAR intensity and duration increased, biomass increased exponentially and the effects of elevated CO_2 and deficit irrigation became more noticeable. During this period, FACE plants grew at a faster rate than controls, with much of the difference attributed to increases in the biomass of stems and nascent reproductive

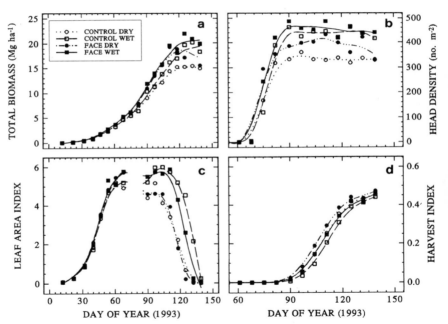

Figure 5 Treatment comparisons of total wheat biomass (Mg ha^{-1}, includes crown tissue but not roots), head density (no. m^{-2}), green leaf area index (GLAI, m^2 m^{-2}), and harvest index (grain biomass divided by total biomass) obtained from 18 weekly samples of wheat in the 1993 FACE experiment. GLAI data were not obtained for plants on sample days 75 and 82. Data are means of four replicates per treatment combination.

tissues (Fig 6). The accelerated rate of development with elevated CO_2 was a biologically significant finding. We observed, for example, that plants exposed to elevated CO_2 reached anthesis about 2 days sooner than controls, regardless of irrigation treatment. Furthermore, wheat plants in the FACE WET treatment matured 1 week earlier than plants in the control WET treatment. In the DRY treatment, grain matured at the about the same time in FACE and control plants. The effect of CO_2 on developmental rate during grain fill may have been suppressed by the improved water status of FACE DRY plants compared with controls.

The processing protocol for the weekly destructive samples provided wheat head numbers and biomass values as soon as awns were visible above the ligule of the flag leaf. These data show that the heads appeared ca. 5 days earlier under FACE (Fig. 5b). After about 2 weeks, most of the heads had emerged in all of the treatments, and numbers stabilized during the grain filling period. Elevated CO_2 appeared to ameliorate the effects of water stress on final head density in the DRY irrigation treatment, but did not affect final head density under well-watered conditions.

GLAI expanded rapidly during February 1993, attaining 2.5 by the end of tillering and reaching maximum levels of more than 5 prior to the booting stage in early-March (Fig 5c). Processing delays spoiled our GLAI data during the 2-week period in March when plants were just beginning to show symptoms of water stress. However, treatment effects were very evident during the second half of the season. Water stress during early grain fill reduced GLAI in the DRY treatment to about 80% of levels observed in the WET treatment. Later in the season, differences among treatments in the senescence of leaves paralleled the pattern of grain maturity and provided a partial explanation for changes in canopy photosynthesis that were discussed earlier. Leaf senescence occurred 2–3 weeks sooner in the DRY treatments than it did under well-watered conditions. Wheat canopies exposed to elevated CO_2 senesced about 1 week earlier than controls.

The evolution of harvest index (HI) (Fig. 5d) and changes in the allocation of biomass to chaff and grain (Fig. 6) show a close correspondence with temporal patterns of head emergence, grain fill, and phenology. Giuntoli and the FACE Team (unpublished data) found that the HI increase rate was independent of CO_2 and irrigation treatment effects during its linear phase. Similar to our findings in cotton, the end of season HI of wheat was greater with water stress and elevated CO_2.

3. Cotton Lint Yields Final marketable yields from the control WET cotton plots (Fig. 7a) met or exceeded statewide averages during each year of our FACE experiments (Mauney *et al.*, 1992, 1994). Lint production was not significantly affected by irrigation treatment in either 1990 or 1991, a result that was partly due to compensatory increases in the harvest index of plants

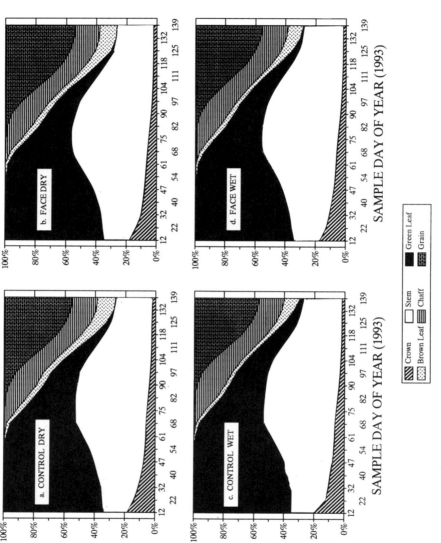

Figure 6 Allocation of biomass (expressed as a percentage of total dry weight) to different wheat plant compartments in the 1992–1993 FACE experiment. Data are shown for crowns (belowground stem and crown tissue without roots), aboveground stems, green leaves, brown or yellow leaves, chaff (nongrain portion of the head), and grain. Data are the means of four

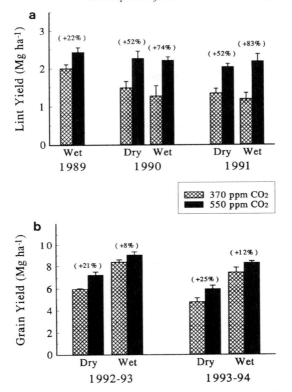

Figure 7 Final marketable yields of cotton lint (a) and wheat grain (b) on a dry weight basis (Mg ha^{-1}) grown in the FACE facility at MAC. [For reference: 1 Mg lint ha^{-1} = 1.86 standard (480 lb.) cotton bales acre^{-1}, and 1 Mg dry grain ha^{-1} = 16.7, 60-lb. bushels acre^{-1} at 12% moisture content.] Hats on bars show 1 standard error above the mean of four replicates. Enhancement effects of the 550 μmol mol^{-1} CO$_2$ (FACE) treatment compared to ambient CO$_2$ (controls) are shown in parentheses.

in the DRY treatment and partly due to the relatively minor water stress to which the plants were actually exposed. Elevated CO_2 caused significant increases in lint yield that were proportionately greater than end of season biomass differentials [see the chapter in this volume by Amthor and Koch (1996)]. The CO_2-related yield enhancement in the WET irrigation treatment was 60% when averaged over the 3 years of the study. Elevated CO_2 stimulated lint yields from the DRY irrigation treatment by 52% in 1990 and 1991.

4. Wheat Grain Yields. Final yields of spring wheat in our FACE studies were obtained from relatively large (~20 m^2) undisturbed areas within each of the subplots. Yields measured for plants grown under well-watered conditions were very high (Fig. 7b). In fact, plants in the control WET treatment during the 1992–1993 season produced 40% more grain than the county average for spring wheat and 10% more than the potential yields for the same cultivar. The effects of treatments on yield were very consistent during both seasons. Grain yields in the DRY treatment were 24 and 32% lower than the well-watered treatments during the first and second years, respectively. After our experiences with cotton, the modest effect (~10%) of CO_2 on final grain yield in the WET treatment was somewhat unexpected. However, when yields were reduced by deficit irrigation, the stimulatory effect of CO_2 doubled. Analyses of variance revealed that the differences between CO_2 treatment means were statistically significant at $p = 0.037$ in 1992–1993 and $p = 0.062$ in 1993–1994.

The CO_2-related stimulation of growth that we observed using FACE technology compared favorably with results from open-top chamber studies conducted in the same field as our FACE experiment in 1992–1993 [A. Frumau and H. Vugts as cited by Kimball et al. (in press)]. However, our findings for a non-water-stressed, wheat agrosystem in an open-field experiment showed a smaller CO_2 response than has been reported for wheat grown in chamber experiments [see the chapter in this volume by Dijkstra et al. (1996) and reviews by Kimball (1983), Poorter (1993), and Idso and Idso (1994)].

Accelerated rates of phenology and postanthesis canopy senescence were surprising phenomena that occurred with elevated CO_2 during both years of our wheat study. Plants in the FACE WET treatment had grain ready for harvest about 1 week before their control counterparts. One possible factor influencing the rates of development may have been related to temperature. For example, Wiegand and Cuellar (1981) showed that the duration of grain filling in wheat was reduced by about 3 days for every 1°C increase in ambient air temperature above 15°C. In our study, however, elevated levels of CO_2 increased average canopy temperatures by an observed 0.6°C over the entire season (Kimball et al., 1994b), probably because of the reduction in g_s in the FACE treatment. Thus, it appeared that tissue temperature by itself did not completely account for the shortened grain filling period. Furthermore, higher ambient temperatures were usually associated with decreased kernel weights (Wardlaw et al., 1980; Wiegand and Cuellar, 1981). Yet in both FACE wheat experiments, kernels from the FACE WET treatment averaged ca. 6% larger than those from control WET. At present we are unable to explain these results, but expect that accelerated developmental rates will provide important clues explaining wheat plant response to elevated CO_2. Regardless of the cause for the decreased grain filling period, the highly deterministic growth patterns in wheat permitted ambient CO_2 treatments to play catch-up at the end of the season, narrowing the differential between the CO_2 treatments in final yields, especially in the WET treatments.

5. Comparison of CO_2 Effects on Cotton and Wheat A crop's response to elevated CO_2 is mediated by ambient temperature (Idso et al., 1987; Long, 1991) and by its ability to develop additional sinks for the accumulation of greater photosynthetic products (Mauney et al., 1978; Stitt, 1991). We thus expected cotton and wheat to be good plant models for our FACE research because they are grown during different seasons of the year and represent extremes in potential sink capacity. Our indeterminate example, cotton, is a warm-season, woody shrub that continues to increase in size and produce fruit as long as growing conditions remain favorable. Although a perennial crop, it is normally cultivated as an annual to minimize problems from overwintering insects. By comparison, wheat is strongly determinate. It is an annual plant with rigid genetic constraints on the rate and extent of sink growth. It also is adapted to grow during cooler temperatures than cotton. Our results showed that the net photosynthesis of both of these crops responded similarly to supraambient CO_2 concentrations as provided by FACE. However, the seasonal trends in biomass and GLAI and the end of season yields revealed important distinctions between the two.

The most important difference between the two crops was related to the effects of CO_2 on the development and persistence of GLAI during the season. CO_2 enhancement of GLAI was greatest for DRY cotton (Fig. 8a), where water stress prevented plants from attaining 100% canopy cover and plants were not source-limited by incident light. The large GLAI coupled with greater light use efficiency explained the greater CO_2-related stimulation of biomass and final lint yield in cotton growing in the DRY treatment. In WET cotton (Fig. 8c), CO_2 effects on GLAI and biomass were more variable. Even though CO_2 enrichment was associated with small (10–20%) reductions in GLAI during the second half of the season, the canopy remained green and CO_2 stimulation of biomass continued until harvest in mid-September.

In wheat, however, the patterns were significantly different (Figs. 8b and 8d). CO_2 caused a small, yet consistent advantage in GLAI during the first part of the season. This disappeared during midseason as the final leaf emerged and expanded. Then, midway into the grain filling period, the wheat leaves in the FACE treatment began to senesce at a much faster rate than controls. The final result was that, during grain filling, the green leaf area duration in FACE was much reduced compared with controls. The CO_2 enhancement factor for biomass gradually increased during the season, reaching a peak around day of year 104 in the WET treatment and about 1 week later in the DRY. Then, as leaf senescence was accelerated in CO_2-

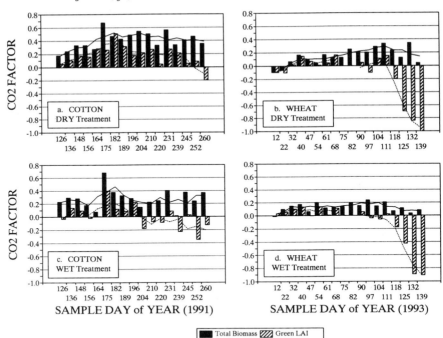

Figure 8 CO_2-related enhancement [(FACE − control)/control] of biomass and green leaf area index (GLAI) for cotton in the 1991 FACE experiment (Figs. 8a and 8c) and wheat in the 1992–1993 FACE experiment (Figs. 8b and 8d). Bars show CO_2 enhancement on the indicated sample day; lines show a three-term running average.

enriched canopies, biomass in control treatments increased relative to the FACE treatments. This caused the CO_2 enhancement factor for biomass to decline in both water treatments. These data support a hypothesis that the determinate growth patterns of wheat render it sink-limited when exposed to supraambient CO_2 levels. They also offer a plausible explanation for the minimal effects of CO_2 on final grain yield under well-watered conditions.

D. Light and Water Utilization Efficiencies

Cotton developed a larger leaf canopy earlier in the season when grown under supraambient CO_2 conditions. This resulted in 15–40% higher absorption of PAR (APAR, MJ m^{-2} day^{-1}) during the first half of the season (Pinter et al., 1994), an important factor affecting potential yield in a crop with indeterminate fruiting behavior. Light use efficiency (LUE) was used to quantify season-long responses of the cotton canopy to CO_2 and irrigation treatments. Defined as dry biomass (g m^{-2} day^{-1}) produced per unit of APAR, LUE was computed for each plant sampling interval and then averaged across the season (Pinter *et al.*, 1994). Results showed that this parameter varied significantly from year to year and reflected the effects of CO_2 and irrigation (Fig. 9a). Elevated CO_2 resulted in an average 26% increase in LUE for the WET irrigation treatment and a 21% increase for the DRY treatment.

Soil water depletion studies showed very similar seasonal ET_c from FACE and control cotton canopies, although marketable yields were increased by elevated levels of CO_2. As a consequence, water use efficiencies [WUE, defined here as kilograms of lint produced per cubic meter of water used

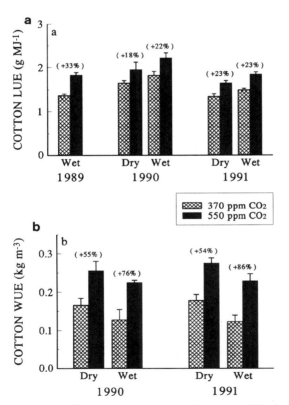

Figure 9 Cotton light use efficiency (panel a, g dry biomass per MJ absorbed PAR) and water use efficiency (panel b, kg lint per m^3 water) measured during the FACE experiments. Cumulative water use was not measured in 1989. Hats on bars show 1 standard error above the mean of four replicates. Enhancement effects of the 550 μmol mol^{-1} CO_2 (FACE) treatment compared to ambient CO_2 (controls) are shown in parentheses.

by the crop (ET_c) during the season] followed the trends in yield and were significantly higher for FACE cotton than for controls (Fig. 9b).

Light and water use efficiencies of wheat grown in FACE are currently being analyzed for comparison with cotton data.

E. Soil Carbon Budget and Sequestration

A rigorous budget of soil carbon has not yet been attempted for our experiments. However, analyses of soil and root samples are continuing, and a few preliminary results are available. First of all, CO_2-related stimulation of total root growth was consistent with observed increases in aboveground biomass [for cotton, see Fig. 3 and also Mauney et al. (1994); for wheat, see Fig. 5]. Elevated CO_2 caused about a 40% increase in cotton roots (Prior et al., 1994) and about a 20% increase in wheat roots (Wechsung and Wechsung, personal communication). Such increases represent greater net input of organic carbon to the soil and, thus, would be expected to influence the rates of many different soil processes.

This indeed seems to be the case. In the cotton experiments, for example, Runion et al. (1994) observed higher microbial activities with elevated CO_2. Supraambient CO_2 also increased soil respiration by an average of 12% during the 1990 and 1991 cotton seasons (Nakayama et al., 1994) and by 19% during the 1992–1993 wheat experiment (Nakayama, 1993). Since these fluxes of CO_2 at the soil surface were averages of measurements obtained during and shortly after the growing season, they do not account for carbon losses that may have occurred between seasons. Notwithstanding, the soil respiration data from cotton imply a net gain in soil C under FACE, while the preliminary data from wheat suggest that soil C inputs and outputs were more balanced.

Support for this hypothesis in cotton comes from two independent studies that were conducted after cotton had been grown under the FACE treatment for several consecutive seasons. In the first study, Wood et al. (1994) observed an overall tendency for soil cores taken from the FACE treatment to have higher total C contents than controls, but the differences were statistically significant only for the 0.1–0.2-m depth increment. In the second study, Leavitt et al. (1994) used a sensitive carbon isotope technique that was possible because the petrochemically derived CO_2 used for FACE enrichment had different $^{13}C/^{12}C$ and $^{14}C/^{12}C$ ratios than current ambient air. Their analyses showed that about 10% of the organic carbon in the FACE soil in 1991 had been derived from the previous 3 years of FACE treatment of the cotton. Moreover, much of this increased C in the soil was in a recalcitrant 6 N HCl resistant fraction, suggesting that elevated CO_2 can indeed increase the sequestration of C in the soil and may mitigate some of the projected rise in atmospheric CO_2 concentration.

IV. Summary and Future Investigations

Growth and yield responses of plants to elevated levels of atmospheric CO_2 have complex interactions with temperature, water, and nutrients that are difficult to quantify by using conventional approaches with closed or open-top chambers. As a result, these responses have not yet been adequately accounted for in simplistic models predicting world food production in the future (viz., Rosenzweig and Parry, 1994). Our series of FACE studies has provided the first interdisciplinary insight into the full-season responses of several important agricultural crops to elevated CO_2 under natural, open-air, field-scale conditions. We believe that our research results will provide unique validation opportunities for simulation models that will be capable of extrapolating the potential impacts of global change in a more realistic fashion [Wall *et al.*, 1994; see the chapter in this volume by Amthor and Loomis (1996)].

Several important conclusions regarding the effects of elevated CO_2 on plants can be drawn from the observations of photosynthesis, transpiration, growth, and yield for cotton and wheat in the FACE experiments.

- In the absence of nutrient limitations, enrichment to 550 μmol mol^{-1} CO_2 caused 20–30% increases in net photosynthesis at individual leaf and canopy levels.
- Midseason accumulation of biomass under the FACE treatment increased from about 20% in wheat to about 40% in cotton.
- CO_2 stimulation of final yields and end of season biomass appears to depend on temperature and the availability of expanding sinks for extra photosynthate.
- Indeterminate crops such as cotton respond more favorably to increased levels of CO_2 than determinate crops such as wheat.
- Photosynthesis, growth, and final grain yields of spring wheat were all stimulated more by elevated CO_2 under water stress conditions than under well-watered conditions.
- Despite significant decreases in stomatal conductance at elevated CO_2, compensatory increases in GLAI resulted in no differences in transpiration per unit ground area in cotton, but there was a slight decrease in wheat.

Although water is perhaps the most important limiting factor affecting the productivity of crops in intensively managed lands, agriculture in less developed regions and plants in most natural ecosystems are additionally restricted by nutrient availability. The consensus of participants in the San Miniato GCTE Workshop in Italy (October, 1993) was that reduced nutrients were important in controlling plant response to CO_2 and probably explained some of the differences that have been observed between managed agricultural systems and more complex, N-limited natural ecosystems. FACE appears to be the most appropriate technology for probing this hypothesis on the ecosystem scale (McLeod, 1993; Schulze and Mooney, 1993). It also confers an "economy of scale" compared with controlled-environment or open-top chambers that makes the cost per unit of scientific opportunity attractive for large, multidisciplinary research projects (Kimball, 1992; Hendrey *et al.*, 1993b).

The logical extension of the research reported here will be to use FACE to determine the CO_2 response of wheat when nutrients are limiting. Beginning in December 1995, wheat will be exposed to ambient and elevated CO_2 and to two levels of soil fertility. Research emphasis will be placed on biogeochemical soil processes, but work will continue on wheat physiology, development, biomass partitioning, and water use. Another area that needs urgent attention is the interaction between CO_2 and atmospheric pollutants such as ozone, which threaten grain productivity in northern, midlatitude agricultural regions (Allen, 1989; Chameides *et al.*, 1994).

Results from such studies are expected to provide meaningful data on the strength of the "CO_2 fertilization effect" on natural terrestrial ecosystems. They will also offer realistic insight into the adequacy of future world food production.

Appendix

Measurements and participants in the FACE experiments with cotton and wheat at the Maricopa Agricultural Center between 1989 and 1994. Acronyms identifying the affiliations of scientists are listed alphabetically at the end of the table.[a]

Experimental parameters	Cotton (1989, 1990, 1991)	Wheat (1992–1993, 1993–1994)
SOIL PROPERTIES		
Carbon isotope analysis	S. Leavitt (TRL); E. Paul (MSU)	S. Leavitt (TRL)
Chemistry	G. Huluka (TU); R. Rauschkolb, H. Cho (MAC); C. W. Wood, H. Rogers, G. Runion, S. Prior (NSDL)	R. Rauschkolb, H. Cho (MAC)
Concentration and fluxes of CO_2 (soil respiration)	F. Nakayama (WCL)	F. Nakayama (WCL)
Decomposition of plant residue	—	D. Akin (RRC); A. Ball (UE); H. W. Hunt, S. Smith (CSU)
Fluxes of CH_4 and N_2O		F. Nakayama (WCL); A. Mosier (SPNR)
Hydraulic conductivity, water retention curve, texture, bulk density	B. Kimball, B. Alexander (WCL); D. Post, A. Warrick (UA); F. Whisler (MISS)	G. Wechsung (HU); F. Adamsen, G. Wall, D. Hunsaker, B. Kimball (WCL)
Mineralization	C. W. Wood, H. Rogers, G. Runion, S. Prior (NSDL)	H. Rogers, S. Prior (NSDL)
Rhizosphere microbes	G. Huluka, R. Ankuma (TU); G. Runion, E. Curl, H. Rogers (NSDL)	—
Volumetric water content	D. Hunsaker (WCL)	D. Hunsaker (WCL)
AGRONOMY		
Population density, biomass components, leaf area, organ counts, heights, yield	J. Mauney (WCRL)	P. Pinter (WCL); A. Giuntoli, F. Miglietta (IATA)
Root biomass and density	H. Rogers, S. Prior (NSDL)	G. Wechsung (HU); F. Wechsung (PIK); G. Wall (WCL)
ANATOMY		
Biomechanical properties of straw	—	H. Rogers, S. Prior (NSDL)
Leaf anatomy	—	D. Akin (RRC)
Stomatal density	S. Malone (GSU)	M. Estiarte, J. Peñuelas (IRTA); A. Giuntoli, F. Miglietta (IATA)

Experimental parameters	Cotton (1989, 1990, 1991)	Wheat (1992–1993, 1993–1994)
BIOCHEMISTRY		
Carbohydrate storage	D. Hendrix (WCRL)	D. Hendrix (WCRL)
C, N in tissues	G. Huluka, D. Hileman, P. Biswas (TU)	F. Wechsung (PIK); G. Wechsung (HU); F. Adamsen (WCL); D. Akin (RRC); T. Sinclair (APL)
Digestibility	D. Akin (of sudangrass planted among cotton, RRC)	D. Akin (RRC)
Lint or grain quality	J. Mauney (WCRL)	C. Morris (WWQL)
Phenolics, antioxidants, secondary compounds	—	M. Estiarte, J. Peñuelas (IRTA); L. Jahnke (UNH); S. Long (UE); M. Badiani, A. Paolacci (UT); D. Clark (ASU)
EVAPOTRANSPIRATION (ET)		
Eddy correlation		A. Fruman, M. Groen, H. Vugts (FUA); Y. Harazono (NIAS)
ET (energy balance and Bowen ratio)	B. Kimball (WCL); D. Fangmeier, M. Yitayew, A. Matthias (UA)	B. Kimball (WCL); A. Frumau, M. Groen, H. Vugts (FUA)
Plant transpiration (stem flow)	W. Dugas, M. Heuer (BRC)	R. Senock, J. Ham (ETL)
Soil evaporation	B. Kimball, M. Johnson (WCL)	B. Kimball, M. Johnson (WCL)
Soil water balance	D. Hunsaker (WCL)	D. Hunsaker (WCL)
MORPHOLOGY		
Organ development, phenology	J. Mauney (WCRL)	G. Wall (WCL); A. Li, A. Trent (UI); M. Bauer, J. Gräfe, P. Michaelis, E. Blum, F. Wechsung (PIK)
PHYSIOLOGY		
Canopy CO_2 flux (eddy diffusivity)		Y. Harazono (NIAS)
Canopy photosynthesis and resistance	D. Hileman, G. Huluka (TU); N. Bhattacharya (WCRL); R. Garcia (WCL)	R. Garcia (WCL)
Chlorophyll concentration	P. Pinter, S. Idso (WCL)	P. Pinter (WCL); F. Wechsung (PIK)
Head and awn photosynthesis	—	F. Wechsung (PIK); R. Garcia (WCL)

Intrinsic water use efficiency	H. Johnson (GSWR); E. Paul (MSU)	H. Johnson (GSWR); E. Paul (MSU)
Leaf fluorescence	J. Burke (PSWC); S. Long (BNL, UE)	S. Long (BNL, UE); A. Giuntoli (IATA)
Leaf photosynthesis, respiration, CO_2 response, and stomatal conductance	D. Hileman (TU); L. Evans (MC); R. Garcia, S. Idso, G. Wall (WCL); N. Bhattacharya, D. Hendrix, J. Radin (WCRL)	R. Garcia, G. Wall (WCL); C. Osborne, S. Long (BNL, UE); M. Badiani, A. Paolacci (UT)
Leaf water potential, osmotic potential	N. Bhattacharya, J. Radin (WCRL); B. Kimball (WCL)	G. Wall, B. Kimball (WCL)
PAR interception and conversion efficiencies	P. Pinter (WCL)	P. Pinter (WCL); A. Frumau (FUA)
Photosynthetic acclimatization, leaf proteins, rubisco activity	—	G. Nie, S. Long (BNL, UE); A. Webber (CEEP); D. Hendrix (WCRL); J. Vu, L. H. Allen (PSP)
ENTOMOLOGY		
Arthropod populations	D. Akey, J. Mauney (WCRL); B. Miller, D. Byrne (UA)	D. Akey (WCRL); P. Pinter (WCL)
COMPARISONS WITH OTHER SYSTEMS		
Open-top chambers	—	A. Frumau, P. Jak, N. de Jong, H. Bleeksmas, J. Rozema (FUA)
OTHER SPECIES		
FACE effects on barley growth, development, photosynthesis	—	M. Bauer, J. Grafe, F. Wechsung (PIK); G. Wall (WCL)
MICROMETEOROLOGY AND REMOTE SENSING		
Leaf, canopy, and soil spectral properties	P. Pinter (WCL)	P. Pinter, T. Clarke (WCL)
Leaf, canopy, and soil temperatures	P. Pinter, B. Kimball (WCL); D. Fangmeier (UA)	P. Pinter, B. Kimball, T. Clarke (WCL); A. Frumau, M. Groen, H. Vugts (FUA)
Solar, PAR, net radiation, soil heat flux, air and soil temperature profiles, vapor pressure, wind speed, precipitation	B. Kimball (WCL); P. Brown (UA)	B. Kimball (WCL); A. Frumau, M. Groen, H. Vugts, J. Worm (FUA); P. Brown (UA)
CULTURAL PRACTICES		
Cultivation, planting	J. Mauney (WCRL)	G. Wall, R. Garcia, P. Pinter, B. Kimball, R. LaMorte, B. Alexander (WCL)
Fertilization	J. Mauney (WCRL)	G. Wall, R. Garcia, R. LaMorte (WCL)
Irrigation timing and amounts	K. Lewin (BNL); J. Mauney (WCRL)	D. Hunsaker, R. LaMorte, P. Pinter (WCL)

Experimental parameters	Cotton (1989, 1990, 1991)	Wheat (1992–1993, 1993–1994)
FACE SYSTEM PERFORMANCE		
CO_2 concentrations	K. Lewin, J. Nagy, G. Hendrey (BNL)	R. LaMorte (WCL); K. Lewin, J. Nagy, G. Hendrey (BNL)
CO_2 flow dynamics	F. Lipfert, G. Alexander (BNL); L. H. Allen (PSP)	—
COMPUTER SIMULATIONS AND MODELING		
CO_2 concentrations and fluxes	—	D. Suarez (SL)
Energy balance	J. Amthor (WCL, WHRC, LLNL); G. Wall, B. Kimball (WCL)	R. Grant (UAL); J. Amthor (WCL, WHRC, LLNL); T. Kartschall, S. Grossman (PIK)
Plant growth	J. Amthor (WCL, WHRC, LLNL); G. Wall, B. Kimball (WCL); S. Maas (CRS)	J. Amthor (WCL, UCD, WHRC, LLNL); R. Grant (UAL); L. A. Hunt (UG); T. Kartschall, S. Grossman, F. Wechsung (PIK); K. Kobayashi (NIAS); J. Müller (AAQ); N. Nikolov, W. Massmann (RMF); T. Sinclair (APL); A. Trent (UI); G. Wall, B. Kimball (WCL)

[a] Research group acronyms: AAQ, Abtellung Agro-ökosystemforschung Quedlinburg, Quedlinburg, Germany; APL, USDA, ARS, Agronomy and Physiology Laboratory, Gainesville, FL; ASU, Arizona State University, Department of Botany, Tempe, AZ; BNL, Brookhaven National Laboratory, Upton, Long Island, NY; BRC, Blackland Research Center, Texas Agricultural Experiment Station, Temple, TX; CEEP, Center for Early Events in Photosynthesis, Arizona State University, Tempe, AZ; CRS, USDA, ARS, Cotton Research Station, Shafter, CA; CSU, Colorado State University, Natural Resource Ecology Laboratory, Boulder, CO; ETL, Evapotranspiration Laboratory, Kansas State University, Manhattan, KS; FUA, Free University of Amsterdam, Amsterdam, Netherlands; GSWR, USDA, ARS, Grassland Soil and Water Research Laboratory, Temple, TX; GSU, Georgia Southern University, Department of Biology, Statesboro, GA; HU, Humbolt University Berlin, Faculty of Agriculture and Horticulture, Berlin, Germany; IATA, Institute of Environmental Analysis and Remote Sensing for Agriculture, CNR, Firenze, Italy; IRTA, Institut de Recerca i Technologia Agroalimentaries, Barcelona, Spain; LLNL, Global Climate Research Division, Lawrence Livermore National Laboratory, Livermore, CA; MAC, University of Arizona, Maricopa Agricultural Center, Maricopa, AZ; MC, Manhattan College, Laboratory of Plant Morphogenesis, The Bronx, NY; MISS, Agronomy Department, Mississippi State University, Mississippi State, MS; MSU, Michigan State University, Department of Crop and Soil Sciences, East Lansing, MI; NIAS, National Institute of Agro-Environmental Sciences, Kannondai, Tsukuba, Japan; NSDL, USDA, ARS, National Soil Dynamics Laboratory, Auburn, AL; PIK, Potsdam Institute for Climate Impact Research, Potsdam Germany; PSP, USDA, ARS, Plant Stress and Protection, Gainesville, FL; PSWC, USDA, ARS, Plant Stress and Water Conservation Research, Lubbock, TX; RMF, Rocky Mountain Forest and Range Experiment Station, Ft. Collins, CO; RRC, USDA, ARS, Russell Research Center, Athens, GA; SL, USDA, ARS, U.S. Salinity Laboratory, Riverside, CA; SPNR, USDA, ARS, Soil and Plant Nutrition Research, Ft. Collins, CO; TRL, University of Arizona, Tree Ring Laboratory, Tucson, AZ; TU, Tuskegee University, Tuskegee, AL; UA, University of Arizona, Tucson, AZ; UAL, University of Alberta, Department of Soil Science, Edmonton, Alberta, Canada; UE, University of Essex, Department of Biology, Colchester, United Kingdom; UCD, University of California, Department of Agronomy and Range Science, Davis, CA; UG, University of Guelph, Crop Science Department, Guelph, Ontario, Canada; UI, University of Idaho, Plant, Soil and Entomological Sciences, Moscow, ID; UNH, University of New Hampshire, Department of Plant Biology, Durham, NH; UT, Universitá della Tuscia, Department of Biochemistry and Agrochemistry, Viterbo, Italy; WCL, USDA, ARS, U.S. Water Conservation Laboratory, Phoenix, AZ; WCRL, USDA,

Acknowledgments

This research was supported by the Agricultural Research Service, U. S. Department of Agriculture, including the U. S. Water Conservation Laboratory (Phoenix, AZ), the Grassland Soil and Water Research Laboratory (Temple, TX), and the Plant Stress and Protection Group (Gainesville, FL). It has also been partially supported by the Carbon Dioxide Research Program of the U. S. Department of Energy, Environmental Sciences Division. Contributions toward operational support were made by the Potsdam Institute for Climate Impact Research (Potsdam, Germany), the NASA Goddard Space Flight Center (Greenbelt, MD), the Natural Resource Ecology Laboratory at Colorado State University (Ft. Collins, CO), and the Department of Soil Science, University of Alberta (Edmonton, Alberta, Canada). We also acknowledge the helpful cooperation of Dr. Roy Rauschkolb and his staff at the University of Arizona, Maricopa Agricultural Center. The FACE apparatus was furnished by Brookhaven National Laboratory, and we are grateful to Mr. Keith Lewin, Dr. John Nagy, and Dr. George Hendrey for their engineering expertise. This work contributes to the Global Change Terrestrial Ecosystem (GCTE) Core Research Programme, which is part of the International Geosphere Biosphere Programme (IGBP). We appreciate the specialized advice provided by the many scientists and graduate students listed in the Appendix and the technical assistance of M. Baker, T. Brooks, O. Cole, M. Gerle, D. Johnson, C. O'Brien, L. Olivieri, J. Olivieri, R. Osterlind, R. Rokey, R. Seay, L. Smith, S. Smith, and K. West. Dr. Gary Richardson, ARS biometrician, provided valuable assistance in the statistical analysis of the data.

References

Allen, L. H., Jr. (1989). Plant responses to rising carbon dioxide and potential interactions with air pollutants. *J. Environ. Qual.* 19, 15–34.

Allen, L. H., Jr. (1992). Free-air CO_2 enrichment field experiments: An historical overview. *Crit. Rev. Plant Sci.* 11, 121–134.

Allen, L. H., Jr., Drake, B. G., Rogers, H. H., and Shinn, J. H. (1992). Field techniques for exposure of plants and ecosystems to elevated CO_2 and other trace gases. *Crit. Rev. Plant Sci.* 11, 85–119.

Amthor, J. S., and Koch, G. W. (1996). Biota growth factor β: Stimulation of terrestrial ecosystem net primary production by elevated atmospheric CO_2. *In* "Carbon Dioxide and Terrestrial Ecosystems" (G. W. Koch and H. A. Mooney, eds.). Academic Press, San Diego.

Amthor, J. S., and Loomis, R. S. (1996). Integrating knowledge of crop responses to elevated CO_2 and temperature with mechanistic simulation models. *In* "Carbon Dioxide and Terrestrial Ecosystems" (G. W. Koch and H. A. Mooney, eds.). Academic Press, San Diego.

Arp, W. J. (1991). Effects of source-sink relations on photosynthetic acclimation to elevated CO_2. *Plant, Cell Environ.* 14, 869–875.

Bauer, A., Smika, D., and Black, A. (1983). Correlation of five wheat growth stage scales used in the Great Plains. *Adv. Agric. Technol.* ISSN0193-3701, 1–17.

Bhattacharya, N. C., Radin, J. W., Kimball, B. A., Mauney, J. R., Hendrey, G. R., Nagy, J., Lewin, K., and Ponce, D. C. (1994). Leaf water relations of cotton in a free-air CO_2-enriched environment. *Agric. Forest Meteorol.* 70, 171–182.

Blum, H. (1993). The response of CO_2-related processes in grassland ecosystems in a three-year field CO_2-enrichment study. *In* "Design and Execution of Experiments on CO_2 Enrichment" (E. D. Schulze and H. A. Mooney, eds.), Pub. EUR 15110 EN, pp. 367–370. Commission of European Communities, Brussels.

Chameides, W. L., Kasibhatla, P. S., Yienger, J., Levy, H., II. (1994). Growth of continental-scale metro-agro-plexes, regional ozone pollution, and world food production. *Science* 264, 74–77.

Dijkstra, P., Nonhebel, S., Grashoff, C., Goudriaan, J., and van de Geijn, S. C. (1996). Response of growth and CO_2 uptake of spring wheat and faba bean to CO_2 concentration under semifield conditions. *In* "Carbon Dioxide and Terrestrial Ecosystems" (G. W. Koch and H. A. Mooney, eds.). Academic Press, San Diego.

Drake, B. G., and Leadley, P. W. (1991). Canopy photosynthesis of crops and native plant communities. *Plant, Cell Environ.* 14, 853–860.

Dugas, W. A., and Pinter, P. J., Jr. (1994). Introduction to the free-air carbon dioxide enrichment (FACE) cotton project. *Agric. Forest. Meteorol.* 70, 1–2.

Dugas, W. A., Heuer, M. L., Hunsaker, D., Kimball, B. A., Lewin, K. F., Nagy, J., and Johnson, M. (1994). Sap flow measurements of transpiration from cotton grown under ambient and enriched CO_2 concentrations. *Agric. Forest. Meteorol.* 70, 231–245.

Estiarte, M., Peñuelas, J., Kimball, B. A., Idso, S. B., LaMorte, R. L., Pinter, Jr., Wall, G. W., and Garcia, R. L. (1994). Elevated CO_2 effects on stomatal density of wheat and sour orange trees. *J. Exp. Bot.* 45, 1665–1668.

Garcia, R. L., Idso, S. B., Wall, G. W., and Kimball, B. A. (1994). Changes in net photosynthesis and growth of *Pinus eldarica* seedlings in response to atmospheric CO_2 enrichment. *Plant, Cell Environ.* 17, 971–978.

Garcia, R. L., Long, S. P., Wall, G. W., Osborne, C. P., Kimball, B. A., Nie, G.-Y., Pinter, P. J., Jr., and LaMorte, R. L. (1995). Photosynthesis and conductance of spring wheat leaves:response to free-air atmospheric CO_2 enrichment. *Plant, Cell Environ.* (submitted).

Grant, R. F., Kimball, B. A., Pinter, P. J., Jr., Wall, G. W., Garcia, R. L., LaMorte, R. L., and Hunsaker, D. J. (1995). CO_2 effects on crop energy balance: Testing *ecosys* with a free-air CO_2 enrichment (FACE) experiment. *Agron. J.* 87:446–457.

Hendrey, G. R. (1992). Introduction. *Crit. Rev. Plant Sci.* 11, 59–60.

Hendrey, G. R., and Kimball, B. A. (1994). The FACE Program. *Agric. Forest Meteorol.* 70, 3–14.

Hendrey, G. R., Lewin, K. F., and Nagy, J. (1993a). Free air carbon dioxide enrichment: development, progress, results. *Vegetatio* 104/105, :17–31.

Hendrey, G., Lewin, K., and Nagy, J. (1993b). Control of carbon dioxide in unconfined field plots. *In* "Design and Execution of Experiments on CO_2 Enrichment" (E. D. Schulze and H. A. Mooney, eds.), Pub. EUR 15110 EN, pp. 309–327. Commission of European Communities, Brussels.

Hendrix, D. L. (1992). Influence of elevated CO_2 on leaf starch of field-grown cotton. *Crit. Rev. Plant Sci.* 11, 223–226.

Hileman, D. R., Bhattacharya, N. C., Ghosh, P. P., Biswas, P. K., Allen, L. L., Jr., Lewin, K. F., and Hendrey, G. R. (1992). Distribution of carbon dioxide within and above a cotton canopy growing in a FACE system. *Crit. Rev. Plant Sci.* 11, 187–194.

Hileman, D. R., Huluka, G., Kenjige, P. K., Sinha, N., Bhattacharya, N. C., Biswas, P. K., Lewin, K. F., Nagy, J., and Hendrey, G. R. (1994). Canopy photosynthesis and transpiration of field-grown cotton exposed to free-air CO_2 enrichment (FACE) and differential irrigation. *Agric. Forest. Meteorol.* 70, 189–207.

Hunsaker, D. J., Hendrey, G. R., Kimball, B. A., Lewin, K. F., Mauney, J. R., and Nagy, J. (1994a). Cotton evapotranspiration under field conditions with CO_2 enrichment and variable soil moisture regimes. *Agric. Forest Meteorol.* 70, 247–258.

Hunsaker, D. J., Kimball, B. A., Pinter, P. J., Jr., LaMorte, R. L., and Wall, G. W. (1994b). Wheat evapotranspiration under free-air CO_2 enrichment and variable soil moisture. *1994 Am. Soc. Agron. Abstr.* 26.

Idso, K. E., and Idso, S. B. (1994). Plant responses to atmospheric CO_2 enrichment in the face of environmental constraints: a review of the past 10 years' research. *Agric. Forest Meteorol.* 69, 153–203.

Idso, S. B., Kimball, B. A., Anderson, M. G., and Mauney, J. R. (1987). Effects of atmospheric CO_2 enrichment on plant growth: The interactive role of air temperature. *Agric. Ecosystem Environ.* 20, 1–10.

Idso, S. B., Kimball, B. A., Wall, G. W., Garcia, R. L., LaMorte, R., Pinter, P. J., Jr., Mauney, J. R., Hendrey, G. R., Lewin, K., and Nagy, J. (1994). Effects of free-air CO_2 enrichment on the light response curve of net photosynthesis in cotton leaves. *Agric. Forest Meteorol.* 70, 183–188.

Kimball, B. A. (1983). Carbon dioxide and agricultural yield: An assemblage and analysis of 430 prior observations. *Agron. J.* 75, 779–788.

Kimball, B. A. (1992). Cost comparisons among free-air CO_2 enrichment, open-top chamber, and sunlit controlled-environment chamber methods of CO_2 exposure. *Crit. Rev. Plant Sci.* 11, 265–270.

Kimball, B. A., and Mauney, J. R. (1993). Response of cotton to varying CO_2, irrigation, and nitrogen: Yield and growth. *Agron. J.* 85, 706–712.

Kimball, B. A., Pinter, P. J., Jr., and Mauney, J. R. (1992). Cotton leaf and boll temperatures in the 1989 FACE Experiment. *Crit. Rev. Plant Sci.* 11, 233–240.

Kimball, B. A., LaMorte, R. L., Seay, R. S., Pinter, P. J., Jr., Rokey, R. R., Hunsaker, D. J., Dugas, W. A., Heuer, M. L., Mauney, J. R., Hendrey, G. R., Lewin, K. F., and Nagy, J. (1994a). Effects of free-air CO_2 enrichment on energy balance and evapotranspiration of cotton. *Agric. Forest Meteorol.* 70, 259–278.

Kimball, B. A., LaMorte, R. L., Seay, R., O'Brien, C., Pinter, P. J., Jr., Wall, G. W., Garcia, R. L., Hunsaker, D. J., and Rokey, R. (1994b). Effects of free-air CO_2 enrichment (FACE) on the energy balance and evapotranspiration of wheat. *USDA, ARS, U.S. Water Conservation Lab. Annu. Res. Rep.* 69–72.

Kimball, B. A., Pinter, P. J., Jr., Wall, G. W., Garcia, R. L., LaMorte, R. L., Jak, P., Frumau, K. F. A., and Vugts, H. F. (1995). Comparisons of responses of vegetation to elevated CO_2 in free-air and open-top chamber facilities. *Adv. CO_2 Effects Res., ASA Special Publication* (in press).

Lawlor, D. W., and Mitchell, A. C., (1991). The effects of increasing CO_2 on crop photosynthesis and productivity: a review of field studies. *Plant, Cell Environ.* 14, 807–818.

Leadley, P. W., and Drake, B. G. (1993). Open top chambers for exposing plant canopies to elevated CO_2 concentration and for measuring net gas exchange. *Vegetatio* 104/105, 3–15.

Leavitt, S. W., Paul, E. A., Kimball, B. A., Hendrey, G. R., Mauney, J. R., Rauschkolb, R., Rogers, H., Lewin, K. F., Nagy, J., Pinter, P. J., Jr., and Johnson, H. B. (1994). Carbon isotope dynamics of free-air CO_2-enriched cotton and soils. *Agric. Forest Meteorol.* 70, 87–101.

Lewin, K. F., Hendrey, G. R., Nagy, J., and LaMorte, R. (1994). Design and application of a free-air carbon dioxide enrichment facility. *Agric. Forest Meteorol.* 70, 15–29.

Lipfert, F., Alexander, Y., Lewin, K., Hendrey, G., and Nagy, J. (1991). A turbulence-driven air fumigation facility for studying air pollution effects on vegetation. *In* "Proceedings of the 7th Symposium of Meteorological Observations and Instrumentation and Special Sessions on Laser Atmospheric Studies," January 14–18, 1991. Amer. Metorol. Soc., Boston, MA.

Lipfert, F., Alexander, Y., Hendrey, G., Lewin, K. F., and Nagy, J. (1992). Performance analysis of the BNL FACE gas injection system. *Crit. Rev. Plant Sci.* 11, 143–163.

Long, S. P. (1991). Modification of the response of photosynthetic productivity to rising temperature by atmospheric CO_2 concentrations: Has its importance been underestimated. *Plant, Cell Environ.* 14, 729–739.

Mauney, J. R., Fry, K. E., and Guinn, G. (1978). Relationship of photosynthetic rate to growth and fruiting of cotton, soybean, sorghum and sunflower. *Crop Sci.* 18, 259–263.

Mauney, J. R., Lewin, K., Hendrey, G. W., and Kimball, B. A., (1992). Growth and yield of cotton exposed to free-air CO_2 enrichment (FACE). *Crit. Rev. Plant Sci.* 11,213–222.

Mauney, J. R., Kimball, B. A., Pinter, P. J., Jr., LaMorte, R. L., Lewin, K. F., Nagy, J., and Hendrey, G. R. (1994). Growth and yield of cotton in response to a free-air carbon dioxide enrichment (FACE) environment. *Agric. Forest Meteorol.* 70, 49–67.

McLeod, A. R. (1993). Open-air exposure systems for air pollutant studies—their potential and limitations. *In* "Design and Execution of Experiments on CO_2 Enrichment" (E. D. Schulze and H. A. Mooney, eds.), Pub. EUR 15110 EN, pp. 353–365. Commission of European Communities, Brussels.

McLeod, A. R., Fackrell, J. E., and Alexander, K. (1985). Open-air fumigation of field crops: Criteria and design for a new system. *Atmos. Environ.* 19, 1639–1649.

Nagy, J., Lewin, K. F., Hendrey, G. R., Hassinger, E., and LaMorte, R. (1994). FACE facility CO_2 concentration control and CO_2 use in 1990 and 1991. *Agric. Forest Meteorol.* 70, 31–48.

Nakayama, F. S. (1993). Soil gas fluxes in CO_2-enriched wheat. *USDA, ARS, U.S. Water Conservation Lab. Annu. Res. Rep.*, 68–69.

Nakayma, F. S., Huluka, G., Kimball, B. A., Lewin, K. F., Nagy, J., and Hendrey, G. R. (1994). Soil carbon dioxide fluxes in natural and CO_2-enriched systems. *Agric. Forest Meteorol.* 70, 131–140.

Nie, G.-Y., Long, S. P., Kimball, B. A., Pinter, P. J., Jr., Wall, G. W., Garcia, R. L., LaMorte, R. L., and Webber, A. N. Free-air CO_2 enrichment effects on the development of the photosynthetic apparatus in wheat as indicated by changes in leaf proteins. *Plant, Cell Environ.* (in press).

Osborne, C. P., Long, S. P., Garcia, R. L., Wall, G. W., Kimball, B. A., Pinter, P. J., Jr., LaMorte, R. L., and Hendrey, G. R. (1995). Do shade and elevated CO_2 have an interactive effect on photosynthesis? An analysis using wheat grown under free-air CO_2 enrichment. *Plant, Cell Environ.* (submitted).

Pearcy, R. W., and Björkman, O. (1983). Physiological effects. *In* "CO_2 and Plants: The Response of Plants to Rising Levels of Atmospheric Carbon Dioxide" (E. R. Lemon, ed.), pp. 65–105. Westview, Boulder, CO.

Pinter, P. J., Jr., Kimball, B. A., Mauney, J. R., Hendrey, G. R., Lewin, K. F., and Nagy, J. (1994). Effects of free-air carbon dioxide enrichment on PAR absorption and conversion efficiency by cotton. *Agric. Forest Meteorol.* 70, 209–230.

Poorter, H. (1993). Interspecific variation in the growth response of plants to an elevated ambient CO_2 concentration. *Vegetatio* 104/105, 77–97.

Prior, S. A., Rogers, H. H., Runion, G. B., and Mauney, J. R. (1994). Effects of free-air CO_2 enrichment on cotton root growth. *Agric. Forest Meteorol.* 70, 69–86.

Rogers, H. H., Prior, S. A., and O'Neil, E. G. (1992). Cotton root and rhizosphere responses to free-air CO_2 enrichment. *Crit. Rev. Plant Sci.* 11, 251–263.

Rosenzweig, C., and Parry, M. L. (1994). Potential impact of climate change on world food supply. *Nature* 367, 133–138.

Runion, G. B., Curl, E. A., Rogers, H. H., Backman, P. A., Rodríguez-Kábana, R., and Helms, B. E. (1994). Effects of free-air CO_2 enrichment on microbial populations in the rhizosphere and phyllosphere of cotton. *Agric. Forest Meteorol.* 70, 117–130.

Schulze, E. D., and Mooney, H. A. (1993). Comparative view on design and execution of experiments at elevated CO_2. *In* "Design and Execution of Experiments on CO_2 Enrichment" (E. D. Schulze and H. A. Mooney, eds.), Pub. EUR 15110 EN, pp. 407–413. Commission of European Communities, Brussels.

Senock, R. S., Ham, J. M., Loughin, T. M., Kimball, B. A., Hunsaker, D. J., Pinter, P. J., Jr., Wall, G. W., Garcia, R. L., and LaMorte, R. L. Free-air CO_2 enrichment (FACE) of wheat: Assessment with sap flow measurements. *Plant, Cell Environ.* (in press).

Stitt, M. (1991). Rising CO_2 levels and their potential significance for carbon flow in photosynthetic cells. *Plant, Cell Environ.* 14, 741–762.

Wall, G. W., and Kimball, B. A. (1993). Biological databases derived from free air carbon dioxide enrichment experiments. *In* "Design and Execution of Experiments on CO_2 Enrichment" (E. D. Schulze and H. A. Mooney, eds.), Pub. EUR 15110 EN, pp. 329–351. Commission of European Communities, Brussels.

Wall, G. W., Amthor, J. S., and Kimball, B. A. (1994). COTCO$_2$: A cotton growth simulation model for global change. *Agric. Forest Meteorol.* 70, 289–342.

Wall, G. W., Kimball, B. A., Hunsaker, D. J., Garcia, R. L., Pinter, P. J., Jr., Idso, S. B., and LaMorte, R. L. (1995). Water potential of leaves of spring wheat grown in a free-air CO$_2$ enriched (FACE) atmosphere under variable soil moisture regimes. *Plant, Cell Environ.* (submitted).

Wardlaw, I. F., Sofield, I., and Cartwright, P. M. (1980). Factors limiting the rate of dry matter accumulation in the grain of wheat grown at high temperatures. *Aust. J. Plant Physiol.* 7, 387–400.

Wiegand, C. L., and Cuellar, J. A. (1981). Duration of grain filling and kernel weight of wheat as affected by temperature. *Crop Sci.* 21, 95–101.

Wood, C. W., Torbert, H. A., Rogers, H. H., Runion, G. B., and Prior, S. A. (1994). Free-air CO$_2$-enrichment effects on soil carbon budget and nitrogen. *Agric. Forest Meteorol.* 70, 103–116.

14

Response of Growth and CO_2 Uptake of Spring Wheat and Faba Bean to CO_2 Concentration under Semifield Conditions: Comparing Results of Field Experiments and Simulations

Paul Dijkstra, Sanderine Nonhebel, Cees Grashoff, Jan Goudriaan, and Siebe C. van de Geijn

I. Introduction

For an analysis of changes in global carbon balance, three aspects of the effects of CO_2 concentration and climate change on the terrestrial ecosystem require study: changes in the area covered by vegetation types (land-use and temperature-induced shifts in ecosystems), changes in C balance of a particular vegetation type and soil, and changes in individual components of the vegetation (species composition) for as much as they affect the vegetation C balance.

Basic responses of plant growth to CO_2 concentration and temperature are quite well understood at the process level for photosynthesis and respiration, although they are less clear for carbohydrate partitioning. However, scaling up from a single plant in laboratory studies to vegetation in the field is more difficult. Field conditions are variable in time and space, interactions with other environmental factors occur, species can differ in their responses to CO_2 concentration, and strong year-to-year variations in weather are found. This makes field studies at the vegetation level essential [Grulke *et al.*, 1990; Drake and Leadley, 1991; Lawlor and Mitchell, 1991;

Chapters 1–5 and 8–12 in this volume (Koch and Mooney, 1996)], but also difficult to interpret. A less complicated ecosystem is the agricultural crop ecosystem: often dense, monospecific, genetically and phenotypically uniform canopies, and grown under well-defined and easy-to-manipulate conditions. Moreover, crop growth models can be used to assess the interaction of weather conditions with CO_2 concentration and climate changes. Furthermore, these models can help to distinguish promising hypotheses for new experimental work, thus contributing to an efficient use of resources.

In this paper, we present the results from a semifield study of the canopy C balance of two crop species at two CO_2 concentrations. In addition, we compare these results with results from simulation studies for both species. The species under study are a determinate grain crop (spring wheat, *Triticum aestivum* L.) and an indeterminate N-fixing legume (faba bean, *Vicia faba* L.).

II. Materials and Methods

A. Experimental Setup

Crops were grown at two CO_2 concentrations, ambient (350 μmol mol^{-1}) and elevated (750 μmol mol^{-1}), under normal agricultural practices from sowing until harvest in two air-conditioned sunlit crop enclosures. The crop enclosures were surrounded by border plants. The enclosures were part of the Wageningen Rhizolab facility, a joint research facility of the AB-DLO and the Wageningen Agricultural University [described in detail by Van de Geijn *et al.*, 1994; Smit *et al.*, 1994].

Spring wheat cv. Minaret was used in 1991; it was sown on April 4, emerged on April 12, and harvested on August 26, 1991. An inbred line of faba bean cv. Minica was used in 1992. This inbred line was selected for high pod retention without insect pollination. Faba bean was sown on April 3, emerged on April 25, and harvested on September 7, 1992. Plants received adequate amounts of water and nutrients with the exception of faba bean plants, which did not receive nitrogen fertilizer, but were inoculated with a compatible rhizobium strain.

The crop enclosures (1.25 × 1.25 m) were made of 5-mm-thick polycarbonate and could be adjusted in height to accommodate plants up to harvest. The system was essentially an open system. Exchange rates of air in each aboveground compartment were typically 50 m^3 hr^{-1} with a recirculation rate of about 800 m^3 hr^{-1}. Temperature was controlled by separate air conditioners for each compartment. Air temperature in the enclosures followed outside temperature, as registered by the weather station linked to the Wageningen Rhizolab (within 0.5°C). CO_2 concentrations of the ingoing and chamber air and airflow rates were monitored every 10 and 20 min, respectively. By using these measurements, CO_2 concentration of the chambers was adjusted by computer-controlled valves, and carbon dioxide exchange was calculated. Part of the air was forced into the soil by a slight overpressure and recovered in an air drain placed in the soil at a 15-cm depth. This prevented the air of the soil compartment from entering the canopy compartment and, at the same time, allowed measurement of the soil–root CO_2 production. The soil–root respiration option was in operation for the 1992 season, while for 1991 only a limited dataset was available. Light intensity was measured in the weather station.

The complete dataset was converted into hourly means. In this paper, data were used from day 188 in 1991 and 1992, selected for their high light intensities in the middle of the growing season. Furthermore, CCER (canopy CO_2 exchange rate) values at maximum light intensity were selected for each day. The daily value of CDR (canopy dark respiration rate) was averaged over the period from 24.00 to 04.00 hr, while SRR (soil root respiration rates) was averaged over the whole 24-hr period. In 1991, root proliferation occurred at a 70-cm depth in association with the presence of root-knot nematodes (*Meloidogyne naasi*) at this depth. This probably did not affect total biomass production.

B. Model Description

In the simulations, growth and production of spring wheat and faba bean were calculated for the present CO_2 concentration of 350 μmol mol^{-1} and for a concentration of 700 μmol mol^{-1}. The models simulated potential production for a crop optimally supplied with water and nutrients and free from pests, diseases, and weeds.

Gross photosynthesis of a crop was calculated on the basis of the photosynthesis–light response curve for individual leaves, characterized by the initial light-use efficiency (EFF) and the maximum rate of leaf gross photosynthesis (AMAX). Both values were affected by the CO_2 concentration. At an average temperature of 20°C, doubling of the CO_2 concentration increased EFF by 15% and AMAX by 100% (Goudriaan and Unsworth, 1990). Integration from leaf photosynthesis to canopy photosynthesis was done by three-point Gaussian integration, based on the leaf photosynthesis–light response curves. The response of the lower leaves to CO_2 concentration was less than that for the upper fully exposed leaves in the canopy. When, for C_3 plants, a doubling of the CO_2 concentration increased AMAX by 100%, the *maximum* rate of crop photosynthesis was raised by 45–65% (due to light extinction) and the *total* daily crop photosynthesis was raised by only 40% (due to the diurnal course of the sun; Goudriaan and Unsworth, 1990).

The model calculated gross photosynthesis for each day as described earlier, using actual values of radiation and temperature and the calculated value of the leaf area index (LAI) of the previous day. Part of the photosynthates produced was used for maintenance respiration of the standing crop (Spitters et al., 1989; Penning de Vries et al., 1989). The remainder was allocated to the various organs, a process controlled by the developmental stage of the crop (van Heemst, 1986). Rate of development was controlled by air temperature. In the organs (roots, stems, seeds, and leaves), the photosynthate was converted into structural dry matter requiring growth respiration, which depended on the chemical composition of the organ (Vertregt and Penning de Vries, 1987). The LAI was calculated from the dry matter growth of leaves and their specific area and was used to calculate the light interception of the next day, thus closing the simulation loop.

This part of the models for spring wheat and faba bean was derived from a general crop growth simulator (Spitters et al., 1989; Penning de Vries et al., 1989). The specific parameters and functions of spring wheat were based mainly on the spring wheat model of van Keulen and Seligman (1987). Characteristics of the faba bean model were derived from experiments with faba bean in The Netherlands (Grashoff, 1990a,b).

Input for the simulation models was daily total solar radiation and maximum and minimum temperature data (Meteostation Haarweg of the Wageningen Agricultural University) during the period 1975–1988. Emergence date for the first runs was fixed at day 90 for spring wheat and day 120 for faba bean. Additional input values, used at the start of the simulation, were values for crop dry matter and leaf area index at emergence.

C. Model Validation

Model validation was done for the ambient CO_2 concentration. The reliability of the model for spring wheat was validated with data from experimental farms in Emmercompascuum and Wieringermeer in The Netherlands (Nonhebel, 1993). The results of the faba bean model proved to be reliable in a comparison with the measured yields of field experiments over 14 years at Wageningen, The Netherlands, and over 2 years at nine locations of the so-called EC-Joint Faba Bean Trials in Denmark, France, Germany, The Netherlands, United Kingdom, and Austria (Grashoff and Stokkers, 1992).

III. Results

A. Field Experiment

For both species, CO_2 concentration significantly ($P < 0.01$) increased the total aboveground biomass (Table I). This increase was larger for faba bean in 1992 than for spring wheat in 1991 (interaction significant $P < 0.01$). Harvestable yield was increased by a similar proportion (34%

Table I Total Aboveground Biomass at the Harvest of Spring Wheat (1991) and Faba Bean (1992) from the Experiment[a]

	Spring wheat (g m^{-2})	Faba bean (g m^{-2})
Ambient	1280 (52)	1302 (68)
Elevated	1705 (106)	2058 (141)
Elevated/ambient	1.33	1.58

[a] SE (in parentheses) was based on within chamber variability ($n = 5$ for spring wheat and $n = 10$ for faba bean).

for spring wheat and 51% for faba bean; Dijkstra *et al.*, 1993). The yield increase for spring wheat was explained mainly by a higher number of ear-bearing tillers and number of seeds per ear. For faba bean, the number of stems and the number of pods per stem contributed equally to the increased yield under elevated CO_2 concentration.

On day 188 of 1991 and 1992 (July 7 and 6, respectively), CCER during the day and CDR during the night were increased by the elevated CO_2 concentration (Fig. 1, Table II). On this clear day, with light intensities over 400 W m^{-2} PAR, CCER at the highest light intensity was increased

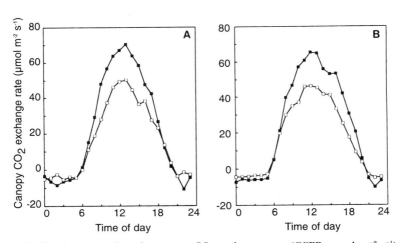

Figure 1 Hourly mean values for canopy CO_2 exchange rate (CCER, μmol m^{-2} s^{-1}) for spring wheat (A) and faba bean (B) as affected by CO_2 concentration (ambient CO_2, open symbols; elevated CO_2, closed symbols).

Table II Effect of CO_2 Concentration on Measured Canopy CO_2 Exchange Rate at Maximum Light Intensity (Hourly Values, $CCER_{max}$) and Canopy Dark Respiration (CDR) for Spring Wheat and Faba Bean on Day 188 of 1991 and 1992, Respectively[a]

	Spring wheat			Faba bean		
	350	750	750/350	350	750	750/350
$CCER_{max}$	50.22	70.38	1.40	45.57	66.85	1.47
CDR	−3.80	−6.03	1.59	−3.85	−6.51	1.69

[a] Values are given in μmol m^{-2} s^{-1}.

slightly more for faba bean than for spring wheat (Table II). This was also found for other parameters of the carbon balance, such as CDR and net fixation over 24 hr. The SRR of faba bean was increased by 50% with elevated CO_2 (from 4.18 to 6.28 μmol CO_2 m^{-2} ground area s^{-1}). Data for spring wheat were not available for this day, but comparable days had similar rates. Of the total carbon fixed during the light, about 8% was lost during the dark due to aboveground respiration, while 23% was lost due to soil–root respiration over 24 hr. This fraction was not affected much by CO_2 concentration; thus, approximately 68% of the carbon fixed over 24 hr was retained. Of course, this fraction will decrease with radiation.

Elevated CO_2 increased $CCER_{max}$ throughout the entire season (Fig. 2) for both crops. The difference between species became evident when $CCER_{max}$ at elevated CO_2 concentration was plotted against $CCER_{max}$ at ambient concentration (Fig. 3). The regression line, forced through the origin, had a slope of 1.45 (± 0.02, $R = 0.992$) for spring wheat and 1.69 (± 0.02, $R = 0.989$) for faba bean. For spring wheat there was an increased responsiveness during the last 2 weeks of growth (not shown). This indicated a direct or an indirect effect of CO_2 concentration on senescence.

We conclude that the responses of yield, biomass, and gas exchange characteristics to CO_2 concentration differed between the two species. Whether this was caused by a physiological difference between the species or by differences in weather conditions between the growing seasons could not be derived from the experimental data.

B. Modeling

One way to analyze the effects of different weather conditions on crop growth is to use crop growth models. By using past weather data from the period 1975–1988, Grashoff et al. (1993) and Nonhebel (1993) simulated the yield and total biomass responses for faba bean and spring wheat to changes in CO_2 concentration and temperature. Over this period, the

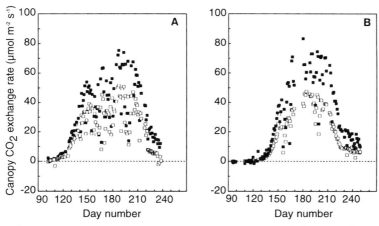

Figure 2 Effect of ambient and elevated CO_2 concentrations on daily values of canopy CO_2 exchange rate ($CCER_{max}$, μmol m^{-2} s^{-1}) at maximum light intensity (hourly values) on spring wheat (A) and faba bean (B) (ambient CO_2, open symbols; elevated CO_2, closed symbols).

simulated mean response to CO_2 doubling for spring wheat was lower than that for faba bean (Table III, boldface type). These results corresponded to those found in the field experiments (Table I). However, in the simulation runs, different emergence dates were used for both crops, while in the field experiments a similar sowing date was applied. Therefore, we studied the effect of sowing and emergence dates on the response of crops to CO_2 elevation.

In The Netherlands, emergence generally takes place at about day 90 for spring wheat and about day 120 for faba bean. These dates were used for the standard crops in the simulation models. To investigate the effect of emergence date on the response to CO_2 concentration for both crops, new simulation runs were done with different emergence dates for both crops. The response ratio to CO_2 doubling increased with a later emergence date, although biomass productivity decreased (Table III). The effect was smaller for faba bean than for spring wheat. This may have been due to the temperature dependence of spring growth in the wheat model. This was absent from the faba bean model.

However, the effect of sowing date could not explain results from the field experiments. Although sowing and emergence dates in the field experiment were almost the same for both crops, the weather conditions during the growing periods differed substantially. Indeed, the mean daily temperatures for May and June 1992 were 6.2 and 4.2°C higher, respectively, than corresponding months in 1991.

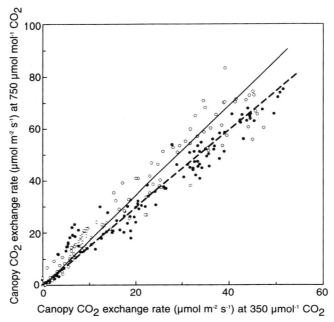

Figure 3 Relation between $CCER_{max}$ values (μmol m^{-2} s^{-1}) of plants grown at ambient and elevated CO_2 concentrations (ambient CO_2, open symbols; elevated CO_2, closed symbols).

These simulation results suggest that a difference between faba bean and spring wheat in the response to CO_2 will be found under normal farming conditions due to a difference in emergence dates.

Table III Simulated Average Annual Potential Aboveground Biomass Production for Spring Wheat and Faba Bean for the Period 1975–1988 Grown in 350 μmol mol^{-1} CO_2 and the Ratio between Production at Elevated CO_2 (700 μmol mol^{-1}) and at 350 μmol mol^{-1} CO_2 for Two Sowing Dates[a]

	Spring wheat		Faba bean	
Emergence (day number)	Biomass (g m^{-2})	Ratio	Biomass (g m^{-2})	Ratio
90 (normal for spring wheat)	**1850**	**1.41**	1425	1.46
	(37.4)	**(0.01)**	(12.9)	(0.01)
120 (normal for faba bean)	1420	1.52	**1285**	**1.52**
	(41.3)	(0.02)	**(14.7)**	**(0.01)**

[a] Boldface numbers indicate the results of runs with standard sowing dates. SEs are in parentheses.

IV. Discussion

A. Differences between Species

Differences between species in the response to CO_2 concentration are often reported. However, the variability within species (between experiments) is often more prominent than the differences between species (Poorter, 1993). This indicates an interaction with experimental conditions and procedures. The increase in total biomass of spring wheat in response to CO_2 elevation in this experiment (33%) was in the range of the results summarized by Rogers and Dahlman (1993), while the response of faba bean (58%) was somewhat higher.

Differences between species in the response of CCER to CO_2 doubling have also been reported (Baker and Allen, 1993). Variations in CCER or CDR can be caused by differences in LAI and/or in the specific rates. For example, soybean CCER was stimulated by 89% (26°C) or 78% (36°C) in response to CO_2 doubling (Campbell *et al.*, 1990), but only by 44% for rice [Baker and Allen, 1993; Baker *et al.*, (this volume)]. This difference in response was attributed to a difference in the canopy structure, which was not completely closed for soybean. The difference in the response of $CCER_{max}$ to CO_2 doubling observed in our studies was attained when canopies were fully closed for both species. Acclimation at the canopy level was not observed for spring wheat (P. Dijkstra, K. Groenwold, and S.C. van de Geijn, unpublished).

B. Environmental Effects: Temperature and Light

For single leaves, and at low values of leaf area index, a large CO_2–light interaction is expected. However, for a closed canopy of spring wheat, the effect of light intensity on the relative response of canopy $CCER_{max}$ was small and nonsignificant, which was corroborated by simulation studies (Dijkstra *et al.*, 1994).

Relative responsiveness to CO_2 enrichment increased with increasing temperatures for *Daucus carota, Raphanus sativa,* and *Gossypium hirsutum* (Idso *et al.*, 1987; Idso and Kimball, 1989). This was related to the effect of temperature on photorespiration (Long, 1991).

The relative response of canopy photosynthesis to CO_2 concentration seemed more dependent on temperature than on light intensity. The important role of environmental factors was confirmed by the following outcomes of the simulation studies: (1) relative response increased with later sowing and emergence dates and, for the Dutch climate, higher temperatures (Table III); (2) the response to CO_2 doubling was higher in the warm year of 1992 than in 1991, in both this field experiment and the simulations (Grashoff and Nonhebel, 1993); and (3) the simulated response to CO_2

concentration was more dependent on temperature than on light intensity for explaining the lower response of spring wheat in 1991 (Grashoff and Nonhebel, 1993).

C. Modeling the Response to CO_2

The models were validated under field conditions for ambient CO_2 concentrations (Grashoff, 1990a,b; Nonhebel, 1993). However, the response of photosynthesis to CO_2 concentration was based on experiments in climate chambers and greenhouses. Doubling of AMAX following a doubled CO_2 concentration, such as that used in the model, is not always found in the literature. Cure and Acock (1986) reported an average stimulation of 52% for a doubling of CO_2 concentration, while Kimball et al. (1993) reported a 66% increase for typical C_3 species. On the other hand, Rozema (1993) found a doubled AMAX for a doubled CO_2 concentration. A similar stimulation was found for faba bean in a greenhouse study (P. Dijkstra, C. S. Pot, and A. H. C. M. Schapendonk, unpublished) and soybean (Clough et al., 1981). The small increments recorded in some pot experiments may be due to resource limitations, for instance, in nitrogen availability (Wong, 1979), which may reduce the effect of CO_2 on AMAX.

It is argued that the short-term response to CO_2 concentration may strongly deviate from the long-term response (Kimball, 1983). Negative feedback mechanisms, such as thicker leaves, resulting in later canopy closure, or leaves with decreased photosynthetic rates after prolonged exposure to high CO_2 (Wong, 1979; Rowland-Bamford et al., 1991), may reduce the positive effect. It is not sufficiently known how important long-term adaptation processes are at the canopy level. Our simulation models did not include such negative feedbacks. The response of canopy photosynthesis to CO_2 concentration was approximately constant in time for spring wheat and faba bean (Dijkstra et al., 1993, 1994), although temporary acclimation of leaf photosynthesis was found (A. Visser, personal communication). This suggested that negative long-term effects were not important or that they were compensated for or hidden at the canopy level. It may not be necessary to include these negative feedbacks in further assessments of the effect of CO_2 increases, at least not for cereals and leguminous crops under optimum nutrient supply.

The outcomes of the field experiments and those of the models showed good agreement: in both cases we concluded that faba bean had a greater response to CO_2 concentration than spring wheat. Results from 1993 indicated that the response of winter wheat to CO_2 elevation was even lower than that of spring wheat (P. Dijkstra et al., unpublished data). This does not automatically lead to the conclusion that the physiological mechanisms operating in the field are as simple as described in the model. Further in-depth validation with field experiments is required, and in fact, the Rhizolab database provides a good opportunity to do so.

D. Species Differences or Temperature Effects

In this paper, we reported the response of biomass productivity and carbon balance to CO_2 concentration for faba bean and spring wheat under semifield conditions. Total aboveground biomass and carbon fixation were increased by the higher CO_2 concentration for both crops. We showed a clear and significant difference in the responses of crop production and canopy photosynthesis to CO_2 enrichment between the two species. We also showed, with simulation models, that the expected response to CO_2 doubling differs between the two species. Emergence date influenced the relative response to CO_2 concentration.

We conclude that the difference in the responses to CO_2 concentration between the two species can, at least partly, be explained by a difference in timing of the growing season, which resulted in exposure to different weather conditions. The lower response of spring wheat in the field experiment was caused by lower temperatures during 1991.

In natural and agricultural ecosystems, phenology is a very important characteristic: at what time does most of the growth occur? At what time is the canopy closed? Phenology determines whether a plant has sufficient water for growth (and yield) and is very important for the outcome of competition between different plant species. It is quite clear from published lab experiments and from the insight in physiological mechanisms of photosynthesis that, under lower temperatures, the response to CO_2 concentration declines (Idso and Kimball, 1989), even to the extent that in early spring the growth response of winter wheat to CO_2 doubling is absent (P. Dijkstra and M. J. H. Jansen, unpublished results). We believe that phenology, and its coincidence with weather conditions, is of crucial importance to the understanding of the response of terrestrial ecosystems to elevated CO_2, perhaps even more so than the often-studied long-term effects of CO_2 elevation on the physiology of the plant.

C_3 species growing in the early and cooler part of a season are expected to exhibit a lower response to CO_2 enrichment than later species. In that way, this influences the competitive strength of the species and, thus, may influence the species composition of a vegetation stand and the carbon balance of canopies.

V. Summary and Conclusions

Spring wheat (1991) and faba bean (1992) were grown under semifield conditions in naturally lit controlled enclosures, with temperatures tracking outside conditions under ambient and elevated CO_2 concentrations. Measurements of carbon balance were made continuously throughout the growing season from sowing until harvest.

(1) Under elevated CO_2, aboveground biomass and canopy photosynthesis increased more for faba bean than for spring wheat. The relative increase in response to a doubling of the CO_2 concentration was greater for the dark respiration rate than for canopy photosynthesis for both crops. The soil–root respiration rate of faba bean increased by about the same factor as the canopy dark respiration rate.

(2) Models, developed independently of the presented experimental data, were used to simulate biomass production for both crops. Simulation of the response to CO_2 concentration with weather data from the period 1975–1988 indicated that normally a larger response of biomass production is expected for faba bean than for spring wheat. This result was comparable to the results from the field experiments. Further modeling indicated that the response to elevated CO_2 increased with later emergence and higher temperatures.

(3) The field experiment with spring wheat was conducted in 1991, while that with faba bean was conducted in 1992. The greater response to CO_2 of faba bean was at least partly caused by the higher temperatures in 1992.

These results suggest that, under farm conditions, a difference between faba bean and spring wheat in the response of aboveground biomass to CO_2 concentration is to be expected. This difference can be at least partly explained simply by the later emergence date, resulting in more favorable weather conditions for faba bean than for spring wheat.

Acknowledgments

We thank Dr. M. van Oijen for his theoretical and technical contributions to the simulation studies and Dr. I. A. J. Haverkort for helpful criticism on the manuscript. We thank Peter van de Glint, Ko Groenwold, Jan van Kleef, and Geurt Versteeg for their essential help in the experiments. The Wageningen Rhizolab is a joint facility of the Research Institute for Agrobiology and Soil Fertility (AB-DLO) and the Wageningen Agricultural University. This research had the financial support of the Dutch National Research Programme on Air Pollution and Climate Change (NOP Project No. 850020) and EC funding in the ENVIRONMENT Programme under Contract CT920169-CROPCHANGE.

References

Baker, J. T., and Allen, L. H., Jr. (1993). Contrasting crop species response to CO_2 and temperature: rice, soybean and citrus. *Vegetatio* 104/105, 239–260.

Baker, J. T., Allen, L. H., Jr., Boote, K. J., and Pickering, N. B. (1996). Assessment of rice responses to global climate change: CO_2 and temperature. *In* "Carbon Dioxide and Terrestrial Ecosystems" (G. N. Koch and H. A. Mooney, eds.). Academic Press, San Diego.

Campbell, W. J., Allen, L. H., Jr., and Bowes, G. (1990). Response of soybean canopy photosynthesis to CO_2 concentration, light and temperature. *J. Exp. Bot.* 41, 427–433.
Clough, J. M., Peet, M. M., and Kramer, P. J. (1981). Effects of high atmospheric CO_2 and sink size on rates of photosynthesis of a soybean cultivar. *Plant Physiol.* 67, 1007–1010.
Cure, J. D., and Acock, B. (1986). Crop responses to carbon dioxide doubling: a literature survey. *Agric. Forest Meteorol.* 38, 127–145.
Dijkstra, P., Schapendonk, A. H. C. M., and Groenwold, K. (1993). Effects of CO_2 enrichment on canopy photosynthesis, carbon economy and productivity of wheat and faba bean under field conditions. *In* "Climate Change: Crops and Terrestrial Ecosystems. *Agrobiological Thema's 9*" (S. C. van de Geijn, J. Goudriaan, and F. Berendse, eds.), pp. 23–42, ISBN 90 73384 17 6. CABO-DLO, Wageningen.
Dijkstra, P., Schapendonk, A. H. C. M., and Van de Geijn, S. C. (1994). Response of spring wheat canopy photosynthesis to CO_2 concentration throughout the growing season: effect of developmental stage and light intensity. *In* "Vegetation, Modelling and Climatic Change Effects" (P. Veroustraete, R. Ceulemans, I. Impens, and J. Van Rensbergen, eds), ISBN 90-5103-090-8, pp. 53–62. SPB Academic Publishing, The Hague.
Drake, B. G., and Leadley, P. W. (1991). Canopy photosynthesis of crops and native plant communities exposed to long-term elevated CO_2. *Plant, Cell Environ.* 14, 853–860.
Goudriaan, J., and Unsworth, M. H. (1990). Implications of increased carbon dioxide and climate change for agricultural productivity and water resources. *In* "Impact of Carbon Dioxide, Trace Gases and Climate Change on Global Agriculture" (B. A. Kimball, N. J. Rosenberg, and L. H. Allen, eds.), pp. 111–130, ASA Special Publication No. 53. Wisconsin.
Grashoff, C. (1990a). Effect of pattern of water supply on *Vicia faba* L. 1. Dry matter partitioning and yield variability. *Netherlands J. Agric. Sci.* 38, 21–44.
Grashoff, C. (1990b). Effect of pattern of water supply on *Vicia faba* L. 2. Pod retention and filling, and dry matter partitioning, production and water use. *Netherlands J. Agric. Sci.* 38, 131–143.
Grashoff, C., and Stokkers, R. (1992). Effect of pattern of water supply on *Vicia faba* L. 4. Simulation studies on yield variability. *Netherlands J. Agric. Sci.* 40, 447–468.
Grashoff, C., and Nonhebel, S. (1993). Effects of CO_2-increase on the productivity of cereals and legumes: model exploration and experimental evaluation. *In* "Climate Change: Crops and Terrestrial Ecosystems. *Agrobiological Thema's 9*" (S. C. van de Geijn, J. Goudriaan, and F. Berendse, eds.), pp. 43–57, ISBN 90 73384 17 6. CABO-DLO, Wageningen.
Grashoff, C., Rabbinge, R., and Nonhebel, S. (1993). Potential effects of global climate change on cool season food legume productivity. *In* "Proceedings of the 2nd International Food Legume Research Conference, 12–16 April 1992, Cairo, Egypt" (F. J. Muehlbauer, ed.). ICARDA.
Grulke, N. E., Riechers, G. H., Oechel, W. C., Hjelm, U., and Jaeger, C. (1990). Carbon balance in tussock tundra under ambient and elevated atmospheric CO_2. *Oecologia* 83, 485–494.
Idso, S. B., and Kimball, B. A. (1989). Growth response of carrot and radish to atmospheric CO_2 enrichment. *Environ. Exp. Bot.* 29, 135–139.
Idso, S. B., Kimball, B. A., Anderson, M. G., and Mauney, R., Jr. (1987). Effects of atmospheric CO_2 enrichment on plant growth: the interactive role of air temperature. *Agric. Ecosystems Environ.* 20, 1–10.
Kimball, B. A. (1983). Carbon dioxide and agricultural yield: an assemblage and analysis of 430 prior observations. *Agron. J.* 75, 779–788.
Kimball, B. A., Mauney, J. R., Nakayama, F. S., and Idso, S. B. (1993). Effects of increasing atmospheric CO_2 on vegetation. *Vegetatio* 104/105, 65–75.
Koch, G. W., and Mooney, H. A. (eds). (1996). "Carbon Dioxide and Terrestrial Ecosystems." Academic Press, San Diego.
Lawlor, D. W., and Mitchell, R. A. C. (1991). The effects of increasing CO_2 on crop photosynthesis and productivity: a review of field studies. *Plant, Cell Environ.* 14, 807–818.

Long, S. P. (1991). Modification of the response of photosynthetic productivity to rising temperature by atmospheric CO_2 concentrations: Has its importance been underestimated? *Plant, Cell Environ.* 14, 729–739.

Nonhebel, S. (1993). Effects of changes in temperature and CO_2 concentration on simulated spring wheat yields in The Netherlands. *Climatic Change* 24, 311–329.

Penning de Vries, F. W. T., Jansen, D. M., ten Berge, H. F. M., & Bakema, A. (1989). Simulation of ecophysiological processes of growth of several annual crops. "Simulation Monographs." Pudoc, Wageningen, 271 pp.

Poorter, H. (1993). Interspecific variation in the growth response of plants to an elevated ambient CO_2 concentration. *Vegetatio* 104/105, 77–97.

Rogers, H. H., and Dahlman, R. C. (1993). Crop responses to CO_2 enrichment. *Vegetatio* 104/105, 117–131.

Rowland-Bamford, A. J., Allen, L. H., Jr., Baker, J. T., and Bowes, G. (1991). Acclimation of rice to changing atmospheric carbon dioxide concentration. *Plant, Cell Environ.* 14, 577–583.

Rozema, J. (1993). Plant responses to atmospheric carbon dioxide enrichment: interactions with some soil and atmospheric conditions. *Vegetatio* 104/105, 172–190.

Smit, A. L., Groenwold, J., and Vos, J. (1994). The Wageningen Rhizolab—a facility to study soil-root-shoot-atmosphere interactions in crops. II. Methods of root observations. *Plant Soil* 161, 289–298.

Spitters, C. J. T., van Keulen, H., and van Kraalingen, D. W. G. (1989). A simple and universal crop growth simulator: SUCROS87. *In* " Simulation and Systems Management in Crop Protection, (R. Rabbinge, S. A. Ward, and H. H. van Laar, eds.), pp. 147–181, Simulation Monographs. Pudoc, Wageningen.

Van de Geijn, S. C., Vos, J., Groenwold, J., Goudriaan, J., and Leffelaar, P. A. (1994). The Wageningen Rhizolab—a facility to study soil-root-shoot-atmosphere interactions in crops. I. Descriptions of the main functions. *Plant Soil* 161, 275–287.

Van Heemst, H. D. J. (1986). The distribution of dry matter during growth of a potato crop. *Potato Res.* 29, 55–66.

Van Keulen, H., and Seligman, N. G. (1987). Simulation of water use, nitrogen nutrition and growth of a spring wheat crop. "Simulation Monographs." Pudoc, Wageningen.

Vertregt, N., and Penning de Vries, F. W. T. (1987). A rapid method for determining the efficiency of biosynthesis of plant biomass. *J. Theoret. Biol.* 128, 109–119.

Wong, S. C. (1979). Elevated atmospheric partial pressure of CO_2 and plant growth. *Oecologia* 44, 68–74.

15

Assessment of Rice Responses to Global Climate Change: CO_2 and Temperature

Jeffrey T. Baker, L. Hartwell Allen, Jr.,
Kenneth J. Boote, and Nigel B. Pickering

I. Introduction

The global atmospheric carbon dioxide concentration ([CO_2]) is currently rising at an increasing rate (Keeling et al., 1989). The current [CO_2] (about 355 μmol mol^{-1}) has been projected to increase to about 670–760 μmol mol^{-1} by the year 2075, mainly due to the continued burning of fossil fuels (Rotty and Marland, 1986). As a result of this rise in [CO_2] and other "greenhouse gases," several atmospheric general circulation models have predicted significant increases in global air temperature (Hansen et al., 1988). Most published studies on [CO_2] and temperature effects on plants have mainly dealt with temperate crop species, while tropical plants have received less attention. Since both [CO_2] and temperature can have large effects on plants, it is important to quantify the effects of these climate variables on food crops such as rice. The objective of this paper is to summarize some of the major research findings on the effects of [CO_2] and temperature on rice (*Oryza sativa* L., cv. IR-30).

II. Materials and Methods: Outdoor, Sunlit, Controlled-Environment Chambers

Allen et al. (1992) compared methods for studying the [CO_2] enrichment of crops and ecosystems, including leaf chambers, greenhouses, open-top or flow through field chambers, controlled-environment cabinets or rooms with artificial lighting, open field release systems, and finally the naturally sunlit, controlled-environment chambers with deep rooting soil bins (soil–plant–atmosphere research units, or SPAR). Of these methods, only the SPAR unit system (Jones et al., 1984) permits continuous, (e.g., 5-min) measurements of whole-season carbon and water balance under conditions that approach the outside environment. Further, the SPAR unit system permits control of $[CO_2]$ at subambient and superambient concentrations, as well as precise control of dry bulb and dew point air temperatures. Improvements to the SPAR facility located at Gainesville, FL are discussed by Pickering et al. (1994a). A similar type of facility is the Wageningen Rhizolab (Van de Geijn, et al. 1994), which additionally features real-time monitoring of root–soil processes. By utilizing the SPAR unit system approach, the season-long effects of a wide range of $[CO_2]$, temperature, and other environmental variables have been studied on a number of crop species, including soybean (Acock et al., 1985; Jones et al., 1985a,b), several species of citrus (Koch et al., 1986; Brakke et al., 1994), rice (*Oryza sativa* L.) (Baker et al., 1990a; Rowland-Bamford et al., 1991), spring wheat (Boone et al., 1990; Wall et al., 1988), and cotton (Reddy et al., 1991, 1992).

Rice (*Oryza sativa*, L., cv. IR-30) plants were grown season long in the six SPAR units described by Jones et al. (1984) and modified for rice (Baker et al., 1990a,b). These chambers were located outdoors and were exposed to natural sunlight. They were constructed of a clear cellulose acetate roof with walls of polyester film (Mylar) (ALMAC Plastics Inc., Atlanta, GA). Aboveground chamber dimensions were 2.0×1.0 m in cross section by 1.5 m in height. The chamber tops were attached to an aluminum vat (1.9×0.8 m in cross section and 0.6 m deep), which was filled with soil to a depth of 0.5 m. This vat provided a watertight, flooded root environment for growing rice in paddy culture. A dedicated computer operating in real time was used to maintain environmental control in each of the chambers and to record plant responses (photosynthesis and evapotranspiration rates) and environmental data (air temperature, humidity, $[CO_2]$, paddy water temperature, and solar radiation) to magnetic disk every 300 s. Signals from an array of sensors in each chamber were monitored at specific time intervals, and control algorithms were used to actuate various devices in order to maintain the desired environmental control. Shown in Table I are summaries of the $[CO_2]$ and temperature set points maintained for the five season-long experiments conducted on rice at this location. Among these experiments, $[CO_2]$ ranged from 160 to 900 μmol CO_2 mol^{-1} in air, while temperature treatments ranged from 25/18/21 to 40/33/3°C (daytime dry bulb air temperature/nighttime dry bulb air temperature/paddy water temperature) (Table 1). Specific details concern-

Table I Treatments for Experiments Conducted on Rice (cv. IR-30) in Outdoor, Naturally Sunlit, Controlled-Environment Chambers

Experiment designation	Planting date	[CO_2] (μmol mol^{-1})	Temperature[a] (°C)	Comments
RICE I	Jan. 22, 1987	160 250 330 500 660 900	31/31/27	Photoperiod extended to >18 h until March 2
RICE II	June 23, 1987	160 250 330 500 660 900	31/31/27	Natural photoperiod
RICE III	Oct. 10, 1988	330 660	28/21/25 34/27/31 40/33/37	Full factorial: CO_2 × temperature
RICE IV	July 14, 1989	660	25/18/21 28/21/25 31/24/28 34/27/31 37/30/34	One additional chamber at 330 μmol mol^{-1} and 28/21/25°C
RICE V	June 8, 1990	330 660	28/21/25	Full factorial: 2CO_2 × 3N fertilizer rates

[a] Temperature treatments are expressed as daytime dry bulb air temperature/nighttime dry bulb air temperature/paddy water temperature.

ing environmental control set points and plant culture for these experiments are given by Baker et al. (1990a,b, 1992a,b).

III. Results and Discussion

A. Growth and Yield

In the RICE I and II experiments, total aboveground biomass, root biomass, tillering, and final grain yield were all increased with increasing [CO_2] treatment, with the greatest increase occurring across the 160–500 μmol mol^{-1} [CO_2] treatment range followed by little additional response to [CO_2] from 500 to 900 μmol mol^{-1} (Baker et al., 1990a). The increased growth under [CO_2] enrichment is often associated with increased tillering in rice, and as a result, CO_2-enriched rice plants often produce greater grain yields due to an increase in the number of panicles (Imai et al., 1985; Baker et al., 1990a).

Figure 1 shows comparisons of average grain yield for the RICE I and II experiments and canopy net photosynthesis (P_n) at 1500 μmol photons m^{-2} s^{-1} on day 61 of RICE II, both normalized to the values obtained in

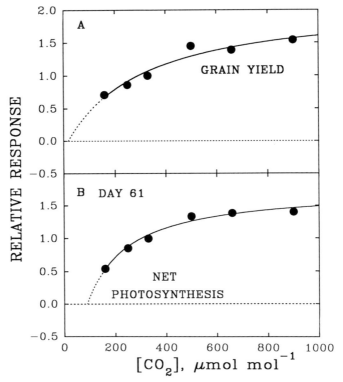

Figure 1 Rice grain yields (A) and canopy net photosynthesis (B) responses to [CO_2] normalized to the values obtained from the 330 μmol mol^{-1} [CO_2] treatment. The percent variation explained by the rectangular hyperbola model was 90 and 96% for grain yield and photosynthesis, respectively. Adapted from Baker and Allen (1993a).

the 330 μmol mol^{-1} [CO_2] treatments. A Michaelis–Menten rectangular hyperbola, modified by adding a y-intercept term, was fit to these data according to Allen et al. (1987) using the NLIN procedure provided by the SAS Institute (Cary, NC):

$$Y = (Y_{max}[CO_2])/([CO_2] + K_m) + Y_i$$

where Y is yield or P_n, [CO_2] is carbon dioxide concentration, Y_{max} is the asymptotic response limit of ($Y - Y_i$) at high [CO_2], Y_i is the intercept on the y axis, and K_m is the value of substrate concentration (e.g., [CO_2] at which ($Y - Y_i$) = $0.5 Y_{max}$). Parameters of $Y_{max} = 2.24$, $K_m = 284$ μmol mol^{-1}, and $Y_i = -0.13$ were obtained for relative grain yield (Fig. 1), with an asymptotic relative yield response ceiling ($Y_{max} + Y_i$) of 2.11 at infinite [CO_2]. For P_n, parameters were 3.96, 70.83 μmol mol^{-1}, and -2.21, for Y_{max}, K_m, and Y_i, respectively, with a relative P_n response ceiling of 1.75 at infinite [CO_2]. Use of these parameters to calculate the percentage increase in relative response at a [CO_2] of 660 μmol mol^{-1} over that obtained at 330 μmol mol^{-1} resulted in 44% (Y = 1.44) and 36% (Y = 1.36) increases for grain yield and P_n, respectively. However, it should be noted that other environmental factors such as temperature, light, and mineral nutrition may strongly influence the shape of the curves shown in Fig. 1.

Across a relatively wide temperature range from 25/18/21 to 37/30/34°C, Baker et al. (1992a) found only a slight apparent temperature optimum for biomass production in the midtemperature ranges. Plants grown at 40/33/37°C were found to be near the upper temperature limit for survival. Temperatures greater than 35°C for more than 1 hr at anthesis induce a high percentage of spikelet sterility (Yoshida 1981). In the 40/33/37°C treatments, plants in the 330 μmol mol^{-1} [CO_2] treatment died during internode elongation, while plants in the 660 μmol mol^{-1} [CO_2] treatment produced small, abnormally shaped panicles that were sterile (Baker et al., 1992b).

Final aboveground biomass, grain yield, and yield components for the five rice experiments are shown in Table II for [CO_2] treatments of 330 and 660 μmol mol^{-1}. While these experiments differ in the time of year they were conducted (Table 1) and utilized different soil types, several trends in plant responses to [CO_2] and temperature are apparent. For both [CO_2] treatments, grain yield was highest in the 28/21/25°C temperature treatment, followed by a decline to zero yield in the 40/33/37°C treatment. [CO_2] enrichment from 330 to 660 μmol mol^{-1} increased grain yield by increasing the number of panicles per plant, while filled grains per panicle and individual seed mass were less affected. As with grain yield, temperature effects on yield components were large and highly significant. Panicles per plant increased while filled grain per panicle decreased sharply with the increasing temperature treatment. Individual seed mass was stable at moderate temperatures, but tended to decline at temperature treatments above 34/27/31°C. Final aboveground biomass and harvest index were increased by [CO_2] enrichment, while the harvest index declined sharply with the increasing temperature treatment. Overall, there were no significant interactive effects between [CO_2] and temperature on grain yield, yield components, or final aboveground biomass (Table II). By utilizing polynomial regression equations to describe the relationship between grain yield and temperature in Table II, Baker and Allen (1993b) found that a linear interpolation across predicted grain yields from 26 to 36°C resulted in grain yield decreases of approximately 9.6 and 10.1% for each 1°C rise in temperature for the 330 and 660 μmol mol^{-1} [CO_2] treatments, respectively.

B. Photosynthesis

The effects of a wide range of long-term [CO_2] enrichment on rice canopy photosynthesis are shown in Fig. 2. Rice photosynthetic rates increased from

Table II Rice Grain Yield, Components of Yield, Total Aboveground Biomass, and Harvest Index for Rice Grown Season Long in Five Separate Experiments[a]

CO_2 (μmol mol^{-1})	Temperature (°C)	Mean air temp.[b] (°C)	Grain yield (Mg/ha)	Panicle plant^{-1} (no./plant)	Filled grain (no./panicle)	Grain mass (mg/seed)	Biomass (g/plant)	Harvest index	Experiment no.
330	28/21/25	24.2	7.9	5.1	34.5	17.4	7.3	0.47	RICE III
	28/21/25	25.1	6.6	3.9	39.6	17.5	6.5	0.44	RICE IV
	28/21/25	25.1	8.0	4.0	47.5	18.5	8.1	0.43	RICE V
	31/31/27	31.0	5.2	5.9	23.0	17.1	5.5	0.42	RICE I
	31/31/27	31.0	4.3	5.4	19.0	17.9	7.2	0.26	RICE II
	34/27/31	30.2	4.2	7.7	15.2	16.2	5.6	0.43	RICE III
	40/33/37	36.2	0.0						RICE III
660	25/18/21	22.1	8.4	4.4	46.0	18.2	7.8	0.47	RICE IV
	28/21/25	24.2	8.4	5.0	37.7	18.0	7.9	0.46	RICE III
	28/21/25	25.1	10.4	4.2	58.3	18.3	8.9	0.50	RICE IV
	28/21/25	25.1	10.1	4.4	54.2	19.0	9.3	0.47	RICE V
	31/31/27	31.0	6.8	6.9	25.1	17.2	7.5	0.40	RICE I
	31/31/27	31.0	6.4	6.0	24.8	18.4	9.3	0.29	RICE II
	34/27/31	30.2	4.8	6.5	18.5	16.7	6.3	0.32	RICE III
	34/27/31	31.1	3.4	7.5	12.9	16.3	8.1	0.18	RICE IV
	37/30/34	34.1	1.0	8.0	3.0	14.2	7.1	0.06	RICE IV
	40/33/37	36.2	0.0						RICE III
F Values
CO_2 concentration			4.2*	4.3*	0.4 NS	0.1 NS	13.1**	4.6*	
Temperature			51.8**	22.6**	51.1**	27.0**	3.2*	32.8**	
CO_2* temperature			1.9 NS	1.7 NS	1.3 NS	0.3 NS	0.2 NS	1.9 NS	

**, *: Significant at the 0.01 and 0.05 probability levels, respectively. NS: not significant.
[a] Adapted from Baker and Allen (1993b).
[b] Mean air temperature is the average of day and night temperatures adjusted for thermoperiod.

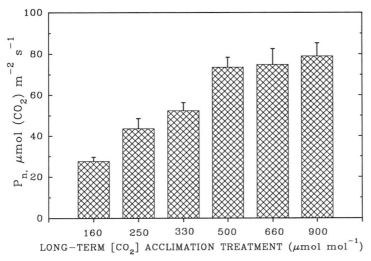

Figure 2 Comparison of rice canopy net photosynthetic rate (P_n) on a land area basis vs long-term [CO_2] acclimation treatment for rice canopies grown and measured at subambient (160, 250), ambient (330), and superambient (500, 660, and 900 μmol CO_2 mol^{-1} in air) [CO_2] treatments 61 days after planting in 1987. The P_n was estimated from linear regression equations of P_n vs photosynthetic photon flux density (PPFD), with PPFD set to 1500 μmol photons m^{-2} s^{-1}. The vertical bars represent 95% confidence intervals. Adapted from Baker et al. (1990b).

160 to 500 μmol mol^{-1}, followed by only small differences among CO_2 treatments from 500 to 900 μmol mol^{-1}. In some crop species, photosynthetic rates initially were stimulated by [CO_2] enrichment, but then subsequently declined with continued [CO_2] enrichment (Aoki and Yabuki, 1977; Kramer, 1981; Wulff and Strain, 1982). Several noncrop species respond similarly, with initial increases in leaf or canopy photosynthetic rates followed by a decline in photosynthesis (Oechel and Strain, 1985; Tissue & Oechel, 1987).

In an extensive review of the literature, Stitt (1991) explored several mechanisms for the often observed gradual decline in photosynthesis following [CO_2] enrichment. One hypothesis put forward to explain this type of acclimation response to elevated [CO_2] is an end product feedback inhibition of photosynthesis, resulting from an imbalance between "source" and "sinks" for photoassimilates. However, for soybean, several experiments have demonstrated that the sinks for photoassimilates are great enough to prevent this decline in individual leaf (Campbell et al., 1988) or whole-canopy (Jones et al., 1984, 1985a) photosynthesis at elevated [CO_2]. A number of studies have reported a decrease in rubisco activity following [CO_2] enrichment (Spencer and Bowes, 1986; Besford and Hand, 1989; Yelle et al., 1989; Besford et al., 1990).

An approach that we have used to detect this type of possible photosynthetic acclimation to long-term [CO_2] is to grow plants for long periods (weeks or months) at two or more [CO_2] and then compare their photosynthetic light responses for brief periods under a common [CO_2]. The hypothesis being tested here is that previous long-term [CO_2] exposure treatment does not affect current short-term photosynthetic rate vs [CO_2] relationships. If this hypothesis is accepted (i.e., we fail to reject), then we may conclude that the photosynthetic rate solely depends on the current [CO_2], and thus there is no evidence for photosynthetic acclimation to long-term [CO_2]. If the different long-term [CO_2] treatments produce different photosynthetic rates when compared at a common short-term [CO_2], we reject the hypothesis. In this case we may conclude that the short-term [CO_2] does not exert sole control over the photosynthetic rate. From this we may infer photosynthetic acclimation to long-term [CO_2]. Additional measurements may then be made to quantify the specific acclimation response. These acclimation responses could conceivably be traced to changes in gross morphology (e.g., changes in leaf thickness or specific leaf weight) or to changes in the photosynthetic biochemistry of the plant (e.g., changes in the amount and/or activity of RuBP carboxylase).

Canopy net photosynthetic rates of rice, measured at a common short-term [CO_2], decreased with increasing long-term [CO_2] growth treatment. To detect this loss of net photosynthetic capacity, Baker et al. (1990b) grew rice at long-term [CO_2] ranging from 160 to 900 μmol mol^{-1} (Fig. 2). The [CO_2] in each chamber was set and controlled for a short period of time to 160, 330, and 660 on days 62, 63, and 64 after planting, respectively (Fig. 3). The canopy net photosynthetic rate decreased with increasing long-term [CO_2] treatment in each of the short- term [CO_2] comparisons. The magnitude of this decrease in the net photosynthetic rate decreased as the short-term [CO_2] increased. Compared at a common short-term [CO_2] of 160 μmol mol^{-1} there was a 3-fold decrease in photosynthesis from the 160 to the 900 μmol mol^{-1} long-term [CO_2] growth treatment (Fig. 3). Figure 4 shows comparisons of canopy *gross* photosynthesis for the same days and short-term [CO_2] treatments as in Fig. 3. Here, canopy gross photosynthesis was calculated by adding the canopy dark respiration rate determined by Baker et al. (1992c) to canopy net photosynthesis. Gross photosynthesis calculated in this way does not allow for photorespiration, nor for the possibility that dark respiration during the day may be different from that at night, although the air temperature in the RICE II experiment was maintained at a constant 31°C both day and night. Comparison of Figs. 3 and 4 indicates that part of the acclimation response of canopy net photosynthesis (Fig. 3) to long-term [CO_2] treatment can be attributed to

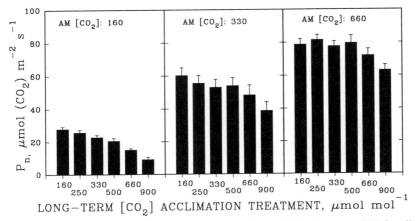

Figure 3 Comparison of rice canopy net photosynthetic rate (P_n) vs long-term [CO_2] acclimation treatment for rice canopies grown at subambient (160, 250), ambient (330), and superambient (500, 660, and 900 μmol CO_2 mol^{-1} in air) [CO_2] treatments in 1987. The P_n estimates were obtained during a short-term [CO_2] changeover study, where the [CO_2] was maintained during the morning hours in all six long-term [CO_2] treatments at 160, 330, and 660 μmol mol^{-1} on days 62, 63, and 64 after planting, respectively. The P_n was estimated from linear regression equations of P_n vs photosynthetic photon flux density (PPFD) with PPFD set to 1500 μmol photons m^{-2} s^{-1}. The vertical bars represent 95% confidence intervals. Adapted from Baker et al. (1990b).

differences in respiration rather than gross photosynthesis (Fig. 4). Expressed on a leaf area basis, this downward acclimation in rice canopy photosynthetic rate was accompanied by a 66% decrease in rubisco activity across the 160–900 μmol mol^{-1} long-term [CO_2] treatment (Rowland-Bamford et al., 1991). A major cause of this decline in rubisco activity was a 32% decrease in the amount of rubisco protein relative to other soluble proteins. Expressed on a leaf area basis, this amounted to a reduction in rubisco activity of 60% with increasing long-term [CO_2] (Rowland-Bamford et al., 1991).

At the leaf level, elevated [CO_2] not only increases photosynthesis but also typically increases the leaf temperature optimum for photosynthesis rate in C_3 plants (Percy and Björkman, 1983; Long, 1991). At the canopy level, apparent photosynthesis has been shown to be relatively insensitive to air temperature across a rather broad air temperature range (Baker et al., 1972; Jones et al., 1985b). Figure 5 shows an example of rice canopy P_n vs PPFD across a daytime air temperature range from 25 to 37°C on day 60 of the RICE IV experiment. While [CO_2] enrichment increased P_n in the 28°C treatment, P_n was very similar among the daytime air temperature treatments for the 660 μmol mol^{-1} [CO_2] treatment. Many studies of canopy

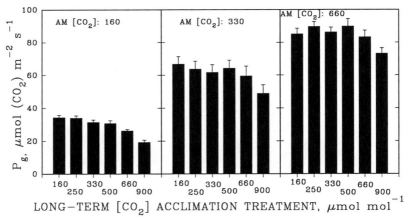

Figure 4 Comparison of rice canopy gross photosynthetic rate (P_g) vs long-term [CO_2] acclimation treatment for rice canopies grown at subambient (160, 250), ambient (330), and superambient (500, 660, and 900 μmol CO_2 mol^{-1} in air) [CO_2] treatments in 1987. The P_g estimates were obtained during a short-term [CO_2] changeover study, where the [CO_2] was maintained during the morning hours in all six long-term [CO_2] treatments at 160, 330, and 660 μmol mol^{-1} on days 62, 63, and 64 after planting, respectively. The P_g was estimated from linear regression equations of P_g vs photosynthetic photon flux density (PPFD) with PPFD set to 1500 μmol photons m^{-2} s^{-1}. The vertical bars represent 95% confidence intervals. Adapted from Baker et al. (1990b, 1992c).

photosynthesis response to air temperature show only small differences across a rather wide range of air temperatures. This may be due in large part to the increase in the evaporative cooling of leaves as air temperature increases.

C. Respiration

While the effects of [CO_2] on photosynthesis have been measured in a number of studies, the response of dark respiration to season-long [CO_2] has received less attention. Baker et al. (1992c) monitored the canopy dark respiration (respiration expressed on a ground area basis, denoted here as R_d) of rice exposed to a wide range of daytime [CO_2] from 160 to 900 μmol mol^{-1}. In this experiment, respiration was measured during the night by periodically flushing each chamber with ambient air, resealing the chambers, and calculating the respiration rate from the canopy [CO_2] efflux. Since R_d was calculated shortly after each chamber was vented with ambient air, the [CO_2] was similar and near ambient in all six season-long daytime [CO_2] treatments.

The seasonal trends in R_d for each of the six daytime [CO_2] treatments are presented in Fig. 6. The trends in R_d both with time during the growing season and with daytime [CO_2] are very similar to those described previously

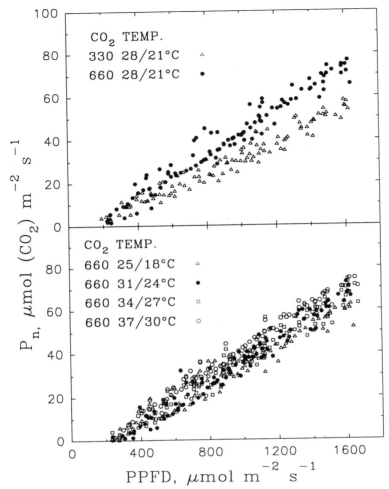

Figure 5 Canopy net photosynthetic rate (P_n) vs incident photosynthetic photon flux density (PPFD) for two daytime [CO_2] treatments and five daytime temperature treatments on day 60 of the RICE IV experiment. Adapted from Baker and Allen (1993b).

for photosynthesis in this experiment (Baker et al., 1990a). The R_d increased with daytime [CO_2] treatment from 160 to 500 μmol mol^{-1} and leveled off across the 500–900 μmol mol^{-1} range. Total aboveground biomass followed a similar trend with [CO_2] treatment.

Specific respiration rate (R_{dw}) was calculated from the data in Fig. 6 by dividing R_d by estimates of plant biomass (D_w) obtained from regression equations of D_w against days after planting (Baker et al., 1992c). The R_{dw} decreased exponentially with time during the growing season in all six

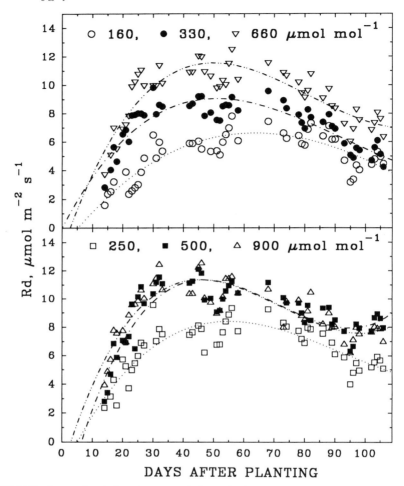

Figure 6 Seasonal variations in canopy dark respiration rate (R_d) of rice exposed to subambient (160, 250), ambient (330), and superambient (500, 660, and 900 μmol mol^{-1}) CO_2 concentrations during the day. Adapted from Baker *et al.* (1992c).

daytime [CO_2] treatments and was higher in the subambient (160 and 250 μmol mol^{-1}) than in the ambient (330 μmol mol) and superambient (500, 660, and 900 μmol mol^{-1}) treatments (Fig. 7). These changes in R_{dw}, both with time during the growing season and across the daytime [CO_2] treatments, are very similar to those found for plant tissue nitrogen concentration. In this experiment, R_{dw} was highly correlated (r = 0.91, P = 0.01) with the total aboveground plant tissue nitrogen concentration ([N]). Baker *et al.* (1992c) concluded that the daytime [CO_2] treatments in this

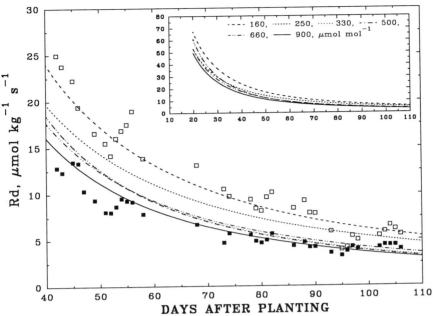

Figure 7 Seasonal trends in specific respiration rate (R_{dw}) for rice exposed to subambient (160, 250), ambient (330), and superambient (500, 600, and 900 μmol mol^{-1}) CO_2 concentrations during the day. Open and closed symbols are for the 160 and 900 μmol mol^{-1} CO_2 concentration treatments, respectively. Adapted from Baker et al. (1992c).

experiment affected R_{dw} mainly by altering the protein composition of the plant tissue.

D. Canopy Assimilation Model

Leaf level sensitivity to [CO_2] and temperature was added to a canopy assimilation model originally developed for row crop geometry (Boote and Loomis, 1991; Pickering et al., 1994b). The model accounts for diffuse and direct beam interception of PPFD after Spitters (1986), computes irradiance on sunlit and shaded leaf classes, and computes the leaf photosynthesis (PGLF) of sunlit and shaded leaf classes with an asymptotic exponential equation after Boote and Loomis (1991):

$$\text{PGLF} = \text{LF}_{max} \text{CO}_{2,max} (1 - e^{(-\text{QE (PPFD)}/\text{LF}_{max}\text{CO}_{2,max})}).$$

Here the temperature and CO_2 effects on quantum efficiency (QE) and light-saturated leaf photosynthesis (LF$_{max}$) are computed by using eq. 16.60a,b of Farquhar and von Caemmerer (1982), assuming limiting RuBP, and temperature effects on the specificity factor of rubisco for CO_2 vs O_2 (GAMST) (Harley et al., 1985) were computed as

$$CO_{2,max} \text{ or } CO_{2,QE} = k(C_i\text{-GAMST})/(4C_i + 8(\text{GAMST}))$$

where k is a scaling factor to set a relative effect of 1.0 at 30°C and 350 μmol mol^{-1} [CO_2] and C_i is internal leaf [CO_2]. Actual QE at any given temperature and [CO_2] is equal to $CO_{2,QE}$ -multiplied by 0.0541 mol mol^{-1} (defined at 30°C and 350 μmol mol^{-1}). Since QE is defined near the light compensation point, C_i was assumed equal to external leaf [CO_2] (C_a) for computing $CO_{2,QE}$. For computing $CO_{2,max}$, a slope of 0.7 was assumed for C_i vs C_a above the CO_2 compensation point (GAMST). These equations give QE responses to [CO_2] and temperature that are very comparable to the data of Ehleringer and Björkman (1977).

Because there were no measurements of single leaf LF_{max}, the canopy assimilation model was used in the solution mode, whereby nonlinear regression was used to solve for LF_{max} and other coefficients (Boote et al., 1993). Using this approach, we evaluated the trends in solved values of LF_{max} and in response to [CO_2] and temperature. For model comparisons, we entered observed leaf area index and 5-min averages of [CO_2], air temperature, and PPFD measured throughout the day for each chamber. A spherical leaf angle distribution and a scattering coefficient of 0.20 were assumed.

This approach was used to solve for single leaf LF_{max} and canopy respiration for individual rice canopies growing at 160, 250, 330, 500, 660, and 900 μmol mol^{-1} on day 60 of the RICE II experiment. Gross assimilation rates were then calculated by subtracting modeled canopy respiration from apparent assimilation values. We then solved for one LF_{max} value to represent all six [CO_2] treatments, with the assumption that CO_2 effects should be accounted for by the $CO_{2,QE}$ and $CO_{2,max}$ terms. The LF_{max} calculated in this fashion reflects the light-saturated rate at 350 μmol mol^{-1} and 30°C. This one solved value for LF_{max} was adequate to predict the canopy response for most [CO_2] treatments, except at lower [CO_2] treatments where the model consistently overestimated assimilation response in high light. Analyses of individual treatments also showed an increasing trend in the LF_{max} with increasing [CO_2]. These two features indicate a problem with the assumption that the complete response to [CO_2] can be modeled on the basis of energy limitation only. The electron transport rate in high light is reportedly inhibited as C_i falls below 300 μmol mol^{-1} (Sharkey et al., 1988; Gerbaud and Andre, 1980). Therefore, an asymptotic equation was added to drive LF_{max} as a function of C_i. This function alternately describes substrate limitation for RuBP regeneration or CO_2 activation of rubisco. The initial slope of this function is proportional to rubisco activity. This addition improved the model fit. As defined, 50% "activation" was achieved at a C_i of 137

μmol mol^{-1}. Figure 8 illustrates the observed and predicted gross assimilation using this modeling approach for day 60 of the RICE II experiment.

IV. Conclusions and Research Recommendations

Photosynthesis, growth, and final grain yield increased with [CO$_2$] from 160 to 500 μmol mol^{-1}, but were very similar from 500 to 900 μmol mol^{-1}. Carbon dioxide enrichment from 330 to 660 μmol mol^{-1} increased grain yield mainly by increasing the number of panicles per plant, while increasing temperature treatments above 28/21/25°C resulted in decreased grain yield largely due to a decline in the number of filled grains per panicle. While [CO$_2$] enrichment always increased photosynthesis, tests for photosynthetic acclimation to long-term [CO$_2$] indicated a loss of some photosynthetic capacity with increasing long-term [CO$_2$] due, at least in part, to a corresponding decrease in RuBP carboxylase content and activity. Specific respiration decreased from the subambient to the ambient and superambient [CO$_2$] and was highly correlated with plant nitrogen concentration. Leaf level sensitivity to [CO$_2$] and temperature was added to a canopy assimilation model originally developed for row crop geometry. By modeling [CO$_2$] effects on both quantum efficiency and light-saturated photosynthesis, good agreement was obtained between observed and predicted rice canopy assimilation across a wide range of [CO$_2$].

Further research on rice is needed to elucidate the [CO$_2$] enrichment effects on carbon partitioning, source–sink relations, and biochemical

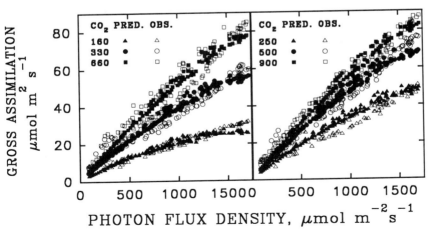

Figure 8 Comparison of observed and predicted gross assimilation for six [CO$_2$] treatments on day 60 of the RICE II experiment. Adapted from Boote et al. (1993).

mechanisms associated with the apparent loss of some photosynthetic capacity under long-term elevated [CO_2]. The paucity of information on respiratory CO_2 efflux of rice under [CO_2] enrichment relative to its importance in the total plant carbon balance also points to an area in need of further study. Additional studies with a wide range of rice cultivars on the interactive effects [CO_2] with other environmental variables, such as temperature, drought, and nutrient stresses, are also needed. Flooded rice fields are a primary anthropogenic source of methane, which is 20–60 times more radiatively active than CO_2 in trapping heat in the Earth's atmosphere. Clearly, further research is needed to elucidate specific physical and physiological mechanisms by which season-long [CO_2] enrichment, through its effects on photosynthesis, tillering, and root biomass production, influences seasonal trends in methane efflux. For predictive purposes, outdoor studies using natural solar irradiance will be especially useful. Finally, the construction and testing of mechanistic, process level simulation models to address climate change effects on plants should continue as a method of extrapolation, prediction, and development of new hypotheses.

V. Summary

Projected future climate changes include a strong likelihood of increased atmospheric carbon dioxide concentration ([CO_2]) and possible increases in air temperatures. We conducted several [CO_2] and temperature experiments on rice (*Oryza sativa*, L., cv. IR-30) in outdoor, naturally sunlit, environmentally controlled plant growth chambers. [CO_2] treatments ranged from 160 to 900 μmol CO_2 mol^{-1} in air, while temperature treatments ranged from 25/18/21 to 40/33/37 °C (daytime dry bulb air temperature/nighttime dry bulb air temperature/paddy water temperature). Photosynthesis, growth, and final grain yield increased with [CO_2] from 160 to 500 μmol mol^{-1}, with much less of a response from 500 to 900 μmol mol^{-1}. Carbon dioxide enrichment from 330 to 660 μmol mol^{-1} increased grain yield mainly by increasing the number of panicles per plant while increasing temperature treatments above 28/21/25°C resulted in decreased grain yield largely due to a decline in the number of filled grains per panicle. Tests for photosynthetic acclimation to long-term [CO_2] indicated that the rice canopy photosynthetic rate measured at a common [CO_2] decreased with increasing long-term [CO_2] growth treatment. Rice-specific respiration (respiration per unit tissue mass) decreased from subambient to ambient and superambient [CO_2] and was associated with a decrease in plant tissue nitrogen content. Leaf level sensitivity to [CO_2] and temperature was added to a canopy assimilation model originally developed for row crop geometry. By modeling [CO_2] effects on both quantum efficiency and light-saturated photosynthesis, good agreement was obtained between observed and predicted rice canopy assimilation across a wide range of [CO_2]. These results indicate that future increases in [CO_2] are likely to benefit rice production by increasing photosynthesis, growth, and grain yield. In warmer areas of the world, possible future increases in air temperature may result in yield decreases.

Acknowledgments

This research is a contribution of the Institute of Food and Agricultural Sciences, University of Florida, and the ARS, U. S. Department of Agriculture, and is supported in part by the U. S. Department of Energy, Carbon Dioxide Research Program, Interagency Agreement No. DE-AI05-88ER69014 with ARS, USDA. This is Florida Agric. Exp. Sta. Journal Series No. R-03627.

References

Acock, B., Reddy, V. R., Hodges, H. F., Baker, D. N. and McKinion, J. M. (1985). *Agron. J.* 77, 942–947.
Allen, L. H., Jr., Boote, K. J., Jones, J. W., Jones, P. H., Valle, R. R., Acock, B., Rogers, H. H. and Dahlman, R. C. (1987). *Global Biogeochem. Cycles* 1, 1–14.
Allen, L. H., Jr., Drake, G. B., Rogers, H. H., and Shinn, J. H. (1992). *Crit. Rev. Plant Sci.* 11, 85–119.
Aoki, M. and Yabuki, K. (1977). *Agric. Meterol.* 18, 475–485.
Baker, J. T., and Allen, L. H., Jr. (1993a). *J. Agric. Meteorol.* 48 (5), 575–582.
Baker, J. T., and Allen, L. H., Jr. (1993b). *Vegetatio* 104/105, 239–260.
Baker, D. N., Hesketh, J. D. and Duncan, W. G. (1972). *Crop Sci.* 12, 431–435.
Baker, J. T., Allen, L. H., Jr., and Boote, K. J. (1990a). *J. Agric. Sci., Camb.* 115, 313–320.
Baker, J. T., Allen, L. H., Jr., Boote, K. J., Jones, P., and Jones, J. W. (1990b). *Agron. J.* 82, 834–840.
Baker, J. T., Allen, L. H., Jr., and Boote, K. J. (1992a). *J. Exp. Bot.* 43, 959–964.
Baker, J. T., Allen, L. H., Jr., and Boote, K. J. (1992b). *Agric. Forest Meterol.* 60, 153–166.
Baker, J. T., Laugel, F., Boote, K. J. and Allen, L. H., Jr. (1992c). *Plant, Cell Environ.* 15, 231–239.
Besford, R. T. and Hand, D. W. (1989). *J. Exp. Bot.* 40, 329–336.
Besford, R. T., Ludwig, L. J., and Withers, A. C. (1990). *J. Exp. Bot.* 41, 925–931.
Boone, M. Y. L., Rickman, R. W., and Whisler, F. D. (1990). *Agron. J.* 82, 718–724.
Boote, K. J., and Loomis, R. S. (1991). The prediction of canopy assimilation. In "Modeling Crop Photosynthesis—from Biochemistry to Canopy" (K. J., Boote, and R. S. Loomis, eds.), CSSA Special Publication No. 19, pp. 109–140. Crop Sci. Soc. Amer., Madison, WI.
Boote, K. J., Pickering, N., Baker, J. T., and Allen, L. H., Jr. (1993). Predicting canopy assimilation of rice in response to carbon dioxide concentration and temperature. In "Research in Photosynthesis" (N. Murata, ed.), pp. 831–834. Proc. IXth International Congress on Photosynthesis.
Brakke, M., Allen, L. H., Jr., Jones, J. W. and Baker, J. T. (1994). *Am. Soc. Hort. Sci.* (submitted).
Campbell, W. J., Allen, L. H., Jr., and Bowes, G. (1988). *Plant Physiol.* 88, 1310–1316.
Ehleringer, J. and Björkman, O. (1977). *Plant Physiol.* 59, 86–90.
Farquhar, G. D., and von Caemmerer, S. (1982). Modelling of photosynthetic response to environment. In "Encyclopedia of Plant Physiology, NS Vol. 12B: Physiological Plant Ecology II" (O. L., Lange, P. S., Nobel, C. B., Osmond, and Ziegler, eds.), pp. 549–587. Springer-Verlag, Berlin.
Gerbaud, A., and Andre, M. (1980). *Plant Physiol.* 66, 1032–1036.

Hansen, J., Fung, I., Lacis, A., Lebedeff, S., Rind, D., Ruedy, R., Russell, G. and Stone, P. (1988). *J. Geophys. Res.* 98(08), 9341–9364.
Harley, P. C., Weber, J. A., and Gates, D. M. (1985). *Planta* 165, 249–263.
Imai, K., Coleman, D. F., and Yanagisawa, T. (1985). *Jpn. J. Crop Sci.* 54, 413–418.
Jones, P., Jones, J. W., Allen, L. H., Jr., and Mishoe, J. W. (1984). *Trans. ASAE* 27, 879–888.
Jones, P., Allen, L. H., Jr., Jones, J. W. and Valle, R. (1985a). *Agron. J.* 77, 119–126.
Jones, P., Allen, L. H., Jr., and Jones, J. W. (1985b). *Agron. J.* 77, 242–249.
Keeling, C. D., Bacastow, R. B., Carter, A. F., Piper, S. C., Whorf, T. P., Heinmann, M., Mook, W. G., and Roeloffzen, H. (1989). A three dimensional model of atmospheric CO_2 transport based on observed winds: Analysis of data. In "Aspects of Climate Variability in the Pacific and the Western Americas" (D. H., Peterson, ed.), Vol. 55, pp. 165–236. Geophysical Monogr., American Geophysical Union, Washington, DC
Koch, K. E., Jones, P. H., Avigne, W. T., and Allen, L. H., Jr. (1986). *Physiol. Plant.* 67, 477–484.
Kramer, P. J. (1981). *BioScience.* 31, 29–33.
Long, S. P. (1991). *Plant, Cell Environ.* 14, 729–739.
Oechel, W. C., and Strain, B. R. (1985). Native species responses to increased atmospheric carbon dioxide concentrations. In "Direct Effects of Increasing Carbon Dioxide on Vegetation" (B. R. Strain, and J.D. Cure, eds.), DOE/ER-0238, pp. 117–154. U. S. Dept. of Energy, Carbon Dioxide Res. Div., Washington, DC.
Pearcy, R. W., and Björkman, O. (1983). Physiological effects. In "Carbon Dioxide and Plants: The Response of Plants to Rising Levels of Atmospheric Carbon Dioxide" (E. R. Lemon, ed.), AAAS Selected Symp. 84, pp. 65–105. Westview Press, Boulder, CO.
Pickering, N. B., Allen, L. H., Jr., Albrecht, S. L., Jones, P., Jones, J. W., and Baker, J. T. (1994a). Environmental plant chambers: control and measurement using CR-10T data loggers. In "Computers in Agriculture" (D. G. Watson, S. Zazueta, and T.V. Harrison, eds.), pp. 29–35. Proc. 5th Intl. Conf., Orlando, Florida, Feb. 5–9. ASAE, St. Joseph, MI.
Pickering, N. B., Jones, J. W., and Boote, K. J. (1994b). Adapting SOYGRO V5.42 for prediction under climate change conditions. In "Climate Change and International Impacts" (C. Rosenzweig, J. W. Jones, and L. H. Allen, Jr., eds.). American Society of Agronomy, Madison, WI (in press).
Reddy, V. R., Baker, D. N., and Hodges, H. F. (1991). *Agron. J.* 83, 699–704.
Reddy, K. R., Hodges, H. F., and Reddy, V. R. (1992). *Agron. J.,* 84, 26–30.
Rotty, R. M., and Marland, G. (1986). Fossil fuel combustion: Recent amounts, patterns, and trends of CO_2. In "The Changing Carbon Cycle: A Global Analysis" (J. R. Trabalka, and D. E. Reichle, eds.), pp. 474–490. Springer-Verlag.
Rowland-Bamford, A. J., Allen, L. H., Jr., Baker, J. T., and Bowes, G. (1991). *Plant, Cell Environ.* 14, 577–583.
Sharkey, T. D., Berry, J. A., and Sage, R. F. (1988). *Planta* 176, 415–424.
Spencer, W., and Bowes, G. (1986). *Plant Physiol.* 82, 528–533.
Spitters, C. J. T. (1986). *Agric. Forest Meteorol.* 38, 231–242.
Stitt, M. (1991). *Plant, Cell Environ.* 14, 741–762.
Tissue, D. L., and Oechel, W. C. (1987). *Ecology* 68, 401–410.
Van de Geijn, S. C., Vos, J., Groenwold, J., Goudriaan, J., and Leffelaar, P. A. (1994). *Plant Soil* 161, 275–287.
Wall, G. W., McKinion, J. M. and Baker, D. N. (1988). ASAE Paper No., 88, 4057, St. Joseph, MI.
Wulff, R., and Strain, B. R. (1982). *Can J. Bot.* 60, 1084–1091.
Yelle, S., Beeson, R. C., Jr., Trudel, M. J. and Gosselin, A. (1989). *Plant Physiol.* 90, 1473–1477.
Yoshida, S. (1981). "Fundamentals of Rice Crop Science." International Rice Research Institute, Los Baños, Philippines.

16

Interactions between CO_2 and Nitrogen in Forests: Can We Extrapolate from the Seedling to the Stand Level?

Dale W. Johnson and J. Timothy Ball

I. Introduction

Nitrogen (N) is unique among nutrients in that it originates from the atmosphere, is intimately tied to organic matter, and rarely accumulates to a significant degree on soil exchange complexes. Other major nutrients (P, S, K, Ca, Mg, Mn) originate primarily from soil minerals and often accumulate in large soil exchangeable pools many times greater than annual vegetation uptake (Johnson, 1992). Soil exchangeable N pools typically are very small and must be replenished several times per year by N mineralization to facilitate vegetation uptake. Thus, it is not surprising that nitrogen deficiency is both widespread and common in terrestrial ecosystems.

Given the widespread occurrence of N deficiency, it must be concluded that the potential for a carbon dioxide (CO_2) "fertilization effect" in terrestrial ecosystems is strongly constrained by nitrogen availability and the effects of CO_2 on nitrogen cycling. Shaver *et al.* (1992) observed three major N cycling constraints on carbon (C) budgets in Arctic ecosystems in Alaska: (i) limits on the variation in C/N ratios in C pools and fluxes, (ii) controls on the distribution of N between plants and soils, and (iii) controls on N inputs and outputs. The first two constraints place limits upon the size of C pools and fluxes for the given amounts of N available, and the third determines the amount of N in the ecosystem. The authors view N budgets and C/N ratios as the key variables through which all other ecosystem perturbations act. They observed that these ecosystems are not responsive to either CO_2 or light on a long-term basis, because neither of these factors affects nitrogen cycling. The authors speculate that the same constraints operate in temperate forest ecosystems, a view that is supported by the frequently observed growth response of forest ecosystems to N fertilization, without any modifications to other environmental factors (Gessel et al., 1973; Heilman and Gessel, 1963; Miller, 1981; Johnson and Todd, 1988; Morrison and Foster, 1977; Pritchett and Comerford, 1982; Raison et al., 1990).

For the purposes of the following discussion, the conceptual model of Shaver et al. (1992) is accepted as a working hypothesis for the forest ecosystem response to CO_2. Thus, it will be assumed that a long-term CO_2 fertilization response in N-deficient forest ecosystems is possible only via changes in the N cycle. (This hypothesis obviously does not apply to that proportion of forest ecosystems that are not N-deficient). Given this premise, then, the question becomes, how will CO_2 affect forest N cycles? The purpose of this paper is to address this question, drawing from the existing literature on N cycling in forests, CO_2 effects on N status in seedlings and saplings, and new results from our ongoing studies of the effects of CO_2 and N on ponderosa and loblolly pines (Pinus ponderosa and Pinus taeda).

II. Nature of Nitrogen Cycling in Forests

Conceptual models of N cycling in forests have been in existence for over 2 decades (Rennie, 1955; Cole et al., 1968; Duvigneaud and Denaeyer-DeSmet, 1970; Curlin, 1970; Miller et al., 1979). While there are some minor variations in the construction of these models, the basic properties are presented in Fig. 1. Only a few of the fluxes shown in Fig. 1 can be measured directly, and others must be calculated using various assumptions. Wet deposition, litter fall, and throughfall can be measured with relative ease, whereas measurements of dry deposition and leaching are made with considerably more difficulty and involve assumptions in order to scale up to an areal (kg ha^{-1}) level. Other fluxes, such as foliar leaching, uptake, requirement, and translocation, must be calculated from direct measurements, including the estimates of dry deposition, which may be subject to considerable error.

The traditional method of calculating uptake, requirement, foliar leaching, and translocation is as follows (Cole and Rapp, 1981):

$$\text{requirement (R)} = \text{nutrients used in new tissue} \quad (1)$$
$$= \text{nutrients in new foliage (F)} + \text{wood} \quad (2)$$
$$= \text{increment (W)} + \text{root increment (RI)}$$

$$\text{uptake (U)} = \text{nutrient return} + \text{nutrient increment in perennial (woody) tissues (W)} \quad (3)$$

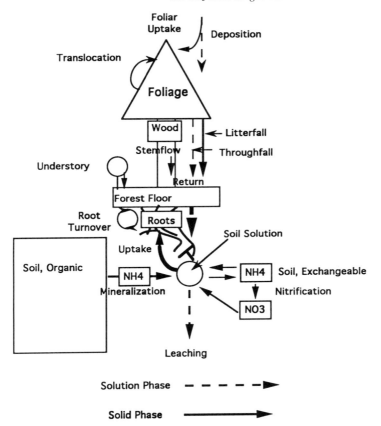

Figure 1 Schematic representation of nutrient cycling in forests.

$$= \text{litter fall (LF)} + \text{foliar leaching} \quad (4)$$
$$\text{(FL)} + \text{root turnover (RT)} + W$$

$$\text{foliar leaching (FL)} = \text{throughfall} + \text{stem flow} - \text{total deposition} \quad (5)$$

$$\text{translocation (T)} = R - U \quad (6)$$

The model described in Eqs. (1)–(6) assumes that the amount of nutrient needed for new growth (foliage, wood, and roots) each year (referred to as requirement) is obtained from recycling via litter fall, foliar leaching, and root turnover and from internal translocation from old to new tissues. The difference between requirement and translocation is defined as uptake; uptake is further explicitly defined as return (litter fall + foliar leaching + root turnover) plus net increment in perennial woody tissues.

Stand age is an important factor determining N uptake and increment rates in forest ecosystems. Uptake and increment rates decline sharply after crown closure when nutrient-rich foliar biomass reaches a steady-state condition (Switzer and Nelson, 1972; Miller, 1981; Turner, 1981). Switzer and Nelson (1972) identified three major stages of nutrient cycling during development: (1) a phase of nutrient accumulation prior to canopy closure, when trees are accumulating nutrient-rich foliar biomass, net annual nutrient increment in biomass is very high, and recycling via the litter fall–decomposition pathway is minimal; (2) a phase of nutrient cycling during which litter fall matches new foliage growth, foliar biomass stabilizes, net annual increment decreases sharply, and the annual uptake of nutrients necessary to supply growing tissues is met largely by recycling; and (3) a stage of decline where senescence ensues and both uptake and increment decline. The second phase is the longest, most nutrient-efficient, and most important on a landscape scale; yet, to date, all studies of tree response to CO_2 have been restricted to only the very earliest parts of the first phase. Forest floor mass and nutrient content accumulate at a rate determined by litter fall and decomposition.

An illustration of the changes in N cycling with stand age is provided by the study of Turner (1981) in a chronosequence of Douglas fir stands (Fig. 2). In these ecosystems, the forest floor becomes an increasingly important N pool with stand age, exceeding vegetation N content by approximately year 70. Within the vegetation component, it is noteworthy that wood becomes the dominant N pool after age 10 and that the foliage N increment becomes zero after year 40 when foliage biomass and N content level off. Uptake increases sharply from years 10 to 25 and then levels off after crown closure. Most of this uptake is satisfied by N return via litter fall and foliar leaching, with only a small proportion needed from either atmospheric or soil N sources to satisfy the N woody increment after age 25.

A complementary and highly relevant dataset on fine root dynamics in age sequences of Douglas fir was presented by Vogt *et al.* (1983). These authors measured fine root biomass in two age sequences of Douglas fir, one on poor sites and one on richer sites. They found that fine root biomass, like foliage, reaches its maximum at canopy closure. In the poor sites, fine root biomass declined after canopy closure, whereas in the richer sites, fine root biomass leveled off, as is the case with foliage biomass.

The high degree to which forests depend upon biogeochemical cycling for their N requirements is illustrated by Cole and Rapp (1981) in their summary of N cycling data from the International Biological Program (IBP). Of the 13 coniferous and 14 deciduous mature forests summarized, the average net N increment was only 20–25% of total uptake (Table I). This implies that factors affecting N recycling (e.g., translocation and decomposition) are of critical importance to the N status of these forest

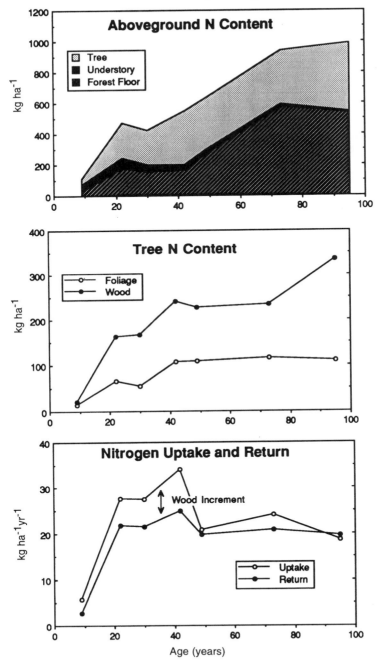

Figure 2 Aboveground N content (top), tree N content (middle), and N uptake and return (bottom) in a chronosequence of Douglas fir stands (data from Turner, 1981).

Table I Nitrogen Requirement, Uptake, Increment, and Atmospheric Input in IBP Forests (Cole and Rapp, 1981)

	Requirement (kg ha^{-1} year^{-1})	Uptake (kg ha^{-1} year^{-1})	Increment (kg ha^{-1} year^{-1})	Deposition (kg ha^{-1} year^{-1})
Coniferous ($n = 13$)	47 ± 17	47 ± 14	11 ± 5	9 ± 8
Deciduous ($n = 14$)	75 ± 18	98 ± 17	14 ± 7	11 ± 6

ecosystems. It is also clear that perennial tissue (wood) N concentration is of critical importance to the requirement for N from outside the biogeochemical cycle, as well as to the efficiency with which this externally derived N is utilized.

The summary of Cole and Rapp (1981) also indicated two other aspects of N cycling in "average" forests that deserve comment here: (1) net translocation was zero (requirement = uptake) (on average) for coniferous species and 24% for deciduous species; (2) atmospheric deposition was approximately equal to N increment (Table I). These average values, however, cannot be taken as representative for all site conditions and species. Turner (1977) showed that N translocation in Douglas fir is substantial and strongly affected by soil N status. He reduced soil N availability by adding carbohydrate and nutrients to the forest floor, on the one hand, and increased soil N availability, on the other hand, by fertilization with urea. With increasing N stress in the carbohydrate plots, translocation increased, causing somewhat decreased litter N concentrations and substantially increased foliage loss to senescence and litter fall. The reverse occurred when N stress was decreased by fertilization.

There are many situations where N increment greatly exceeds atmospheric N input, especially in young, rapidly growing stands that are building N-rich foliar biomass (reviewed by Johnson, 1992). Two illustrations of such situations are given in Table II: an aspen–hardwood stand in Wisconsin (Pastor and Bockheim, 1984) and a eucalyptus plantation in Australia (Turner and Lambert, 1983). In both cases, N increment exceeds atmospheric N deposition by severalfold, necessitating the "mining" of N from soils in order to meet stand N needs. In each of these two cases, N leaching is quite low. In contrast, atmospheric N deposition greatly exceeds N increment in the senescent red spruce forests in the Great Smoky Mountains, and N leaching is substantial (Johnson et al., 1991; Table II).

Table II Nitrogen Deposition, Increment, and Leaching in Some Forest Ecosystems

	Deposition (kg ha^{-1} year^{-1})	Increment (kg ha^{-1} year^{-1})	Leaching (kg ha^{-1} year^{-1})
Aspen–hardwood[a] (Wisconsin)	5	25	<0.5
Eucalyptus[b] (Australia)	1	30	2
Red spruce[c] (North Carolina)	27	1	21

[a] Pastor and Bockheim, 1984.
[b] Turner and Lambert, 1983.
[c] Johnson et al., 1991.

III. Potential Effects of Elevated CO_2 on Nitrogen Cycling

Given the great differences in the nature of N cycles in seedlings vs mature forests, we must ask whether it is possible to make meaningful extrapolations from seedling or sampling studies to mature forests. One question that immediately arises is how will elevated CO_2 affect wood N concentrations and, therefore, N increment in mature forests? Do wood N data from seedling–sapling studies have any relevance to this question? If so, the data we have now from 18-month-old ponderosa pine seedlings suggest that little gain in N use efficiency will be realized in woody tissues: we find increasing rather than decreasing N concentrations in stem tissue with increasing CO_2 [Johnson et al. (1996) and Norby et al. (1996) in this volume]. Seedling stem tissue is not equivalent to wood N tissue from mature trees, however, so that this question must, for now, remain unanswered.

The often-noted reduction in foliar N concentration with elevated CO_2 [Norby et al., 1986a,b; Campagna and Margolis, 1989; Brown, 1991; see also the review by Johnson and Henderson (1994)], if it is sustained in mature forests, might cause a number of changes in N cycling, including uptake, litter fall, decomposition, and primary production.

Research to date on the effects of CO_2 upon litter quality and decomposition are meager [O'Neill and Norby, 1996 (this volume)]; however, there is as yet no evidence that CO_2 will have a major effect upon either. Norby et al., (1986b) concluded that the effects of increased CO_2 on the chemical composition of senesced leaves from white oak (*Quercus alba*) seedlings were too small to significantly affect litter decomposition rates. Furthermore, Couteaux et al., (1991) demonstrated that initial indices of litter quality, such as C/N or lignin/N ratios, are not accurate predictors of decomposition rates over the long term. Our initial studies of the effects of CO_2 on the decomposition of aspen leaves have also shown that decomposition and N mineralization are very dynamic processes and that the effects of treatment are very transient (Johnson and Henderson, 1994). O'Neill and Norby (1996, this volume) reviewed their current results, as well as other literature on the effects of elevated CO_2 on litter quality, and concluded that the evidence is as yet inconclusive, but most field results suggest that decomposition will not be significantly affected by elevated CO_2.

There is some evidence that elevated CO_2 can affect soil carbon and nitrogen mineralization through rhizosphere effects. Körner and Arnone (1992) found a reduction in soil C and an increase in soil respiration and nitrate leaching in an artificial tropical ecosystem subjected to elevated CO_2. They attributed this to increased soil organic matter decomposition in the rhizosphere. Similarly, Zak et al. (1993) found increased microbial biomass, C, and N mineralization in the rhizosphere soils of *Populus grandidentata* seedlings subjected to elevated CO_2. Curtis et al. (1996, this volume) present results from additional studies supporting this model. These findings, if they hold true in general and in mature forest ecosystems, have considerable implications for the ability of rapidly growing forests to "mine" soil N in times of high N demand. On the other hand, if elevated CO_2 causes rhizodeposition of labile C with low C/N ratio, it may cause increased microbial N demand and immobilization rather than mineralization of available N. Diaz et al. (1993) report just such a situation, where elevated CO_2 caused increased root exudation from native herbaceous plants, which in turn caused microbial immobilization rather than mineralization of N and other nutrients, potentially causing nutrient feedback in the opposite direction of that posed by Zak et al. (1993). As a third point on the rhizosphere response curve, we have as yet found no effect of CO_2 on N availability in ponderosa pine (Henderson, unpublished data). Thus, the issue of CO_2 effects on rhizosphere processes remains an open one, and more research is needed to understand these important interactions.

Elevated CO_2 may have profound effects on N cycling on an entirely different scale of response, namely, through species change. Bazazz et al. (1990) found greatly differing responses to elevated CO_2 among seven cooccurring tree species of the northeastern United States. American beech (*Fagus grandifolia*), paper birch (*Betula papyrifera*), black cherry (*Prunus serotina*), sugar maple (*Acer saccharum*), and eastern hemlock (*Tsuga canadensis*) exhibited significantly increased growth in response to elevated CO_2, whereas growth responses in red maple (*Acer rubrum*) and white pine (*Pinus strobus*) were marginal. The authors noted that the largest grown responses were in the shade-tolerant species. Changes of species in response to elevated CO_2, especially changes from coniferous to deciduous or vice versa (see Table I), will have profound effects upon nutrient distribution and cycling in forest ecosystems.

The mechanisms and potential feedback by which increased atmospheric CO_2 could affect nutrient cycling are very complex, making predictions very difficult, especially in view of the lack of information on an ecosystem scale. Studies to date have been limited to seedlings, and effects of CO_2 on nutrient cycling in mature forest ecosystems must be deduced from small-scale, process-level studies (e.g., through simulation modeling).

IV. What Do Seedling–Sapling Studies Tell Us about Ecosystem-Level Response?

Studies of nutrient responses to CO_2 have, until now, necessarily been conducted at a single-tree level using seedlings. The great majority of these studies have been conducted in greenhouses using potted plants. We have found that the disturbance effect on soils during potting has the potential to seriously compromise pot studies in terms of their applicability to the field (Johnson et al., 1995). In field studies using open-top chambers, we found positive growth responses of ponderosa pine to both CO_2 and N over the first 18 months of growth [Johnson et al., 1996 (this volume)]. In a controlled-environment study using the same soil (a typic haplohumult) and the same treatments, we found significant negative growth responses to N. The difference was apparently due to the stimulation of N mineralization upon soil potting, as revealed by soil solution sampling: NO_3^- concentrations in the unfertilized pots (peaking at 5000 μmol l^{-1}) were 3 orders of magnitude greater than are typical of either the Placerville site or forest ecosystems in general (Johnson and Lindberg, 1991). Soil solution NO_3^- concentrations in the unfertilized chambers at the Placerville site were typically <20 μmol l^{-1}. The degree to which disturbance-caused N mineralization has affected other pot studies is unknown, but may have caused some misinterpretations concerning the potential of supposedly N-limited systems to respond to CO_2.

Aside from the potential problems with soil disturbance in pot studies, there are several reasons why these studies must fall short of addressing nutritional issues in the real world. First and foremost, as has been mentioned, mature forests recycle 70–90% of the nutrients they take up to meet their annual growth requirements; thus, an understanding of CO_2 effects upon forest nutrient cycling is fundamental to understanding potential nutrient limitations to CO_2 response. Studies of isolated plants in greenhouses or open-top chambers, however illuminating in terms of physiological responses, simply cannot adequately address potential nutritional constraints because nutrient cycles do not exist in these circumstances.

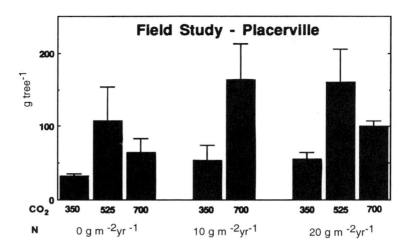
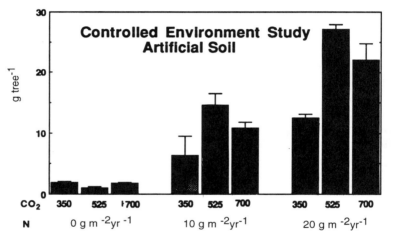
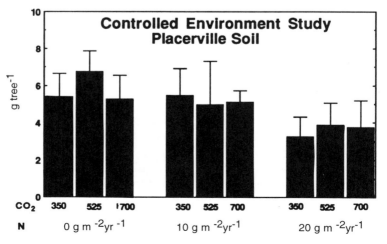

Although the need for information at the stand level is great, there are serious technical, fiscal, and even conceptual problems with conducting CO_2 experiments on mature forest ecosystems that must be addressed before any such experiments are proposed. Free Air CO_2 Experiment-(FACE) type experiments offer a potential means of stand-level experimentation; for example, they offer the possibility of directly testing the null hypothesis that no CO_2 response can occur in N-deficient forests after "root closure" because the soil is fully exploited. [Corollaries to this hypothesis are that (1) increases in N use efficiency will be transient and inadequate to support a response to CO_2 without additional N uptake, and (2) there will be no increase in rhizosphere N mineralization.] Capital costs of FACE studies are high (at least by ecological standards), however, and technical questions remain at this time. Conceptual problems exist with regard to a forest's "memory": is it meaningful to impose square-wave increases in CO_2 on forest ecosystems that have developed under an entirely different regime? Developing FACE experiments at Duke forest will add considerably to our knowledge in all these areas. Meanwhile, we are faced with extrapolation from seedling–sapling to mature forest ecosystems, whether it is justifiable to do so or not.

What can be said about the relevance of seedling–sapling-scale studies to ecosystem-level response? First, if there is indeed a growth response on the seedling–sapling level, we can presume that the "memory" of such a response, if not the response itself, will carry over into the mature ecosystem. Forest fertilization studies have demonstrated conclusively that seedling-stage growth responses carry into the mature stages by simply advancing the stage of stand development, even in cases where the fertilization effect itself is long past (Miller, 1981). By analogy, then, we might assume that fertilization by CO_2 will have the same effect, facilitating increased N uptake in the early stages of development that carries into the future even if there is no further CO_2 response after canopy closure.

Second, the effects of CO_2 on rhizosphere C and N dynamics will have considerable significance to stand-level responses, if the same effects do in fact occur in mature forests. Soils constitute the largest N pool in forest ecosystems, usually exceeding 85% of total ecosystem capital even in the most N-deficient ecosystems (Cole and Rapp, 1981). Yet most soil N is inert and unavailable for either uptake or leaching, with only a rather loosely

Figure 3 Biomass of seedlings in the Placerville field study at 72 weeks (top), in the first controlled-environment study with low N artificial soil at 46 weeks (middle), and in the second controlled-environment study using Placerville soil at 52 weeks (bottom). Standard errors are shown (data from Johnson *et al.*, 1994, 1995, 1996).

defined "mineralizable" pool being biologically active (Aber et al., 1989). This mineralizable pool, the size of which is typically estimated by incubating soils or litter, is that portion of soil N for which heterotrophs (decomposers), plants, and nitrifiers (auto- and heterotrophic) compete. Increases (e.g., Zak et al., 1993) or decreases (e.g., Diaz et al., 1993) in the size of this pool will have considerable significance for potential forest response to CO_2. On the other hand, if CO_2 has no effect upon the mineralizable N pool [as is the case in our studies of ponderosa pine; Johnson et al., 1996 (this volume)], only increases in the efficiency of utilization of this pool through increased root exploration can facilitate greater N uptake and a positive growth response.

Finally, we can address some fundamental questions with regard to forest response to CO_2 under various degrees of N limitation using seedling–sapling studies. We have learned from our studies to date that N deficiency should be thought of as a continuum, with the possibilities for growth response to CO_2 decreasing as N deficiency worsens. Our CO_2 studies on ponderosa pine with differing soils and N treatments illustrate this (Fig. 3). In the Placerville field study, there was a significant positive response to elevated CO_2 without N fertilization because there was enough native N (0.09% N) in the soil to allow such a response (Fig. 3, top) [Johnson et al., 1996 (this volume)]. There was a slight positive response to N fertilization in the Placerville field studies, suggesting that N was suboptimal, but not to the extent that it precluded a response to CO_2. In contrast, the extreme N deficiency imposed by using an artificial soil (0.02% N) precluded a growth response to CO_2 in the first controlled-environment study (Fig. 3, middle) (Johnson et al., 1994). Growth response to N fertilization in this case was positive and highly significant. At the other end of the spectrum, soil disturbance caused excessive N mineralization in the second controlled-environment study using Placerville soil, so that response to CO_2 was negligible and response to additional N was negative.

V. Conclusions

Results from seedling and sapling studies cannot be directly extrapolated to the field because some important ecosystem-level properties and processes simply are not present at the seedling and sapling stages. Specifically, there is usually no nutrient cycle at the seedling and sapling stages, crown and root closure typically have not occurred, and competition among species is usually absent or limited. It may be possible, however, to extrapolate the results of some specific process-level studies from seedlings and saplings to mature forest ecosystems. For example, effects of elevated CO_2 on soil N mineralization (or lack of an effect) have important implications for the responses of mature forests to elevated CO_2. Controlled-environment studies must, however, take into account the possibility of introducing significant artifacts by soil disturbance and consequent increases in N mineralization.

Information on the response of some processes to elevated CO_2 in mature forests—such as increased rooting and soil exploration—will require FACE-type experiments in mature stands. However, even in the event that a growth response occurs only at the seedling or sampling stages, the effects may carry into later stages simply through advancement of stand development.

It can safely be concluded at this stage that growth response to CO_2 is constrained by N limitation, as Shaver *et al.* (1992) hypothesize, but that response to CO_2 is not necessarily precluded by suboptimal N conditions. N deficiency should be thought of as a continuum, with responses to elevated CO_2 precluded at the extreme, but increasing with increasing N availability, rather than as an on–off type response. In that a range of N conditions exists in mature forests throughout the world, it seems safe to say that growth responses to CO_2 have occurred and will likely continue to occur on average, assuming that atmospheric CO_2 levels continue to increase.

Acknowledgments

Research was supported by the Electric Power Research Institute (RP3041-02) and the Nevada Agricultural Experiment Station, University of Nevada, Reno, NV. Technical assistance by Valerie Yturiaga, Carol Johnson, Peter Ross, and Greg Ross is greatly appreciated.

References

Aber, J. D., Nadelhoffer, K. J., Streudler, P., and Melillo, J. M. (1989). Nitrogen saturation in northern forest ecosystems. *Bioscience* 39, 378–386.

Bazzaz, F. A., Coleman, J. S., and Morse, S. R. (1990). Growth responses of seven major co-occurring tree species of the northeastern United States to elevated CO_2. *Can. J. Forest Res.* 20, 1479–1484.

Brown, K. R. (1991). Carbon dioxide enrichment accelerates the decline in nutrient status and relative growth rate of Populus tremuloides Michx. seedlings. *Tree Physiol.* 8, 161–173.

Campagna, M. A., and Margolis, H. A. (1989). Influence of short-term atmospheric CO_2 enrichment on growth, allocation patterns, and biochemistry of black spruce seedlings at different stages of development. *Can. J. Forest Res.* 19, 773–782.

Cole, D. W., and Rapp, M. (1981). Elemental cycling in forest ecosystems. *In* "Dynamic Properties of Forest Ecosystems" (D. E. Reichle, ed.), pp. 341–409. Cambridge University Press, London.

Cole, D. W., Gessel, S. P., and Dice, S. F. (1968). Distribution and cycling of nitrogen, phosphorus, potassium, and calcium in a second-growth Douglas-fir forest. *In* "Primary Production and Mineral Cycling in Natural Ecosystems" (H. E. Young, ed.), pp. 197–213. University of Maine Press, Orono, ME.

Couteaux, M.-M., Mousseau, M., Celkerier, M.-L., and Bottner, P. (1991). Increased atmospheric CO_2 and litter quality decomposition of sweet chestnut litter with animal food webs of different complexities. *Oikos* 61, 54–64.

Curlin, J. W. (1970). Nutrient cycling as a factor in site productivity and forest fertilization. In "Tree Growth and Forest Soils" (C. T. Youngberg and C. R. Davey, eds.) pp. 313–326. Oregon State University Press, Corvallis, OR.

Curtis, P. S., Zak, D. R., Pregitzer, K. S., Lussnehop, J., and Teeri, J. A. (1996). Linking above- and belowground responses to rising CO_2 in northern deciduous forest species. In "Carbon Dioxide and Terrestrial Ecosystems" (G. W. Koch and H. A. Mooney, eds.). Academic Press, San Diego.

Diaz, S., Grime, J. P., Harris, J., and McPherson, E. (1993). Evidence of feedback mechanism limiting plant response to elevated carbon dioxide. *Nature* 364, 616–617.

Duvigneaud, P., and Denaeyer-DeSmet, S. (1970). Biological cycling of minerals in temperate deciduous forests. In "Analysis of Forest Ecosystems" (D. E. Reichle, ed.), pp. 199–255. Springer-Verlag, New York.

Gessel, S. P., Cole, D. W., and Steinbrenner, E. C. (1973). Nitrogen balances in forest ecosystems of the Pacific Northwest. *Soil Biol. Biochem.* 5, 19–34.

Heilman, P. E., and Gessel, S. P. (1963). Nitrogen requirements and the biological cycling of nitrogen in Douglas-fir stands in relationship to the effects of nitrogen fertilization. *Plant Soil* 18, 386–402.

Johnson, D. W. (1992). Nitrogen retention in forest soils. *J. Environ. Qual.* 21, 1–12.

Johnson, D. W., and Lindberg, S. E. (eds.) (1991). "Atmospheric Deposition and Forest Nutrient Cycling: A Synthesis of the Integrated Forest Study," Ecological Series 91. Springer-Verlag, New York.

Johnson, D. W., and Henderson, P. (1994). Effects of Forest Management and Elevated Carbon Dioxide on Soil Carbon Storage. *Adv. Soil Sci.* (in press).

Johnson, D. W., and Todd, D. E. (1988). Nitrogen fertilization of young yellow-poplar and loblolly pine plantations at differing frequencies. *Soil Sci. Soc. Am. J.* 52, 1468–1477.

Johnson, D. W., Van Miegroet, H., Lindberg, S. E., Harrison, R. B., and Todd, D. E. (1991). Nutrient cycling in red spruce forests at the Great Smoky Mountains. *Can. J. Forest Res.* 21, 769–787.

Johnson, D. W., Ball, J. T., and Walker, R. F. (1994). Effects of CO_2 and nitrogen on nutrient uptake in ponderosa pine seedlings. *Plant Soil* (in press).

Johnson, D. W., Walker, R. F., and Ball, J. T. (1995). Lessons from lysimeters: Nitrogen mineralization from soil disturbance compromises controlled environment study. *Ecol. Appl.* (in press).

Johnson, D. W., Henderson, P. H., Ball, J. T., and Walker, R. F. (1996). Effects of CO_2 and N on growth and N dynamics in ponderosa pine. In "Carbon Dioxide and Terrestrial Ecosystems" (G. W. Koch and H. A. Mooney, eds.). Academic Press, San Diego.

Körner, C., and Arnone, J. A. (1992). Responses to elevated carbon dioxide in artificial tropical ecosystems. *Science* 257, 1672–1675.

Miller, H. G. (1981). Forest fertilization: Some guiding concepts. *Forestry* 54, 157–167.

Miller, H. G., Cooper, J. M., Miller, J. D., and Pauline, O. J. L. (1979). Nutrient cycles in pine and their adaption to poor soils. *Can. J. Forest Res.* 9, 19–26.

Morrison, I. K., and Foster, N. W. (1977). Fate of urea fertilizer added to a boreal forest Pinus banksiana Lamb stand. *Soil Sci. Soc. Am. J.* 41, 441–448.

Norby, R. J., O'Neill, E. G., and Luxmoore, R. J. (1986a). Effects of atmospheric CO_2 enrichment on the growth and mineral nutrition of Quercus alba seedlings in nutrient-poor soil. *Plant Physiol.* 82, 83–89.

Norby, R. J., Pastor, J., and Melillo, J. M. (1986b). Carbon-nitrogen interactions in CO_2-enriched white oak: physiological and long-term perspectives. *Tree Physiol.* 2, 233–241.

Norby, R. J., Wullschleger, S. D., and Gunderson, C. A. (1996). Tree Responses to Elevated CO_2 and Implications for Forests. In "Carbon Dioxide and Terrestrial Ecosystems" (G. W. Koch and H. A. Mooney, eds). Academic Press, San Diego.

O'Neill, E. G., and Norby, R. J. (1996). Litter quality and decomposition rates of foliar litter produced under CO_2 enrichment. *In* "Carbon Dioxide and Terrestrial Ecosystems" (G. W. Koch and H. A. Mooney, eds.). Academic Press, San Diego.

Pastor, J. J., and Bockheim, J. G. (1984). Distribution and cycling of nutrients in an aspen-mixed-hardwood-Spodosol ecosystem in northern Wisconsin. *Ecology* 65, 339–353.

Pritchett, W. L., and Comerford, N. B. (1982). Long-term response to phosphorus fertilization on selected southeastern coastal plain soils. *Soil Sci. Soc. Am. J.* 46, 640–644.

Raison, R. J., Khanna, P. K., Connell, M. J., and Falkner, R. A. (1990). Effects of water availability and fertilization on nitrogen cycling in a stand of *Pinus radiata*. *Forest Ecol. Manage.* 30, 31–43.

Rennie, P. J. (1955). The uptake of nutrients by mature forest growth. *Plant Soil* 7, 49–95.

Shaver, G. R., Billings, W. D., Chapin, F. S., III, Giblin, A. E., Nadelhoffer, K. J., Oechel, W. C., and Rastetter, E. B. (1992). Global change and the carbon balance of arctic ecosystems. *Bioscience* 42, 433–441.

Switzer, G. L., and Nelson, L. E. (1972). Nutrient accumulation and cycling in loblolly pine (Pinus taeda L.) plantation ecosystems: The first twenty years. *Soil Sci. Soc. Am. Proc.* 36, 143–147.

Turner, J. (1977). Effect of nitrogen availability on nitrogen cycling in a Douglas-fir stand. *Forest Sci.* 23, 307–316.

Turner, J. (1981). Nutrient cycling in an age sequence of western Washington Douglas-fir stands. *Ann. Bot.* 48, 159–169.

Turner, J., and Lambert, M. J. (1983). Nutrient cycling within a 27-year-old Eucalyptus grandis plantation in New South Wales. *Forest Ecol. Manage.* 6, 155–168.

Vogt, K. A., Moore, E. E., Vogt, D. J., Redlin, M. J., and Edmonds, R. L. (1983). Conifer fine root and mycorrhizal root biomass within the forest floors of Douglas-fir stands of different ages and site productivities. *Can. J. Forest Res.* 13, 429–437.

Zak, D. R., Pregitzer, K. S., Curtis, P. S., Teeri, J. A., Fogel, R., and Randlett, D. L. (1993). Elevated atmospheric CO_2 and feedback between carbon and nitrogen cycles. *Plant Soil* 151, 105–117.

17

Protection from Oxidative Stress in Trees as Affected by Elevated CO_2 and Environmental Stress

Andrea Polle

I. Introduction

Increasing atmospheric CO_2 concentrations can stimulate the growth of trees (Mousseau and Saugier, 1993). However, in natural habitats this potentially positive response may be affected by adverse conditions such as climatic stresses (high light intensities, low temperatures, drought stress), biotic stresses (wounding, pathogens), anthropogenic pollutants (sulfur dioxide, ozone, nitrogen oxides), or nutritional limitations. In the temperate zone, drought stress is a major limiting factor to tree growth (Aussenac, 1978). Rising concentrations of CO_2 and other radiatively active gases are expected to increase the frequency and severity of drought, which in turn has the potential to affect the distribution and productivity of woody plants (Wigley et al., 1984). The tropospheric ozone concentration is also expected to rise in parallel with that of CO_2 (Chameides et al., 1994; Penkett, 1984). Ozone is probably one of the most important air pollutants because it can cause large-scale damage to forest trees and significant yield loss of field crops (Cowling et al., 1990; Heck et al., 1982; Johnson and Taylor, 1989; Krause et al., 1985; Chameides et al., 1994). Because of the great economic and ecological significance of environmental constraints such as ozone and drought, it is important to understand the mode of action of these stress factors and the protection measures that exist in plants.

At the cellular level, unfavorable environmental conditions such as drought stress or ozone can cause oxidative stress by increased production of reactive oxygen species, including $^1O_2$, $O_2^{\cdot-}$, H_2O_2, and $OH^{\cdot}$ [for a review, see Polle and Rennenberg (1993a,b)]. Ozone itself is a strong oxidant (2.07 eV) and, therefore, can damage plants directly by the oxidation of proteins, lipids, or other cellular components or indirectly by the formation of secondary reactive compounds in aqueous media (Grimes *et al.*, 1983; Heath, 1987). In ozone-injured leaves, free radicals and organic peroxides have been detected (Mehlhorn *et al.*, 1990; Hewitt *et al.*, 1990). Electron spin resonance studies have shown that drought stress also results in the increased production of free radicals in leaves (Price and Hendry, 1991; Quartacci and Navari-Izzo, 1992). As the formation of reactive oxygen species is an inevitable event in an oxygen-containing environment, aerobic organisms constitutively contain protective systems that normally keep the concentrations of activated oxygen low. Oxidative stress will, however, cause damage to cellular structures, if increased production rates of oxidants are not adequately compensated. In the present report, the localization and function of antioxidants in plants will be reviewed with regard to the question of whether elevated CO_2 might protect plants from ozone- or drought-induced injury by providing more substrate for detoxification and repair.

II. Detoxification of Reactive Oxygen Species

In plants, the antioxidative system consists of low-molecular-weight compounds (antioxidants) and protective enzymes, which either are directly involved in the removal of harmful oxygen species (superoxide dismutase, catalase, peroxidase) or are necessary for the regeneration of the reduced state of antioxidative substrates (monodehydroascorbate radical reductase, dehydroascorbate reductase, glutathione reductase). The antioxidants, ascorbate (vitamin C), glutathione (a small tripeptide composed of glutamic acid, cysteine, and glycine), and α-tocopherol (a lipophilic compound), participate as substrates in enzymatic reactions or scavenge oxidants in nonenzymatic reactions. The subcellular localization of antioxidants and protective enzymes is shown in Fig. 1.

In mature leaf tissue, chloroplasts are a major production site of activated oxygen species (Asada and Takahashi, 1987). For instance, in isolated thylakoids approximately 5–15% of the electrons produced at the reducing site of photosystem I were transferred to molecular oxygen, yielding superoxide radicals (Hodgson and Raison, 1991). A similar estimation for the photosynthetic electron flow to O_2 was obtained by using intact spinach leaves in which CO_2 assimilation and photorespiration were inhibited (Wu *et al.*, 1991). It should be emphasized that the actual extent and rate of superoxide production *in situ* under stress conditions are not known. Still, even a small leakage of electrons to molecular oxygen will result in a considerable O_2^{-} production rate, on the order of 100 μmol per liter of

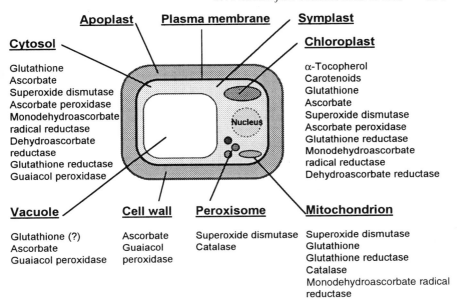

Figure 1 Subcellular localization of antioxidants and protective enzymes in a typical plant cell [adapted from Polle and Junkermann (1994b)].

chloroplast volume per second. The concentration of $O_2^{\cdot -}$ is kept low by the activity of superoxide dismutase. In spruce needles, the chloroplastic concentration of superoxide dismutase is sufficient to diminish the steady-state concentration of superoxide radicals by more than 3 orders of magnitude (Kröniger et al., 1992). The product of superoxide dismutase is H_2O_2. The accumulation of H_2O_2 must be prevented because H_2O_2 oxidizes thiol-containing enzymes of the Calvin cycle, thereby inhibiting photosynthesis (Tanaka et al., 1982). Removal of H_2O_2 is achieved by a powerful detoxification system, the "ascorbate–glutathione pathway" (Fig. 2). In this pathway, ascorbate is used for the reduction of H_2O_2 by an ascorbate-specific peroxidase. Ascorbate peroxidase has a high specificity for ascorbate, whereas its activity with guaiacol, an artifical substrate commonly used to assay "unspecific" peroxidase activity, amounts to only about 3% of its maximum activity (Asada, 1992). Ascorbate peroxidase produces monodehydroascorbate radicals and water. Further reactions in the ascorbate–glutathione pathway serve to regenerate reduced ascorbate (Fig. 2). The final step, which couples the detoxification of reactive oxygen species to photosynthetically produced reductant, NADPH, is accomplished by glutathione reductase (Foyer and Halliwell, 1976). The reactions of this pathway can sustain electron transport to some extent when the normal electron acceptor of

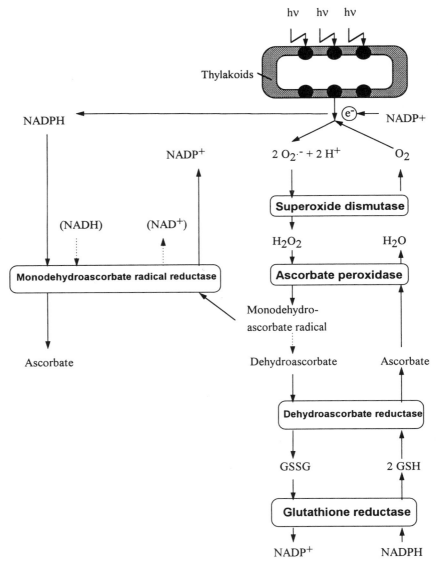

Figure 2 Detoxification of superoxide radicals and hydrogen peroxide in chloroplasts and cytosol [after the work of Foyer and Halliwell (1976); Groden and Beck, (1979); Hossain et al., (1984)].

photosynthesis, NADP+, is limiting. The components of the superoxide dismutase–ascorbate–glutathione pathway are also present in the cytosol (Fig. 1).

An important mechanism to prevent overreduction of the photosynthetic apparatus is afforded by photorespiration. In chloroplasts, CO_2 and O_2 compete for the same binding site of ribulose-1,5-bisphosphate carboxylase/oxygenase (rubisco). When oxygen binds to rubisco, ribulose 1,5-bisphosphate is oxidized, yielding phosphoglycerate and phosphoglycolate. After dephosphorylation, glycolate is transported into peroxisomes and oxidized to glyoxylate. This reaction produces H_2O_2, which is immediately removed by peroxisomal catalase (Tolbert and Essner, 1981). In further metabolic steps, photorespiration gives rise to CO_2 and NADP+ (Zelitch, 1972), thereby helping to maintain high photosynthetic capacity in the mesophyll under conditions of limited external CO_2 supply, for instance, when the stomata are closed under drought stress (Cornic et al., 1989; Cornic and Briantais, 1991).

In plants grown in elevated CO_2 concentrations, reductions in rubisco activity, which are frequently associated with reductions in rubisco protein content, have been reported, even though photosynthetic performance was not affected (Bowes, 1991, 1993). Estimations suggest that the observed decreases in rubisco can be expected to decrease photorespiration by up to 50% (Sharkey, 1988). Increased atmospheric CO_2 concentrations may further reduce photorespiration by favoring carboxylation over oxygenation. A decreased rate of photorespiration and an increased availability of CO_2 for photosynthesis might imply that the need for detoxification of reactive oxygen species is reduced in plants grown in elevated atmospheric CO_2.

III. Interactions of Environmental Stresses and Elevated CO_2

A. Ozone

CO_2 and O_3 are expected to interact at several levels, i.e., affect the stomatal conductance of leaves and, thereby, modulate the dose of ozone taken up into the leaf, change the balance of carboxylation and oxygenation reactions of rubisco, and affect other enzymatic activities (Long, 1994). To date, few studies have addressed the question of whether an elevated atmospheric concentration of CO_2 can ameliorate negative ozone effects in plants. The effect of combined exposure to CO_2 (750 ppm) and O_3 (80 ppb) was studied in radish for 27 days (Barnes and Pfirrmann, 1992) and in Norway spruce after one growth season from April to October (Polle et al., 1993). When ozone and CO_2 were applied as single variables, typical responses were observed in both species, i.e., reductions in photosynthesis and growth in the presence of elevated ozone concentrations and stimulation of photosynthesis and growth in the presence of elevated CO_2 concentrations (Barnes and Pfirrmann, 1992; Polle et al., 1993).

In Norway spruce, ozone caused significant reductions in superoxide dismutase and catalase activities on the basis of protein as well as dry or fresh weight (Fig. 3; cf. Polle et al., 1993). A reduction in superoxide dismutase activity after ozone exposure of spruce trees has also been reported by Hausladen et al. (1990). Matters and Scandalios (1987) observed a reduction in catalase activity in ozone-exposed maize leaves. A problem in assessing the significance of the observed changes in activities of antioxidative enzymes is that the primary mode of action of ozone *in planta* is not understood. At the level of photosynthesis, one of the initial effects of ozone is a reduction in carboxylation efficiency (Farage et al., 1991) caused by reductions in rubisco protein (Pell et al., 1992). A low amount of rubisco would cause a reduction in the rate of photorespiration and, thus, explain the ozone-induced reduction in catalase activity (Polle et al., 1993; Matters and Scandalios, 1987).

Superoxide dismutase has been implicated in ozone defense (Bennett et al., 1984). However, a literature survey shows that there is no clear response of superoxide dismutase activity to ozone: increases, decreases, and no effect of ozone on superoxide dismutase activity have been reported (Bowler et al., 1992). In most cases investigated, transgenic plants expressing increased superoxide dismutase activity were not more ozone-resistant than the wild type (Rennenberg and Polle, 1994). It has also been shown that the antiozonant ethylenediurea (EDU) did not act through increased superoxide dismutase activity (Pitcher et al., 1992). These observations suggest that ozone-mediated alterations in superoxide dismutase activity are downstream events, indicating either damage or adaptation to metabolic requirements, but are not directly related to degrading ozone.

Needles of Norway spruce trees exposed to elevated concentrations of CO_2 showed significant reductions in both superoxide dismutase and catalase activities (Polle et al., 1993; cf. Fig. 3). A reduction in catalase activity was also observed in the leaves of tomato plants grown in elevated CO_2 concentrations (Havir and McHale, 1989). Since catalase is essential for the removal of H_2O_2 formed in photorespiration, a reduction in catalase activity is consistent with the idea of a decreased rate of photorespiration in plants exposed to an elevated CO_2 level.

The effects of ozone and elevated CO_2 were additive, since the lowest activities of antioxidative enzymes were found in the needles of spruce trees grown at elevated concentrations of both gases (Fig. 3). In the previous year's needles, ozone caused pigment reduction and chlorotic spots regardless of whether elevated CO_2 was present (Fig. 3). Young spruce trees cultivated outdoors for another season after ozone–CO_2 pretreatment ex-

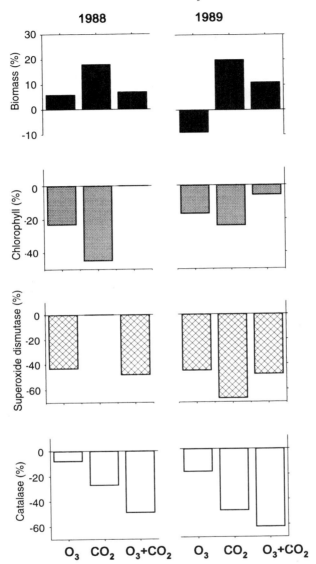

Figure 3 Relative changes in biomass, pigments, and catalase and superoxide dismutase activities in the current and previous year's needles of Norway spruce trees (*Picea abies*, L.) grown in elevated atmospheric CO_2 (750 ppm) or in elevated ozone (80 ppb), as compared to controls grown in ambient CO_2 and ozone concentrations (350 ppm and 20 ppb, respectively) [modified after Polle *et al.* (1993)]. Calculations were performed with data on a fresh weight basis.

hibited severe needle loss in the age class that had been produced in the previous season in the presence of elevated ozone, irrespective of exposure to elevated CO_2 (Polle and Pfirrmann, unpublished results). Apparently, the "ozone memory effect," initially reported by Sandermann and coworkers (1990), was not prevented at elevated atmospheric CO_2. The metabolic basis of the "ozone memory effect" is not known.

Elevated ozone partially counteracted positive growth effects in spruce and radish (Polle et al., 1993; cf. Fig. 3; Barnes and Pfirrmann, 1992). However, in radish elevated CO_2 provided transient protection since ozone-mediated reductions in photosynthesis were initially prevented in plants exposed to a combination of ozone and CO_2. At the end of the experiment, the reduction in photosynthesis was similar in radish plants exposed either to ozone as single variable or to a combination of both gases (Barnes and Pfirrmann, 1992).

Little is known regarding the induction and metabolic bases of ozone tolerance in plants (Kangasjärvi et al., 1994). Some studies indicate that high antioxidative capacity in plants confers high tolerance to oxidative stress (Shaaltiel et al., 1988; Malan et al., 1990). Tobacco plants with high concentrations of ascorbate as a result of "preconditioning" in high light were less ozone-sensitive than tobacco plants preconditioned in low light and, thus, containing less ascorbate (Menser, 1964). These observations suggest that a low antioxidative capacity, as observed in leaves of spruce and oak trees grown in elevated atmospheric CO_2 (Figs. 3 and 4), might

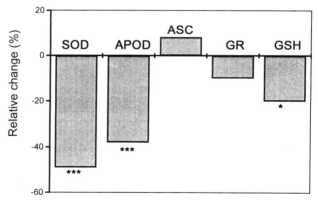

Figure 4 Relative changes in antioxidant enzymes and their substrates in the leaves of young oak seedlings (*Quercus robur*, L.) grown at an elevated CO_2 concentration (700 ppm), as compared to controls grown in ambient atmospheric CO_2. Similar results are obtained regardless of whether the data are expressed on the basis of the protein content of leaf extracts or on the basis of the dry or fresh weight of leaves (Schwanz and Polle, unpublished results). Abbreviations: SOD, superoxide dismutase; APOD, ascorbate peroxidase; ASC, reduced ascorbate; GR, glutathione reductase; GSH, glutathione.

be disadvantageous in an ozone-polluted environment, provided that ozone uptake rates are unaffected.

The flux of ozone into the leaf is controlled by the stomata (Dobson *et al.*, 1990; Tingey and Hogsett, 1985; Wieser and Havranek, 1993), whereas cuticles provide an almost impermeable barrier to ozone (Kerstiens and Lendzian, 1989). Since plants grown in elevated CO_2 generally display reduced stomatal conductance, a reduction in ozone uptake rates might be expected. The experiments reported earlier suggest that ozone-mediated damage may be delayed, but that principal patterns of ozone injury remain unchanged. In order to understand ozone–CO_2 interactions in plants, it will be necessary to identify primary sites of ozone action. Measurements of ozone uptake into leaves indicate that the concentration of ozone drops to zero either in the gaseous phase of the substomatal cavity or in the aqueous phase of the cell wall (Chameides, 1989; Laisk *et al.*, 1989). Ascorbate, which is present in the cell walls of herbaceous and woody plants [spruce needles, Castillo *et al.* (1987) and Polle *et al.* (1990); *Sedum album*, Castillo and Greppin (1988); spinach, Takahama and Oniki (1992)], has been implicated in the degradation of and protection from ozone (Chamaides, 1989; Polle and Rennenberg, 1993a). However, this view has been challenged. Luwe *et al.* (1993) found that apoplastic ascorbate was oxidized, but not regenerated at an appreciable rate in ozone-fumigated spinach leaves. Enzymes mediating the regeneration of ascorbate, such as monodehydroascorbate radical reductase or dehydroascorbate reductase and glutathione reductase, have not been detected in the apoplast (Polle *et al.*, 1990). Thus, at least in spinach, the protection afforded by apoplastic ascorbate was limited and insufficient (Luwe *et al.*, 1993).

Volatile hydrocarbons emitted by plants such as terpenes, isoprene, and ethylene are also potential reaction partners for ozone (Hewitt *et al.*, 1990; Mehlhorn and Wellburn, 1987). In air, the products of these reactions are peroxides (Hewitt and Kok, 1991; Simonaitis *et al.*, 1991). The observation that organic peroxides were found in isoprene-emitting leaves after exposure to ozone suggests that ozone–isoprene reactions may already take place inside the leaf (Hewitt *et al.*, 1990). The toxicity of organic peroxides has been documented *in vitro* (Marklund, 1971; Polle and Junkermann, 1994a) and may also be important in the field (Polle and Junkermann, 1994a). Especially in tree species, the emission of volatile hydrocarbons is an important sink of photosynthetically fixed carbon (Monson *et al.*, 1991). Because these emissions are affected by elevated CO_2 on the one hand (Sharkey *et al.*, 1991), and might mediate ozone injury on the other (Hewitt *et al.*, 1990), it will be important to investigate these interactions in order to understand factors mediating ozone injury and to enhance plant protection in a changing environment.

B. Drought Stress

In order to avoid dehydration, plants restrict water loss by reducing stomatal conductance. Under these conditions, net CO_2 fixation is limited; however, this reduction in photosynthesis is not associated with a decrease in the capacity for assimilation (Jones, 1985). There is evidence that photorespiration is an important protection mechanism, maintaining full mesophyll capacity in plants subjected to drought stress (Cornic et al., 1989; Cornic and Briantais, 1991; Wu et al., 1991). As dehydration becomes more severe, the mesophyll capacity for CO_2 assimilation is also decreased. The biochemical reasons responsible for mesophyll limitation are not fully understood, but damage to the photosynthetic apparatus does not seem to be involved (Kaiser, 1987; Cornic et al., 1989; Quick et al., 1992). The leaf water potentials at which metabolic limitations to CO_2 become apparent are species-dependent (cf. Grieu et al., 1988). In oak trees, ecophysiologically characterized as a drought-tolerant species (Dreyer et al., 1992), drought-induced reductions in photosynthesis were related to an increase in nonchemical energy dissipation, which was fully reversible upon rehydration (Epron et al., 1992). Nonchemical energy quenching has been correlated with zeaxanthin levels (Demmig-Adams, 1990). Since ascorbate is necessary for the production of zeaxanthin from violaxanthin in the xanthophyll cycle, ascorbate might be important in protecting the photosynthetic apparatus from photodamage (Foyer, 1993; Polle and Rennenberg, 1993b). It has been frequently been observed that drought stress caused increases in the activities of antioxidants and protective enzymes in the leaves of herbaceous plants or mosses and that drought tolerance was correlated with the presence of a high capacity of the antioxidative system (Smirnoff, 1993).

The effect of drought stress on photosynthesis and growth of plants in elevated CO_2 has been reviewed (Chaves and Pereira, 1992). Since elevated CO_2 increases leaf-level water-use efficiency (Mooney et al., 1991), it can potentially increase drought resistance. We investigated the role of antioxidants in mediating drought tolerance in oak seedlings raised in ambient and elevated atmospheric CO_2 (700 ppm). In the leaves of well-watered plants grown at elevated CO_2, the activities of antioxidative enzymes were significantly lower than those in the leaves of control plants raised in ambient atmospheric CO_2 (Fig. 4). Moderate drought stress, which had no effect on the relative leaf water content and the water potential ($\Psi > -1$ MPa), but caused an approximately 2-fold reduction in stomatal conductance for water vapor (C. Picon and E. Dreyer, personal communication), resulted in opposite responses in the antioxidant system in leaves of oak seedlings grown either in ambient or elevated CO_2 concentrations (Table I). Drought caused a significant reduction in superoxide dimutase and catalase activities and an increase in ascorbate in leaves from oak seedlings

Table I Response of Antioxidants in the Leaves of Oak Plants Subjected to Drought Stress[a]

Antioxidant	$1 \times CO_2$	$2 \times CO_2$
Superoxide dismutase	↓	↑
Ascorbate peroxidase	=	↑
Glutathione reductase	=	=
Ascorbate	↑	↑
Glutathione	=	=
Catalase	↓	=

[a] Seedlings were raised in a greenhouse in the presence of either ambient or elevated CO_2 concentrations (700 ppm). Five-month-old plants were subjected to drought stress. After 3 weeks, leaves were harvested and analyzed. The relative water content of leaves was similar in plants grown in ambient or elevated atmospheric CO_2 and was not affected by withholding water (Schwanz and Polle, unpublished results). Arrows indicate increases (↑) or decreases (↓) compared to well-watered control plants.

grown in ambient atmospheric CO_2. In contrast, drought had no effect on catalase activity, but caused significant increases in superoxide dismutase and ascorbate peroxidase activities in plants grown in elevated atmospheric CO_2. The activities of antioxidative enzymes in plants grown in elevated CO_2 concentrations and subjected to drought stress were similar to those found in well-watered plants grown in ambient atmospheric CO_2. These observations suggest that plants grown in elevated CO_2 fall back to "business as usual" while experiencing drought. The increase in reduced ascorbate in response to drought may provide protection by scavenging singlet oxygen or mediating zeaxanthin formation (Polle and Rennenberg, 1993b). These results show that protection from oxidants was improved in young oak trees subjected to drought stress in an environment with elevated atmospheric CO_2 concentrations. In order to predict ecosystem reponses, the competitiveness of other, less drought-tolerant ecotypes needs to be studied.

IV. Summary

Light energy captured by the antennae pigments is normally converted to chemical energy and used for the production of reductant. If environmental stresses limit assimilation, excess light energy must be dissipated into waste pathways. Antioxidants are involved in protection from oxidative stress at different levels, i.e., by mediating nonchemical energy dissipation in the reaction centers, by scavenging reactive oxygen species produced in light-driven electron transport, and by removing hydrogen peroxide formed in photorespiration (Fig. 5). In the leaves of young oak seedlings and in the needles of Norway spruce trees grown in elevated atmospheric CO_2, significant reductions in antioxidant capacity were observed. These observations suggest that plants exposed to high atmospheric CO_2 experience less oxidative stress.

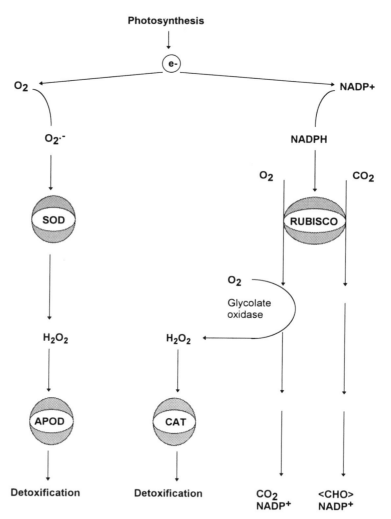

Figure 5 Photosynthetic electron flux to oxygen and carbon dioxide. Abbreviations: SOD, superoxide dismutase; CAT, catalase; APOD, ascorbate peroxidase; rubisco, ribulose-1,5-bisphosphate carboxylase/oxygenase; <CHO>, carbohydrate.

However, in natural environments plants frequently will encounter adverse conditions that may affect the positive responses to elevated CO_2 concentrations. The interactions of elevated CO_2 with two important environmental variables, i.e., drought and ozone, have been investigated. In laboratory studies, elevated CO_2 improved the performance of young oak seedlings during drought stress. In contrast, ozone-mediated injury was not prevented in plants grown in elevated atmospheric CO_2, but may occur with some delay. It is recognized that the effective dose of ozone causing damage to plants is controlled by stomatal uptake and, therefore, (1) is not simply related to atmospheric ozone concentrations and (2) is affected by CO_2. The negative effects of ozone as a single variable on photosynthesis are relatively well-characterized. However, a large gap exists in our knowledge regarding the nature of primary ozone reactions in leaves and the resulting products, which are thought to mediate the negative effects of ozone. These reactions and their implications in CO_2 × ozone interactions need to be characterized. The importance of increasing atmospheric CO_2 concentrations for productivity is likely to be overestimated if interactions with other important environmental variables are not taken into account.

Acknowledgments

Collaboration with Dr. T. Pfirrmann and Dr. H. Payer (GSF-Forschungszentrum für Gesundheit und Umwelt, München, Germany), who provided ozone–CO_2-fumigated plant material, and with Dr. E Dreyer and J. M. Guehl (INRA, Nancy, France), who provided CO_2-exposed, drought-stressed plant material, is gratefully acknowledged. The contributions of P. Schwanz (Institut für Forstbotanik und Baumphysiolgie, Freiburg, Germany) and S. Chakrabarti (Fraunhofer Institut für Atmosphärische Umweltforschung, Garmisch-Partenkirchen, Germany) are appreciated. I am grateful to Prof. Dr. H. Rennenberg for his constant support, valuable discussions, and critical reading of the manuscript. The work reported here was conducted at the Fraunhofer Institut für Atmosphärische Umweltforschung and at the Institut für Forstbotanik und Baumphysiologie, Professur für Baumphysiologie. Financial support was provided by the Bayerisches Staatsministerium für Landsentwicklung und Umweltfragen and the by European Community (EC EV5V-CT92-0093).

References

Asada, K. (1992). Ascorbate peroxidase—a hydrogen peroxide scavenging enzyme in plants. *Physiol. Plant.* 85, 235–241.
Asada, K., and Takahashi, M. (1987). Production and scavenging of active oxygen in photosynthesis. In "Photoinhibition" (D. J. Kyle, C. B. Osmond, and R. Arntzen, eds.), pp. 227–287. Elsevier, Amsterdam.
Aussenac, G. (1978). La sécheresse de 1976: influence de déficits hydriques sur la croissance des arbres forestière. *Rev. Forest Fr.* 30, 103–114.
Barnes, J., and Pfirrmann, T. (1992). The influence of CO_2 and O_3, singly and in combination, on gas exchange, growth and nutrient status of radish (*Raphanus sativus*, L.). *New Phytol.* 121, 403–412.
Bennett, J., Lee, E., and Heggestad, H. (1984). Biochemical aspects of plant tolerance to ozone and oxyradicals: superoxide dismutase. In "Gaseous Pollutants and Plant Metabolism" (M. Koziol, and F. Whatley, eds.), pp. 413–423. Butterworths.

Bowes, G. (1991). Growth at elevated CO_2: photosynthetic responses mediated through Rubisco. *Plant, Cell Environ.* 14, 795–806.
Bowes, G. (1993). Facing the inevitable. Plants and increasing atmospheric CO_2. *Annu. Rev. Plant Physiol. Plant Mol. Biol.* 44, 309–333.
Bowler, C., Van Montagu, M., and Inzé, D. (1992). Superoxide dismutase and stress tolerance. *Annu. Rev. Plant Physiol. Plant Mol. Biol.* 43, 83–116.
Castillo, F. J., and Greppin, H. (1988). Extracellular ascorbic acid and enzyme activities related to ascorbic acid metabolism in *Sedum album*, L., leaves after ozone exposure. *Exp. Environ. Bot.* 28, 231–238.
Castillo, F. J., Miller, P. R., and Greppin, H. (1987). 'Waldsterben' Extracellular biochemical markers of photochemical air pollution damage to Norway spruce. *Experientia* 43, 111–115.
Chameides, W. L. (1989). The chemistry of ozone deposition to plant leaves: role of ascorbic acid. *Environ. Sci. Technol.* 23, 595–600.
Chameides, W. L., Kasibhatla, P. S., Yienger, J., and Levy, H. (1994). Growth of continental-scale metro-agro-plexes, regional ozone pollution and world food production. *Science* 264, 74–77.
Chaves, M. M., and Pereira, J. (1992). Water, CO_2 and climate change. *J. Exp. Bot.* 43, 1131–1139.
Cornic, G., and Briantais, J. M. (1991). Partitioning of photosynthetic electron flow between CO_2 and O_2 reduction in a C_3 leaf (*Phaseolus vulgaris* L.) at different CO_2 concentrations and during drought stress. *Planta* 183, 178–184.
Cornic, G., LeGouallec, J. L., Briantais, J. M., and Hodges, M. (1989). Effect of dehydration and high light on photosynthesis of two C3 plants (*Phaseolus vulgaris* L. and *Elatostema repens* (Lour.) Hall f.). *Planta* 177, 84–90.
Cowling, E., Shriner, J., Barnard, A., Lucier, A., Johnson, A., and Kiester, A. (1990). Air borne chemicals and forest health in the United States. *In* "Report from 19th IUFRO World Congress," Vienna, pp. 25–36.
Demmig-Adams, B. (1990). Carotenoids and photoprotection in plants: a role for the xanthophyll zeaxanthin. *Biochim. Biophys. Acta* 1020, 1–20.
Dobson, M., Taylor, G., and Freer-Smith, P. (1990). The control of ozone uptake by *Picea abies*, (L.) Karst. and *P. sitchensis* (Bong.) Carr. during drought and interacting effects on shoot water relations. *New Phytol.* 116, 465–474.
Dreyer. E., Granier, A., Breda, N., Cochard, H., Epron, D., and Aussenac, G. (1992). Oak trees under drought constraints: ecophysiological aspects. *In* "Recent Advances in Studies on Oak Dieback, Proceedings, Bari," pp. 1–18.
Epron, D., Dreyer, E., and Breda, N. (1992). Photosynthesis of oak trees (*Quercus petraea* (Matt.) Liebl.) during drought under field conditions: diurnal course of net CO_2 assimilation and photochemical efficiency of photosystem II. *Plant, Cell Environ.* 15, 809–820.
Farage, P., Long, S., Lechner, E., and Baker, N. (1991). The sequence of change within the photosynthetic apparatus of wheat following short-term exposure to ozone. *Plant Physiol.* 95, 529–535.
Foyer, C. (1993). Ascorbic acid. *In* "Antioxidants in Higher Plants" (R. Alscher and J. Hess, eds.), pp. 31–58. CRC Press, Boca Raton, FL.
Foyer, C., and Halliwell, B. (1976). The presence of glutathione and glutathione reductase in chloroplasts: a proposed role in ascorbic acid metabolism. *Planta* 133, 21–25.
Grieu, P., Guehl, J. M., and Aussenac, G. (1988). The effects of soil and atmospheric drought on photosynthesis and stomatal control of gas exchange in three coniferous species. *Physiol. Plant.* 73, 97–104.
Grimes, H., Perkins, K., and Boss, W. (1983). Ozone degrades into hydroxyradicals under physiological conditions. *Plant Physiol.* 72, 1016–1020.
Groden, D., and Beck, E. (1979). H_2O_2 destruction by ascorbate dependent systems from chloroplasts. *Biochim. Biophys. Acta* 546, 426–435.
Hausladen, A., Madamachani, N., Fellows, S., Alscher, R., and Amundson, R. (1990). Seasonal changes in antioxidants in red spruce as affected by ozone. *New Phytol.* 115, 447–456.

Havir, E., and McHale, N. A. (1989). Regulation of catalase activity in leaves of *Nicotiana sylvestris* by high CO_2. *Plant Physiol.* 89, 952–957.

Heath, R. (1987). The biochemistry of ozone attack on the plasma membrane of plant cells. *Adv. Phytochem.* 21, 29–54.

Heck, W. W., Taylor, O. C., Adams, R., Bingham, G., Miller, J., Preston, E., and Weinstein, L. (1982). Assessment of crop loss from ozone. *J. Air Pollut. Contr. Assoc.* 32, 353–361.

Hewitt, N., and Kok, G. (1991). Formation and occurrence of organic hydroperoxides in the troposphere: Laboratory and field observation. *J. Atm. Chem.* 12, 181–194.

Hewitt, N., Kok, G., and Fall, R. (1990). Hydroperoxide in plants exposed to ozone mediates air pollution damage to alkene emitters. *Nature* 344, 56–58.

Hodgson, R., and Raison, J. (1991). Superoxide production by thylakoids during chilling and its implication in the susceptibility of plants to chilling-induced photoinhibition. *Planta* 183, 222–228.

Hossain, M., Nakano, K., and Asda, K. (1984). Monodehydroascorbate reductase in spinach chloroplasts and its participation in the regeneration of ascorbate for scavenging hydrogen peroxide. *Plant Cell Physiol.* 61, 385–395.

Johnson, D., and Taylor, G. E. (1989). Role of air pollution in forests decline in eastern North America. *Water Air Soil Pollut.* 48, 21–43.

Jones, H. G. (1985). Partitioning stomatal and non-stomatal limitations to photosynthesis. *Plant, Cell Environ.* 8, 95–104.

Kaiser, W. (1987). Effect of water deficit on photosynthetic capacity. *Physiol. Plant.* 71, 142–149.

Kangasjärvi, J., Talvinen, J., Utriainen, M., and Karjalainen, R. (1994) Plant defense systems induced by ozone: a review. *Plant, Cell Environ.* 17, 783–794.

Kerstiens, G., and Lendzian, K. (1989). Interactions between ozone and plant cuticles. I. Ozone deposition and permeability. *New Phytol.* 112, 13–19.

Krause, G. H., Jung, K. D., and Prinz, B. (1985). Experimentelle Untersuchungen zur Aufklärung der Waldschäden in der Bundesrepublik Deutschland. *VDI-Berichte* 560, 627–655.

Kröniger, W., Rennenberg, H., and Polle, A. (1992). Purification of two superoxide dismutase isozymes and their subcellular localization in needles and roots of Norway spruce (*Picea abies*, L.). *Plant Physiol.* 100, 334–340.

Laisk, A., Kull, O., and Moldau, H. (1989). Ozone concentration in leaf intracellular spaces is close to zero. *Plant Physiol.* 90, 1163–1167.

Long, S. P. (1994) The potential effects of concurrent increases in temperature, CO_2, and O_3 on net photosynthesis, as mediated by RUBISCO. *In* "Plant Reponses to the Gaseous Environment" (R. G. Alscher and A. R. Wellburn, eds.), pp. 21–38. Chapman & Hall, New York.

Luwe, M., Takahama, U., and Heber, U. (1993). Role of ascorbate in detoxifying ozone in the apoplast of spinach (*Spinacia oleracea*, L.) leaves. *Plant Physiol.* 101, 969–976.

Malan, C., Greyling, M., and Gressel, J. (1990). Correlation between CuZn-superoxide dismutase and glutathione reductase and environmental and xenobiotic stress tolerance in maize inbreds. *Plant Sci.* 69, 157–166.

Marklund, S. (1971). Hydroxymethyl hydroperoxide as inhibitor and peroxide substrate of horseradish peroxidase. *Eur. J. Biochem.* 21, 348–354.

Matters, G., and Scandalios, J. (1987). Synthesis of isozymes of superoxide dismutase in maize leaves in response to O_3, SO_2, and elevated O_2. *J. Exp. Bot.* 38, 842–852.

Mehlhorn, H., and Wellburn, A. (1987). Stress ethylene formation determines plant sensitivity to ozone. *Nature* 327, 417–418.

Mehlhorn, H., Tabner, B., and Wellburn, A. (1990). Electron spin resonance evidence for the formation of free radicals in plants exposed to ozone. *Physiol. Plant.* 79, 377–383.

Menser, A. (1964). Response of plants to air pollutants III. A relation between ascorbic acid levels and ozone susceptibility of light-preconditioned tobacco plants. *Plant Physiol.* 39, 564–567.

Monson, R., Guenther, A., and Fall, R. (1991). Physiological reality in relation to ecosystem- and global-level estimates of isoprene emission. In "Trace Gas Emissions by Plants" (T. Sharkey, E. Holland, and H. Mooney, eds.), Physiological Ecology Series, pp.185–208. Academic Press, New York.

Mooney, H., Drake, B., Luxmoore, W., Oechel, W., and Pitelka, L. F. (1991). Predicting ecosystem response to elevated CO_2 concentrations. *Bioscience* 41, 96–1904.

Mousseau, M., and Saugier, B. (1993). The direct of effect increased CO_2 on gas exchange and growth of forest tree species. *J. Exp. Bot.* 43, 1121–1130.

Pell, E. J., Eckhardt, N., and Enyedi, A. (1992). Timing of ozone stress and resulting status of ribulose bisphosphate carboxylase/oxygenase and associated net photosynthesis. *New Phytol.* 120, 397–403.

Penkett, S. A. (1984). Ozone increases in ground-level European air. *Nature* 311, 14–15.

Pitcher, L., Brennan, E., and Zilinskas, B. (1992). The antiozonant ethylenediurea does not act via superoxide dismutase induction in bean. *Plant Physiol.* 99, 1388–1392.

Polle, A., and Rennenberg, H. (1993a). Significance of antioxidants in plant adaptation to environmental stress. In "Plant Adaption to Environmental Stress" (L. Fowden, T. Mansfield, and J. Stoddard, eds.), pp. 263–273. Chapman & Hall, London.

Polle, A., and Rennenberg, H. (1993b). Photooxidative stress in trees. In "Causes of Photooxidative Stress in Plants and Amelioration of Defense Systems" (C. Foyer and P. Mullineaux, eds.), pp. 199–218. CRC Press, Boca Raton, FL.

Polle, A., and Junkermann, W. (1994a). Inhibition of apoplastic and symplastic peroxidase activity from Norway spruce by the photo-oxidant hydroxymethyl hydroperoxide. *Plant Physiol.* 104, 617–621.

Polle, A., and Junkermann, W. (1994b). Does atmospheric H_2O_2 contribute to forest decline? *Environ. Sci. Technol.* 28, 812–815.

Polle, A., Chakrabarti, K., Schürmann, W., and Rennenberg, H. (1990). Composition and properties of hydrogen peroxide decomposing systems in extracellular and total extracts from needles of Norway spruce (*Picea abies*, L., Karst). *Plant Physiol.* 94, 312–319.

Polle, A., Pfirrmann, T., Chakrabarti, S., and Rennenberg, H. (1993). The effects of enhanced ozone and enhanced carbon dioxide concentrations on biomass, pigments and antioxidative enzymes in spruce needles (*Picea abies*, L.). *Plant, Cell Environ.* 16, 311–316.

Price, A., and Hendry, G. A. (1991). Iron-catalysed oxygen radical formation and its possible contribution to drought damage in nine native grasses and three cereals. *Plant, Cell Environ.* 14, 477–484.

Quartacci, M., and Navari-Izzo, F. (1992). Water stress and free radical mediated changes in sun flower seedlings. *J. Plant Physiol.* 139, 621–625.

Quick, P., Chaves, M. M., Wendler, R., David, M., Rodrigues, M., Passarinho, J., Pereira, J., Acock, M., Leegood, R., and Stitt, M. (1992). The effect of water stress on photosynthetic carbon metabolism in four species grown under field conditions. *Plant, Cell Environ.* 15, 25–35.

Rennenberg, H., and Polle, A. (1994). Protection from oxidative stress in transgenic plants. *Biochem. Soc. Trans.* 22, 936–940.

Sandermann, H., Langebartels, C., and Heller, W. (1990). Ozonstreß bei Pflanzen. Frühe und Memory-Effekte von Ozon bei Nadelbäumen. *Ztsch. f. Umweltchem. Ökotox.* 2, 14–15.

Shaaltiel, Y., Glazer, A., Bozion, P., and Gressel, J. (1988). Cross tolerance to herbicidal and environmental oxidants of plant biotypes tolerant to paraquat, sulfur dioxide and ozone. *Pest. Biochem. Physiol.* 31, 13–33.

Sharkey, T. (1988). Estimating the rate of photorespiration in leaves. *Physiol. Plant.* 73, 147–152.

Sharkey, T., Loreto, F., and Delwiche, C. (1991). High carbon dioxide and sun/shade effects on isoprene emission from oak and aspen tree leaves. *Plant, Cell Environ.* 14, 333–338.

Simonaitis, R., Olszyna, K., and Meagher, J. (1991). Production of hydrogen peroxide and organic peroxides in the gas phase reactions of ozone with natural alkenes. *Geophys. Res. Lett.* 18, 9–12.

Smirnoff, N. (1993). The role of active oxygen in the response of plants to water deficit and desiccation. *New Phytol.* 125, 27–58.
Takahama, U., and Oniki, T. (1992). Regulation of peroxidase dependent oxidation of phenolics in the apoplast of spinach leaves by ascorbate. *Plant Cell Physiol.* 33, 379–387.
Tanaka, K., Otsubo T., and Kondo, N. (1982). Participation of hydrogen peroxide in the inactivation of Calvin cycle SH-enzymes in SO_2-fumigated spinach leaves. *Plant Cell Physiol.* 23, 999–1007.
Tingey, D., and Hogsett, W. (1985). Water stress reduces ozone injury via a stomatal mechanism. *Plant Physiol.* 77, 944–947.
Tolbert, N., and Essner, E. (1981). Microbodies: peroxisomes and glyoxysomes. *J. Cell. Biol.* 91, 271s–283s.
Wieser, G., and Havranek, W. (1993). Ozone uptake in the sun and shade crown of spruce: quantifying the physiological effects of ozone exposure. *Trees* 7, 227–232.
Wigley, T. M., Briffa, K. R., and Jones, P. D. (1984). Prediction plant productivity and water resources. *Nature* 312, 102–103.
Wu, J., Neimanis, S., and Heber, U. (1991). Photorespiration is more effective than the Mehler reaction in protecting the photosynthetic apparatus against photoinhibition. *Bot. Ac.* 104, 281–291.
Zelitch, J. (1972). The photooxidation of glyoxylate by envelope-free spinach chloroplasts and its relation to photorespiration. *Arch. Biochem. Biophys.* 150, 698–707.

18
Integrating Knowledge of Crop Responses to Elevated CO_2 and Temperature with Mechanistic Simulation Models: Model Components and Research Needs

Jeffrey S. Amthor and Robert S. Loomis

> Primary causes are unknown to us, but are subject to simple and constant laws that may be discovered by observation.
> —Jean Baptiste Joseph Fourier (1768–1830).

I. Introduction

Atmospheric CO_2 concentration, $[CO_2]_a$, increased from ca. 280 to ca. 360 ppm during the past 200 years, and it is expected that $[CO_2]_a$ will continue to increase for many decades. A concern is that increasing atmospheric concentrations of CO_2 and other infrared radiation absorbing gases will increase Earth's surface temperature and alter precipitation patterns due to an enhanced "greenhouse effect" (Manabe, 1983; Manabe and Stouffer, 1994; n.b., Fourier's "hothouse" analogy was mechanistically incorrect). Temperature increases of just a few degrees Celsius can significantly affect crop growth and productivity. For example, grain production by wheat (*Triticum* spp.) and rice (*Oryza* spp.), arguably the two most important plant genera on earth, is negatively related to temperature (Fischer, 1983; Evans, 1993).

In addition to influencing climate, elevated $[CO_2]_a$ can affect plants in more direct ways. For example, elevated $[CO_2]_a$: (1) generally enhances leaf and canopy CO_2 assimilation rates in C_3 plants (Drake and Leadley, 1991) because CO_2 is a substrate of photosynthesis and is also a competitive inhibitor of photorespiration; (2) may slow or increase plant respiration, depending in part on the basis of expressing the respiration rate and whether short-term or long-term elevated CO_2 treatments are used (Wullschleger *et al.*, 1994; Amthor, 1996); (3) often reduces stomatal aperture and increases the ratio of CO_2 assimilated to water transpired in C_3 and C_4 plants (Eamus, 1991; Gunderson *et al.*, 1993); (4) may alter the partitioning of photoassimilate among plant organs (Rogers *et al.*, 1994); (5) can affect plant chemical composition [often nitrogen concentration is smaller and C/N ratio is greater in elevated $[CO_2]_a$] (Wong, 1979; Norby *et al.*, 1986, 1992; Curtis *et al.*, 1989, 1990); and (6) usually (Lawlor and Mitchell, 1991; Bowes, 1993), but not always (Hunt *et al.*, 1993), increases plant growth and crop productivity. It is important to quantify these and other effects of increasing $[CO_2]_a$ on crops and pastures. In particular, there is a need to:

(A) summarize or integrate knowledge of crop responses to elevated $[CO_2]_a$ (in combination with changes in temperature, air-pollution levels, precipitation amount and timing, nitrogen inputs, and other factors) and
(B) predict future crop productivity with respect to potential global environmental changes.

Models can be used to help meet these needs because they can achieve explicit integrations of knowledge. Moreover, it is easier to increase $[CO_2]_a$ and change temperature in the data input to a model than it is to change $[CO_2]_a$ and temperature in experiments, especially in the field. This is *not* to say that models can replace experiments, but that models should be an important and complementary component of elevated CO_2 research. Indeed, many experiments are needed, but most of the desired field experiments cannot be conducted with the available resources. Models provide an alternative for studying responses to environmental change, especially for plant–environment combinations not easily studied by experiment. Models also provide a means of extrapolating the results of specific experiments—and general concepts—across spatial and temporal scales. It is important to note, however, that models not adequately meeting need (A) should not be expected to adequately meet need (B). That is, models that do not encompass present knowledge, and most models do not, will not provide the best available predictions of plant and ecosystem responses to elevated $[CO_2]_a$ and altered climate.

Some effects of $[CO_2]_a$ and temperature that must be included in models used to predict crop responses to environmental changes are outlined in this chapter. Knowledge gaps and model weaknesses are also considered. In many respects, the articulation of points of ignorance is more important than the summation of knowledge because future experiments should address knowledge gaps. Although the focus herein is on crops, most of the factors discussed also apply to other terrestrial ecosystems.

II. Models

All science involves models, but their nature and uses vary widely among disciplines and individual scientists. In general, building a model is a powerful means for integrating knowledge (need (A)). Herein, the word *model* refers to an explicit quantitative mathematical representation of a crop system, with the emphasis placed on plant physiology and growth, but also including physical processes in soil and atmosphere. Underlying concepts, rationales, limits, uses, and structures of plant and crop models have been outlined by, for example, de Wit (1970), de Wit *et al.* (1978), Loomis *et al.* (1979), Hesketh and Jones (1980), Penning de Vries (1983), Wisiol and Hesketh (1987), Penning de Vries *et al.* (1989), Thornley and Johnson (1990), and Jorgensen (1994). Because no (or only a very few) existing models are capable of predicting crop responses to elevated $[CO_2]_a$ with the needed certainty [and models of other ecosystems are even more limited (Jarvis, 1993)], this chapter focuses on properties of models that could be used to predict crop responses to elevated $[CO_2]_a$ rather than reviewing specific existing models. While it is not important, or even possible, to include everything that is known about plants in models, for the present purposes it *is* important to incorporate all plant processes that have a quantitative importance to growth and productivity and that are likely to be significantly affected by elevated $[CO_2]_a$ and climatic change.

Crop models contain *state variables,* such as leaf area, root length, or nonstructural carbohydrate mass, and *rate variables,* such as the rate of canopy photosynthesis or root growth. Rate variables quantify the rate of change of state variables, although conservation laws constrain growth and should be central to models. That is, the substrates used in one process, e.g., cell wall biosynthesis or respiration, cannot be used in another. The model is acted on by *driving* or *forcing variables*—which can be either state or rate variables—such as solar irradiance, photosynthetic photon flux area density (PPFD; 400–700 nm), $[CO_2]_a$, air temperature, humidity, and sowing date. Driving variables are input parameters used by equations that define rates such as photosynthesis and respiration or are used to initialize state variables. Rate variables can also be functions of state variables or of state variables *and* driving variables. In *dynamic* simulation models, state variables are updated at each iteration or time step (e.g., each minute, hour, or day) according to values of rate variables. Rate variables are generally considered to be constant during a time step, so that time step length is determined in part by the time constants of processes being simulated and whether the time course of the affected state variables is important in the simulation. Many crop models use inappropriately long time steps (e.g., 1 day) in order to limit computer-use time and model complexity. With daily time steps, important diel variations in temperature, PPFD, $[CO_2]_a$, and other factors are neglected. The present availability of fast, inexpensive computers has eliminated some of the rationale for long time steps, and the need for complex models to predict crop responses to global environmental change (see the following) has eliminated the rest.

Although prediction and explanation have been the main goals of modelers, many have found modeling to be equally important as a guide to research because gaps in knowledge become apparent during the process of model building. For example, when formulating rate equations for leaf growth, it becomes evident that knowledge of mechanisms relating carbohydrate level to leaf growth and development is limited. Model building, therefore, can directly assist experimental programs by setting priorities for research. Furthermore, finished models that integrate knowledge are research tools and effective vehicles for the interchange of information.

A. Mechanism versus Empiricism

Plants are plastic—across many space and time scales—in response to environment. To predict crop behavior in new environments, the responses of rate variables to driving variables must mimic those of real plants. This is best achieved with a hierarchic model structure. For crop plants, a typical hierarchy of organizational levels is as follows.

Entity	Level of organization
Landscape	$n + 2$
Crop	$n + 1$
Whole plant	n
Organ	$n - 1$
Tissue	$n - 2$
Cell	$n - 3$
Organelle	$n - 4$

A hierarchy is used to explain processes at one level, say a single whole plant (level $= n$), by describing processes at lower levels, say leaves, cells, organelles, or enzymes (levels $< n$). In turn, whole-plant and crop processes are scaled or integrated to the levels of crops and landscapes, respectively. Lower level processes are themselves constrained by higher level states and rates. Also, process controls at a given level can come from higher or lower levels, as they do in nature. Some implications of hierarchies for research in plant physiology beyond simulation modeling were discussed by Passioura (1979).

A hierarchic model structure allows realistic adjustments in lower level processes, such as leaf photosynthetic capacity or root hydraulic conductivity, as environmental conditions change. Adjustments in lower levels can lead to feedback at higher levels, resulting in plasticity similar to that possessed by real plants. Thus, a hierarchic model structure is the basis of a *mechanistic*, i.e., explanatory, crop model. Specifically, a mechanistic model simulates a process at a given level of organization by quantifying and integrating various physical, biochemical, or physiological processes at lower or higher levels of organization. For example, a mechanistic model of leaf photosynthesis is based on the knowledge of processes that take place at the lower levels of the leaf cell, organelle, or macromolecule and on the receipt of light and CO_2 by the leaf, as determined by the higher levels of the canopy and landscape. It is also possible to hypothesize a mechanism (assume a lower level action giving rise to a higher level observation without observations of the lower level action) and include that hypothesis in a hierarchic model, but observed and proposed lower level processes should be distinguished.

Because a mechanistic model can explain system behavior, it has the potential to make predictions outside the envelope of present experience, i.e., to extrapolate. Because $[CO_2]_a$ is now at levels that exceed those estimated for other times since the inception of agriculture (Fig. 1), and because $[CO_2]_a$ is continuing to increase, the task of predicting effects of future $[CO_2]_a$ on crops is one of extrapolation. Models used for this task must be mechanistic.

Hierarchic models contrast with *empirical*, i.e., single-level, models that describe processes without making reference to lower or higher levels of organization. Empirical models are often statistical descriptions or summaries of data that generally fail to explain rates or states. That is to say, they do not explain results from underlying mechanisms. An empirical model can be used for interpolative predictions (i.e., within the range of conditions that gave rise to previous observations), and such models can be accurate and useful, but they cannot provide reliable predictions outside the envelope of the data from which they were constructed.

At some low level, all models must involve empirical descriptions (see the opening quotation of Fourier). There is no mechanistic explanation of a phenomenon at the lowest level of study, because there are then no lower levels upon which to base explanations. Indeed, explanation and description are themselves level-of-organization dependent. For example, an observed increase in leaf photosynthesis (CO_2 assimilation) rate caused by an increase in $[CO_2]_a$ can be explained by the kinetics of ribulose-1,5-bisphosphate carboxylase/oxygenase (rubisco), the enzyme that catalyzes the assimilation of CO_2 in C_3 photosynthesis. The kinetics of rubisco, however, are empirically derived (e.g., Laing *et al.*, 1974; Jordan and Ogren,

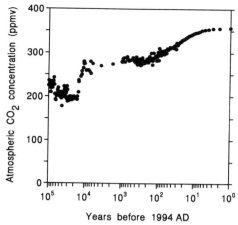

Figure 1 Atmospheric CO_2 concentration during the most recent 100,000 years derived from measurements of air enclosed in Greenlandic and Antarctic ice (Neftel *et al.*, 1985, 1988; Friedli *et al.*, 1986; Barnola *et al.*, 1987; Siegenthaler *et al.*, 1988; Etheridge *et al.*, 1988; Wahlen *et al.*, 1991; Leuenberger *et al.*, 1992) and of ambient air at Mauna Loa Observatory, Hawaii (Keeling and Whorf, 1994). [Barnola *et al.* (1991) present "slightly lower" CO_2 concentrations between 50,000 and 10,000 years before the present compared with Barnola *et al.* (1987), and Raynaud *et al.* (1993) discuss measurement methods and data interpretation.] Crop plant domestication seems to have emerged independently in several geographic regions on the order of 10,000 years ago (Evans, 1993). This roughly corresponds with the beginning of the present interglacial period and a ca. 40% increase in CO_2 concentration that occurred during a few thousand years. In broad terms, atmospheric CO_2 increased from ca. 190–200 ppm during the last glacial maximum (ca. 18,000 years ago) to ca. 280 ppm prior to the Industrial Revolution and has since increased to almost 360 ppm (note the logarithmic time scale).

1984), although some aspects of substrate–enzyme interactions are mechanistically understood.

In addition to a contrast between mechanistic and empirical models—and in fact, most models are made up of mechanistic and empirical submodels, so that the distinction is often gray—model attibutes of *realism* and *accuracy* should be considered. A realistic model is one that predicts the right crop responses to a change in environment for the right reasons. Realism can be attained with hierarchic models, whereas single-level models often fail to respond to new environments in the right way or for the right reasons. Accuracy can be attained by any model, that is to say, both hierarchic and single-level models can make accurate predictions, but often for different reasons. The present need is for models that can be applied across environments with confidence, but without tuning for each new environment. To do that, the explanations must be realistic, i.e., the explanations and mechanisms encoded in the model must be correct. Accuracy is important, of course, but accuracy across environments—including future environments with elevated $[CO_2]_a$—is no more important than realism and will probably be possible only with realistic models.

Because empirical models generally are easier to build and use than mechanistic models, they have been (mis)used more frequently than mechanistic models to predict effects of elevated $[CO_2]_a$ on crop productivity. The simplest empirical models of crop response to $[CO_2]_a$ can be developed by examining the relationship between historic increases in $[CO_2]_a$ and grain yield. For example, the relationship between $[CO_2]_a$ and wheat grain yield in the United Kingdom (UK) has been positive—and strongly so—during the past several centuries (Fig. 2). This correlation is mostly spurious, however, because even though an increase in $[CO_2]_a$ generally stimulates C_3 photosynthesis, recent increases in crop productivity have been mainly due to nitrogen fertilization, cultivar selection, pest control, irrigation, and other management practices (Loomis and Connor, 1992; Evans, 1993). One can be confident that the 800–1500% increase in grain yield of UK wheat that accompanied the ca. 25% increase in $[CO_2]_a$ during the past few centuries was not primarily due to a stimulation of photosynthesis by elevated $[CO_2]_a$ *because* a quantitative mechanistic understanding of the relationships between CO_2 and C_3 photosynthesis and between nitrogen supply and plant growth exists. A further increase in $[CO_2]_a$ of 70–80 ppm could not, by itself, cause another 10-fold increase in wheat yield! A similar relationship between historic $[CO_2]_a$ and rice grain yield in China is evident (Fig. 2B), but again, the increase in $[CO_2]_a$ does not fully explain the increase in grain yield during recent history. The moral: simple, empirical (single-level) models may be seriously misleading when used to (1) extrapolate beyond their historic base of data or (2) identify cause-and-effect relationships between environment and crop productivity.

B. Model Testing

Untested models may be useful in summarizing knowledge, evaluating and developing new theory, and quantifying a potential consequence to a crop due to a change in an environmental variable or a specific physiological process. Crop models are no more than suppositions about real events, however, until tested with specific data or general observations on growth and productivity of field-grown crops. Three types of crop model tests can be considered: (a) comparisons to whole-crop data obtained under a range of conditions, (b) behavioral tests, and (c) comparisons to the rates and states of lower level processes measured in the field. A powerful (a)-type test is one in which growth and yield are predicted for crops sown at different densities or treated with a range of nutrient or water (where suitable) amounts. A related (a)-type test that puts proper pressure on development submodels is to vary the sowing date. Behavioral tests involve

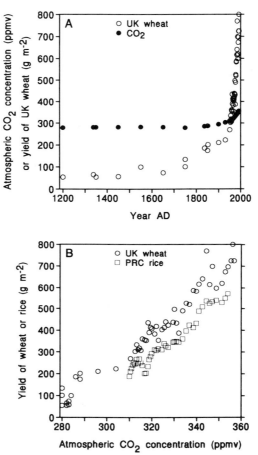

Figure 2 (A) Yield of wheat in the United Kingdom (UK) and global atmospheric CO_2 concentrations from AD 1200 to the present. Yield values from AD 1200 to 1930 are from Stanhill (1976). More recent data are from FAO (1955–1993). Atmospheric CO_2 concentrations are from sources listed in legend to Fig. 1; some have been interpolated to correspond with years for which wheat yield estimates exist. (B) Yield of UK wheat (from A) and People's Republic of China (PRC) rice (Colby *et al.*, 1992) as a function of atmospheric CO_2 concentration (from Fig. 1) during the years for which those yield estimates exist.

similar perturbations, but in a conceptual way that is easier to conduct than new field experiments. For example, it is well-known that root/shoot ratios are affected by drought, and a model should predict this response. Additionally, the predicted root/shoot response to drought should have the same form as that of real crops, e.g., concave up or down with respect to soil water content, and should be based on rational reasons, i.e., present knowledge. Important (c)-type tests include comparing predicted rates and states such as plant chemical composition, organ photosynthetic or respiratory rate, leaf temperature, root density with depth in soil, and leaf nonstructural carbohydrate level to newly measured or literature values. These (c)-type tests directly check the explanatory power of a model.

Experiments (measurements) are needed for model testing, and experiments designed to test models will often result in more useful data than experiments designed without model testing in mind (Lawlor and Mitchell, 1991). Models should be tested with *independent* data whenever possible. Independent data are data that are not used to develop or parametrize a model. If a model relies on the measurement of a particular process for parametrization in order to make a "prediction" about that same process, that model is simply an elaborate form of curve fitting and may not be useful for predicting plant responses to a future environment because there is no access to future measurements. It is difficult to test predictions of crop responses to elevated $[CO_2]_a$ because crops can be exposed to elevated $[CO_2]_a$ only in special circumstances. For example, free-air CO_2 enrichment (FACE) technology [Allen *et al.*, 1992; based on, for example, Greenwood *et al.*, 1982; McLeod *et al.*, 1985] can provide CO_2 enrichment over significant areas (10^1–10^3 m^2), although stomatal responses to elevated $[CO_2]_a$ and the resulting effects on crop energy balance and transpiration are not necessarily well studied with FACE technology because of "island" or "oasis" effects (Section III.D). Open-top chambers (e.g., Allen *et al.*, 1992) can be used only on the spatial scale of ca. 10^1 m^2, and open-top chambers usually alter microclimate. Treatment of native vegetation with elevated temperature has not been satisfactorily achieved in the field, although elevated temperature is a frequent side effect of open-top chamber experiments. For crops, higher or lower temperatures can be "applied" by changing location or sowing date (e.g., Gilmore and Rogers, 1958). It is noteworthy that a crop model—indeed, any model—can never be proven correct, no matter how well it is tested, but it can be falsified or shown to be incorrect by a single accurate datum (e.g., Popper, 1958; Penning de Vries *et al.*, 1989; Thornley and Johnson, 1990). Nonetheless, confidence in the predictions of a model increases when a model passes tests. Progress is made when a new model is shown to produce better results than current models (Thornley and Johnson, 1990). A goal of crop modeling is better models, rather than the final or ultimate model—which can never be written.

Model testing is the bridge between summarizing or integrating existing knowledge within a model (need (A)) and predicting plant responses to a new environment (need (B)). A model that does not pass independent tests of its predictions or is unrealistic need not be taken seriously; such a model should be revised or abandoned. In our opinion, most existing models fail most tests with independent data, i.e., predictions differ from measurements, and also fail behavioral tests, i.e., predictions are unrealistic. This has not, however, generally led to the reexamination of model structures and underlying assumptions with an eye toward model correction before further model use. Rather, models have been tuned *a posteriori* to fit available data or, worse, used to predict crop response to an environment without being corrected in any way. Suffice it to say at this time that there have been few tests of models with data from CO_2 enrichment experiments in the field, so that evaluations of the usefulness of present models to predict productivity in elevated $[CO_2]_a$ must be based on examinations of their structure. Sadly, this is sufficient to cast serious doubt on many crop models as tools for predicting effects of proposed future $[CO_2]_a$ and climatic factors on growth and productivity. Nonetheless, tested comprehensive hierarchic models can be used for meaningful extrapolative predictions. Indeed, they should provide the best available estimates of crop responses to a range of $[CO_2]_a$ values and climatic conditions.

III. Processes That Should Be Included in Models Used To Predict Crop Responses to Elevated CO_2

Processes that are known or suspected to be quantitatively important to plant physiology and growth, and that are affected directly or indirectly by CO_2 enrichment, must be included in models designed to predict crop responses to elevated $[CO_2]_a$. Most of the primary plant physiological processes fall into this category, so that appropriate models will be comprehensive. (Note, however, that the previous section indicated that most models are too weak to provide clear guidance on the issue of which processes are quantitatively important. The notion that most "primary plant physiological process" are important to crop responses to elevated CO_2 therefore is partly conjecture.) Because the models should also account for potentially important feedback among physiological processes and feedback between crop and environment, they will also be complex.

Models addressing the impacts of elevated $[CO_2]_a$ and climatic change on crops also must include soil water and heat states and fluxes, but these will not be considered herein (see, for example, van Bavel and Ahmed, 1976; Campbell, 1985). Soil nutrient uptake, assimilation, and use also must be included in models (see, for example, van Keulen *et al.*, 1975; Penning de Vries *et al.*, 1989).

A. Acclimation and Adaptation

Crop organs and plants continually acclimate (phenotypically respond to environmental changes) to daily and seasonal changes in temperature,

PPFD, vapor pressure, and other environmental factors. Some aspects of acclimation can be simulated directly, e.g., diurnal patterns of osmotic adjustment. Others can be simulated by making lower level rates and states functions of environmental history. Indeed, accounting for acclimation is an important reason for using multilevel models, although few existing crop models account for acclimation.

Whereas acclimation is important to many crop processes on the time scale of days to weeks, adaptation (genetic response to an environmental change) may alter model parameters over a longer time scale, e.g., many years. According to Bowes (1993), "our knowledge of how plant populations adapt as they face selection pressures from increasing CO_2 is negligible." For crops, the complication of human, in addition to evolutionary, selection of genotypes makes predictions of future crop processes all the more difficult.

Uncertainty about the changing responses of plants to increasing $[CO_2]_a$ and temperature through acclimation, adaptation, and new cultivars represents a possibly important limitation to predicting future crop responses to elevated $[CO_2]_a$ and climatic change.

B. Photosynthesis and Photorespiration

1. C_3 Photosynthesis At present and near-future atmospheric CO_2 levels, daytime CO_2 uptake by the leaves of C_3 plants is strongly and positively related to CO_2 concentration. There are two main reasons for this relationship. First, at present $[CO_2]_a$, CO_2 concentration in the chloroplasts of C_3 leaves is less than the Michaelis–Menten constant of rubisco for CO_2. Second, CO_2 competitively inhibits photorespiration (i.e., ribulose 1,5-bisphosphate oxygenation), which is also catalyzed by rubisco. In addition, CO_2 is required as a coactivator of rubisco, but this CO_2–rubisco interaction is probably of limited significance with respect to the stimulation of C_3 photosynthesis by elevated CO_2. Wheat, rice, barley (*Hordeum vulgare*), oat (*Avena sativa*), rye (*Secale cereale*), soybean (*Glycine max*), peanut (*Arachis hypogaea*), pea (*Pisum sativum*), bean (*Phaseolus vulgaris*), sunflower (*Helianthus annuus*), rapeseed (*Brassica* spp.), potato (*Solanum tuberosum*), cassava (*Manihot esculenta*), sweet potato (*Ipomoea batatas*), sugarbeet (*Beta vulgaris*), ryegrass (*Lolium perenne*), alfalfa or lucerne (*Medicago sativa*), clovers (*Trifolium* spp.), cotton (*Gossypium* spp.), and most horticultural fruit and nut species are C_3 crops.

Quantitative mechanistic models that are relevant summaries of knowledge of C_3 photosynthesis and photorespiration have been developed. They are generally based on the well-known model first published by Farquhar *et al.* (1980) and reviewed, modified, and extended many times (e.g., Farquhar and von Caemmerer, 1982; Sharkey, 1985; Farquhar, 1989; Collatz *et al.*, 1990, 1991; Harley and Sharkey, 1991; Sage and Reid, 1994). This

Farquhar model of C_3 photosynthesis encompasses most of what is known about short-term interactions among photosynthesis, photorespiration, and CO_2 concentration. It is based on the biochemical kinetics of rubisco, which can be studied in laboratory experiments, although only approximate values of the rubisco kinetic constants are known (von Caemmerer and Evans, 1991). The mathematical equations of the model are presented and reviewed in, for example, Farquhar and von Caemmerer (1982) and Sage (1990).

Effects of temperature on C_3 photosynthesis are also included in the Farquhar model, in part empirically (e.g., Collatz *et al.*, 1991). One potentially significant effect of elevated CO_2 is an increase in the optimal temperature for CO_2 assimilation, which is predicted by the model (Long, 1991) and observed in experiments (Berry and Björkman, 1980). Effects of leaf nitrogen content on CO_2 assimilation can also be accommodated in this model (Amthor, 1994b).

The Farquhar model of C_3 photosynthesis and photorespiration provides an excellent compromise between simplicity and mechanistic detail. It also has been relatively well tested and is quite accurate in comparison to other photosynthesis models. It is a good choice for inclusion in C_3 crop models designed to assess the effects of increasing $[CO_2]_a$ on photosynthesis and growth, although other (more complex or simpler) photosynthesis models might also be appropriate in some circumstances.

In spite of the generally positive relationship between CO_2 concentration and C_3 leaf CO_2 assimilation rate, and the success of the Farquhar photosynthesis model, there are several reports of a negative relationship between CO_2 level and leaf photosynthesis when the intercellular CO_2 concentration, $[CO_2]_i$, exceeds a few hundred parts per million in short-term experiments (e.g., Woo and Wong, 1983; Sage *et al.*, 1990). Indeed, the Farquhar model has been extended to account for such a relationship on biochemical grounds (Harley and Sharkey, 1991), although this phenomenon may be an exception rather than a rule.

Difficulty can be encountered in the parametrization of the kinetic constants of the Farquhar model for individual crops in space and time. One question is whether plants grown in present $[CO_2]_a$, from which most data have been obtained, are representative of future plants growing in elevated CO_2. Also, although no change is apparent in the C_3 photosynthesis apparatus due to crop breeding during the past 100 years—or due to evolution during the last two hundred million years, during which $[CO_2]_a$ has greatly varied—changes might occur in breeding during the coming decades that will in turn require modifications to the Farquhar model or its parametrizations. Indeed, the model can be used to evaluate the implications of changes in rubisco kinetic parameters.

2. C_4 Photosynthesis Because of the CO_2-concentrating mechanism associated with bundle sheath cells in C_4 plant leaves, effects of CO_2 enrichment above present levels on C_4 photosynthesis are small. Nonetheless, models of C_4 crops should contain rational models of C_4 photosynthetic metabolism and its relationships to CO_2 concentration. One such quantitative model of C_4 photosynthesis has been described by Collatz et al. (1992). For reasons discussed there, this model is an appropriate component of models of C_4 crops, although other C_4 photosynthesis models might also be suitable. In spite of the near saturation of C_4 photosynthesis by the present $[CO_2]_a$ level, C_4 crop productivity can be significantly enhanced by elevated $[CO_2]_a$ (Lawlor and Mitchell, 1991). Maize (*Zea mays*), *Sorghum* spp., sugarcane (*Saccharum officinarum*), and *Panicum* spp. are important C_4 crops.

"Scaling" CO_2–photosynthesis relationships from the level of chloroplasts in C_3 or C_4 leaves to whole-crop canopies is a significant consideration for models of crop photosynthesis. Some aspects of such scaling have been discussed by, for example, Norman (1993) and Amthor (1994b). It is also important to consider the time step of photosynthesis calculations because a time step longer than 1 hr, or even longer than a few minutes, may limit the accuracy of photosynthesis predictions (Wang et al., 1992).

C. Carbohydrate Supply and Growth

A key issue in modeling crop responses to elevated CO_2 is the plant carbon source and sink balance. Carbon assimilation occurs mostly in leaves (sources), but much of that carbon is used for growth and respiration in other organs (sinks). In addressing source–sink relationships, it is useful to account for the numbers and activities of meristems, i.e., the crop's morphogenetic template (see, e.g., Mutsaers, 1984; Denison and Loomis, 1989). Also, the "structure" of sink activity is important: plants produce leaves at stem nodes, shoot meristems must produce new internodes before a new leaf can be formed, axillary buds exist that can grow based on a range of circumstances, and a "switch" must be made from vegetative to reproductive growth in meristems before flowers and fruits can be initiated. This information should be included in models designed to predict source–sink relations and growth in different CO_2 environments. In models of sources, sinks, and partitioning, it is proper to differentiate structural tissue from the temporary accumulation of nonstructural carbohydrate and to define growth as the synthesis of new structural tissue at the expense of temporary pools of carbohydrates and other compounds such as amino acids (see Penning de Vries et al., 1979). In addition to carbon assimilation and use, crop growth is strongly controlled by nitrogen acquisition and partitioning, so that models must account for the effects of both carbon and nitrogen on growth.

A large body of experimental data on the responses of partitioning to environmental conditions—including elevated $[CO_2]_a$ (Rogers et al., 1994)—and stage of development exists, and this can be used to hypothesize underlying mechanisms that can be incorporated into models. Many models, however, use "fixed" partitioning coefficients. Such models probably will not properly simulate the effects of elevated $[CO_2]_a$ on partitioning.

Any consideration of carbon sources and sinks recalls old questions: When does source activity, and when does sink activity, limit growth? Simple answers to these questions do not exist, in part because both source activity and sink activity may limit growth, but at different times, and in part because a mechanistic basis for predicting plant growth does not exist. General observations that crop growth is stimulated by elevated $[CO_2]_a$ (Lawlor and Mitchell, 1991; Rogers et al., 1994) do indicate that carbohydrate supplies limit growth, at least during part of a crop's life cycle. Effects of elevated $[CO_2]_a$ on source–sink relationships may vary among crops. For example, the growth of indeterminate crops such as cotton may be more responsive to CO_2 enrichment than that of determinate crops (Radin et al., 1987; Lawlor and Mitchell, 1991). This is probably associated with favorable source–sink interactions. In agriculture, some flexibility exists with respect to managing source–sink relationships through sowing density and other management practices, so that models should be sensitive to plant density, and other factors, both as predictors of future productivity and as tools for present crop management.

To the extent that growth is limited by carbohydrate supply at ambient CO_2 levels, rather than another substrate such as nitrogen, increased photosynthesis, i.e., carbohydrate supply, due to elevated CO_2 should increase growth. It may also be the case that increased nonstructural carbohydrate levels per se stimulate growth and growth respiration by inducing various growth and growth-related process (Farrar and Williams, 1991a; Amthor, 1994a). Thus, increased nonstructural carbohydrate levels resulting from enhanced photosynthesis in elevated $[CO_2]_a$ may give rise to increased growth by a positive feedforward mechanism.

There is also negative feedback between source and sink activities. For example, laboratory manipulations of plants indicate that elevated carbohydrate levels can result in repression or reduction of photosynthetic enzymes and pigments (e.g., Sheen, 1994; Van Oosten et al., 1994; Webber et al., 1994). Such down-regulation responses to elevated $[CO_2]_a$ may be less apparent in field-grown plants than in pot-grown plants (e.g., Bowes, 1991), however, so that it is too simplistic to merely reduce photosynthetic capacity in a model when $[CO_2]_a$ is increased. Hierarchic models can (and should) consider a multiplicity of interactions among growth, photosynthesis, and carbohydrate accumulation, but knowledge of such interactions is limited.

Root–shoot partitioning of carbon, which can be affected by elevated $[CO_2]_a$ (Rogers et al., 1994), must be included in models. Several root–shoot partitioning models related to a "functional equilibrium" (Brouwer, 1983) have been developed (e.g., Reynolds and Thornley, 1982; Thornley and Johnson, 1990). Such models often work well in young plants, but may be less successful (without modification) for crops with established root systems.

The ratio of carbon to nitrogen (or protein) in plant biomass may be greater in elevated than in ambient $[CO_2]_a$, perhaps more so in C_3 than in C_4 crop species (e.g., Wong, 1979). This is often attributed to nitrogen dilution due to the accumulation of "extra" carbon in nonstructural carbohydrates, to a reduction in photosynthetic proteins, or to both. Models that correctly predict carbohydrate production, growth (including the composition of tissue formed), respiration, and nitrogen acquisition should also correctly predict the effects of $[CO_2]_a$ on nitrogen dilution. But, if elevated $[CO_2]_a$ results in a reduction in plant content of rubisco or other photosynthetic proteins (e.g., Sheen, 1994), the carbon to nitrogen ratio can be affected independently of nonstructural carbohydrate accumulation, which is a plant response requiring additional model complexity.

D. Stomates, Transpiration, and Crop Energy Balance

Stomatal aperture and leaf surface conductance often are negatively related to short-term changes in $[CO_2]_a$. On the other hand, there are several cases, including some long-term CO_2-enrichment experiments, in which stomatal conductance is not strongly related to CO_2 concentration (e.g., Gunderson *et al.*, 1993). When stomata do respond to CO_2 concentration, apparently it is $[CO_2]_i$ that is of significance, rather than the ambient or leaf surface CO_2 concentrations (Mott, 1988). Stomata in C_3 and C_4 leaves may be about equally sensitive to a given change in $[CO_2]_a$ or $[CO_2]_i$ (e.g., Morison and Gifford, 1983). The response of conductance to $[CO_2]_i$ cannot be predicted *a priori*, however, and must be determined by experiment.

Stomatal conductance can be linearly related to photosynthetic rate, and the ratio of $[CO_2]_i$ to $[CO_2]_a$ then is about constant over a range of $[CO_2]_a$ values (e.g., Wong *et al.*, 1985). On the basis of this observation, several models link conductance to photosynthesis (e.g., Farquhar and Wong, 1984; Collatz *et al.*, 1991; Aphalo and Jarvis, 1993; Amthor, 1994b).

The decline in stomatal conductance and the increase in photosynthesis with a short-term increase in $[CO_2]_a$ lead to a marked increase in the instantaneous ratio of CO_2 assimilated to water transpired in C_3 plants. This ratio is also increased in C_4 plants by elevated $[CO_2]_a$ due to stomatal closure. Indeed, when the growth of C_4 crops is enhanced by CO_2 enrichment, it may be due in part to an improved water balance (Lawlor and Mitchell, 1991).

Reduced stomatal conductance can increase leaf temperature due to reduced latent heat exchange. Thus, even without global warming, crop temperature may be elevated by increased $[CO_2]_a$, and models used to predict effects of global environmental change on crop physiology and growth should include a full canopy (and soil) energy balance. This requires the use of micrometeorological and soil physical relationships, including effects of atmospheric vapor pressure on transpiration. These physical relationships can be considered mechanistic from the viewpoint of crop modeling and should be directly applicable to predictions of crop responses to increased $[CO_2]_a$—to the extent that stomatal conductance can be predicted accurately—and climatic change.

Stomatal density (stomata per square meter of leaf) may be greater at subambient than at present ambient $[CO_2]_a$ as a result of more stomata per epidermal cell or smaller epidermal cells with the same number of stomata per epidermal cell. This reponse has been observed in experiments involving subambient and ambient $[CO_2]_a$'s, as well as in leaves of herbarium specimens collected during the last 200–250 years (Woodward, 1987; Peñuelas and Matamala, 1990). At supraambient $[CO_2]_a$'s, however, stomatal density may be about equal to values at present ambient $[CO_2]_a$ (Woodward, 1987), so that further increases in $[CO_2]_a$ may have little effect on stomatal density, although conductance (aperture) is likely to be reduced.

Reduced stomatal conductance has important implications for crop water use, the amount of precipitation or irrigation required by a crop, and even the locations at which crops can be grown. Regional reductions in stomatal conductance may also affect atmospheric evaporative demand. If regional latent heat exchange decreases, then sensible heat exchange increases (for a constant net radiation, although canopy longwave radiation emission increases with an increase in canopy temperature). "This combination of changes . . . will result in an increase of the humidity deficit (rising temperature, decreasing moisture content), which will enhance" transpiration (Jacobs and de Bruin, 1992). Plants in controlled environment chambers containing constant vapor pressure (controlled by processes other than evapotranspiration) will also show a greater change in transpiration for a given change in leaf conductance than would a regional landscape. The negative feedback from regional transpiration to the driving force for transpiration in the planetary boundary layer is an important, but mostly overlooked aspect of modeling plant, ecosystem, and landscape responses to elevated $[CO_2]_a$. Other relationships between elevated $[CO_2]_a$ and crop water use that should be considered in (or predicted by) models have been discussed by Eamus (1991).

E. Responses to Increased Temperature

Increased crop temperature brought about by an enhanced greenhouse effect or by an altered canopy energy budget will affect most plant processes and, therefore, model rate variables. Effects of temperature on growth and development are especially in need of research, and many model shortcomings (i.e., inaccurate or unrealistic predictions) can probably be traced back to an incomplete understanding of growth and developmental responses to temperature. Effects of temperature on developmental rate have been reviewed by Bonhomme *et al.* (1994; for maize) and Slafer and Rawson (1994; for wheat).

Developmental rate strongly responds to temperature, a fact that has important implications for predicting crop responses to global warming. Not only is developmental response to temperature nonlinear, but its slope is positive at low temperatures and negative at high temperatures with a relatively narrow temperature optimum: ca. 30–35°C for warm-season crops (e.g., Fig. 3) and a lower temperature for cool-season crops (e.g., Loomis and Connor, 1992). Equations useful for simulating the effects of temperature on developmental rate have been presented by, for example, Sharpe and DeMichele (1977) and Thornley and Johnson (1990).

Clearly, reliance on thermal units, unless they are calculated nonlinearly and over the diel cycle (i.e., hourly or more frequently), can introduce large errors into predictions of crop development over a growing season. Many models, unfortunately, do not account for the nonlinearity of developmental response to temperature over diel cycles. Moreover, most models use air temperature rather than plant temperature (the proper measure)

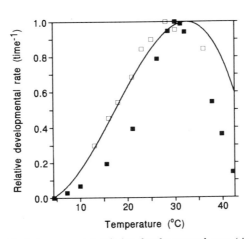

Figure 3 Relationship between maize relative developmental rate (time^{-1}; normalized to maximum value of unity) and temperature. Developmental rate was assessed as the inverse of time from planting to anthesis [solid box, Gilmore and Rogers, 1958, using Lehenbauer's data; open box, Warrington and Kanemasu, 1983a (cultivar XL45)] and as the number of leaf primordia initiated each day [line, Warrington and Kanemasu, 1983b (regression equation.)]

to calculate developmental rate. Thus, when any factor, such as crop water status and transpiration rate, causes a deviation in plant temperature from air temperature, air temperature will no longer be the correct model-driving variable in calculations of developmental rate equations (e.g., Seligman *et al.*, 1983). (The developmental rate of some crops also depends on day length. Although daylength will not change as a result of elevated $[CO_2]_a$, the time of year or latitude at which specific crops are grown may change if the temperature changes. This is a point to consider in crop models, although crop breeding may match the photoperiod responses of crops that will be grown at new latitudes or times of year, as a result of climatic change, to their new photoperiod lengths.)

Temperature can differentially affect rates of development and growth, which may be particularly important during grain filling in crops such as wheat. An increase from "low" to "moderate" temperature may stimulate wheat development and photosynthesis about equally, so that even though the duration of grain filling is reduced, final grain yield may be unaffected (e.g., Sofield *et al.*, 1977). An increase from "moderate" to "high" temperature, however, may further shorten the duration of grain filling, but without a compensating increase in photosynthesis or grain filling rates (e.g., Wardlaw *et al.*, 1980). The lack of growth stimulation at high temperature, while the developmental rate is accelerated, may be caused by several physiological or biochemical factors. For example, above ca. 30°C, wheat grain starch-producing enzymes may become unstable (Keeling *et al.*, 1993). In addition, photosynthesis typically responds less strongly to temperature than does development at temperatures several degrees Celsius below the optimum for development, so that increased temperature often results in a larger relative increase in developmental rate than in photosynthetic rate, which in turn limits the amount of photosynthesis that can occur during the grain filling period. With today's $[CO_2]_a$ and climate, "high" temperature during the grain-filling period may be the major environmental limitation to the productivity of wheat (Fischer, 1983) and other C_3 crops, but perhaps not for C_4 crops in high light (Evans, 1993). In elevated CO_2, however, more photosynthesis can occur during a time interval (at least in C_3 crops), so that the balance of accelerated development due to increased temperature and accelerated photosynthesis due to elevated CO_2 might result in increased, decreased, or no change in grain production.

Some carbohydrate used during grain filling arises from photosynthate stored prior to the grain-filling period, but again, elevated temperature may reduce the time available for such photosynthesis and storage, whereas elevated CO_2 may increase the potential for photosynthesis and carbohydrate storage during a given time interval. The interplay among development, growth, and photosynthesis will likely be as complicated in the future as it is at present. Many interactions and feedbacks, both positive and negative, are significant and crop models should include them all.

Plants in warm environments often have small levels of nonstructural carbohydrates, whereas plants in cool environments often have large carbohydrate levels. This has been attributed to a stronger positive response to elevated temperature by growth compared with photosynthesis, which results in a depletion of carbohydrate pools (Farrar and Williams, 1991b). Elevated $[CO_2]_a$ often results in elevated levels of nonstructural carbohydrates, so that effects of the combination of elevated $[CO_2]_a$ and elevated temperature on carbohydrate levels are complex. Indeed, crop carbohydrate balance is a complicated business in any CO_2 environment. Because carbohydrate balances are subject to many input and output processes (Evans, 1975), relationships among $[CO_2]_a$, temperature, and carbohydrate balance can be best studied with a model.

Warming may cause a decrease in root/shoot ratio, but observations are variable (Farrar and Williams, 1991b). Effects of elevated $[CO_2]_a$ on crop root/shoot ratio have also been variable, with perhaps more increases than decreases being observed (Lawlor and Mitchell, 1991; Rogers *et al.*, 1994). There are no known direct effects of temperature or $[CO_2]_a$ on root–shoot interactions or partitioning, so that elevated temperature and $[CO_2]_a$ do not necessitate new model components or relationships, although most existing models require improvements in their general treatment of partitioning. In addition, there is no reason that elevated temperature or $[CO_2]_a$ should have a consistent effect on root/shoot ratio, because there are so many indirect responses to those environmental variables. Models to be used in any environment should account for effects of temperature and $[CO_2]_a$ on carbohydrate level, growth rate, photosynthesis, and photosynthate partitioning and their interactions.

Rates of photosynthesis and respiration are affected by short-term changes in temperature, but it is well-known that photosynthesis (Berry and Björkman, 1980) and respiration (Amthor, 1994c) can acclimate to temperature change. There is no quantitative mechanistic model of acclimation to temperature, however, so it must be accounted for empirically (e.g., Säll and Pettersson, 1994). At present, few models account for the acclimation of photosynthesis or respiration to seasonal changes in temperature. Adaptation to temperature, both natural and as a result of crop breeding, should also be considered in models of crop growth in a warmer world.

If a model properly treated responses of development, growth, photosynthesis, partitioning, and respiration to temperature, it would be a powerful tool in the study of crop responses to climatic change. Unfortunately, most models do not have such traits, and the prediction of crop responses to elevated $[CO_2]_a$ in combination with elevated temperature is problematic.

F. Respiration

In models, respiration is generally considered as the sum of two or more functional components. Each respiratory component is linked to a different process, such as growth of new cells, maintenance of existing cells, and transport of ions from soil solution into roots. These processes are supported by respiration through the production of ATP, reductant, or carbon skeleton intermediates, with the pathways of respiration and related catabolism assumed to be the same regardless of the process supported. This is a reasonable and useful approach.

The most common distinction is between growth respiration (respiration linked to growth, including ion uptake from soil) and maintenance respiration (respiration linked to maintenance processes). There is a significant body of theory, based in large part on the biochemical model of Penning de Vries *et al.* (1974), that leads to a rather strict stoichiometry between growth rate and growth respiration rate for a given tissue composition. An increase in plant growth associated with elevated $[CO_2]_a$, or any other environmental change, is therefore expected to give rise to a concomitant increase in growth respiration (Amthor, 1996). A change in the chemical composition of growing plant tissue brought about by elevated $[CO_2]_a$, however, can alter the stoichiometry between rates of growth and growth respiration. Thus, it is not sufficient to calculate structural dry matter gain by plants as a function of $[CO_2]_a$ in models; the composition of the tissue formed also must be simulated. That is, models should relate growth respiration rate to growth rate and to the composition of the tissue being formed, and both can be affected by $[CO_2]_a$.

If the amount of ions taken up from soil is calculated independently of growth, and the metabolic costs of ion uptake are known, the amount of respiration required to support that ion uptake can be calculated separately from growth respiration (Johnson, 1990). The relationship between ion uptake respiration and ion uptake is understood mechanistically, but the specific costs of uptake (energy per ion) are not generally known, so that present models must rely on estimates (hypotheses) of the stoichiometries between ion uptake and ion uptake respiration.

Models that relate maintenance respiration rate to the rate of maintenance processes are based largely on the pioneering work of Penning de Vries (1975). In these models, maintenance respiration rate is generally linked to the rates of processes such as protein turnover and transport to counteract leaks within cells, say from cytosol to vacuole. This theory is not, however, fully quantitative; prediction of maintenance respiration rate is difficult because the controls of maintenance processes and their specific metabolic costs are not understood in quantitative terms (Amthor, 1994d). Thus, maintenance respiration must be simulated semiempirically. Because maintenance respiration rate may be related to plant or organ protein content (e.g., Ryan, 1991; Li and Jones, 1992; but see Byrd *et al.*, 1992), specific maintenance respiration rate (specific rate of x means rate of x per unit mass of tissue performing x) can be affected by changes in plant composition caused by elevated $[CO_2]_a$ or temperature.

Many effects of elevated $[CO_2]_a$ and temperature on respiration rate, i.e., the indirect effects (sensu Amthor, 1996), can be explained by changes in growth rate or plant composition. Models that calculate growth, maintenance, and ion uptake respiration as a function of growth rate and composition will therefore simulate these responses to $[CO_2]_a$ and temperature. But, there is also evidence that CO_2 inhibits respiration directly (Amthor, 1996). Because direct effects of CO_2 on respiration have not been well explained, it is necessary to empirically model direct effects of CO_2 on respiration.

IV. Knowledge Needs

Several information gaps that limit further model development were explicitly and implicitly identified earlier. It is easier, however, to outline what is known about plant responses to elevated $[CO_2]_a$ than what is unknown because the list is shorter. In our opinion, what is known is relatively simple. Elevated $[CO_2]_a$ stimulates leaf-level C_3 photosynthesis and may cause stomatal closure. Nearly all plant processes are dependent on photosynthesis. Photosynthesis may become acclimated to elevated $[CO_2]_a$ during short-term (days to months) experiments. Adaptation, natural or human-caused, to long-term (decades to centuries) elevated $[CO_2]_a$ might occur, but our knowledge of this is limited. Elevated $[CO_2]_a$ may directly inhibit respiration. It also may indirectly stimulate *or* slow respiration. Elevated $[CO_2]_a$ often stimulates growth, probably via increased carbohydrate supply. Elevated $[CO_2]_a$ can affect plant composition, including the amount of rubisco. It also can alter partitioning patterns. Note that most of these statements—concerning what *is* known—are qualified.

For crops, and more so for other vegetation, additional research and synthesis concerning the controls exerted by CO_2 (or carbohydrates) and temperature on growth and development in the field are needed. Much remains to be learned about these processes in all CO_2 environments—not just in elevated CO_2. Present models include a good deal of empiricism, i.e., simulation at a single level of organization, although a framework for the proper treatment of growth is available. That is to say, it is known that plants are composed of cells and that different organs and tissues have different cell types. The ranges of cell wall thickness (structural carbohydrate contents) and protein contents are known for different cell types.

The role of cell division in growth is understood. Locations of cell division are known. It is also known that cell expansion is needed for growth and where that expansion occurs. Cellular differentiation is known to be important, and its general features have been quantified and localized. Cation–anion balances can be quantified. Many of the laws (sensu Fourier) controlling growth are also understood. For example, the behavior of apical meristems is well understood, the role of indole-3-acetic acid is known, secondary cambia are always secondary, and many flowering and fruiting "rules" are clear. Most of this knowledge, however, is not used in models. There also is a vast amount of literature on morphogenesis that is only rarely considered by modelers. These examples highlight gaps between what is known and what knowledge most modelers use.

Nonetheless, a mechanistic understanding of the controls on growth exerted by carbohydrates does not exist. Furthermore, the important effects of temperature on both development and growth are inadequately understood (Evans, 1993), particularly in comparison with knowledge of the effects of $[CO_2]_a$ on photosynthesis. Because of this, the emphasis in elevated CO_2 research, including model building, must shift from descriptions of short-term photosynthetic responses to $[CO_2]_i$, PPFD, and temperature to a program explicitly examining the underlying mechanisms of the control of growth, development, and partitioning. That is, knowledge must be advanced so that growth responses to the environment and to photosynthate supply can be stated explicitly and in quantitative terms. These are knowledge gaps between what is known and what needs to be known.

But should—rather than can—modeling continue? Yes, modeling should continue in spite (or perhaps because) of the many knowledge gaps, because models are excellent vehicles for integrating and communicating information and for identifying additional important knowledge gaps and indicating appropriate research. Models can also be used to answer many "what if?" questions and to quantify *potential* effects of elevated $[CO_2]_a$ on crops. Used in this way, models are research tools. For example, a model can be used to assess the effects on growth of direct inhibition of leaf respiration by elevated $[CO_2]_a$. A model could also be used to quantify implications for whole-crop carbon balance and productivity of different stoichiometries between ion uptake from soil and ion uptake respiration. Modeling should continue also because models can be employed to make sense out of various responses to temperature and $[CO_2]_a$ as they interact to determine yield; combinations of temperature and $[CO_2]_a$ that might occur in the future can be used as model input to predict effects on developmental rate, grain growth rate, and final yield. Such an exercise might identify particularly important combinations of temperature and $[CO_2]_a$ that could then be studied in detail by experiment. After all, it is easier to conduct a large series of $[CO_2]_a$ × temperature experiments with a model than in the field, and a model might be used to circumvent some expensive, but relatively unneeded, experiments by highlighting probable important environmental controls on productivity. Models can also be used to compare the consequences of alternative hypotheses, allowing some hypotheses to be shown to be unacceptable if not outright false.

Most experimental data indicate an increase in crop growth with an increase in $[CO_2]_a$, and models should (and usually do) predict this response when other environmental conditions remain unchanged. Few data (or model simulations) are available from nutrient-limited crops, however. For natural vegetation (ecosystems), nutrient limitations may lessen plant growth responses to elevated $[CO_2]_a$ [reviewed by Field et al. (1992)], and the same might be expected for low-input crops. This issue deserves further experimental study, and proper crop models can be helpful in studying potential nutrient–CO_2 interactions and, therefore, in guiding related field research.

V. Conclusions

Short-term responses of photosynthesis and photorespiration to changes in $[CO_2]_a$ and temperature are well understood, and mechanistic models of leaf CO_2 assimilation are now in common use. Aside from photosynthesis and photorespiration, however, mechanistic understanding of physiological responses to elevated $[CO_2]_a$ and temperature is limited, although stomatal response to $[CO_2]_a$ and temperature often can be predicted semiquantitatively and effects of $[CO_2]_a$ and temperature on growth, respiration, and development can be predicted with some realism. This is a clear cue that elevated $[CO_2]_a$ and temperature research programs should shift the present emphasis from studies of leaf photosynthesis and empirical descriptions of growth to explanations of development, partitioning, and growth.

Because knowledge of mechanisms of control of crop physiological and growth processes is incomplete, models must be combinations of mechanism and empiricism; fully mechanistic models are not possible. Indeed, the limited number of mechanistic crop models is indicative of important knowledge gaps. But, in addition to a gap between what is known and what needs to be known for successful modeling, another type of knowledge gap exists: the gap between present knowledge and the use of that knowledge in models. Simple models cannot encapsulate present knowledge. Greater model complexity is necessary to close the gap between what is known and what knowledge is used in models. Obtaining a useful balance between mechanism and empiricism and understanding the limitations to extrapolative prediction are the tasks at hand for modelers.

Realistic models that can be applied across environments without tuning are needed, and a greater emphasis should be placed on model testing. Models that must be tuned for each new situation will not be helpful in predicting crop responses to global environmental change, although they may be suited to other applications. Realism in models of complex systems such as a soil–crop–atmosphere association follows only from a hierarchic structure. For accuracy, the lower levels of organization must be simulated properly, and lower level model accuracy stands to be significantly improved following additional research on fundamental controls of plant growth and development and their links to the physical environment. With respect to predicting crop productivity in a future, higher CO_2 world, models should integrate knowledge at levels of organization lower than the crop, allow the expression of interactions within and among levels, and arrive at explainable predictions of whole-crop responses to the environment.

Most quantitative predictions made to date of plant (and especially ecosystem) responses to elevated $[CO_2]_a$ and temperature are equivocal. Nonetheless, we are cautiously optimistic that, during the coming decade, models will significantly improve our understanding of crop (and other terrestrial ecosystem) responses to the changing environment. Progress will depend in part on a shift in experimental programs from descriptions of photosynthesis to explanations of growth and development and their controls. It will also depend on a shift from empirically based to mechanistically based models.

References

Allen, L. H., Jr., Drake, B. G., Rogers, H. H., and Shinn, J. H. (1992). Field techniques for exposure of plants and ecosystems to elevated CO_2 and other trace gases. *Crit. Rev. Plant Sci.* 11, 85–119.

Amthor, J. S. (1994a). Higher plant respiration and its relationships to photosynthesis. In "Ecophysiology of Photosynthesis" (E.-D. Schulze and M. M. Caldwell, eds.), pp. 71–101. Springer-Verlag, Berlin.

Amthor, J. S. (1994b). Scaling CO_2-photosynthesis relationships from the leaf to the canopy. *Photosynthesis Res.* 39, 321–350.

Amthor, J. S. (1994c). Plant respiratory responses to the environment and their effects on the carbon balance. In "Plant–Environment Interactions" (R. E. Wilkinson, ed.), pp. 501–554. Marcel Dekker, New York.

Amthor, J. S. (1994d). Respiration and carbon assimilate use. In "Physiology and Determination of Crop Yield" (K. J. Boote, J. M. Bennett, T. R. Sinclair, and G. M. Paulsen, eds.), pp. 221–250. Amer. Soc. Agron., Madison, WI.

Amthor, J. S. (1996). Plant respiratory responses to elevated CO_2 partial pressure. In "Advances in Carbon Dioxide Effects Research" (L. H. Allen, M. B. Kirkham, D. M. Olszyk, and C. Whitman, eds.). Amer. Soc. Agron. Madison, WI (in press).

Aphalo, P. J., and Jarvis, P. G. (1993). An analysis of Ball's empirical model of stomatal conductance. *Ann. Bot.* 72, 321–327.

Barnola, J. M., Raynaud, D., Korotkevich, Y. S., and Lorius, C. (1987). Vostok ice core provides 160,000-year record of atmospheric CO_2. *Nature* 329, 408–414.

Barnola, J.-M., Pimienta, P., Raynaud, D., and Korotkevich, Y. S. (1991). CO_2-climate relationship as deduced from the Vostok ice core: a re-examination based on new measurements and on a re-evaluation of the air dating. *Tellus* 43B, 83–90.

Berry, J., and Björkman, O. (1980). Photosynthetic response and adaptation to temperature in higher plants. *Annu. Rev. Plant Physiol.* 31, 491–543.

Bonhomme, R., Derieux, M., and Edmeades, G. O. (1994). Flowering of diverse maize cultivars in relation to temperature and photoperiod in multilocation field trials. *Crop Sci.* 34, 156–164.

Bowes, G. (1991). Growth at elevated CO_2: photosynthetic responses mediated through Rubisco. *Plant Cell Environ.* 14, 795–806.

Bowes, G. (1993). Facing the inevitable: plants and increasing atmospheric CO_2. *Annu. Rev. Plant Physiol. Plant Mol. Biol.* 44, 309–332.

Brouwer, R. (1983). Functional equilibrium: sense or nonsense? *Neth. J. Agric Sci.* 31, 335–348.

Byrd, G. T., Sage, R. F., and Brown, R. H. (1992). A comparison of dark respiration between C_3 and C_4 plants. *Plant Physiol.* 100, 191–198.

Campbell, G. S. (1985). "Soil Physics with BASIC." Elsevier, Amsterdam.

Colby, W. H., Crook, F. W., and Webb, S.-E. H. (1992). "Agricultural Statistics of the People's Republic of China, 1949–90." Agricultural and Trade Analysis Division, Economic Research Service, U.S. Department of Agriculture, Washington, DC.

Collatz, G. J., Berry, J. A., Farquhar, G. D., and Pierce, J. (1990). The relationship between the Rubisco reaction mechanism and models of photosynthesis. *Plant Cell Environ.* 13, 219–225.

Collatz, G. J., Ball, J. T., Grivet, C., and Berry, J. A. (1991). Physiological and environmental regulation of stomatal conductance, photosynthesis and transpiration: a model that includes a laminar boundary layer. *Agric. Forest Meterol.* 54, 107–136.

Collatz, G. J., Ribas-Carbo, M., and Berry, J. A. (1992). Coupled photosynthesis-stomatal conductance model for leaves of C_4 plants. *Aust. J. Plant Physiol.* 19, 519–538.

Curtis, P. S., Drake, B. G., and Whigham, D. F. (1989). Nitrogen and carbon dynamics in C_3 and C_4 estuarine marsh plants grown under elevated CO_2 in situ. *Oecologia* 78, 297–301.

Curtis, P. S., Balduman, L. M., Drake, B. G., and Whigham, D. F. (1990). Elevated atmospheric CO_2 effects on belowground processes in C_3 and C_4 estuarine marsh communities. *Ecology* 71, 2001–2006.

Denison, R. F., and Loomis, R. S. (1989). "An Integrative Physiological Model of Alfalfa Growth and Development." Univ. California, Div. Agric. Natural Resources, Publ. 1926.

de Wit, C. T. (1970). Dynamic concepts in biology. In "Prediction and Measurement of Photosynthetic Productivity" (I. Setlik, ed.), pp. 17–23. Pudoc, Wageningen, The Netherlands.

de Wit, C. T., *et al.* (1978). "Simulation of Assimilation, Respiration, and Transpiration of Crops." Pudoc, Wageningen.

Drake, B. G., and Leadley, P. W. (1991). Canopy photosynthesis of crops and native plant communities exposed to long-term elevated CO_2. *Plant Cell Environ.* 14, 853–860.

Eamus, D. (1991). The interaction of rising CO_2 and temperatures with water use efficiency. *Plant Cell Environ.* 14, 843–852.

Etheridge, D. M., Pearman, G. I., and de Silva, F. (1988). Atmospheric trace-gas variations as revealed by air trapped in an ice core from Law Dome, Antarctica. *Ann. Glaciol.* 10, 28–33.

Evans, L. T. (1975). The physiological basis of crop yield. In "Crop Physiology: Some Case Histories" (L. T. Evans, ed.), pp. 327–355. Cambridge Univ. Press, Cambridge, UK.

Evans, L. T. (1993). "Crop Evolution, Adaptation and Yield." Cambridge Univ. Press, Cambridge, UK.

FAO (annual). "Production Yearbook." Food and Agriculture Organization of the United Nations, Rome.

Farquhar, G. D. (1989). Models of integrated photosynthesis of cells and leaves. *Philes. Trans. R. Soc. London B* 323, 357–367.

Farquhar, G. D., and von Caemmerer, S. (1982). Modeling photosynthetic response to environmental conditions. In "Encyclopedia of Plant Physiology," New Series, Vol. 12B (O. L. Lange, P. S. Nobel, C. B. Osmond, and H. Ziegler, eds.), pp. 549–587. Springer-Verlag, Berlin.

Farquhar, G. D., and Wong, S. C. (1984). An empirical model of stomatal conductance. *Aust. J. Plant Physiol.* 11, 191–210.

Farquhar, G. D., von Caemmerer, S., and Berry, J. A. (1980). A biochemical model of photosynthetic CO_2 assimilation in leaves of C_3 species. *Planta* 149, 78–90.

Farrar, J. F., and Williams, J. H. H. (1991a). Control of the rate of respiration in roots: compartmentation, demand and the supply of substrate. In "Compartmentation of Plant Metabolism in Nonphotosynthetic Tissues" (M. J. Emes, ed.), pp. 167–188. Cambridge Univ. Press, Cambridge, UK.

Farrar, J. F., and Williams, M. L. (1991b). The effects of increased atmospheric carbon dioxide and temperature on carbon partitioning, source-sink relations and respiration. *Plant Cell Environ.* 14, 819–830.

Field, C. B., Chapin, F. S., III, Matson, P. A., and Mooney, H. A. (1992). Responses of terrestrial ecosystems to the changing atmosphere: a resource-based approach. *Annu. Rev. Ecol. Syst.* 23, 201–235.

Fischer, R. A. (1983). Wheat. In "Potential Productivity of Field Crops under Different Environments," pp. 129–154. International Rice Research Institute, Los Baños, Philippines.

Friedli, H., Lötscher, H., Oeschger, H., Siegenthaler, U., and Stauffer, B. (1986). Ice core record of the $^{13}C/^{12}C$ ratio of atmospheric CO_2 in the past two centuries. *Nature* 324, 237–238.

Gilmore, E. C., Jr., and Rogers, J. S. (1958). Heat units as a method of measuring maturity in corn. *Agron. J.* 50, 611–615.

Greenwood, P., Greenhalgh, A., Baker, C., and Unsworth, M. (1982). A computer-controlled system for exposing field crops to gaseous air pollutants. *Atmos. Environ.* 16, 2261–2266.

Gunderson, C. A., Norby, R. J., and Wullschleger, S. D. (1993). Foliar gas exchange responses of two deciduous hardwoods during 3 years of growth in elevated CO_2: No loss of photosynthetic enhancement. *Plant Cell Environ.* 16, 797–807.

Harley, P. C., and Sharkey, T. D. (1991). An improved model of C_3 photosynthesis at high CO_2: reversed O_2 sensitivity explained by lack of glycerate reentry into the chloroplast. *Photosynth. Res.* 27, 169–178.

Hesketh, J. D., and Jones, J. W. (eds.) (1980). "Predicting Photosynthesis for Ecosystem Models," Vols. I and II. CRC Press, Boca Raton, FL.

Hunt, R., Hand, D. W., Hannah, M. A., and Neal, A. M. (1993). Further responses to CO_2 enrichment in British herbaceous species. *Funct. Ecol.* 7, 661–668.

Jacobs, C. M. J., and de Bruin, H. A. R. (1992). The sensitivity of regional transpiration to land-surface characteristics: significance of feedback. *J. Climate* 5, 683–698.

Jarvis, P. G. (1993). Prospects for bottom-up models. In "Scaling Physiological Processes: Leaf to Globe" (J. R. Ehleringer and C. B. Field, eds.), pp. 115–126. Academic Press, San Diego.

Johnson, I. R. (1990). Plant respiration in relation to growth, maintenance, ion uptake and nitrogen assimilation. *Plant Cell Environ.* 13, 319–328.

Jordan, D. B., and Ogren, W. L. (1984). The CO_2/O_2 specificity of ribulose 1,5-bisphosphate carboxylase/oxygenase. *Planta* 161, 308–313.

Jorgensen, S. E. (1994). "Fundamentals of Ecological Modelling," 2nd ed. Elsevier, Amsterdam.

Keeling, C. D., and Whorf, T. P. (1994). Atmospheric CO_2 records from sites in the SIO air sampling network. In "Trends '93: A Compendium of Data on Global Change" (T. B. Boden, D. P. Kaiser, R. J. Sepanski, and F. W. Stoss, eds.), ORNL/CDIAC-65, pp. 16–26. Carbon Dioxide Information Analysis Center, Oak Ridge National Laboratory, Oak Ridge, TN.

Keeling, P. L., Bacon, P. J., and Holt, D. C. (1993). Elevated temperature reduces starch deposition in wheat endosperm by reducing the activity of soluble starch synthase. *Planta* 191, 342–348.
Laing, W. A., Ogren, W. L., and Hageman, R. H. (1974). Regulation of soybean net photosynthetic CO_2 fixation by the interaction of CO_2, O_2, and ribulose 1,5-diphosphate carboxylase. *Plant Physiol.* 54, 678–685.
Lawlor, D. W., and Mitchell, R. A. C. (1991). The effects of increasing CO_2 on crop photosynthesis and productivity: a review of field studies. *Plant Cell Environ.* 14, 807–818.
Leuenberger, M., Siegenthaler, U., and Langway, C. C. (1992). Carbon isotope composition of atmospheric CO_2 during the last ice age from an Antarctic ice core. *Nature* 357, 488–490.
Li, M., and Jones, M. B. (1992). Effect of nitrogen status on the maintenance respiration of C_3 and C_4 *Cyperus* species. *In* "Molecular, Biochemical and Physiological Aspects of Plant Respiration" (H. Lambers and L. H. W. van der Plas, eds.), pp. 509–514. SPB Academic Publ. The Hague.
Long, S. P. (1991). Modification of the response of photosynthetic productivity to rising temperature by atmospheric CO_2 concentrations: has its importance been underestimated? *Plant Cell Environ.* 14, 729–739.
Loomis, R. S., and Connor, D. J. (1992). "Crop Ecology: Productivity and Management in Agricultural Systems." Cambridge Univ. Press, Cambridge, UK.
Loomis, R. S., Rabbinge, R., and Ng, E. (1979). Explanatory models in crop physiology. *Annu. Rev. Plant Physiol.* 30, 339–367.
Manabe, S. (1983). Carbon dioxide and climatic change. *Adv. Geophys.* 25, 39–82.
Manabe, S., and Stouffer, R. J. (1994). Multiple-century response of a coupled ocean-atmosphere model to an increase of atmospheric carbon dioxide. *J. Climate* 7, 5–23.
McLeod, A. R., Fackrell, J. E., and Alexander, K. (1985). Open-air fumigation of field crops: criteria and design for a new experimental system. *Atmos. Environ.* 19, 1639–1649.
Morison, J. I. L., and Gifford, R. M. (1983). Stomatal sensitivity to carbon dioxide and humidity. A comparison of two C_3 and two C_4 grass species. *Plant Physiol.* 71, 789–796.
Mott, K. A. (1988). Do stomata respond to CO_2 concentrations other than intercellular: *Plant Physiol.* 86, 200–203.
Mutsaers, H. J. W. (1984). KUTUN: a morphogenetic model for cotton (*Gossypium hirsutum* L.). *Agric. Syst.* 14, 229–257.
Neftel, A., Moor, E., Oeschger, H., and Stauffer, B. (1985). Evidence from polar ice cores for the increase in atmospheric CO_2 in the past two centuries. *Nature* 315, 45–47.
Neftel, A., Oeschger, H., Staffelbach, T., and Stauffer, B. (1988). CO_2 record in the Byrd ice core 50,000–5,000 years BP. *Nature* 331, 609–611.
Norby, R. J., O'Neill, E. G., and Luxmoore, R. J. (1986). Effects of atmospheric CO_2 enrichment on the growth and mineral nutrition of *Quercus alba* seedlings in nutrient-poor soil. *Plant Physiol.* 82, 83–89.
Norby, R. J., Gunderson, C. A., Wullschleger, S. D., O'Neill, E. G., and McCracken, M. K. (1992). Productivity and compensatory responses of yellow-poplar trees in elevated CO_2. *Nature* 357, 322–324.
Norman, J. M. (1993). Scaling processes between leaf and canopy levels. *In* "Scaling Physiological Processes: Leaf to Globe" (J. R. Ehleringer and C. B. Field, eds.), pp. 41–76. Academic Press, San Diego.
Passioura, J. B. (1979). Accountability, philosophy and plant physiology. *Search* 10, 347–350.
Penning de Vries, F. W. T. (1975). The cost of maintenance processes in plant cells. *Ann. Bot.* 39, 77–92.
Penning de Vries, F. W. T. (1983). Modeling of growth and production. *In* "Encyclopedia of Plant Physiology," New Series Vol. 12D (O. L. Lange, P. S. Nobel, C. B. Osmond, and H. Ziegler, eds.), pp. 117–150. Springer-Verlag, Berlin.

Penning de Vries, F. W. T., Brunsting, A. H. M., and van Laar, H. H. (1974). Products, requirements and efficiency of biosynthesis: A quantitative approach. *J. Theor. Biol.* 45, 339–377.

Penning de Vries, F. W. T., Witlage, J. M., and Kremer, D. (1979). Rates of respiration and of increase in structural dry matter in young wheat, ryegrass and maize plants in relation to temperature, to water stress and to their sugar content. *Ann. Bot.* 44, 595–609.

Penning de Vries, F. W. T., Jansen, D. M., ten Berge, H. F. M., and Bakema, A. (1989). "Simulation of Ecophysiological Processes of Growth in Several Annual Crops." Pudoc, Wageningen, The Netherlands.

Peñuelas, J., and Matamala, R. (1990). Changes in N and S leaf content, stomatal density and specific leaf area of 14 plant species during the last three centuries of CO_2 increase. *J. Exp. Bot.* 41, 1119–1124.

Popper, K. R. (1958). "The Logic of Scientific Discovery." Hutchinson, London.

Radin, J. W., Kimball, B. A., Hendrix, D. L., and Mauney, J. R. (1987). Photosynthesis of cotton plants exposed to elevated levels of carbon dioxide in the field. *Photosynth. Res.* 12, 191–203.

Raynaud, D., Jouzel, J., Barnola, J. M., Chappellaz, J., Delmas, R. J., and Lorius, C. (1993). The ice record of greenhouse gases. *Science* 259, 926–934.

Reynolds, J. F., and Thornley, J. H. M. (1982). A shoot:root partitioning model. *Ann. Bot.* 49, 585–597.

Rogers, H. H., Runion, G. B., and Krupa, S. V. (1994). Plant responses to atmospheric CO_2 enrichment with emphasis on roots and the rhizosphere. *Environ. Pollut.* 83, 155–189.

Ryan, M. G. (1991). Effects of climate change on plant respiration. *Ecol. Appl.* 1, 157–167.

Sage, R. F. (1990). A model describing the regulation of ribulose-1,5-bisphosphate carboxylase, electron transport, and triose phosphate use in response to light intensity and CO_2 in C_3 plants. *Plant Physiol.* 94, 1728–1734.

Sage, R. F., and Reid, C. D. (1994). Photosynthetic response mechanisms to environmental change in C3 plants. *In* "Plant–Environment Interactions" (R. E. Wilkinson, ed.), pp. 413–499. Marcel Dekker, New York.

Sage, R. F., Sharkey, T. D., and Pearcy, R. W. (1990). The effect of leaf nitrogen and temperature on the CO_2 response of photosynthesis in the C_3 dicot *Chenopodium album* L. *Aust. J. Plant Physiol.* 17, 135–148.

Säll, T., and Pettersson, P. (1994). A model of photosynthetic acclimation as a special case of reaction norms. *J. Theor. Biol.* 166, 1–8.

Seligman, N. G., Loomis, R. S., Burke, J., and Abshahi, A. (1983). Nitrogen nutrition and canopy temperature in field-grown spring wheat. *J. Agric. Sci. Camb.* 101, 691–697.

Sharkey, T. D. (1985). Photosynthesis in intact leaves of C_3 plants: physics, physiology, and rate limitations. *Bot. Rev.* 51, 53–105.

Sharpe, P. J. H., and DeMichele, D. W. (1977). Reaction kinetics of poikilotherm development. *J. Theor. Biol.* 64, 649–670.

Sheen, J. (1994). Feedback control of gene expression. *Photosynth. Res.* 39, 427–438.

Siegenthaler, U., Friedli, H., Loetscher, H., Moor, E., Neftel, A., Oeschger, H., and Stauffer, B. (1988). Stable-isotope ratios and concentration of CO_2 in air from polar ice cores. *Ann. Glaciol.* 10, 151–156.

Slafer, G. A., and Rawson, H. M. (1994). Sensitivity of wheat phasic development to major environmental factors: a re-examination of some assumptions made by physiologists and modellers. *Aust. J. Plant Physiol.* 21, 393–426.

Sofield, I., Wardlaw, I. F., Evans, L. T., and Zee, S. Y. (1977). Nitrogen, phosphorus and water contents during grain development and maturation in wheat. *Aust. J. Plant Physiol.* 4, 799–810.

Stanhill, G. (1976). Trends and deviations in the yield of the English wheat crop during the last 750 years. *Agro-Ecosystems* 3, 1–10.

Thornley, J. H. M., and Johnson, I. R. (1990). "Plant and Crop Modelling: A Mathematical Approach to Plant and Crop Physiology." Oxford Univ. Press, Oxford, UK.

van Bavel, C. H. M., and Ahmed, J. (1976). Dynamic simulation of water depletion in the root zone. *Ecol. Model.* 2, 189–212.

van Keulen, H., Seligman, N. G., and Goudriaan, J. (1975). Availability of anions in the growth medium to roots of an actively growing plant. *Neth. J. Agric. Sci.* 23, 131–138.

Van Oosten, J.-J., Wilkins, D., and Besford, R. T. (1994). Regulation of the expression of photosynthetic nuclear genes by CO_2 is mimicked by regulation by carbohydrates: a mechanism for the acclimation of photosynthesis to high CO_2? *Plant Cell Environ.* 17, 913–923.

von Caemmerer, S., and Evans, J. R. (1991). Determination of the average partial pressure of CO_2 in chloroplasts from leaves of several C_3 plants. *Aust. J. Plant Physiol.* 18, 287–305.

Wahlen, M., Allen, D., Deck, B., and Herchenroder, A. (1991). Initial measurements of CO_2 concentrations (1530 to 1940 AD) in air occluded in the GISP 2 ice core from central Greenland. *Geophys. Res. Lett.* 18, 1457–1460.

Wang, Y.-P., McMurtrie, R. E., and Landsberg, J. J. (1992). Modelling canopy photosynthetic productivity. *In* "Crop Photosynthesis: Spatial and Temporal Determinants" (N. R. Baker and H. Thomas, eds.), pp. 43–67. Elsevier Science Publishers, Amsterdam.

Wardlaw, I. F., Sofield, I., and Cartwright, P. M. (1980). Factors limiting the rate of dry matter accumulation in the grain of wheat grown at high temperature. *Aust. J. Plant Physiol.* 7, 387–400.

Warrington, I. J., and Kanemasu, E. T. (1983a). Corn growth response to temperature and photoperiod. I. Seedling emergence, tassel initiation, and anthesis. *Agron. J.* 75, 749–754.

Warrington, I. J., and Kanemasu, E. T. (1983b). Corn growth response to temperature and photoperiod. II. Leaf-initiation and leaf-appearance rates. *Agron. J.* 75, 755–761.

Webber, A. N., Nie, G.-Y., and Long, S. P. (1994). Acclimation of photosynthetic proteins to rising atmospheric CO_2. *Photosynth. Res.* 39, 413–425.

Wisiol, K., and Hesketh, J. D. (eds.) (1987). "Plant Growth Modeling for Resource Management," Vols. 1 and 2. CRC Press, Boca Raton, FL.

Wong, S. C. (1979). Elevated atmospheric partial pressure of CO_2 and plant growth. I. Interactions of nitrogen nutrition and photosynthetic capacity in C_3 and C_4 plants. *Oecologia* 44, 68–74.

Wong, S.-C., Cowan, I. R., and Farquhar, G. D. (1985). Leaf conductance in relation to rate of CO_2 assimilation. I. Influences of nitrogen nutrition, phosphorus nutrition, photon flux density, and ambient partial pressure of CO_2 during ontogeny. *Plant Physiol.* 78, 821–825.

Woo, K. C., and Wong, S. C. (1983). Inhibition of CO_2 assimilation by supraoptimal CO_2: effect of light and temperature. *Aust. J. Plant Physiol.* 10, 75–85.

Woodward, F. I. (1987). Stomatal numbers are sensitive to increases in CO_2 from pre-industrial levels. *Nature* 327, 617–618.

Wullschleger, S. D., Ziska, L. H., and Bunce, J. A. (1994). Respiratory responses of higher plants to atmospheric CO_2 enrichment. *Physiol. Plant.* 90, 221–229.

19
Progress, Limitations, and Challenges in Modeling the Effects of Elevated CO_2 on Plants and Ecosystems

James F. Reynolds, Paul R. Kemp, Basil Acock, Jia-Lin Chen, and Daryl L. Moorhead

I. Introduction

Our ability to predict the potential effects of climate change and elevated atmospheric CO_2 concentration (CO_2) on plants and ecosystems clearly depends on the development of mechanistic simulation models (Mooney et al., 1991). Climatic factors and CO_2 affect plants and ecosystems both directly and indirectly in complex ways. Increased atmospheric CO_2 has immediate and direct effects on photosynthesis, stomatal response, and respiration, which are then translated to varying (and often unknown) degrees to whole-plant growth, allocation, and morphology (Strain, 1985; Bazzaz, 1990; Grodzinski, 1992). Changes in whole-plant structure and function, in turn, affect lower hierarchical levels, causing, for example, changes in leaf photosynthetic capacity (Harley et al., 1992), and are translated to higher hierarchical levels, potentially altering population dynamics and species interactions (Bazzaz and McConnaughay, 1992), ecosystem processes such as trophic interactions (Fajer et al., 1991; Ayres, 1993), and carbon and nutrient cycling (Woodward et al., 1991). Many of these responses are further modified by environmental variables such as nutrients, temperature, and rainfall, which are also expected to change in the future.

A wholly empirical approach to predicting ecosystem responses to elevated CO_2 is not only impractical but potentially misleading. It is impossible to design more than a small fraction of the many different combinations of experiments that would be needed to sort out complex interactions (Reynolds and Acock, 1985). The size and expense of equipment restrict studies to relatively small spatial scales (e.g., a square meter). Also, the experiments are necessarily short, so that it is impossible to observe long-term changes that may lead to new feedbacks, homeostasis, or shifts in species (Strain and Thomas, 1992; Pacala and Hurtt, 1993). To date, there is considerable imbalance in the representation of species examined, with the majority being cultivated species (Körner, 1993), which may not be representative of wild species. Furthermore, most studies have relied on potted plants, which may behave differently from species in the field (Thomas and Strain, 1991).

Mechanistic models enable us to complement specific data with more general knowledge of plant and ecosystem behavior. However, with so few empirical data, we still do not understand many of the processes that control ecosystem responses, especially the responses to high CO_2. In the absence of knowledge, hypotheses proliferate and these largely untested hypotheses become incorporated in mechanistic models. It is, therefore, important to assess the limitations and uncertainties of the numerous mechanistic models that have been proposed for forecasting ecosystem, regional, and global responses to climate change and elevated CO_2 (e.g., Solomon, 1986; Pastor and Post, 1988; Schimel *et al.*, 1990; Running and Nemani, 1991; Botkin and Nisbet, 1992; Monserud *et al.*, 1993; Baker *et al.*, 1993, Rastetter *et al.*, 1991). Different models represent similar phenomena in different ways with unknown consequences. The underlying assumptions are often overly simplistic, e.g., the use of linear relations to represent effects of CO_2 on ecosystems in spite of considerable experimental evidence suggesting nonlinear responses (Bazzaz, 1990; Woodward *et al.*, 1991). In addition, models must consider (or ignore) processes that operate at various levels of biological organization, as well as the translation of effects across hierarchical scales (Wessman, 1992; Reynolds *et al.*, 1993a). In most cases, model testing (validation) is inadequate or impossible (Oreskes *et al.*, 1994).

In this chapter, we review the progress that has been made at the various hierarchical levels (leaf to the globe) in modeling the responses of plants and ecosystems to elevated CO_2, and we examine the current limitations and uncertainties of these models. Our review is intended to be representative, not exhaustive. Many of the limitations in models represent gaps in our collective knowledge about the responses of plants and ecosystems to elevated CO_2. We therefore emphasize them in an effort to help direct further research and, particularly, to link experimental and modeling research. With a tight coupling of modeling and experimentation, the limited body of empirical knowledge can be greatly extended through state-of-the-art model simulations that capture complex combinations of environmental–biotic interactions (Reynolds *et al.*, 1993b).

II. Leaf-Level Models

A. Photosynthesis

The most significant effect of CO_2 at the leaf level is on photosynthesis. Elevated CO_2 directly affects photosynthesis through interactions at the biochemical and physiological levels of carboxylation (Bowes, 1991; Stitt, 1991). In addition, there may be direct or indirect effects of elevated CO_2 at the leaf level on stomatal regulation (Strain, 1985), cellular respiration (Amthor, 1991), carbohydrate metabolism (Farrar and Williams, 1991; Stitt 1991; Woodward *et al.*, 1991), and secondary carbon chemistry (Johnson and Lincoln, 1991; Lambers, 1993). Photosynthesis is well understood, and the short-term effects of elevated CO_2 can be modeled with relatively high confidence by using the mechanistic model of Farquhar and von Caemmerer (1982; see Harley and Tenhunen, 1991; Harley and Sharkey, 1991; Harley *et al.*, 1992).

Limitations and Uncertainties: Long-Term Adjustments of Photosynthesis to Elevated CO_2 (i.e., Acclimation); Interactions with Other Environmental Limitations Although mechanistic models of leaf photosynthesis readily describe short-term responses of photosynthesis to increasing CO_2 (Harley and Tenhunen, 1991), there are considerable complications in the application of these models to the prediction of long-term responses to elevated CO_2 and in scaling up from leaf photosynthesis to the whole plant. In the short term, elevated CO_2 increases the leaf photosynthetic rate, but in the long term photosynthesis may decline, even approaching its value at ambient CO_2 (Tissue and Oechel, 1987). A reduction in the photosynthetic rate from the short-term potential rate is termed acclimation and may result from changes in leaf anatomy and/or biochemistry. There are large differences among species in their long-term responses to elevated CO_2, even among similar life forms (Oberbauer *et al.*, 1985; Reekie and Bazzaz, 1989; Sage *et al.*, 1989; El Kohen *et al.*, 1993), and some studies have found little evidence of acclimation of photosynthesis (Ziska *et al.*, 1991; Bunce, 1992).

One form of acclimation results from changes in the amounts of key photosynthetic enzymes and chlorophyll (von Caemmerer and Farquhar, 1984; Sage *et al.*, 1989). This has been modeled empirically (Harley *et al.*, 1992; Long, 1991) by adjusting model parameters (V_{cmax}, V_{omax}, and J_{max}) to account for reduced enzyme activity or reduced enzyme content, perhaps in response to changing nitrogen (N) allocation patterns within the leaf (Tissue *et al.*, 1993) and plant (Larigauderie *et al.*, 1988). Another aspect of long-term reduction in photosynthetic response to elevated CO_2 may be related to feedback inhibition on photosynthesis due to starch accumulation or even damaged chloroplast structure from large starch grains (Sasek *et al.*, 1985).

Some insights into the question of acclimation of leaf-level responses to elevated CO_2 have come from conceptual (Field and Mooney, 1986) and mathematical (Hilbert and Reynolds, 1991) models. Hilbert and Reynolds (1991) coupled a mechanistic photosynthesis model with an allocation model, and the combined model predicted that the acclimation of photosynthesis results from reallocation of N away from carboxylation enzymes. This model prediction was corroborated by the experimental findings of Tissue *et al.* (1993), who showed a decline in rubisco content in elevated CO_2. They also found that light reaction enzymes were increased. Thus, while elevated CO_2 may initially result in an accumulation of photosynthetic intermediates because of insufficient ATP and NADPH from the light reactions (Eamus and Jarvis, 1989), ultimately, reallocation of N within the leaf would lead to an increase in enzyme systems in the light reactions, an alleviation of end product accumulation, and a balance among the components of the photosynthetic system (Field and Mooney, 1986; Tissue *et al.*, 1993).

Experimental and modeling studies clearly suggest that the long-term photosynthetic responses of species will depend on the interaction of photosynthetic metabolism with leaf-level and plant-level allocation patterns. Species that maintain sink strength and have abundant external nutrient supplies should exhibit less long-term acclimation of photosynthesis. Models must also account for changes in leaf anatomy induced by high CO_2, especially where these changes can be related to acclimation and whole-plant responses to elevated CO_2.

Some of the differences in long-term responses reported in various studies may arise from the interactions of elevated CO_2 with other environmental variables, mainly, nutrients, temperature, and water (Strain and Cure, 1985; Bazzaz, 1990; Long, 1991). There is considerable uncertainty with regard to how model parameters might be affected in the long term by water or nutrient stress. However, some responses to environmental variables presently are incorporated directly into the Farquhar and von Caemmerer (1982) photosynthesis model, and these aspects of the model provide a means of theoretically investigating interactions of elevated CO_2 with other environmental factors. For example, one modification of the photosynthesis model included the effect of inorganic phosphate recycling on carboxylation (Harley *et al.*, 1992), and this model predicted that at high CO_2 phosphate limitation (in the habitat) could limit photosynthesis. This model prediction was corroborated by Conroy *et al.* (1990), who showed that photosynthesis enhancement at high CO_2 was decreased in phosphorus-deficient conditions. Long (1991) used the Farquhar and von Caemmerer (1982) model to explore the effects of elevated CO_2 on temperature and light responses of leaf photosynthesis. The model predicted that elevated CO_2 caused changes in photorespiration, quantum yield, light compensation point, and light-saturated photosynthesis, and the changes were much greater with increased leaf temperature. Long's results suggested that, while the enhancement of photosynthesis in light-saturated leaves may be subject to down-regulation (acclimation), the enhancement in shaded leaves was less likely to decline. This has important implications for leaves in canopies and shade plants.

B. Stomatal Conductance

The mechanisms by which CO_2 interacts with the regulation of stomata remain elusive. An empirical model that relates stomatal conductance to photosynthesis and ambient humidity (Ball *et al.*, 1987) has been assumed to apply under conditions of elevated CO_2 (Harley and Tenhunen, 1991; Harley *et al.*, 1992). Mechanistic models of stomatal function may be more appropriate in future model development, particularly with regard to the interaction of stomatal regulation under interacting effects of CO_2 and water stress (e.g., Mansfield *et al.*, 1990).

Limitations and Uncertainties: Effects of CO_2 on Stomata   The lack of understanding of the direct and indirect effects of CO_2 on stomatal function clearly limits the development of process models (Mott, 1990). Stomatal conductance is often found to diminish by 25–40% for a number of species exposed to doubled atmospheric CO_2 (Morison, 1985; Eamus and Jarvis, 1989; Lawlor and Mitchell, 1991), but other studies have found no decrease in stomatal conductance (Tolley and Strain, 1985; Hollinger, 1987; Bunce, 1992). There is little evidence of acclimation of stomatal conductance at elevated CO_2 (Eamus, 1991), although there are reports of changes in stomatal density with long-term exposure to elevated CO_2 (Woodward and Bazzaz, 1988; Oberbauer *et al.*, 1985). Elevated CO_2 may also interact with other environmental variables in altering the responses of stomata. Stomatal sensitivity to vapor pressure deficit was increased in some species and decreased in another grown in elevated CO_2 (Hollinger, 1987). Stomatal sensitivity to elevated CO_2 may be increased (Morison and Gifford, 1983) or decreased (Beable *et al.*, 1979) by increased light. Thus, further studies on the effects of elevated CO_2 on stomata and the interactions with other environmental factors are needed for the development of mechanistic models of stomatal response to elevated CO_2.

III. Plant-Level Models

While there are models dealing with nearly all aspects of plant structure and function, we focus here on modeling several processes that can be strongly influenced by CO_2 and that are ultimately important in determining plant development, the acquisition of resources, and the interaction of plants with one another and with other organisms at various trophic levels. These processes are canopy photosynthesis, transpiration, and allocation of carbon and nitrogen (N) to various vegetative and reproductive structures. Allocation affects plant morphology, the ability to compete for light, water, and nutrients, the number and viability of seed produced, and, hence, the species composition of ecosystems.

A. Canopy Photosynthesis and Transpiration

Integration from the individual leaf to the plant canopy for estimating whole-plant photosynthesis and transpiration has been accomplished with both detailed (integrated) and simplified (scaled) canopy models [see Norman (1993) for a discussion of integration versus scaling; see also Reynolds *et al.* (1992) and McMurtrie and Wang (1993)]. Integrated canopy models include elements of energy and gas exchange for a large number of leaf classes over a large number of canopy layers and can predict radiation interception and canopy photosynthesis and transpiration that compare favorably with measured values for grass (Norman, 1993) and shrub canopies (Reynolds *et al.*, 1992). Both Reynolds *et al.* (1992) and Norman (1993) investigated simplified, or scaled, approaches to modeling the plant canopy and concluded that one or a few canopy layers, with leaves all in the same average arrangement, are sufficient to accurately represent the plant canopy in most cases, so long as the leaves are divided into sunlit and shaded classes. Reynolds *et al.* (1992) concluded that scaling from the leaf to the canopy depended upon the accurate characterization leaf temperatures (i.e., leaf energy exchange) and leaf N within the canopy and that the scaled canopy model would reasonably predict short-term responses of canopy photosynthesis and transpiration to elevated CO_2. Baldocchi (1993) further concluded that detailed turbulent transfer calculations could also be eliminated from canopy models with minimal effect on predicted canopy gas exchange, at least for relatively rough canopies.

Limitations and Uncertainties: Accounting for Secondary Effects of Elevated CO_2 on Canopy Architecture and Reallocation of Nutrients within the Canopy Simplified canopy models, consisting of only one or a few leaf layers, may not adequately account for changes in canopy architecture or redistribution of N within the canopy at high CO_2. For example, elevated CO_2 may induce changes in specific leaf area, node number, or stem biomass that could, in turn, alter the energy budget and gas exchange of the canopy (Bazzaz, 1990). Elevated CO_2 may alter leaf senescence patterns and, thus, leaf shading within the canopy (Bazzaz, 1990). A further secondary effect of elevated CO_2, predicted by the model study of Chen *et al.* (1993), is the reallocation of N within the canopy to maintain the photosynthetic efficiency of both sun and shade leaves (see also Hirose et al., 1988). Further studies must be undertaken to evaluate the consequences of various canopy model simplifications in predicting canopy responses to elevated CO_2, and these model results will need to be validated against the canopy responses of field-grown plants at elevated CO_2.

B. Carbon and Nitrogen Allocation among Vegetative Organs

Elevated CO_2 may lead to a number of changes in allocation patterns among plant organs (Sionit et al., 1985; Luxmoore et al., 1986; Norby et al., 1986b; Larigauderie et al., 1988). Many of the allocation shifts observed at the whole-plant level in enriched CO_2 are consistent with the maintenance of a balance between the rate of supply of carbon from leaves and the rate of supply of nutrients from roots (Davidson, 1969; Chapin, 1980). Models of allocation based on this balanced activity or functional equilibrium have been used to successfully predict plant growth responses for several species under a variety of environmental conditions (Reynolds and Thornley, 1982; McMurtrie and Wolf, 1983; Hilbert and Reynolds, 1991; Luo et al., 1994).

Hilbert et al. (1991) used a combination cost benefit–balanced activity model to predict optimal leaf N concentration, allocation patterns between leaf and root, and growth of a hypothetical plant in response to elevated CO_2. This relatively simple model reproduced many of the observed responses to changes in the resource environment of the shoots, including the increase in leaf N for plants grown in high light and the decrease in leaf N for plants grown in elevated CO_2. Decreased leaf N in elevated CO_2 (resulting in decreased leaf photosynthesis) was produced in the model as a result of maintaining the optimal balance between roots and shoots to maximize growth. Although the model is based on a teleonomic goal (to maximize growth rate), it includes mechanistic relationships between physiological processes in the plant (photosynthesis, nutrient uptake) and morphology (allocation to structure). More research clearly is needed to uncover the mechanisms governing the allocation and partitioning of carbon and N among organs (Dewar, 1993). However, model results such as those of Hilbert et al. (1991) provide hypotheses for experimentalists to test and should further aid in the establishment of some general paradigms for plant allocation responses in an environment of elevated CO_2.

Limitations and Uncertainties: Knowledge of Mechanisms of Allocation and Long-Term Allocation Patterns in Elevated CO_2 A significant weakness in the modeling of plant growth is the lack of understanding of the mechanisms of allocation. The sink strength of various organs appears to be an important factor governing allocation and, particularly, the long-term response of plants to elevated CO_2 (Thomas and Strain, 1991). However, there are many unanswered questions about the process. What controls sink strength? Does elevated CO_2 affect sink strength? Are there some generalizations regarding the control of sink strength during the growing season? Are there similarities within life form groups? A second set of questions relates to the long-term nature of the allocation response. The theory of "balanced activity" allocation was based on plants undergoing exponential growth (i.e., early growth of annual plants). Is there a general theory that governs the allocation of perennials? While experimental research ultimately will be required to provide answers to these questions, we believe that insight into the effects of elevated CO_2 on allocation can be gained by exercising newly developed mechanistic translocation models, such as the Muensch translocation model (Dewar, 1993), as well as phenomenological allocation models, such as that by Hilbert and Reynolds (1991).

C. Carbon and Nitrogen Allocation to Reproductive Structures

Changes in allocation to reproductive structures could be an important factor affecting population persistence in future environments (Bazzaz and Garbutt, 1988). Agronomic studies consistently show an increase in the yield of seeds and/or fruits in response to elevated CO_2 (Kimball, 1983). In fact, CO_2 enrichment of greenhouses is well documented for increasing the yields of selected species (Enoch and Kimball, 1986). However, studies on the effects of CO_2 on reproductive allocation in naturally occurring species have produced a wide range of results, including decreases (Bazzaz, 1990). To date there has been no application of a reproductive allocation model in the prediction of the response of a naturally occurring species to elevated CO_2. The use of existing models seems to offer little promise for general application for prediction or even for the study of elevated CO_2 effects on reproductive allocation, since they are quite species specific and are based on age- and environmental-controlled developmental phases, which may undergo rather unpredictable changes in elevated CO_2 (Bazzaz, 1990).

Limitations and Uncertainties: Lack of Generality in Reproductive Response to Elevated CO_2 The inconsistency in results obtained with managed and unmanaged species demands explanation, as does the wide variation in responses of wild species. The simplest paradigm that we can propose is that (at least some of) the variation is related to growth form and reproductive sink strength. Most agronomic species are annuals and have been bred for high reproductive allocation. Wild annuals also rely on high reproductive allocation for survival, whereas perennials can invest either in seed production or in the parts of the plant that persist from one growing season to the next. Thus, annual species have a relatively large and continuous sink for reproductive allocation (once initiated), which apparently responds positively to increased carbon flow under elevated CO_2. However, for perennials the trade-off in investment in vegetative versus reproductive structure apparently is much more variable (and currently unpredictable) in response to elevated CO_2. It is clear that much more research is needed before advanced models of reproductive allocation response can be developed. However, we believe that this is one arena in which a joint program of modeling and experimentation may help to uncover elements of generality in the response of plants to elevated CO_2.

D. Types of Plant-Level Models Needed

Modeling of plant responses to elevated CO_2 revolves around two somewhat different goals. One is the *study* of how the direct and indirect effects of elevated CO_2 on plant morphology and physiology will be translated throughout the plant to produce changes in growth patterns. The second goal is the *prediction* of plant response to elevated CO_2 within the context of larger scale predictions of population, community, and ecosystem responses to elevated CO_2 (see also Wang *et al.*, 1992). These two goals require two kinds of models. In the former case, models are needed that are fully based on the underlying biochemical, physiological, and morphological processes. Such models must be able to account for the significant primary, secondary, and tertiary effects of elevated CO_2 on plant response (Strain, 1985). They must be well integrated with experimental findings on the responses to elevated CO_2 of the leaf, organ, and whole plant. They may be combined with detailed abiotic models to explore the interaction of elevated CO_2 with temperature, light, water, humidity, and nutrients. Highly mechanistic models of this type will serve to test hypotheses of plant response to elevated CO_2 and, in corroboration with the experimental database, to identify the key attributes that will be incorporated into the second type of model: the elevated CO_2 response model. This model will be robust in its capacity to predict whole-plant growth responses to the effects of elevated CO_2 and climate change, but will lack the mechanistic details at the cellular and leaf levels that can lead to mechanistic regress or scaling errors (O'Neill, 1988; Sharpe, 1990; Wang *et al.*, 1992; Reynolds *et al.*, 1993a).

IV. Population, Community, and Stand Models

There are a great number of models of plant populations that employ a variety of approaches, including those that emphasize (a) interactions among plants through indirect effects on resource availability and requirements (i.e., resource-use models; Tilman, 1988, 1990), (b) the occurrence of openings, or gaps, in the community and the changing environment (particularly light) within the gap (i.e., gap or succession models; Shugart, 1984; Shugart *et al.*, 1992), (c) explicit interactions with neighbors (i.e., interference models; Weiner and Conte, 1981; Pacala and Silander, 1985, 1989; Bonan, 1988), and (d) probabilities of individuals being present at particular locations through time (i.e., Markov models; Horn, 1975; van Hulst, 1979; van Tongeren and Prentice, 1986). We briefly discuss the advantages and disadvantages of these approaches with respect to their application to predicting population structure and dynamics in response to elevated CO_2. We then discuss some uncertainties and limitations that apply in general to the application of population models to predicting responses to increasing CO_2.

Models of resource use offer the most direct step in translating the effects of CO_2 from the whole-plant level to the population and community, since they do not require any additional information about plant–plant interactions or spacing. Instead, they predict indirect interactions among plants via effects on resource availability (Field *et al.*, 1992). Changes in resource distributions and availability may account for much of the indirect effect of CO_2 at the community level, leading to changes in species interactions and composition (Tilman, 1993). This could be true whether CO_2 is acting as a limiting nutrient or as a controlling physical factor within an ecosystem—a factor that has yet to be determined (Tilman, 1993). As a controlling physical factor, elevated CO_2 could affect an individual plant's properties, such assimilate translocation, leaf area, and root production, which secondarily affects the plant's resource acquisition and, thus, the resource distribution and availability to neighbors. Resource-use models also offer a rather straightforward way to link whole-plant responses with ecosystem-level processes, as they may be incorporated directly as an element or a module of an ecosystem process model. However, models of this type do not consider the explicit spatial distribution of plants and resources and ignore aspects of the biology and life history of populations that may be subject to changes under higher CO_2 and different climates (Bazzaz, 1990; Mooney *et al.*, 1991; Woodward *et al.*, 1991).

The gap models of forest succession have some features of both resource-use models and individual-based interference models. These models follow the fate of all individuals but provide an intermediate degree of spatial averaging since the exact spatial location of individuals is not known; the plot or patch is chosen to be of the size that potentially involves interactions among all individuals in the patch. Individuals both affect and are affected by the aboveground (mainly) and belowground resources of the patch. These models have been used to simulate population dynamics in a number of forest systems (see Shugart, 1984; Shugart *et al.*, 1992), as well as a grassland ecosystem (Coffin and Lauenroth, 1989). These models have also been used to predict responses of forests to climate changes (e.g., Solomon,

1986; Pastor and Post, 1988; Urban and Shugart, 1989; Botkin and Nisbet, 1992). The success of gap models in predicting forest responses to long-term prehistoric climate shifts (Solomon et al., 1980, 1981; Solomon and Webb, 1985) suggests that these models may be useful for further development and exploration of CO_2–climate change effects.

Two limitations of gap models need to be addressed before their predictions can be considered robust. First, there is a high degree of circularity with respect to modeled climate responses. The environmental response functions for limiting potential growth generally are derived from correlations of current tree distributions with environmental conditions across the distributional range, rather than from independent tests for growth response. For example, the low and high temperature limits to growth are based on temperatures at the northern and southern limits of distribution, respectively (Shugart, 1984). Thus, model predictions of species distributions as a function of climate should closely match actual distributions by construct. However, as Pacala and Hurtt (1993) point out, determination of a species response based upon its actual distribution (i.e., its realized niche) represents only a compressed response or a fraction of the potential response of the tree in the absence of competitors (i.e., its fundamental niche). A species' realized niche (distribution) could change considerably under conditions of changing competition (such as with a different ensemble of species) due to the effects of elevated CO_2 and climate change.

The second limitation of gap models is that the effects of elevated CO_2 on tree response are not readily included, since most models have relatively simple growth functions and do not consider many aspects of a tree's growth that could be affected by elevated CO_2, (e.g., photosynthesis, canopy structure, changing allocation patterns). Nevertheless, as effects of CO_2 on tree growth are determined experimentally, they may be incorporated into the models. Linking gap models with mechanistic models of nutrient cycling (such as in Pastor and Post, 1986) will also be important to account for changes in the nutrient resource structure of the stand with increasing CO_2.

Individual-based interference models may consider any of a number of attributes of plant life history (e.g., germination, dispersal, fecundity, survival) and population structure (e.g., age, size, density) and how they are affected by neighbors of the same (Pacala and Silander, 1985) or different species (Pacala and Silander, 1989). Models of this type may provide a way to explore how plant-to-plant interactions are affected by elevated CO_2 and could be directly related to experimental studies involving plants distributed in various groupings, spacing, and neighborhood arrangements (examples cited in Bazzaz, 1990). Such a model–experiment interaction is laudably called for by Pacala and Hurtt (1993). However, a limitation of most individual-based models is that the life history attributes are functions of interference coefficients, which are determined from field experiments. Presently there is no theoretical basis, and little experimental data, for determining how the coefficients might change under elevated CO_2. Furthermore, the way in which these kinds of models can be incorporated in larger scale models (ecosystem or regional) is not as clear as with the resource-use or gap models.

Similar limitations apply to the use of Markov models in the study and prediction of population response to changing CO_2. The probabilities of occurrence of individuals and transitions to other individuals are based upon observations of these occurrences in natural or experimental situations, of which there will be very few in the case of elevated CO_2. Thus, Markov models would seem to have little application in the prediction of general population response to elevated CO_2, although they may have use for studying facets of population response such as thresholds for CO_2-induced transitions among competitors or sensitivities of competitors to CO_2-induced transitions.

Limitations and Uncertainties: Scaling of Plant Responses to the Population and Community and Generalizing the Responses of Plant Communities to Elevated CO_2 Because the effects of CO_2 on plant growth vary from species to species, it has been predicted that elevated CO_2 will bring about changes in plant communities by preferentially promoting the growth of some species over others (Zangerl and Bazzaz, 1984; Bazzaz, 1990; Long and Hutchin, 1991). But this variable species response also complicates model development based on generality. Experimental studies of CO_2 enrichment of artificial communities of two or more species (Carter and Peterson, 1983; Bazzaz and Carlson, 1984; Zangerl and Bazzaz, 1984; Bazzaz and Garbutt, 1988; Bazzaz et al., 1989) indicate that elevated CO_2 may increase total community productivity (though not always), change the relative contributions of different species to community production, and change the reproductive output of plants, potentially influencing future community structure. These studies further reveal that the community-level responses to elevated CO_2 are not simple linear combinations of the individual responses to CO_2 (Bazzaz et al., 1989). Thus, the outputs of the whole-plant models cannot be used directly to predict population and community response to CO_2, since intra- and interspecific competition affect the way in which individual plants respond to CO_2. These considerations argue in favor of individual-based interference models for the study and prediction of community response to elevated CO_2, particularly for aspects such as species composition, plant density, and size distributions. On the other hand, the distributions (limitations) of resources in the community can affect the response of plants to CO_2 (Zangerl and Bazzaz, 1984; Woodward et al., 1991) and may ultimately determine community composition and plant density (Tilman, 1988). This consideration argues in favor of resource-use models. The development of models that combine the dynamics of individuals with physiological responses to resources [e.g., LINKAGES, Pastor and Post (1986); HYBRID, Friend et al. (1993); WHORL, Sorrensen-Cothern et al. 1993)] should prove quite useful in the study of population and community response to elevated CO_2.

V. Ecosystem Models

A large number of models have been developed during the last decade to study the structure and function of ecosystems (Pastor and Post, 1986; Parton et al., 1987), evaluate ecosystem response to management plans (McMurtrie et al., 1990), forecast trends in ecosystem productivity or other states, particularly in response to perturbation (Miller et al., 1984; Aber and Federer, 1992), and study linkages of the ecosystem with larger scale processes [Schimel et al., 1990; see Ågren et al. (1991) for a review of ecosystem models]. There is now considerable interest in the potential for using these models to make predictions about ecosystem response to global change (Pastor and Post, 1988; Rastetter et al., 1991; Baker et al., 1993; Ryan et al., 1994). However, while existing ecosystem models might make reasonable predictions regarding ecosystem response to climate change (i.e., those models that include significant detail regarding the climatic response of ecosystems), they do not, presently, include sufficient detail or validation for the purpose of predicting ecosystem response to elevated CO_2. Nevertheless, existing ecosystem models may provide specific model elements for use in developing CO_2-responsive models (e.g., Comins and McMurtrie, 1993), as well as providing general guidelines for structure, scaling, and simplification in further ecosystem model development.

Because of the paucity of ecosystem-level experiments, the development of ecosystem models responsive to CO_2 will rely on the integration of results from plant-level experiments and plant models such as growth, leaf production, and litter production with ecosystem processes such as decomposition, nutrient cycling, hydrology, and herbivory. The physiological models that encompass plant responses to CO_2 operate at scales of seconds, minutes, and days, while ecosystem processes are generally modeled at scales of days, months, or years. The complications in scaling point out the need for a nested modeling approach that utilizes the various types of models at the scale for which they are best suited, but translates this information to other hierarchical levels. Reynolds et al. (1993a) used a detailed physiological model to develop a response surface for the relative growth rate of chaparral shrubs as a function of N, solar radiation, and CO_2. This response surface was then used in a stand-level population model to investigate the response of chaparral ecosystems to N, solar radiation, and CO_2 under different fire regimes. Along similar lines, Comins and McMurtrie (1993) developed the G'DAY forest ecosystem model, which is composed of three nested models: a detailed physiological model of canopy photosynthesis, a less complex carbon balance model, and an aggregated scheme that links soil and plant processes. Running and Coughlan (1988) approached the scaling problem in FOREST-BGC by separating processes into fast (calculated daily), e.g., canopy gas exchange, hydrology, and energy budgets, and slow (calculated annually) ones, e.g., allocation of carbon and nitrogen and decomposition.

A. Primary Production/Biomass/Carbon Cycling

Ryan *et al.* (1994) compared seven ecosystem-level models in their predictions of productivity and standing biomass response to elevated CO_2 for two coniferous forest stands—*Pinus sylvestris* in Sweden and *Pinus radiata* in Australia. The models differed substantially in their approaches, processes modeled, and complexity. Most, but not all, of the models were able to match biomass and productivity data reasonably well under the ambient environment over a 4 (Sweden) or 11 (Australia) year period. However, with simulated elevated temperature ($+4°C$) and/or elevated CO_2 (700 ppm), the models differed substantially in predictions of stem biomass, aboveground production, and soil carbon storage over a long-term simulation (60 years for *P. radiata* and 120 years for *P. sylvestris*). Most of the models predicted large increases in stem biomass after long-term exposure to elevated CO_2. These increases occurred despite moderate (in *P. radiata*) to strong (in *P. sylvestris*) nitrogen limitations on growth and contrast with other studies suggesting that the response of nutrient-limited ecosystems to elevated CO_2 would be minimal (Rastetter *et al.*, 1991; Comins and McMurtrie, 1993). For example, Comins and McMurtrie (1993), using the G'DAY model, found that a 27% increase in forest production immediately following a doubling of CO_2 was eliminated in just 3 or 4 years because of limitations to productivity by nitrogen cycling through the ecosystem, resulting in only a 5% increase in productivity over the long term. Leadley and Reynolds (1992) developed a simple process–phenomenological model to predict total plant production and accumulation of carbon in an Arctic tussock tundra. This model also suggests that elevated CO_2 and increased temperature would have little effect on the biomass of the ecosystem because it is primarily limited by N.

Limitations and Uncertainties: Model Development and Testing (Validation) of Predictions; Generality of Ecosystem Responses The model comparisons of Ryan *et al.* (1994) reveal the degree to which current models are in error. After 1 year of simulated growth at doubled CO_2, productivity was predicted to increase variously among models, from 28 to 123% for *P. radiata* stands and from 20 to 46% for *P. sylvestris* stands. After 60 or 120 years, stem biomass was predicted to increase variously from about 2 to 200% for *P. radiata* and from about 5 to 60% for *P. sylvestris*. Since there are no elevated CO_2 experiments against which to compare these results, we cannot know which model(s), if any, might be correct. All but two of the models were developed for purposes other than predicting ecosystem responses to elevated CO_2 and have received little development and virtually no testing with respect to elevated CO_2 predictions. Thus, it is not too surprising that they differ so much with regard predictions for the same ecosystems.

There are several ecosystems in which CO_2 is being elevated experimentally. These systems provide the opportunity for testing prototype models for predicting responses to elevated CO_2. Leadley and Reynolds (1992) developed a simple model for predicting the response of tussock tundra to perturbations, including elevated CO_2. The simulated growth of tussock tundra compared favorably with the experimental results of Tissue and Oechel (1987), who showed little enhancement in production and standing biomass after 3 years of growth under doubled CO_2. However, CO_2 enrichment experiments of the same duration in two other ecosystems [salt marsh, Arp *et al.* (1993); tallgrass prairie, Owensby *et al.* (1993)] have shown continued response of production and biomass to elevated CO_2. Thus, establishment of the generality of the modeled CO_2 responses will require the validation of models in multiple ecosystem types under a variety of conditions of nutrient and stress limitations, in addition to enriched atmospheric CO_2.

B. Ecosystem Physiognomy and Species Composition

Many ecosystem models primarily follow flows between pools of carbon, nitrogen, water, etc., without an explicit consideration of the species composition of the ecosystem (e.g., Parton *et al.*, 1987; Aber and Federer, 1992). While the models, to be sure, base flows and pools on the characteristics of dominant species or groups of species in the ecosystem, they do not readily account for species transitions, without changes in model parameters. However, Clark (1993) points out that there are numerous examples of climate changes during the last several thousand years inducing shifts in species composition that had important implications with regard to litter quality, N cycling, water use, hydrology, and landscape heterogeneity within the ecosystem. The failure of ecosystem models to account for the species composition changes would probably result in the complete failure to correctly predict general energy and matter fluxes, as well as feedback and interaction with climate states in larger scale models of atmosphere–biosphere interactions. Thus, successful ecosystem modeling must involve a consideration, if not direct incorporation, of plant species composition changes. In this context, a satisfactory approach may prove to be in the linking of community–stand models, such as gap models with ecosystem flow models, as has been done with LINKAGES (Pastor and Post, 1986) and HYBRID (Friend *et al.*, 1993).

Limitations and Uncertainties: Incorporating Plant Community Structure into Ecosystem Models; CO_2 Effects on Migration and Dispersal Just how much population–community structure should be included within ecosystem models and other large-scale models will depend on two factors, irrespective of the general modeling approach. One is the nature of the specific question being explored. In some cases, the explicit species compositions of a region must be followed because of questions about changes in species diversity or changes in a species with particular significance, such as commercial importance, threat of extinction, or keystone status in a community. The second factor that will govern how much population–community structure must be included in ecosystem models relates to the more general questions of how much community structure and species composition may alter under elevated CO_2 and climatic change and how such shifts may be translated to ecosystem processes, such as biomass accumulation, primary productivity, carbon storage, decomposition, and nutrient cycling. These processes are only just now beginning to be explored both experimentally and through models. Plant community structure and species composition may be judiciously simplified within larger scale models if, for example, they are found to have limited impact upon ecosystem energy and matter fluxes. Such an approach is being taken by Running and Hunt (1993) in the development of a generic biome simulation model (BIOME-BGC). Clark (1993) suggests that many aspects of population structure (density, area occupied, self-thinning) may be expressed in terms of a few variables or parameters associated with either individual plant growth or overall stand growth (age). However, experimentation will be necessary to validate model assumptions and simplifications with respect to responses to elevated CO_2.

A second major limitation to modeling long-term species composition responses to climate change and elevated CO_2 is accounting for dispersal and plant migration (Pacala and Hurtt, 1993). Dispersal and migration are poorly understood for most species. Modeled dispersal will depend on the scale or distance modeled, temporal distribution of dispersal agents, dormancy and seed bank patterns, and most importantly, the species. While there are many attributes of the plant community and ecosystem that can be accounted for through the simplification and aggregation of species into life form groups, seed dispersal is not likely to be adequately accounted for by traditional life form groupings, since there is such great variability in dispersal within a life form group (e.g., deciduous trees have seed weight differences of more than 2 orders of magnitude—and some have seeds with special adaptations for dispersal). Thus, species may have to be considered individually or new aggregations may have to be developed to treat dispersal. However, this will present difficulties when merging with the traditional life form groupings within an existing model. If some members of a life form group arrive at a region in a modeled system, to what extent is the whole life form group represented? What would be the nature of a region if only a depleted portion of a life form class is present?

C. Decomposition and Nutrient Cycling

A potential effect of elevated CO_2 in ecosystems could be the modification of nutrient cycling (Pastor and Post, 1988). The very close coupling of nitrogen and carbon cycles within the plant and the ecosystem (Chapin, 1980; Conroy *et al.*, 1992; Field *et al.*, 1992; Rastetter and Shaver, 1992; van de Geijn and van Veen, 1993) indicates that there may be many avenues for interactions and feedback as one or the other cycle is altered via elevated CO_2 and climate change. Nutrient cycles may be altered through changing litter decomposition rates, plant nutrient uptake, or internal cycling of nutrients within plants, any or all of which may occur under elevated CO_2 (Graham *et al.*, 1990). The potential for increased carbon acquisition by plants under elevated CO_2 can be limited by the availability of soil nutrients (Larigauderie *et al.*, 1988; Conroy *et al.*, 1992; Woodin *et al.*, 1992; Silvola and Ahlholm, 1993), which, in turn, is controlled by decomposition. While the process of decomposition is relatively well-known from the standpoint of the effects of moisture, temperature, and nutrient quality of the litter (e.g., Meentemeyer, 1978; Melillo *et al.*, 1982), the effects of changing CO_2 levels on decomposition and nutrient cycling have been studied only recently (O'Neill, 1994).

One of the main avenues for the interaction of elevated CO_2 with decomposition may be via effects on litter quality. Litter characteristics, such as lignin and nutrient contents, strongly influence decay patterns (Meentemeyer, 1978; Melillo *et al.*, 1982). The overall litter quality of the ecosystem may be altered by elevated CO_2, either by changes induced directly in the litter produced or by changes in species composition of plant communities and associated litter characteristics (Pastor and Post, 1988). The former case seems plausible since, for many species, C/N ratios in plant tissues increase with CO_2 enrichment (e.g., Larigauderie *et al.*, 1988; Couteaux *et al.*, 1991). However, the C/N ratios of senescent tissues may not reflect those of living tissue. Curtis *et al.* (1989) reported that the litter C/N ratio in the rush, *Scirpus olneyi*, was unaffected by CO_2 enrichment, and Kemp *et al.* (1994) found little difference in C/N ratios of grass litter grown at ambient and twice-ambient CO_2. Lignin/N ratios of litter for some deciduous tree species also were unaffected by elevated CO_2 (Norby *et al.*, 1986a). Clearly, more data are required before any generalizations can be made.

Simulation studies have examined the effects of changing climate on decomposition and nutrient cycles (e.g., Pastor and Post, 1988, Rastetter *et al.*, 1991; Moorhead and Reynolds, 1993; Ojima *et al.*, 1993). Moorhead and Reynolds (1993) suggested that decomposition and nitrogen dynamics in tussock tundra soil would be sensitive to climate change. Pastor and Post (1988) evaluated the potential effects of climate changes on long-term forest productivity and species distribution and noted that vegetation composition depends in large part on soil water and nitrogen availability. In turn, changes in vegetation composition can alter soil nitrogen availability.

Limitations and Uncertainties: Knowledge of the Effects of Elevated CO_2 on Plant Litter and Belowground Production and Turnover The greatest challenge in applying models of decomposition to a changing CO_2 world is predicting the changes in litter quality and rates of litter input into the ecosystem that may occur under elevated CO_2. Other factors must also be considered. For example, the results of a temperate grassland model suggest that combined CO_2 and N fertilization have synergistic effects on ecosystem carbon dynamics (Thornley *et al.*, 1991), and increasing amounts of labile carbon in plant rhizospheres associated with elevated CO_2 may stimulate microbial activities, influencing nutrient availability (Norby *et al.*, 1986a; O'Neill *et al.*, 1987; Conroy *et al.*, 1990; Comins and McMurtrie, 1993).

D. Hydrology

Responses of ecosystem hydrology to climate change and/or elevated CO_2 have often been modeled somewhat independently of the rest of the ecosystem (Aston, 1984; Idso and Brazel, 1984; Gleick, 1986; Rosenberg *et al.*, 1989). However, the potentially complex and nonlinear responses of plants to elevated CO_2 suggest that hydrologic models not coupled in some manner to plant responses will be inaccurate. Simple "adjustments" of modeled transpiration to reflect increased stomatal resistance or increased air temperature (e.g., Aston, 1984; Idso and Brazel, 1984) will not account for changes in transpiration that may result from CO_2-induced changes in leaf area, plant density, and root growth that could affect water loss and uptake (Tyree and Alexander, 1993). Furthermore, stomatal resistance and leaf transpiration are not directly related to evapotranspiration from a canopy or vegetated surface (Eamus, 1991; Morison, 1993). Increased stomatal resistance increases the leaf temperature and decreases the vapor pressure within the air around the canopy, both of which increase the vapor pressure deficit and, thus, offset increased stomatal resistance. Morison (1993) showed that a 60% increase in leaf stomatal resistance results in a decrease in total evapotranspiration of only 10–15% over a range of leaf area indexes from 1 to 6. While we propose that hydrologic models must be linked to plant process (ecosystem) models, it will be equally important for ecosystem models to be linked to detailed hydrologic (and microclimatic) models in order to predict climate-induced changes in the soil environment that could limit or interact with plant responses to elevated CO_2 (Bonan, 1993), especially in systems with limited water.

Limitations and Uncertainties: General Effects of Elevated CO_2 on Ecosystem Hydrologic Responses and Translation of Effects to the Large Scale Meaningful predictions of changes in ecosystem and watershed hydrology as a function of changing CO_2 remain elusive. A rather significant complication is the change in climate that may accompany elevated CO_2. Changes in hydrology are apt to be most affected by changes in precipitation (Nemec and Schaake, 1982; Gleick, 1986), for which there is great uncertainty in the future. Changes in temperature and particularly precipitation could offset the changes in plant transpiration induced by elevated CO_2. For example, Revelle and Waggoner (1983) suggested that CO_2-induced climate changes will increase evapotranspiration and reduce runoff on watersheds in the western United States by 40–75%. This contrasts with the results of Idso and Brazel (1984) and Aston (1984), who suggested that the antitranspirant effect of CO_2 would increase stream runoff by 40–60% in Arizona. Gifford (1988) suggests that there may be little or no change in evapotranspiration when all effects of elevated CO_2 are taken into account. Clearly, experimental results are needed from field studies in different ecosystems to establish trends and generalities in terms of the effects of elevated CO_2 on hydrology. Second, more studies are needed to understand the linkages and interactions between plant transpiration, the plant boundary layer, and large-scale hydrology. Much of this understanding is based upon studies in agricultural systems with monocultures or limited species diversity. The process of evapotranspiration is less known for ecosystems with water limitations and diverse species having different seasonal patterns of water use.

E. Plant–Animal Interactions

There has been considerable speculation about the ramifications of elevated CO_2 on plant chemical makeup and the consequences for plant herbivores (Strain and Bazzaz, 1983; Bryant *et al.*, 1983; Lincoln *et al.*, 1984; Lambers, 1993). Some suggest that secondary metabolites may accumulate and repel herbivores (Strain and Bazzaz, 1983), whereas others suggest that decreased tissue N levels may increase herbivory (Lincoln *et al.*, 1984). Most studies of insect herbivores support the notion of increased rates of consumption (Lincoln *et al.*, 1984; Akey *et al.*, 1988; Lincoln and Couvet, 1989; Johnson and Lincoln, 1990), presumably due to the decreased nutritional quality (associated with increased C/N) of plant tissues produced under elevated CO_2 (Lincoln and Couvet, 1989; Lincoln *et al.*, 1993). However, other features such as leaf toughness and starch content may also be involved (Lincoln *et al.*, 1993). For the most part, elevated CO_2 has not been shown to increase secondary compounds that act as general herbivore repellents (Fajer *et al.*, 1989; Lincoln and Couvet, 1989; Johnson and Lincoln, 1990), although under low nutrient levels secondary compounds may increase (Johnson and Lincoln, 1991; Lambers, 1993). While increases in herbivory may compensate for the lower nutritional quality of CO_2-enriched plants, there are examples of decreased growth rate (Lindroth *et al.*, 1993) and increased mortality (Fajer *et al.*, 1989) in some insect species.

As yet, there have been no models developed to explore plant–animal responses to elevated CO_2. However, the hypothesis that herbivores are nitrogen rather than carbon limited (Mattson, 1980), coupled with changes in plant C/N ratios and secondary metabolites under elevated CO_2, offers one avenue for the development of simple models of insect herbivory. Other aspects of changes in insect developmental patterns, densities, and life histories (Fajer *et al.*, 1989; Lindroth *et al.*, 1993) may be added as the effects of CO_2 and climate changes become known.

Limitations and Uncertainties: Lack of Understanding and Experimental Data The primary limitation of building models to predict animal–plant interactions to elevated CO_2 lies with the lack of mechanistic understanding of relationships between plants and herbivores. There are still questions about the effects of elevated CO_2 on plant secondary metabolism (Lambers, 1993), animal consumption patterns (Lincoln *et al.*, 1993), and long-term changes in insect populations. Virtually nothing is known about long-term plant population responses to CO_2-induced changes in herbivore activities. Furthermore, the effects of elevated CO_2 on herbivory by animal groups other than arthropods are just beginning to be studied (Owensby *et al.*, 1990). Other herbivore groups may respond differently than insects. For example, the rate-limiting step for consumption in ruminants may be fermentation in the rumen, in which case there would not be an increase in plant harvesting to compensate for lower tissue quality (Owensby *et al.*, 1990), although there could be a shift in the selection of more nutritious plant species.

VI. Regional and Global Models

Ultimately, the goal of much of the model development at the leaf, plant, population, and ecosystem levels is to provide understanding and predictions that may be used either directly or indirectly in the development of models capable of predicting regional or global responses of ecosystems and biomes to climate and CO_2 changes. At this scale, two approaches can be employed: bottom-up and top-down. In fact, we believe that both approaches should be employed to take advantage of different strengths and weaknesses. Consistency between approaches would lend confidence to model predictions. Since there has been little model development with respect to elevated CO_2 at the regional and global scale, we focus our discussion to a general consideration of the approaches and some important limitations of each.

A. Bottom-up Approaches

As the name conveys, bottom-up models employ information and relationships from below the scale of focus and are generally based on processes or mechanisms that link or relate various elements to one another. Use of these mechanistic models presumably will yield much information about the degree to which CO_2 and climate effects at the organismal level are ultimately translated through the community to different trophic levels and to larger scales. However, a model of landscape, regional, and ultimately global responses to CO_2 and climate forcing may not follow from an averaging or summation of the modeled organism and ecosystem responses. One reason is that most ecosystem models are, to a large extent, abstractions of the ecosystem (Running and Hunt, 1993; Sharpe and Rykiel, 1991). While they are useful tools for studying *potential* responses, as well as identifying ecosystem processes that are sensitive to CO_2 and climate forcing, they may not predict the actual responses of landscapes and regions, which will be affected by large-scale processes such as atmosphere–biosphere interactions, landscape heterogeneity, land-use patterns, and other anthropogenic modifications not included in the patch-scale ecosystem models. Thus, at a minimum, ecosystem predictions will require careful interfacing with geographical information systems and considerations, if not models, of large-scale processes, constraints, and feedback. A second problem in predicting large-scale patterns and responses directly with the mechanistic ecosystem models is mechanistic regress (Sharpe, 1990) or the transmutation of scale (O'Neill *et al.*, 1986). Most ecosystem models described in the previous section include processes that operate over scales of a few meters and on the order of days or even hours. An attempt to incorporate these processes directly into large-scale models results in compounding of errors associated with multiple linkages connecting processes across hierarchies (mechanistic regress) or, similarly, in the loss of mechanism because the control by processes at one scale is overridden by controlling processes at a larger scale (transmutation). Mechanistic models involving many processes also become quite complex, difficult to interpret, and tend to be unstable for long-term simulations (Landsberg, 1986; O'Neill, 1988; Reynolds *et al.*, 1993a). Thus, successful models of landscapes, regions, and the globe will involve careful integration of the mechanism of indirect effects of CO_2 on ecosystems with large-scale correlative or phenomenological modeling approaches.

B. Top-down Approaches

A fundamental difference in the top-down, compared to the bottom-up, approach is not so much the lack of mechanism, but rather the extent to which information is used from hierarchies below the level of focus. In a top-down model, processes that occur below the scale of focus are subsumed into derived variables that relate or correlate with the variable of focus. To the extent that these newly derived relationships are representative of processes, interactions, and feedback among real or abstract parts, we can say that the large-scale model is, to a degree, mechanistic. However, the large-scale model may also consist of (either solely or in conjunction with one another) relationships that are phenomenological, correlative (empirical), or inverted (defined and constrained by inputs from a scale greater than the one of focus). For example, a number of correlative models have been developed to forecast regional and global responses of vegetation to climate change. These models are based on the correlations between various climatic variables and the distribution of vegetation types (Emanuel *et al.*, 1985; Cramer and Leemans, 1993; Monserud *et al.*, 1993) or plant life forms (Cramer and Leemans, 1993). The principal limitation of these models is in their ability to extrapolate beyond the range over which the correlation has been established. For example, these models generally develop climate-–vegetation distribution correlations based on existing climate–vegetation boundaries. However, current boundaries represent complex interactions due to plant competition, soil and nutrient conditions, herbivore and pathogen pressures, historical conditions (and inertia), and, only partly, climatic response. Climatic change and elevated CO_2 may bring about changes in any or all of these interactions, with the result that actual vegetation shifts could be quite different from those projected with simple climate correlation models. A second difficulty in the development of correlative models for forecasting vegetation changes is incorporating the effects of elevated CO_2. There is only a small experimental base from which to build correlations and responses of vegetation units. However, Woodward (1993) has demonstrated convincingly that a consideration of the effects of elevated CO_2 in concert with climate change results in a fundamentally different outcome than predictions based solely on climate. Therefore, we conclude that forecasts using simple climate correlation models without the incorporation of CO_2 effects and some consideration of abiotic and biotic interactions are premature and potentially misleading.

While there is little experimental data that can be drawn on for correlations between CO_2 and vegetation or plant distributions, several avenues may provide information for developing correlations between CO_2 and plant distributions. These include studies of prehistoric climate/CO_2–vegetation patterns and plant distributions along gradients having naturally elevated CO_2, such as near springs that vent large amounts of CO_2 (Woodward et al., 1991). Finally, an important element of the large-scale models will be the coupling of models of climate/CO_2 effects with models and/or data dealing explicitly with spatial information about land form, soils, and land use. These models ultimately will be coupled with climate models to produce interactions at the global biosphere–atmosphere level.

VII. Future Challenges

There are fundamental limitations of models that must be kept in mind when evaluating their results and considering their potential contributions. Models are based on assumptions and conceptualizations about how systems behave and what are important factors, relationships, and interactions that govern the structure and function of that system. The growing diversity among model predictions of plant and ecosystem responses to changing climate and CO_2 requires that we carefully examine model details, particularly model assumptions. The simple schematic diagram shown in Fig. 1 illustrates that the pattern of carbon flow depends upon allocation patterns at numerous levels within the plant. These allocation patterns are controlled by plant genetics, hormone levels, and internal nutrient and carbon concentration gradients, as well as external environmental factors. Furthermore, these allocation patterns in plants will determine interactions between plants and among plants and organisms at other trophic levels. This will lead to changes in the habitat resource structure, with further feedback on plant allocation patterns. Models of plants and ecosystems make a variety of assumptions about how plants allocate carbon and nutrients and about how resources are utilized by plants. Model comparisons that identify differences due to specific assumptions will lead to a greater understanding of how assumptions affect modeled responses to elevated CO_2 and will help in identifying assumptions that are invalid or need further refinement.

Model assumptions must also be tested against independent data sets from field experiments that are representative of diverse natural systems. Many of the limitations of models at present are due to gaps in our knowledge base. There are many aspects of the long-term effects of elevated CO_2 that are poorly understood. Consequently, it is difficult to include factors in a model that account for these effects. There will be genetic, physiological, and behavioral adjustments of organisms to climate and CO_2 changes. Many of these responses have yet to be studied, let alone be included in models. While most of the experimental research has focused on the growth and physiology of single organisms, processes such as acclimation and speed

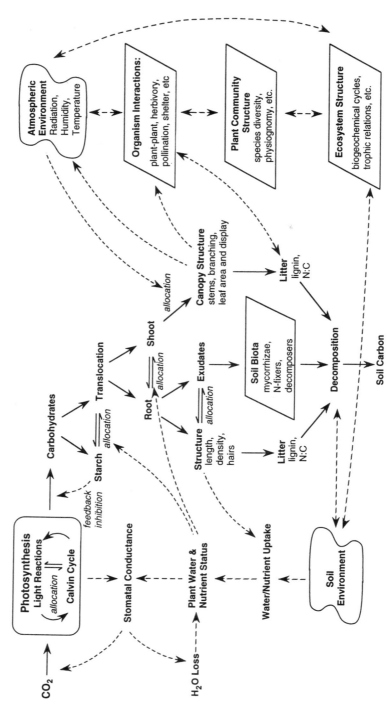

Figure 1 Conceptual model of carbon flow (solid arrows) through a plant and the important feedbacks and interactions (dashed arrows) with other environmental factors, as well as with other plants and trophic levels.

of migration may be more important in determining the composition of vegetation in the future (Bradshaw and McNeilly, 1991). Many of the complex feedbacks within an ecosystem are just beginning to be explored, and model results are helping to identify important feedbacks that should be studied (Comins and McMurtrie, 1993).

Another general limitation of most mechanistic models of communities and ecosystems is that spatial heterogeneity is not considered (Clark, 1993). Spatial heterogeneity may be due to completely unrelated phenomena operating at considerably different spatial and temporal scales. For example, in herbaceous communities upon uniform landscapes the spatial pattern may be related to plant–plant interactions, whereas in tundra communities intricate vegetation patterns develop in association with drainage patterns that result from physical freeze–thaw processes. Long-term responses of plant communities to elevated CO_2 will depend in part upon complex factors related to spatial processes. For communities in which spatial patterns are related to biotic factors, there may be indirect effects of CO_2 due to changes in biotic interactions. For communities in which spatial patterns are more strongly determined by abiotic factors, there may be strong point-to-point differences in CO_2 effects because of differential responses to CO_2. Human activities are spatially diverse and must also be accounted for in model forecasts, as they could override many climate/CO_2 effects on plant communities.

Lastly, a critical weakness in current models, particularly at hierarchies from the whole plant and above, is the lack of sufficient testing against independent data sets (Reynolds *et al.*, 1993b). The modeling of CO_2/climate effects on organismal interactions will require experimental data collected from larger scale experiments, and final validation of models must be based upon field conditions if we are to be confident in their formulations and predictions. Usually most, if not all, of a database is utilized in the development of the model, leaving no set of independent data for validation. Two approaches for achieving elevated CO_2 over relatively large areas of quasi-natural vegetation are the open-top chamber (Leadley and Drake, 1993) and free-air CO_2 enrichment (FACE) systems (Hendrey *et al.*, 1993). These experimental systems offer an important source of data for independent model testing. Ultimately, validation of ecosystem models regarding long-term effects of CO_2 will probably be accomplished with experimental data from FACE systems.

Acknowledgments

This research was supported by DOE Grant DE-FG05-92ER61493 and Interagency Agreement DE-AI05-88ER69014. We thank B. Schlesinger, J. Clark, and two anonymous reviewers for many helpful comments.

References

Aber, J. D., and Federer, C. A. (1992). A generalized, lumped-parameter model of photosynthesis, evapotranspiration and net primary production in temperate and boreal forest ecosystems. *Oecologia* 92, 463–474.

Ågren, G. I., McMuririe, R. E., Parton, W. J., Pastor, J., and Shugart, H. H. (1991). State of the art of models of production-decomposition linkages in conifer and grassland ecosystems. *Ecol. Appl.* 1, 119–138.

Akey, D. H., Kimball, B. A., and Mauney, J. R. (1988). Growth and development of the Pink Bollworm, *Pectinophora gossypiella* (Lepidoptera: Gelechiidae), on bolls of cotton grown in enriched carbon dioxide atmosphere. *Environ. Entomol.* 17, 42–45.

Amthor, J. S. (1991). Respiration in a future, higher CO_2 world. *Plant Cell Environ.* 14, 13–20.

Arp, W. J., Drake, B. G., Pockman, W. T., Curtis, P. S., and Whigham, D. F. (1993). Interactions between C_3 and C_4 salt marsh plant species during four years of exposure to elevated atmospheric CO_2. *Vegetatio* 104/105, 133–143.

Aston, A. R. (1984). The effect of doubling atmospheric CO_2 on streamflow: a simulation. *J. Hydrol.* 67, 273–280.

Ayres, M. P. (1993). Plant defense, herbivory, and climate change. *In* "Biotic Interactions and Global Change" (P. M. Kareiva, J. G. Kingsolver, and R. B. Huey, eds.), pp. 75–94. Sinauer Associates Inc., Sunderland, MA.

Baker, B. B., Hanson, J. D., Bourdon, R. M., and Eckert, J. B. (1993). The potential effects of climatic change on ecosystem processes and cattle production on U.S. rangelands. *Clim. Change* 25, 97–117.

Baldocchi, D. D. (1993). Scaling water vapor and carbon dioxide exchange from leaves to a canopy: rules and tools. *In* "Scaling Physiological Processes" (J. Ehleringer and C. Field, eds.), pp. 77–114. Academic Press, San Diego, CA.

Ball, J. T., Woodrow, I. E., and Berry, J. A. (1987). A model predicting stomatal conductance and its contribution to the control of photosynthesis under different environmental conditions. *In* "Progress in Photosynthesis Research, Vol. IV, No. 5. Proceedings of the VII International Congress on Photosynthesis" (I. Biggins, ed.), pp. 221–224. Dr. W. Junk, The Hague, The Netherlands.

Bazzaz, F. A. (1990). The response of natural ecosystems to the rising global CO_2 levels. *Annu. Rev. Ecol. Syst.* 21, 167–196.

Bazzaz, F. A., and Carlson, R. W. (1984). The response of plants to elevated CO_2. I. Competition among an assemblage of annuals at two levels of moisture. *Oecologia* 62, 196–198.

Bazzaz, F. A., and Garbutt, K. (1988). The response of annuals in competitive neighborhoods: Effects of elevated CO_2. *Ecology* 69, 937–946.

Bazzaz, F. A., and McConnaughay, K. D. M. (1992). Plant-plant interactions in elevated CO_2 environments. *Aust. J. Bot.* 40, 547–563.

Bazzaz, F. A., Garbutt, K., Reekie, E., and Williams, G. (1989). Using growth analysis to interpret competition between a C_3 and a C_4 annual under ambient and elevated CO_2. *Oecologia* 79, 223–233.

Beable, C. L., Jarvis, and P. G., Neilson, R. E. (1979). Leaf conductances as related to xylem water potential and carbon dioxide concentration in Sitka spruce. *Physiol. Plant.* 52, 391–400.

Bonan, G. B. (1988). Analysis of neighborhood competition among annual plants: implications of a plant growth model. *Ecol. Model.* 65, 123–136.

Bonan, G. B. (1993). Do biophysics and physiology matter in ecosystem models? *Clim. Change* 24, 281–285.

Botkin, D. A., and Nisbet, R. A. (1992). Projecting the effects of climate change on biological diversity in forests. *In* "Global Warming and Biological Diversity" (R. Peters and T. Lovejoy, eds.), pp. 277–293. Yale Univ. Press, New Haven, CT.

Bowes, G. (1991). Growth at elevated CO_2: Photosynthetic responses mediated through Rubisco. *Plant Cell Environ.* 14, 795–806.

Bradshaw, A. D., and McNeilly, T. (1991). Evolutionary response to global climatic change. *Ann. Bot.* 67 (Suppl.), 5–14.

Bryant, J. P., Chapin, F. S., and Klein, D. R. (1983). Carbon/nutrient balance of boreal plants in relation to vertebrate herbivory. *Oikos* 40, 357–368.

Bunce, J. A. (1992). Stomatal conductance, photosynthesis and respiration of temperate deciduous tree seedlings grown outdoors at an elevated concentration of carbon dioxide. *Plant Cell Environ.* 15, 541–549.

Carter, D. R., and Peterson, K. M. (1983). Effects of CO_2-enriched atmosphere on the growth and competitive interaction of a C_3 and C_4 grass. *Oecologia* 58, 188–193.

Chapin, F. S., III (1980). The mineral nutrition of wild plants. *Annu. Rev. Ecol. Syst.* 11, 233–260.

Chen, J.-L., Reynolds, J. F., Harley, P. C., and Tenhunen, J. D. (1993). Coordination theory of leaf nitrogen distribution in a canopy. *Oecologia* 93, 63–69.

Clark, J. S. (1993). Scaling the population level: effects of species composition and population structure. *In* "Scaling Physiological Processes: Leaf to Globe" (J. F. Ehleringer and C. B. Field, eds.), pp. 255–285. Academic Press, San Diego, CA.

Coffin, D. P., and Lauenroth, W. K. (1989). A gap dynamics simulation model of succession in a semiarid grassland. *Ecol. Model.* 49, 229–266.

Comins, H. N., and McMurtrie, R. E. (1993). Long-term response of nutrient-limited forests to CO_2 enrichment; equilibrium behavior of plant-soil models. *Ecol. Appl.* 3, 666–681.

Conroy, J. P., Milnam, P. H., Reed, M. L., and Barlow, E. W. (1990). Increases in phosphorus requirements for CO_2-enriched pine species. *Plant Physiol.* 92, 977–982.

Conroy, J. P., Milham, P. J., and Barlow, E. W. R. (1992). Effect of nitrogen and phosphorus availability on the growth response of *Eucalyptus grandis* to high CO_2. *Plant Cell Environ.* 15, 843–847.

Couteaux, M.-M., Mousseau, M., Celerier, M.-L., and Bottner, P. (1991). Increased atmospheric CO_2 and litter quality: decomposition of sweet chestnut leaf litter with animal food webs of different complexities. *Oikos* 61, 54–64.

Cramer, W. P., and Leemans, R. (1993). Assessing impacts of climate change on vegetation using climate classification systems. *In* "Vegetation Dynamics & Global Change" (E. A. Solomon and H. H. Shugart, eds.), pp. 190–217. Chapman and Hall, New York.

Curtis, J. D., Drake, B. G., and Whigham, D. F. (1989). Nitrogen and carbon dynamics in C_3 and C_4 estuarine march plants grown under elevated CO_2 *in situ*. *Oecologia* 78, 297–301.

Davidson, R. L. (1969). Effect of root/leaf temperature differentials on root/shoot ratios in some pasture grasses and clover. *Ann. Bot.* 33, 561–569.

Dewar, R. C. (1993). A root-shoot partitioning model based on carbon-nitrogen-water interactions and Müench phloem flow. *Funct. Ecol.* 7, 356–368.

Eamus, D. (1991). The interaction of rising CO_2 and temperature with water use efficiency. *Plant Cell Environ.* 14, 843–852.

Eamus, D., and Jarvis, P. G. (1989). The direct effects of increase in global atmospheric CO_2 concentration on natural and commercial temperate trees and forests. *Adv. Ecol. Res.* 19, 1–55.

El Kohen, A., Venet, L., and Mousseau, M. (1993). Growth and photosynthesis of two deciduous forest species at elevated carbon dioxide. *Funct. Ecol.* 7, 480–486.

Emanuel, W. R., Shugart, H. H., and Stevenson, M. P. (1985). Climatic change and the broadscale distribution of terrestrial ecosystem complexes. *Clim. Change* 7, 29–43.

Enoch, H. Z., and Kimball, B. A. (eds.) (1986). "Carbon Dioxide Enrichment of Greenhouse Crops, Volume II: Physiology, Yield and Economics." CRC Press, Inc., Boca Raton, FL.

Fajer, E. D., Bowers, M. D., and Bazzaz, F. A. (1989). The effect of enriched CO_2 atmospheres on plant/insect herbivore interactions. *Science* 243, 1198–1200.

Fajer, E., Bowers, M. D., Bazzaz, F. A. (1991). The effects of enriched CO_2 atmospheres on the buckeye butterfly, *Junonia coenia*. *Ecology* 72, 751–754.

Farquhar, G. D., and von Caemmerer, S. (1982). Modeling of photosynthetic response to environment. *In* "Encyclopedia of Plant Physiology, New Series, Vol. 12B, Physiological Plant Ecology II," pp. 549–587. Springer-Verlag, New York.

Farrar, J. F., and Williams, M. L. (1991). The effects of increased atmospheric carbon dioxide and temperature on carbon partitioning, source-sink relations and respiration. *Plant Cell Environ.* 14, 819–830.

Field, C., and Mooney, H. A. (1986). The photosynthesis-nitrogen relationship in wild plants. *In* "On the Economy of Plant Form and Function" (T. G. Givnish, ed.), pp. 25–55. Cambridge Univ. Press, Cambridge, England.

Field, C. B., Chapin, F. S., III, Matson, P. A., and Mooney, H. A. (1992). Responses of terrestrial ecosystems to the changing atmosphere: A resource-based approach. *Annu. Rev. Ecol. Syst.* 23, 201–235.

Friend, A. D., Shugart, H. H., and Running, S. W. (1993). A physiology-based gap model of forest dynamics. *Ecology* 74, 792–797.

Gifford, R. M. (1988). Direct effects of higher carbon dioxide concentration on vegetation. *In* "Greenhouse—Planning for Climate Change" (G. I. Pearman, ed.), pp. 506–519. CSIRO Publ., Melbourne.

Gleick, P. H. (1986). Methods for evaluating the regional hydrologic impacts of global climatic change. *J. Hydrol.* 88, 97–116.

Graham, R. L., Turner, M. G., and Dale, V. H. (1990). How increasing CO_2 and climate change affect forests. *BioScience* 40, 575–587.

Grodzinski, B. (1992). Plant nutrition and growth regulation by CO_2 enrichment. *Bioscience* 42, 517–525.

Harley, P. C., and Sharkey, T. D. (1991). An improved model of C_3 photosynthesis at high CO_2: Reversed O_2 sensitivity explained by lack of glycerate reentry into the chloroplast. *Photosynth. Res.* 27, 169–178.

Harley, P.C., and Tenhunen, J. D. (1991). Modeling the photosynthetic response of C_3 leaves to environmental factors. *In* "Modeling Crop Photosynthesis: From Biochemistry to Canopy" (K. J. Boote and R. S. Loomis, eds.), Spec. Publ. No. 19, pp. 17–39. Amer. Soc. Agron., Madison, WI.

Harley, P. C., Thomas, R. B., Reynolds, J. F., and Strain, B. R. (1992). Modeling photosynthesis of cotton grown in elevated CO_2. *Plant Cell Environ.* 15, 271–282.

Hendrey, G. R., Lewin, K. F., and Nagy, J. (1993). Free air carbon dioxide enrichment: development, progress, results. *Vegetatio* 104/105, 17–31.

Hilbert, D. W., and Reynolds, J. F. (1991). A model allocating growth among leaf proteins, shoot structure, and root biomass to produce balanced activity. *Ann. Bot.* 68, 417–425.

Hilbert, D. W., Larigauderie, A., and Reynolds, J. F. (1991). Influence of carbon dioxide and daily photon-flux density on optimal leaf nitrogen concentration and root:shoot ratio. *Ann. Bot.* 68, 365–376.

Hirose, T., Werger, W. J. A., Pons, T. L., and van Rheenen, J. W. A. (1988). Canopy structure and leaf nitrogen distribution in a stand of *Lysimachia vulgaris* L. as influenced by stand density. *Oecologia* 77, 145–150.

Hollinger, D. Y. (1987). Gas exchange and dry matter allocation responses to elevation of atmospheric CO_2 concentration in seedlings of three tree species. *Tree Physiol.* 3, 193–202.

Horn, H. S. (1975). Markovian properties of forest succession. *In* "Ecology and Evolution of Communities" (M. L. Cody and J. M. Diamond, eds.), pp. 196–211. Belknap, Cambridge, MA.

Idso, S. B., and Brazel, A. J. (1984). Rising atmospheric carbon dioxide concentrations may increase streamflow. *Nature* 32, 51–53.

Johnson, R. H., and Lincoln, D. E. (1990). Sagebrush and grasshopper responses to atmospheric carbon dioxide concentration. *Oecologia* 84, 103–110.
Johnson, R. H., and Lincoln, D. E. (1991). Sagebrush carbon allocation patterns and grasshopper nutrition: the influence of carbon dioxide enrichment and soil mineral limitation. *Oecologia* 87, 127–134.
Kemp, P. R., Waldecker, D. G., Owensby, C. E., Reynolds, J. F., and Virginia, R. A. (1994). Effects of elevated CO_2 and nitrogen fertilization pretreatments on decompostion of tallgrass prairie leaf litter. *Plant Soil* 165, 115–127.
Kimball, B. A. (1983). Carbon dioxide and agricultural yield: an assemblage and analysis of 770 prioir observations. WCL Report 14. U. S. Water Cons. Lab., Phoenix, AZ.
Körner, C. (1993). CO_2 fertilization: The great uncertainty in future vegetation development. *In* "Vegetation Dynamics and Global Change" (A. M. Solomon and H. H. Shugart, eds.), pp. 53–70. Chapman and Hall, New York.
Lambers, H. (1993). Rising CO_2, secondary plant metabolism, plant-herbivore interactions and litter decomposition—theoretical considerations. *Vegetatio* 104/105, 263–271.
Landsberg, J. J. (1986). "Physiological Ecology of Forest Production." Academic Press, New York.
Larigauderie, A., Hilbert, D. W., Oechel, W. C. (1988). Effect of CO_2 enrichment and nitrogen availability on resource acquisition and resource allocation processes in a grass, *Bromus mollis*. *Oecologia* 77, 544–549.
Lawlor, D. W., and Mitchell, A. C. (1991). The effects of increasing CO_2 on crop photosynthesis and productivity: a review of field studies. *Plant Cell Environ.* 14, 807–818.
Leadley, P. W., and Reynolds, J. F. (1992). Long-term response of an Arctic sedge to climate change—a simulation study. *Ecol. Appl.* 2, 323–340.
Leadley, P. W., and Drake, B. G. (1993). Open top chambers for exposing plant canopies to elevated CO_2 concentration and for measuring net gas exchange. *Vegetatio* 104/105, 3–15.
Lincoln, D. E., and Couvet, D. (1989). The effects of carbon supply on allelochemicals and caterpillar consumption of peppermint. *Oecologia* 78, 112–114.
Lincoln, D. E., Sionit, N., and Strain, B. R. (1984). Growth and feeding response of *Pseudoplusia includens* to host plants grown in controlled carbon dioxide atmospheres. *Environ. Entomol.* 13, 1527–1530.
Lincoln, D. E., Fajer, E. D., and Johnson, R. H. (1993). Plant-insect herbivore interactions in elevated CO_2 environments. *Trends Ecol. Evol.* 8, 64–68.
Lindroth, R. L., Kinney, K. K., and Platz, C. L. (1993). Responses of deciduous trees to elevated atmospheric CO_2: productivity, phytochemistry, and insect performance. *Ecology* 74, 763–777.
Long, S. P. (1991). Modification of the response of photosynthetic productivity to rising temperature by atmospheric CO_2 concentrations: Has its importance been underestimated? *Plant Cell Environ.* 14, 729–739.
Long, S. P., and Hutchin, P. R. (1991). Primary production in grasslands and coniferous forests with climate change: an overview. *Ecol. Appl.* 1, 139–156.
Luo, Y., Mooney, H., and Field, C. B. (1994). Predicting responses of photosynthesis and root fraction to elevated CO_2: Interaction among carbon, nitrogen and growth. *Plant Cell Environ.* 17, 1195–1204.
Luxmoore, R. J., O'Neill, E. G., Ells, J. M., and Rogers, H. H. (1986). Nutrient-uptake and growth responses of Virginia pine to elevated atmospheric CO_2. *J. Environ. Qual.* 15, 244–251.
Mansfield, T. A., Hetherington, A. M., and Atkinson, C. J. (1990). Some current aspects of stomatal physiology. *Annu. Plant Physiol. Plant Mol. Biol.* 41, 55–75.
Mattson, W. (1980). Nitrogen and herbivory. *Annu. Rev. Ecol. Syst.* 11, 119–162.
McMurtrie, R., and Wolf, L. (1983). A model of competition between trees and grass for radiation, water and nutrients. *Ann. Bot.* 52, 449–458.

McMurtrie, R., and Wang, Y.-P. (1993). Mathematical models of the photosynthetic response of tree stands to rising CO_2 concentrations and temperatures. *Plant Cell Environ.* 16, 1–13.

McMurtrie, R. E., Rook, D. A., and Kelliher, F. M. (1990). Modelling the yield of *Pinus radiata* on a site limited by water and nitrogen. *Forest Ecol. Manage.* 30, 381–413.

Meentemeyer, V. (1978). Macroclimate and lignin control of litter decomposition rates. *Ecology* 59, 465–472.

Melillo, J. M., Aber, J. D., and Muratore, J. F. (1982). Nitrogen and lignin control of hardwood leaf litter decomposition dynamics. *Ecology* 63, 621–626.

Miller, P. C., Miller, P. M., Blake-Jacobsen, M., Chapin, F. S., Everett, K. R., Hilbert, D. W., Kummerow, J., Linkins, A. E., Marion, G. M., Oechel, W. C., Roberts, S. W., and Stuart, L. (1984). Plant–soil processes in *Eriophorum vaginatum* tussock tundra in Alaska: a systems modelling approach. *Ecol. Monogr.* 54, 361–405.

Monserud, R. A., Tchebakova, A. M., and Leemans, R. (1993). Global vegetation change predicted by the modified BUDYKO model. *Clim. Change* 25, 59–83.

Mooney, H. A., Drake, B. G., Luxmoore, R. J., Oechel, W. C., and Pitelka, L. F. (1991). Predicting ecosystem responses to elevated CO_2 concentrations. *BioScience* 41 (2), 96–104.

Moorhead, D. L., and Reynolds, J. F. (1993). Effects of climate change on decomposition in arctic tussock tundra: A modeling synthesis. *Arctic Alpine Res.* 25, 403–412.

Morison, J.I. L. (1985). Sensitivity of stomata and water use efficiency to high CO_2. *Plant Cell Environ.* 8, 467–474.

Morison, J. I. L. (1993). Response of plants to CO_2 under water limited conditions. *Vegetatio* 104/105, 193–209.

Morison, J. I. L., and Gifford, R. M. (1983). Stomatal sensitivity to carbon dioxide and humidity. A comparison of two C_3 and two C_4 species. *Plant Physiol.* 71, 789–796.

Mott, K. A. (1990). Sensing of atmospheric CO_2 by plants. *Plant Cell Environ.* 13, 731–737.

Nemec, J., and Schaake, J. (1982). Sensitivity of water resources to climate variations. *Hydrol. Sci. J.* 27, 327–343.

Norby, R. J., O'Neill, E. G., and Luxmoore, R. J. (1986a). Effects of atmospheric CO_2 enrichment on the growth and mineral nutrition of *Quercus alba* seedlings in nutrient poor soil. *Plant Physiol.* 82, 83–89.

Norby, R. J., Pastor, J., and Melillo, J. M. (1986b). Carbon-nitrogen interactions in CO_2-enriched white oak: physiological and long-term perspectives. *Tree Physiol.* 2, 233–41.

Norman, J. M. (1993). Scaling processes between leaf and canopy levels. *In* "Scaling Physiological Processes: Leaf to Globe" (J. F. Ehleringer and C. B. Field, eds.), pp. 41–76. Academic Press, San Diego, CA.

Oberbauer, S. F., Strain, B. R., and Fetcher, N. (1985). Effect of CO_2 enrichment on seedling physiology and growth of two tropical species. *Physiol. Plant.* 65, 352–356.

Ojima, D. S., Parton, W. J., Schimel, D. S., Scurlock, J. M. O., and Kittel, T. G. F. (1993). Modeling the effects of climatic and CO_2 changes on grassland storage of soil-C. *Water Air Soil Pollut.* 70, 643–657.

O'Neill, E. G. (1994). Responses of soil biota to elevated atmospheric carbon dioxide. *Plant Soil* 165, 55–65.

O'Neill, E. G., Luxmoore, R. J., and Norby, R. J. (1987). Elevated atmospheric CO_2 effects on seedling growth, nutrient uptake, and rhizosphere bacterial populations of *Liriodendron tulipifera* L. *Plant Soil* 104, 3–11.

O'Neill, R. V. (1988). Hierarchy theory and global change. *In* "Scales and Global Change" (T. Rosswall, R. G. Woodmansee, and P. G. Risser, eds.). John Wiley & Sons, New York.

O'Neill, R. V., DeAngelis, D. L., Waide, J. B., and Allen, T. F. H. (1986). "A Hierarchical Concept of Ecosystems." Princeton Univ. Press, Princeton, NJ.

Oreskes N., Shrader-Frechette, K., and Belitz, K. (1994). Verification, validation, and confirmation of numerical models in the earth sciences. *Science* 263, 641–646.

Owensby, C. E., Coyne, P. I., Auen, L. M., and Sionit, N. (1990). Rangeland-plant response to elevated CO_2. "Response of Vegetation to Carbon Dioxide." No. 59. U. S. Dept. of Energy, Washington, DC.

Owensby, C. E., Coyne, P. I., Ham, J. M., Auen, L. M., and Knapp, A. K. (1993). Biomass production in a tallgrass prarie ecosystem exposed to ambient and elevated CO_2. *Ecol. Appl.* 3 (4), 644–653.

Pacala, S. W., and Silander, J. A., Jr. (1985). Neighborhood models of plant population dynamics. 1. Single-species models of annuals. *Am. Nat.* 125, 385–411.

Pacala, S. W., and Silander, J. A., Jr. (1989). Tests of neighborhood population dynamic models in field communities of two annual weed species. *Ecol. Monogr.* 60, 113–134.

Pacala, S. W., and Hurtt, G. C. (1993). Terrestrial vegetation and climate change: integrating models and experiments. *In* "Biotic Interactions and Global Change" (P. M. Kareiva, J. G. Kingsolver, and R. B. Huey, eds.), pp. 57–74. Sinauer Assoc., Sunderland, MA.

Parton, W. J., Schimel, D. S., Cole, C. V., and Ojima, D. S. (1987). Analysis of factors controlling soil organic matter levels in Great Plains grasslands. *Soil Sci. Soc. Am. J.* 51, 1173–1179.

Pastor, J., and Post, W. M. (1986). Influence of climate, soil moisture, and succession on forest carbon and nitrogen cycles. *Biogeochemistry* 2, 3–27.

Pastor, J., and Post, W. M. (1988). Response of northern forests to CO_2 induced climate change. *Nature* 343, 51–53.

Rastetter, E. B., and Shaver, G. R. (1992). A model of multiple-element limitation for acclimating vegetation. *Ecology* 73, 1157–1174.

Rastetter, E. B., Ryan, M. G., Shaver, G. R., Melillo, J. M., Nadelhoffer, K. J., Hobbie, J. E., and Aber, J. D. (1991). A general biogeochemical model describing the responses of the C and N cycles in terrestrial ecosystems to changes in CO_2, climate, and N deposition. *Tree Physiol.* 9, 101–126.

Reekie, E. G., and Bazzaz, F. A. (1989). Competition and patterns of resource use among seedlings of five tropical trees grown at ambient and elevated CO_2. *Oecologia* 79, 212–222.

Revelle, R. R., and Waggoner, P. E. (1983). Effects of a carbon dioxide-induced climatic change on water supplies in the western United States. *In* "Changing Climate, Report of the Carbon Dioxide Assessment Committee, National Research Council," pp. 419–432. National Academy Press, Washington, DC.

Reynolds, J. F., and Thornley, J. H. M. (1982). A shoot:root partitioning model. *Ann. Bot.* 49, 585–597.

Reynolds, J. F., and Acock, B. (1985). Predicting the response of plants to increasing carbon dioxide: a critique of plant growth models. *Ecol. Model.* 29, 107–129.

Reynolds, J. F., Chen, J.-L., Harley, P. C., Hilbert, D. W., Dougherty, R. L., and Tenhunen, J. D. (1992). Modeling the effects of elevated CO_2 on plants: extrapolating leaf response to a canopy. *Agric. Forest Meteorol.* 61, 69–94.

Reynolds, J. F., Hilbert, D. W., and Kemp, P. (1993a). Scaling ecophysiology from the plant to the ecosystem: a conceptual framework. *In* "Scaling Physiological Processes" (J. Ehleringer and C. Field, eds.), pp. 127–140. Academic Press, San Diego, CA.

Reynolds, J. F., Acock, B., and Whitney, R. (1993b). Linking CO_2 experiments and modelling. *In* "Design and Execution of Experiments on CO_2 Enrichment" (E.-D. Schulze and H. A. Mooney, eds.), Ecosystems Research Report No. 6, pp. 93–106. Commission of the European Communities, Brussels, Belgium.

Rosenberg, N. J., McKenney, M. S., and Martin, P. (1989). Evapotranspiration in a greenhouse warmed world: a review and a simulation. *Agric. Forest Meteorol.* 47, 303–320.

Running, S. W., and Coughlan, J. C. (1988). A general model of forest ecosystem processes for regional applications. I. Hydrologic balance, canopy gas exchange, and primary production processes. *Ecol. Model.* 42, 125–154.

Running, S. W., and Nemani, R. R. (1991). Regional hydrologic and carbon balance responses of forests resulting from potential climate change. *Clim. Change* 19, 349–368.

Running, S. W., and Hunt, E. R., Jr. (1993). Generalization of a forest ecosystem process model for other biomes, BIOME-BGC, and application for global-scale models. *In* "Scaling Physiological Processes" (J. Ehleringer and C. Field, eds.), pp. 141–158. Academic Press, San Diego, CA.

Ryan, M. G., Hunt, E. R., Jr., Ågren, G. L., Friend, A. D., Pulliam, W. M., Linder, S., McMurtrie, R. E., Aber, J. D., Rastetter, E. B., and Raison, R. J. (1994). Comparing models of ecosystem function for temperate conifer forests. I. Model description and validation. II. Simulations and effect of climate change. *In* "Effects of Climate Change on Forests and Grasslands, SCOPE" (H. A. Mooney, *et al.*, eds.). John Wiley & Sons, New York.

Sage, R. F., Sharkey, T. D., and Seemann, J. R. (1989). Acclimation of photosynthesis to elevated CO_2 in five C_3 species. *Plant Physiol.* 89, 590–596.

Sasek, T. W., DeLucia, E. H., and Strain, B. R. (1985). Reversibility of photosynthetic inhibition in cotton after long-term exposure to elevated CO_2 concentrations. *Plant Physiol.* 78, 619–622.

Schimel, D. S., Parton, W. J., Kittel, T. G. F., Ojima, D. S., and Cole, C. V. (1990). Grassland biogeochemistry: Links to atmospheric processes. *Clim. Change* 17, 13–25.

Sharpe, P. J. H. (1990). Forest modeling approaches: Compromises between generality and precision. *In* "Process Modelling of Forest Growth Responses to Environmental Stress" (R. K. Dixon, R. S. Meldahl, G. A. Ruark, and W. G. Warren, eds.), pp. 180–190. Timber Press, Portland, OR.

Sharpe, P. J. H., and Rykiel, E. J., Jr. (1991). Modelling integrated response of plants to multiple stresses. *In* "Responses of Plants to Multiple Stresses" (H. A. Mooney, W. E. Winner, and E. J. Pell, eds.), Physiological Ecology Series, pp. 205–224. Academic Press, New York.

Shugart, H. H. (1984). "A Theory of Forest Dynamics." Springer-Verlag, New York.

Shugart, H. H., Smith, T. M., and Post, W. M. (1992). The potential for application of individual-based models for assessing the effects of global change. *Annu. Rev. Ecol. Syst.* 23, 15–38.

Silvola, J., and Ahlholm, U. (1993). Effects of CO_2 Concentration and nutrient status on growth, growth rhythm and biomass partitioning in a willow, *Salix phylicifolia*. *Oikos* 67, 227–234.

Sionit, N., Strain, B. R., Hellmers, H., Riechers, G. H., and Jaeger, C. H. (1985). Long-term atmospheric CO_2 enrichment effects and the growth and development of *Liquidambar styaciflua* and *Pinus taeda* seedlings. *Can. J. Forest Res.* 15, 468–471.

Solomon, A. M. (1986). Transient response of forests to CO_2-induced climate change: simulation experiments in eastern North America. *Oecologia* 68, 567–579.

Solomon, A. M., and Webb, T., III. (1985). Computer-aided reconstruction of late Quaternary landscape dynamics. *Annu. Rev. Ecol. Syst.* 16, 63–84.

Solomon, A. M., Delcourt, H. R., West, D. C., and Blasing, T. J. (1980). Testing a simulation model for reconstruction of prehistoric forest stand dynamics. *Quat. Res.* 14, 275–293.

Solomon, A. M., West, D. C., and Solomon, J. A. (1981). Simulating the role of climate change and species immigration in forest succession. *In* "Forest Succession: Concepts and Application" (D. C. West, H. H. Shugart, and D. B. Botkin, eds.), pp. 154–177. Springer-Verlag, New York.

Sorrensen-Cothern, K. A., Ford, E. D., and Sprugel, D. G. (1993). A model of competition incorporating plasticity through modular foliage and crown development. *Ecol. Monogr.* 63, 277–304.

Stitt, M. (1991). Rising CO_2 levels and their potential significance for carbon flow in photosynthetic cells. *Plant Cell Environ.* 14, 741–762.

Strain, B. R. (1985). Physiological and ecological controls on carbon sequestering in ecosystems. *Biogeochemistry* 1, 219–232.

Strain, B. R., and Bazzaz, F. A. (1983). Terrestrial plant communities. *In* "CO_2 and Plants: The Response of Plants to Rising Levels of Atmospheric Carbon Dioxide" (E. R. Lemon, ed.), pp. 177–222. Westview Press, Inc., Boulder, CO.

Strain, B. R., and Cure, J. (eds.) (1985). "Direct Effects of Increasing Carbon Dioxide on Vegetation," Publication ER-0238. United States Department of Energy, Washington, DC.
Strain, B. R., and Thomas, R. B. (1992). Field measurements of CO_2 enhancement and climate change in natural vegetation. *Water, Air Soil Pollut.* 64, 45–60.
Thomas, R. B., and Strain, B. R. (1991). Root restriction as a factor in photosynthetic acclimation of cotton seedlings grown in elevated carbon dioxide. *Plant Physiol.* 96, 627–634.
Thornley, J. H. M., Fowler, D., and Cannell, M. G. R. (1991). Terrestrial carbon storage resulting from CO_2 and nitrogen fertilization in temperate grasslands. *Plant Cell Environ.* 14, 1007–1011.
Tilman, D. (1988). "Plant Strategies and the Dynamics and Structure of Plant Communities." Princeton Univ. Press, Princeton, NJ.
Tilman, D. (1990). Mechanisms of plant competition for nutrients: the elements of a predictive theory of competition. *In* "Perspectives on Plant Competition" (J. B. Grace and D. Tilman, eds.), pp. 117–141. Academic Press, San Diego, CA.
Tilman, D. (1993). Carbon dioxide limitation and potential direct effects of its accumulation on plant communities. *In* "Biotic Interactions and Global Change" (P. M. Kareiva, J. G. Kingsolver, and R. B. Huey, eds.), pp. 333–346. Sinauer Associates Inc., Sunderland, MA.
Tissue, D. T., and Oechel, W. C. (1987). Response of *Eriophorum vaginatum* to elevated CO_2 and temperature in the Alaskan tussock tundra. *Ecology* 68, 401–410.
Tissue, D. T., Thomas, R. B., and Strain, B. R. (1993). Long-term effects of elevated CO_2 and nutrients on photosynthesis and rubisco in loblolly pine seedlings. *Plant Cell Environ.* 16, 859–865.
Tolley, L. C., and Strain, B. R. (1985). Effects of CO_2 enrichment and water stress on gas exchange of *Liquidambar styraciflua* and *Pinus taeda* seedlings grown under different levels of irradiance. *Oecologia* 65, 166–172.
Tyree, M. T., and Alexander, J. D. (1993). Plant water relations and the effects of elevated CO_2: A review and suggestions for future research. *Vegetatio* 104/105, 47–62.
Urban, D. L., and Shugart, H. H. (1989). Forest response to climatic change: a simulation study for southeastern forests. *In* "The Potential Effects of Global Climate Change on the United States: Appendix D—Forests" (J. B. Smith and D. A. Tirpak, eds.), U.S. Environmental Protection Agency, Office of Policy, Planning, Evaluation, Washington, DC.
van de Geijn, S. C., and van Veen, J. A. (1993). Implications of increased carbon dioxide levels for carbon input and turnover in soils. *Vegetatio* 104, 283–292.
van Hulst, R. (1979). On the dynamics of vegetation: Markov chains as models of succession. *Vegetatio* 40, 3–14.
van Tongeren, O., and Prentice, I. C. (1986). A spatial simulation model for vegetation dynamics. *Vegetatio* 65, 163–173.
von Caemmerer, S., and Farquhar, G. D. (1984). Effects of partial defoliation, changes of irradiance during growth, short-term water stress and growth enhanced p(CO_2) on the photosynthetic capacity of *Phaseolus vulgaris* L. *Planta* 160, 320–329.
Wang, Y.-P., McMurtrie, R. E., and Landsberg, J. J. (1992). Modelling canopy photosynthetic productivity. *In* "Crop Photosynthesis: Spatial and Temporal Determinants, Vol. 12, Topics in Photosynthesis" (N. R. Baker and H. Thomas, eds.), pp. 43–67. Elsevier Science Publ., New York.
Weiner, J., and Conte, P. T. (1981). Dispersal and neighborhood effects in an annual plant competition model. *Ecol. Model.* 13, 131–147.
Wessman, C. A. (1992). Spatial scales and global change: Bridging the gap from plots to GCM grid cells. *Annu. Rev. Ecol. Syst.* 23, 175–200.
Woodin, S., Graham, B., Killick, A., Skiba, U., and Cresser, M. (1992). Nutrient limitation of the long term response of heather (*Calluna vulgaris* (L.) Hull) to CO_2 Enrichment. *New Phytol.* 122, 635–642.

Woodward, F. I. (1993). The lowland-to-upland transition—modelling plant responses to environmental change. *Ecol. Appl.* 3, 404–408.

Woodward, F. I., and Bazzaz, F. A. (1988). The responses of stomatal density to CO_2 partial pressure. *J. Exp. Bot.* 39, 1771–1781.

Woodward, F. I., Thompson, G. B., and McKee, I. F. (1991). The effects of elevated concentrations of carbon dioxide on individual plants, populations, communities and ecosystems. *Ann. Bot.* 67, 23–38.

Zangerl, A. R., and Bazzaz, F. A. (1984). The response of plants to elevated CO_2. II. Competitive interactions among individual plants under varying light and nutrients. *Oecologia* 62, 412–417.

Ziska, L. H., Hogan, K. P., Smith, A. P., and Drake, D. G. (1991). Growth and photosynthetic response of nine tropical species with long-term exposure to elevated carbon dioxide. *Oecologia* 86, 383–389.

20

Stimulation of Global Photosynthetic Carbon Influx by an Increase in Atmospheric Carbon Dioxide Concentration

Yiqi Luo and Harold A. Mooney

I. Introduction

Atmospheric CO_2 concentration (C_a) is rapidly and unambiguously increasing (Siegenthaler and Sarmiento, 1993; Thorning et al., 1989), rising from 280 ppm in preindustrial times to nearly 360 ppm in 1994 and possibly doubling in the next century. Rising C_a could substantially stimulate global photosynthetic carbon influx from the atmosphere to the biosphere (Allen et al., 1986; Melillo et al., 1993), potentially resulting in terrestrial ecosystems storing 1–2 Gt (1 Gt = 10^{15} g) of carbon per year (Tans et al., 1990; Wigley and Raper, 1992; Gifford, 1994). Quantification of such stimulation on the global scale, therefore, is crucial for our understanding of global carbon cycling in a changing C_a environment.

Estimation of global photosynthetic carbon influx has been exceedingly difficult, utilizing either experimental or modeling approaches. Available techniques only allow us to make leaf-level and small-scale ecosystem measurements (Field and Mooney, 1990). Measurements on both scales indicate that photosynthetic responses to elevated CO_2 are extremely variable (Luo et al., 1994; Sage, 1994). The biochemical capacity for leaf photosynthesis increases for some species and decreases for others under elevated CO_2 (Sage et al., 1989; Stitt, 1991; Sage, 1994). Net ecosystem carbon assimilation in elevated CO_2 changed little in Artic tundra (Grulke et al., 1990; Tissue and Oechel, 1987), but increased by 80% in salt marsh ecosystems on the Chesapeake Bay [Drake and Leadley, 1991; Drake et al., 1996 (this volume)], depending on species composition and species-specific CO_2 responses

(Mooney et al., 1991; Luo, 1995). Variable responses of photosynthesis at the leaf and ecosystem levels make it difficult to extrapolate small-scale studies to a global estimation of photosynthetic carbon influx.

Early modeling studies of global carbon cycling employed a single parameter, i.e., the biotic growth factor β, to account for terrestrial carbon content changes with C_a (Bacastow and Keeling, 1973). The β factor can be defined as a fractional change in net primary productivity (NPP) with a fractional change in C_a (Gates, 1985):

$$\beta = \left(\frac{\Delta \text{NPP}}{\text{NPP}}\right)\left(\frac{C_a}{\Delta C_a}\right) \tag{1}$$

Bacastow and Keeling (1973) stated that β could be as low as 0.05, but their results indicate a likely range of 0.2–0.6 based on the observed atmospheric increase from 1959 to 1969 (Gates, 1985). Amthor and Koch (1996, this volume), by using a different formulation for β, conclude from the scant experimental data available that β (NPP) is probably greater than zero in many ecosystems, but acknowledge a very wide range of values between, and even within, ecosystems. Limited understanding of the biological basis of the β factor circumscribes its applications in global carbon cycling studies (Harvey, 1989).

Modeling studies have integrated more biological information into predicting global terrestrial carbon exchanges (Prentice and Fung, 1990; Smith et al., 1992; Melillo et al., 1993; Potter et al., 1993). These models usually use geographical maps of world vegetation and soils and characteristic parameters of each vegetation and soil type. CO_2 effects on carbon uptake are generally based on experimental results at the leaf or small ecosystem scales (Melillo et al., 1993). Environmental heterogeneity in global ecosystems and diverse species characteristics, however, still hinder our understanding of CO_2 influences on carbon fluxes between the biosphere and atmosphere.

Here we have employed an analytical approach in an attempt to overcome difficulties associated with environmental heterogeneity and species characteristics in studying global photosynthetic carbon influx. We examine leaf photosynthesis, focusing on its relative response to a small change in atmospheric CO_2 concentration (the leaf-level $\mathcal{L}$ factor). From simple mathematical manipulations of a mechanistic model of leaf photosynthesis (Farquhar et al., 1980), we find that the $\mathcal{L}$ factor is an approximate constant for any C_3 plant, regardless of the geographical location and canopy position. We explore the biochemical basis of the $\mathcal{L}$ factor being an approximate constant and discuss the possibility of extrapolating the $\mathcal{L}$ factor across spatial and temporal scales to estimate the additional amount of global photosynthetic carbon influx (P_G) stimulated by a small C_a increase. We also develop a relationship between the $\mathcal{L}$ factor and the biotic growth factor β.

II. The Model

Annual global photosynthetic carbon influx [P_G, Gt ($=10^{15}$ g) yr^{-1}], i.e., gross primary productivity, is the sum of carbon influx from total leaf area within canopies (x) over the global surface (y) over the period of a year (t) [$P(x,y,t)$, g m^{-2} s^{-1}]. Mathematically, it can be expressed as

$$P_G = \int_{t=\text{year}} \int_{y=\text{globe}} \int_{x=\text{canopy}} P(x, y, t) \, dx \, dy \, dt \qquad (2)$$

For simplicity, $P(x,y,t)$ is abbreviated as P hereafter. For 1 unit change in the global atmospheric CO_2 concentration (C_a, ppm), the rate of P_G change (Gt ppm^{-1} yr^{-1}) is

$$\frac{dP_G}{dC_a} = \iiint \frac{dP}{dC_a} \, dx \, dy \, dt$$
$$= \iiint (\mathcal{L} P) \, dx \, dy \, dt \qquad (3)$$

where $\mathcal{L}$ is a leaf-level factor (ppm^{-1}) defined as

$$\mathcal{L} = \frac{1}{P} \frac{dP}{dC_a} \qquad (4)$$

The $\mathcal{L}$ factor denotes the relative leaf photosynthetic response to a 1 ppm C_a change.

The vast majority of all terrestrial plants share a C_3 photosynthetic pathway (Bowes, 1993). Leaf photosynthesis of C_3 plants, which is usually limited either by electron transport or by ribulose-1,5-bisphosphate carboxylase/oxygenase (rubisco) activity, is predicted by the well-established Farquhar et al. (1980) model. When photosynthesis is limited by electron transport, leaf photosynthetic carbon influx (P_l) is

$$P_l = J \frac{C_i - \Gamma}{4.5 C_i + 10.5 \Gamma} \qquad (5)$$

where J is the electron transport rate (μmol m^{-2} s^{-1}), C_i is the intercellular CO_2 concentration (ppm), and Γ is the CO_2 compensation point without dark respiration (ppm). With the assumption that C_i is proportional to C_a as,

$$C_i = \alpha C_a, \quad 0 < \alpha < 1, \qquad (6)$$

the corresponding $\mathcal{L}$ factor is

$$\mathcal{L}_1 = \frac{15 \alpha \Gamma}{(\alpha C_a - \Gamma)(4.5 \alpha C_a + 10.5 \Gamma)} \qquad (7)$$

Parameter J is eliminated from Eq. (7) because $\mathcal{L}_1$ is a measure of relative response.

When photosynthesis is limited by rubisco activity, leaf carbon influx (P_2) is

$$P_2 = V_{cmax}\frac{C_i - \Gamma}{C_i + K} \qquad (8)$$

where V_{cmax} is the maximum carboxylation rate (μmol m^{-2} s^{-1}) and K is a coefficient (ppm) associated with enzyme kinetics. The corresponding $\mathcal{L}$ factor is

$$\mathcal{L}_2 = \frac{\alpha(K + \Gamma)}{(\alpha C_a - \Gamma)(\alpha C_a + K)} \qquad (9)$$

Parameter V_{cmax} is eliminated from Eq. (9) because $\mathcal{L}_2$ is also a measure of relative response. Because the physiological process of electron transport is less sensitive to CO_2 concentration than is carboxylation, $\mathcal{L}_1$ and $\mathcal{L}_2$ define the lower and upper limits, respectively, of the $\mathcal{L}$ factor.

Equations (7) and (9) suggest that both the lower and upper limits of the $\mathcal{L}$ factor are independent of the light-related parameter J and enzyme-related parameter V_{cmax} and vary with C_a, α, Γ, and K. With the assumption that the four parameters are constant (the validity is discussed in the following), the $\mathcal{L}$ factor is a constant regardless of plant species, vertical position in a canopy, geographical location on the earth, and time of year. Thus, Eq. (3) becomes

$$\frac{dP_G}{dC_a} = \mathcal{L} P_G \qquad (10)$$

Equation (10) indicates that the rate of P_G change relative to C_a can be calculated simply from $\mathcal{L}$ and P_G. It follows that the additional amount of annual photosynthetic carbon influx (ΔP_G, Gt yr^{-1}), stimulated by a yearly increase in atmospheric CO_2 concentration (ΔC_a), can be estimated by

$$\begin{aligned}\Delta P_{G,1} &= \mathcal{L}_1\, P_G\, \Delta\, C_a \\ \Delta P_{G,2} &= \mathcal{L}_2\, P_G\, \Delta\, C_a\end{aligned} \qquad (11)$$

where $\Delta P_{G,1}$ and $\Delta P_{G,2}$ are the lower and upper limits of ΔP_G, respectively. When the photosynthesis of global vegetation is mainly limited by light, biochemical processes related to electron transport limit photosynthesis, and the additional global photosynthetic carbon influx is close to the lower limit. Otherwise, it may approach the upper limit.

III. Results and Discussion

A. Relative Photosynthetic Response to CO_2 ($\mathcal{L}$ Factor)

Photosynthesis in the light-limited environment increases by 0.183% ($\mathcal{L}_1$) for C_a at 280 ppm, 0.115% at 357 ppm, and 0.077% at 440 ppm (Fig. 1),

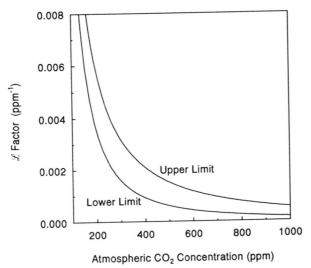

Figure 1 Lower ($\mathcal{L}_1$) and upper ($\mathcal{L}_2$) limits of the $\mathcal{L}$ factor (relative leaf photosynthetic response to a 1-ppm CO_2 change) within a range of atmospheric CO_2 concentrations from 100 to 1000 ppm, predicted by Eqs. (7) and (9) with $\alpha = 0.70$, $\Gamma = 35$ ppm, and $K = 650$ μmol m^{-2} s^{-1}.

with $\alpha = 0.70$ and $\Gamma = 35$ ppm when the atmospheric CO_2 concentration (C_a) increases by 1 ppm [Eq. (7)]. In the enzyme-limited condition, photosynthesis increases by 0.352% ($\mathcal{L}_2$) for C_a at 280 ppm, 0.248% at 357 ppm, and 0.183% at 440 ppm (Fig. 1), with $\alpha = 0.70$, $\Gamma = 35$ ppm, and $K = 650$ ppm due to a 1-ppm CO_2 increase [Eq. (9)].

Relative photosynthetic response to CO_2 (the $\mathcal{L}$ factor) varies with parameters, α, Γ, and K. The C_i/C_a ratio (α) is sensitive to water status and CO_2 concentration but fairly constant among species when plants grow in their natural environments (Pearcy and Ehleringer, 1984; Evans and Farquhar, 1991). When α decreases by 0.10 from 0.70, $\mathcal{L}_1$ increases by 15% and $\mathcal{L}_2$ by 7% (Fig. 2A). When α increases by 0.10, $\mathcal{L}_1$ decreases by 12% and $\mathcal{L}_2$ by 6%. The CO_2 compensation point (Γ) varies little among species, but depends strongly on temperature (Jordan and Ogren, 1984; Brooks and Farquhar, 1985). A value of $\Gamma = 35$ ppm chosen here corresponds to that at 20°C, which is about 4°C higher than the average earth surface temperature (Schlesinger, 1991) to account for most photosynthetic machinery distributed in warmer tropic and temperate zones. As globally averaged temperature changes by ±5°C, leading to approximately ±7 ppm changes in Γ from 35 ppm, $\mathcal{L}_1$ varies by 19% and $\mathcal{L}_2$ by 4% (Fig. 2B). The enzyme kinetic parameter K is variable among species (Evans and Seemann, 1984; Harley

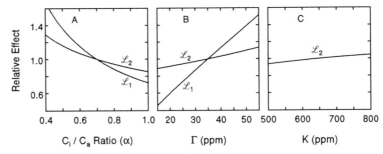

Figure 2 Effects of (A) C_i/C_a ratio (α), (B) Γ (CO_2 compensation point), and (C) K (enzyme kinetic parameter) on the lower and upper limits of $\mathcal{L}$. Parameter values are the same as in Fig. 1.

and Tenhunen, 1991), but only slightly affects the upper limit of the $\mathcal{L}$ factor (Fig. 2C).

The Farquhar et al. (1980) model describes two general biochemical processes of leaf photosynthesis: ribulose 1,5-bisphosphate (RuBP) regeneration driven by light and carbon fixation catalyzed by ribulose-1,5-bisphosphate carboxylase/oxygenase (rubisco). Effects of light, nutrient availability, and plant characteristics on photosynthesis are reflected by variations in J and V_{cmax} values, and the values of these two parameters vary greatly (Wullschleger, 1993). Parameter V_{cmax} ranges from 6 μmol m^{-2} s^{-1} for the coniferous species *Picea abies* (Benner et al., 1988) to 194 μmol m^{-2} s^{-1} for the agricultural species *Beta vulgaris* (Taylor and Terry, 1984) and averages 64 μmol m^{-2} s^{-1} for 109 species (Wullschleger, 1993). Parameter J increases with light in a rectangular hyperbolic shape and reaches a maximum, J_{max} (Farquhar et al., 1980). The latter also varies greatly among species (Wullschleger, 1993). Both V_{cmax} and J_{max} vary with nutrient availability (Field, 1983; Harley et al., 1992). Our mathematical derivation eliminates parameters J and V_{cmax}, leading to the $\mathcal{L}$ factor being independent of plant characteristics, light, and nutrient environment. The resultant $\mathcal{L}$ factor is only a function of α, Γ, K, and C_a. Since α, Γ, and K only slightly affect the $\mathcal{L}$ factor (Fig. 2), and C_a varies little over different geographical locations (Conway and Tans, 1989) and canopy positions (Monteith and Unsworth, 1990), the $\mathcal{L}$ factor is virtually a constant across ecosystems, but a function of time-associated changes in C_a.

B. Biochemical Basis of the $\mathcal{L}$ Factor

That the $\mathcal{L}$ factor is an approximate constant at a given C_a is rooted in the nature of the biochemical reactions of photosynthesis. Photosynthesis (i.e., carboxylation of RuBP) and photorespiration (i.e., oxygenation of RuBP) are both catalyzed by rubisco (Andrews and Lorimer, 1987). The rubisco reaction with molecular carbon dioxide via carboxylation of RuBP leads to carbohydrate synthesis in the photosynthetic carbon reduction (PCR) cycle (Fig. 3). The rubisco reaction with molecular oxygen via oxygenation of RuBP leads to carbohydrate anabolism in the photorespiratory carbon oxidation (PCO) cycle with resultant release of CO_2 (Fig. 3). An increased CO_2 concentration competes with O_2 and decreases the oxygenase activity of rubisco (Farquhar *et al.*, 1980; Stitt, 1991; Lawlor, 1993), leading to an increased ratio of carboxylation to oxygenation.

The carboxylation acceptor, RuBP, is regenerated in the PCR cycle, which is driven by light energy. When light limits photosynthesis, regeneration of RuBP controls the photosynthetic rate (Farquhar *et al.*, 1980). With a certain amount of regenerated RuBP, a portion of RuBP binds molecular carbon dioxide in the PCR cycle to produce carbohydrate, and a portion of RuBP is used to bind molecular oxygen to release CO_2 in the PCO cycle (Fig. 3). Despite different light levels, leading to different electron transport rates and regeneration rates of RuBP (von Caemmerer and Edmonson, 1986; Andrew and Lorimer, 1988; Sage *et al.*, 1990), light does not change the fraction of RuBP involved in carboxylation versus oxygenation. An increasing CO_2 concentration does not change the regeneration rates of RuBP, but increases the portion of RuBP for carboxylation in the PCR cycle and decreases the portion of RuBP for oxygenation in the PCO cycle. Carboxylation efficiency (carboxylation rate per unit of photosynthetic machinery) increases. Thus, the relative photosynthetic response to CO_2 (the $\mathscr{L}$ factor) is not dependent on RuBP regeneration rate and light level, but is dependent on changes in CO_2 concentration.

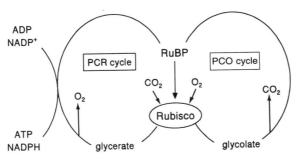

Figure 3 Illustration of the biochemical reaction of photosynthesis. Rubisco (ribulose-1,5-bisphosphate carboxylase/oxygenase) catalyzes both the carboxylation and oxygenation of RuBP (ribulose 1,5-bisphosphate). RuBP reacts with CO_2 (carboxylation) in the photosynthetic carbon reduction (PCR) cycle to produce carbohydrate and with O_2 (oxygenation) in the photorespiratory carbon oxidation (PCO) cycle to release CO_2. RuBP is regenerated in the PCR cycle, consuming energy and electrons generated in light reactions. The ratio of carboxylation to oxygenation (the $\mathscr{L}$ factor) is regulated by the CO_2/O_2 ratio and temperature, but is independent of light-driven regeneration of RuBP and plant-specific content of rubisco.

When light does not limit photosynthesis and RuBP is saturated, rubisco controls the photosynthetic rate (Farquhar et al., 1980). With a fixed amount of rubisco, a portion of rubisco binds molecular carbon dioxide to produce carbohydrate in the PCR cycle, and a portion of rubisco binds molecular oxygen to release CO_2 in the PCO cycle (Fig. 3). A small increase in C_a does not change the rubisco amount, but increases the fraction of rubisco binding with molecular carbon dioxide and decreases the portion of rubisco binding molecular oxygen. It follows that the relative photosynthetic response to CO_2 (the $\mathscr{L}$ factor) is independent of rubisco content but varies with the CO_2 concentration.

The ability of rubisco to bind CO_2 versus O_2, i.e., the CO_2/O_2 specificity of rubisco, depends on temperature (Farquhar et al., 1980; Jordan and Ogren, 1984). This temperature dependence is reflected in the CO_2 compensation point (G), which slightly influences the $\mathscr{L}$ factor (Fig. 2B). Although plant water status influences the CO_2/O_2 ratio at the reaction sites, homeostatic adjustments through photosynthetic rate and stomatal opening lead to a fairly constant intercellular CO_2 concentration and CO_2/O_2 ratio (Pearcy and Ehleringer, 1984; Evans and Farquhar, 1991). Thus, parameter $\alpha(= C_i/C_a)$ varies within a narrow range and slightly affects the $\mathscr{L}$ factor. In short, a small increase in atmospheric CO_2 leads to an increase in CO_2/O_2, the carboxylation/oxygenation ratio, and the rate of carboxylation, the fractional increase of which (the $\mathscr{L}$ factor) is independent of light-driven regeneration of RuBP (J) and plant-specific content of rubisco (V_{cmax}), is slightly influenced by temperature and water stress.

C. Spatial Extrapolation of the $\mathscr{L}$ Factor to the Global Scale

Mathematical derivation and biochemical examination both indicate that the relative photosynthetic response to CO_2 (the $\mathscr{L}$ factor) is independent of light, nutrient environment, and species characteristics, but is a function of atmospheric CO_2 concentration and is slightly influenced by temperature and water. This property of the $\mathscr{L}$ factor provides the possibility of extrapolating the leaf-level $\mathscr{L}$ factor to the global scale for estimating the additional P_G stimulated by C_a increase. Indeed, small-scale measurements can be extrapolated to the global scale, provided that the parameter in question is an approximate constant relative to environmental and biological variables. An example is atmospheric CO_2 concentration, which is independent of temperature, moisture, and other atmospheric factors and is only slightly influenced by biospheric activities, i.e., an approximate constant across spatial scales at a given time.

Along with the approximate global constancy of C_a, several other factors make it possible to extrapolate the $\mathscr{L}$ factor across spatial scales to estimate carbon influx from the atmosphere to the biosphere. First, photosynthesis is the almost exclusive pathway through which terrestrial ecosystems take up carbon from the atmosphere (Mooney et al., 1987). Second, the vast majority of plants in terrestrial ecosystems share the C_3 photosynthetic pathway (Bowes, 1993). Third, photosynthesis of C_3 plants can be well described by the Farquhar et al. (1980) model. Since the $\mathscr{L}$ factor is derived simply from the Farquhar model, its approximate constancy at a given C_a propagates to the majority of terrestrial plants in the earth system.

Additional global photosynthetic carbon influx estimated from extrapolating the $\mathscr{L}$ factor to the global scale [Eq. (11)] is between 0.21 and 0.45 Gt yr^{-1} in 1993 compared to that in 1992 (Fig. 4), with $C_a = 357$ ppm, $\Delta C_a = 1.5$ ppm (Thorning et al., 1989), and $P_G = 120$ Gt yr^{-1} (Olson et al., 1983). Stimulation of global carbon influx by increasing C_a diminishes from 0.25 in 1970 to 0.17 Gt yr^{-1} in 2020 for the lower limit and from 0.52 to 0.38 Gt yr^{-1} for the upper limit during the same period (Fig. 4), if C_a increases by 1.5 ppm each year.

The predicted ΔP_G varies with α, Γ, and K in a parallel manner for the $\mathscr{L}$ factor (Fig. 2). In addition, C_a over different geographical locations and at different positions in a canopy generally varies by less than 20 ppm (Conway and Tans, 1989; Monteith and Unsworth, 1990). Thus, global variations of CO_2 concentration at leaf surfaces may be much less than 40 ppm. That would cause a slight change in the estimated ΔP_G. Although CO_2 concentration in the forest floor may be as high as 420 ppm (Bazzaz

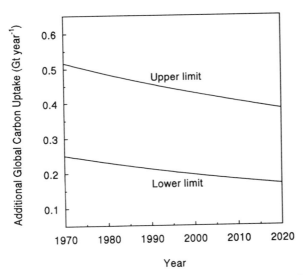

Figure 4 Lower ($\Delta P_{G,1}$) and upper ($\Delta P_{G,2}$) limits of the additional amount of global photosynthetic carbon influx stimulated by a 1.5-ppm increase in atmospheric CO_2 concentration per year, predicted by Eq. (11) with $\alpha = 0.70$, $\Gamma = 35$ ppm, $K = 650$ ppm, $P_G = 120$ Gt yr^{-1}, and $\Delta C_a = 1.5$ ppm.

and Williams, 1991), its effect on ΔP_G is negligible because understory plants contribute little to P_G. Estimated P_G varies by 20–30 Gt yr^{-1} (Olson, et al., 1983). When a different P_G is used, ΔP_G could vary by up to 20%. The additional carbon influx (ΔP_G) could be slightly lowered by the presence of C_4 plants, whose photosynthesis is less sensitive to C_a change than the C_3 species (Collatz et al., 1992), and some C_3 plants, whose photosynthesis is limited by phosphate regeneration, which desensitizes the CO_2 response (Sharkey, 1985; Wullschleger, 1993).

D. Applying the $\mathscr{L}$ Factor for Long-Term Studies

Our model predicts additional global photosynthetic carbon influx induced by a small increase in C_a. The prediction is based on the assumption that a few ppm CO_2 change in the atmosphere does not alter parameters J and V_{cmax}. Plants grown under twice the current ambient CO_2 concentration, however, have shown both increases and decreases in photosynthetic capacity (see, for example, Sage et al., 1989; Wong, 1990; Ryle et al., 1992). Changes in photosynthetic capacity result from variations in J_{max} (the maximum J) and V_{cmax}. When J and V_{cmax} vary with CO_2 concentration, the $\mathscr{L}$ factor is modified as

$$\mathscr{L}_1' = \mathscr{L}_1 + \frac{1}{J}\frac{dJ}{dC_a}$$
$$\mathscr{L}_2' = \mathscr{L}_2 + \frac{1}{V_{cmax}}\frac{dV_{cmax}}{dC_a} \quad (12)$$

Consequently, the lower and upper limits of the additional carbon influx are

$$\Delta P'_{G,1} = \Delta P_{G,1} + \Delta C_a \int\int\int \left(\frac{P}{J}\frac{dJ}{dC_a}\right) dx\, dy\, dt$$
$$\Delta P'_{G,2} = \Delta P_{G,2} + \Delta C_a \int\int\int \left(\frac{P}{V_{cmax}}\frac{dV_{cmax}}{dC_a}\right) dx\, dy\, dt \quad (13)$$

Equation 13 indicates that, if globally averaged J and V_{cmax} decrease by 10% in the 700-ppm CO_2 concentration in comparison to those in 350 ppm, $\Delta P_G'$, on average, should be smaller by 0.0514 Gt yr^{-1} than ΔP_G, with $P_G = 120$ Gt yr^{-1} and $\Delta C_a = 1.5$ ppm each year. If globally averaged J and V_{cmax} increase by 10% in the 700-ppm CO_2 concentration, $\Delta P_G'$ on average, should be larger by 0.0514 Gt yr^{-1} than ΔP_G.

Two reviews reveal that leaf photosynthetic capacity aggregated across a variety of species varies little with CO_2 (Sage, 1994; Luo, 1995). Based on theoretical interpretations of A/C_i (assimilation versus intercellular CO_2 concentration) curves, Sage (1994) concluded that growth in elevated CO_2 leads to a higher photosynthetic capacity than in ambient CO_2 for 12 out of 27 species with plants grown in pots and for 2 out of 3 species with plants grown in the field. Luo (1995) used nitrogen–photosynthesis relationships (Field, 1983; Harley *et al.*, 1992) and predicted that J_{max} and V_{cmax} decreased by 2.1% for all 33 species surveyed from published papers, by 4.1% for a subgroup of 11 crop species, and by 1.1% for a subgroup of 22 wild species with a doubled CO_2. Although ecosystem carbon fluxes depend on canopy structure, effects of elevated CO_2 on canopy development are largely unknown (Luo and Mooney, 1995).

With the assumption that globally averaged J and V_{cmax} do not change with C_a, the $\mathcal{L}$ factor can be directly used to estimate cumulative additional carbon influx. The atmospheric CO_2 concentration increased by 42 ppm from 1958 to 1993. That results in 5.6–12.1% more carbon influx in 1993 than in 1958, equaling an additional 6.7–14.5 Gt yr^{-1} with $P_G = 120$ Gt yr^{-1} (Table I). A 77-ppm C_a increase from preindustrial times to 1993 could stimulate global carbon influx by 11.8–25.5%. Doubling of C_a from 350 to 700 ppm would lead to a 23.8–70.1% increase in global photosynthetic carbon influx (Table I).

Use of the $\mathcal{L}$ factor to estimate the long-term stimulation of global carbon influx by a cumulative C_a increase requires caution due to two major issues. One is the possibility of CO_2-induced adjustments in leaf and ecosystem photosynthetic properties and, thus, the parameters V_{cmax} and J_{max}. The other is the degree of limitation of global photosynthetic stimulation and global NPP by nutrient limitation under rising CO_2 (Luo and Mooney, 1995). Limited evidence supports the idea that the parameters V_{cmax} and J_{max}, averaged across a group of species, vary little with CO_2 concentration. We recognize, however, that this is based largely on data for herbaceous species, whereas most of the global terrestrial carbon fixation is by long-lived woody species, for which there is limited data. The second, and related, point of caution concerns the influence of elevated CO_2 on interactions between the carbon and nitrogen cycles (Johnson *et al.*, 1996, O'Neill and Norby, 1996, and Curtis *et al.*, 1996, all in this volume) and the role of nitrogen deposition from the atmosphere in meeting increased nitrogen demand under elevated CO_2 (Schindler and Bayley, 1993; Hudson *et al.*,

Table I Changes in Global Photosynthetic Carbon Influx in Four Periods of Time

Period	ΔC_a (ppm)	ΔP_G (%)	$\Delta P_G{}^a$ (Gt yr^{-1})
1992–1993	1.5	0.17–0.37	0.21–0.45
1958–1993	42	5.6–12.1	6.7–14.5
Preindustrial times to 1993	77	11.8–25.5	14.1–30.6
1988 to 21st century	350	23.8–70.1	28.6–84.1

^{a} With $P_G = 120$ Gt yr^{-1} (1 Gt = 10^{15} g).

1994). At this time, uncertainties in these areas only serve to emphasize the tentative nature of our conclusions regarding the long-term stimulation of carbon influx by a cumulative increase in C_a.

E. Photosynthetic β Factor Defined from the $\mathscr{L}$ Factor

The $\mathscr{L}$ factor developed here can be used to define the photosynthetic β factor (β_p) as

$$\beta_p = C_a \mathscr{L} = \left(\frac{dP}{P}\right)\left(\frac{C_a}{dC_a}\right) \qquad (14)$$

The lower and upper limits of β_p are 0.51 and 0.99, respectively, when C_a is at 280 ppm, 0.41 and 0.89 at 357 ppm, and 0.22 and 0.65 at 700 ppm (Fig. 5). The prediction is consistent with observed data for numerous species (Fig. 5). All of the 18 data points from 5 species are within the predicted lower and upper limits.

Equation (14) is similar to Eq. (1). In Eq. (1), the growth factor β represents biomass production changes against C_a changes (Bacastow and Keeling, 1973), whereas β_p in Eq. (14) describes photosynthesis changes against C_a changes. Indeed, the growth β factor is closely related to β_p

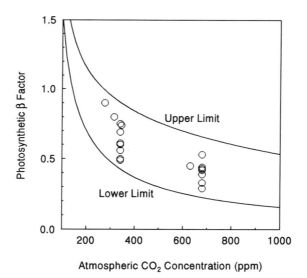

Figure 5 Lower and upper limits of photosynthetic β factor within a range of atmospheric CO_2 concentrations from 100 to 1000 ppm, predicted by Eq. (14) with $\alpha = 0.70$, $\Gamma = 35$ ppm, and $K = 650$ ppm. Symbols represent data from species *Acer saccharum* (Jurick et al., 1985), *Glycine max* (Allen et al., 1986), *Populus deltoides* (Regehr et al., 1975), and *Populus grandidentata* and *Quercus rubra* (Jurik et al., 1985).

(Gates, 1985; Allen et al., 1986). Measured β (growth) for *Glycine max* was about 15% lower than measured β_p (Allen et al., 1986). Values of β used in a global carbon cycling model by Bacastow and Keeling (1973) ranged from 0.2 to 0.6 for the observed C_a increase from 1959 to 1969. Goudriaan and Ketner (1984) used a β value of 0.5 to model the observed C_a increase from 1958 to 1980. Gifford (1980) found a β value of 0.6 to be necessary in his global carbon model. Those values of the growth β factor are well within the range of the photosynthetic β factor predicted from the $\mathcal{L}$ factor. A close match between values of the growth β factors used in various global carbon cycling models and predicted β_p ($\mathcal{L}C_a$) provides more support for the global application of the $\mathcal{L}$ factor.

IV. Summary

We have used leaf-level physiology to estimate the additional amount of global terrestrial carbon influx (P_G) stimulated by an increase in the atmospheric CO_2 concentration (C_a). We examined leaf photosynthesis (P), focusing on its relative response to a small change in C_a [$=dP/(PdC_a)$]. Although the response of P to C_a (dP/dC_a) varies greatly with light, nutrients, and species, normalization of dP/dC_a against P eliminates their effects. As a result, the leaf-level $\mathcal{L}$ factor is independent of light, nutrient environment, and species characteristics of C_3 plants, but rather is a function of C_a and is slightly influenced by temperature and water. Since the $\mathcal{L}$ factor is derived simply from the Farquhar et al. (1980) model, which predicts the photosynthesis of C_3 plants, which are the vast majority of terrestrial plants in global ecosystems, its property of being an approximate constant at a given C_a propagates to the majority of terrestrial plants in the earth system. Thus, we are able to extrapolate from the $\mathcal{L}$ factor to estimate the additional amount of P_G as stimulated by a small C_a increase. That is, 0.21–0.45 Gt ($=10^{15}$ g) yr^{-1} with $P_G = 120$ Gt yr^{-1} in 1993, compared with that in 1992, due to a 1.5-ppm C_a increase in 1993. Application of the $\mathcal{L}$ factor for long-term studies is valid when ecosystem photosynthetic properties do not change with C_a and increased carbon assimilation can be matched by nutrient supply through aerial deposition or perhaps increased nutrient use efficiency. Limited available data substantiate these two prerequisites. In this case, P_G increases by 11.8–25.5% for a 77-ppm C_a increase from preindustrial times to 1993 and by 23.8–70.1% for a C_a increase from 350 to 700 ppm. In addition, we defined the product $\mathcal{L}C_a$ as a *photosynthetic β factor* (β_p). Values of the biotic growth β factor used in a variety of global carbon cycling models are well within the predicted β_p range.

Appendix

A list of symbols and abbreviations.

Name	Description
C_a	Atmospheric CO_2 concentration
C_i	Intercellular CO_2 concentration
J	Electron transport rate
J_{max}	Maximum electron transport rate
K	Coefficient associated with enzyme kinetics
$\mathscr{L}$	Relative photosynthetic responses to 1-ppm CO_2 change
$\mathscr{L}_1$	Lower limit of $\mathscr{L}$
$\mathscr{L}_2$	Upper limit of $\mathscr{L}$
NPP	Net primary productivity
P	Photosynthetic rate
P_1	Photosynthetic rate with limitation of electron transport
P_2	Photosynthetic rate with limitation of rubisco activity
P_G	Global photosynthetic carbon influx
$P_{G,1}$	Lower limit of P_G
$P_{G,2}$	Upper limit of P_G
PCR	Photosynthetic carbon reduction
PCO	Photorespiratory carbon oxidation
rubisco	Ribulose-1,5-bisphosphate carboxylase/oxygenase
RuBP	Ribulose 1,5-bisphosphate
V_{cmax}	Maximum carboxylation rate
α	Ratio of intercellular to ambient CO_2 concentration
β	Biota growth factor
β_p	Biota photosynthetic factor
Γ	CO_2 compensation point

Acknowledgments

We thank G. D. Farquhar, C. B. Field, P. M. Vitousek, R. S. Loomis, F. S. Chapin, III, and G. W. Koch for their suggestions and comments. This work is supported by a grant from the National Science Foundation (BSR-90-20347).

References

Allen, L. H., Jr., Boote, K. L., Jones, J. W., Jones, P. H., Valle, R. R., Acock, B., Rogers, H. H., and Dalhman, R. C. (1986). Response of vegetation to rising carbon dioxide: Photosynthesis, biomass, and seed yield of soybean. *Global Biogeochem. Cycles* 1, 1–14.

Amthor, J. S., and Koch, G. W. (1996). Biota Growth Factor β. *In* "Carbon Dioxide and Terrestrial Ecosystems" (G. W. Koch and H. A. Mooney, eds.). Academic Press, San Diego.

Andrews, T. J., and Lorimer, G. H. (1987). Rubisco: structure, mechanisms, and prospects for improvement. *In* "The Biochemistry of Plants, A Comprehensive treatise" (P. K. Stumpf and E. E. Conn, eds.), pp. 131–218. Academic Press, San Diego.

Bacastow, R., and Keeling, C. D. (1973). Atmospheric carbon dioxide and radiocarbon in the natural carbon cycle: II. Changes from A.D. 1700 to 2070 as deduced from a geochemical model. *In* "Carbon and the Biosphere, Proceedings of the 24th Brookhaven Symposium in Biology" (G. M. Woodwell and E. V. Pecan, eds.), AEC Symp. Ser. Vol. 30, pp. 86–135. Atomic Energy Commission, Upton, NY.

Bazzaz, F. A., and Williams, W. E. (1991). Atmospheric CO_2 concentrations within a mixed forest: Implications for seedling growth. *Ecology* 72, 12–16.

Benner, P., Sabel, P., and Wild, A. (1988). Photosynthesis and transpiration of healthy and diseased spruce trees in the course of three vegetation periods. *Trees* 2, 223–232.

Bowes, G., (1993). Facing the inevitable: Plants and increasing atmospheric CO_2. *Annu. Rev. Plant Physiol. Plant Mol. Biol.* 44, 309–332.

Brooks, A., and Farquhar, G. D. (1985). Effect of temperature on the CO_2/O_2 specificity of ribulose-1,5-bisphosphate carboxylase/oxygenase and the rate of respiration in the light. Estimates from gas-exchange measurements on spinach. *Planta* 165, 397–406.

Collatz, G. J., Ribas-Carbo, M., and Berry, J. A. (1992). Coupled photosynthesis-stomatal conductance model for leaves of C_4 plants. *Aust. J. Plant Physiol.* 19, 519–538.

Conway, T. J., and Tans, P. P. (1989). "Atmospheric CO_2 Concentrations—The NOAA/Geophysical Monitoring for Climate Change (GMCC) Flask Sampling Network," Rep. NDP-005/R1. Carbon Dioxide Info. and Anal. Cent., Oak Ridge Nat. Lab., U.S. Dept. of Energy, Oak Ridge, TN.

Curtis, P. S., Zak, D. R., Pregitzer, K. S., Lussenhop, J., and Teeri, J. A. (1996). Linking Above-1 and Belowground Responses to Rising CO_2 in Northern Deciduous Forest Species. *In* "Carbon Dioxide and Terrestrial Ecosystems" (G. W. Koch and H. A. Mooney, eds.). Academic Press, San Diego.

Drake, B. G., and Leadley, P. W. (1991). Canopy photosynthesis of crops and native plant communities exposed to long-term elevated CO_2. *Plant, Cell Environ.* 14, 853–860.

Drake, B. G., Peresta, G., Geugeling, E., and Matamala, R. (1996). Long-Term Elevated CO_2 Exposure in a Chesapeake Bay Wetland. *In* "Carbon Dioxide and Terrestrial Ecosystems" (G. W. Koch and H. A. Mooney, eds.). Academic Press, San Diego.

Evans, J. R., and Farquhar, G. D. (1991). Modeling canopy photosynthesis from the biochemistry of the C_3 chloroplast. *In* "Modeling Crop Photosynthesis—from Biochemistry to Canopy" (K. J. Boote and R. S. Loomis, eds.), pp. 1–15. Crop Science Society of America, American Society of Agronomy, Madison, WI.

Evans, J. R., and Seemann, J. R. (1984). Differences between wheat genotypes in specific activity of ribulose-1,5-bisphosphate carboxylase/oxygenase and the relationship to photosynthesis. *Plant Physiol.* 74, 759–765.

Farquhar, G. D., von Caemmerer, S., and Berry, J. A. (1980). A biochemical model of photosynthetic CO_2 assimilation in leaves of C_3 species. *Planta* 149, 79–90.

Field, C. B. (1983). Allocating leaf nitrogen for the maximization of carbon gain: leaf age as a control on the allocation program. *Oecologia* 56, 341–347.

Field, C. B., and Mooney, H. A. (1990). Measuring photosynthesis under field conditions: Past and present approaches. *In* "Measurement Techniques in Plant Science" (Y. Hashimoto, et al., eds.), pp. 185–205. Academic Press, San Diego.

Gates, D. M. (1985). Global biospheric response to increasing atmospheric carbon dioxide concentration. *In* "Direct Effects of Increasing Carbon Dioxide on Vegetation" (B. R. Strain and J. D. Cure, eds.), DOE/ER-0238, pp. 171–184. U.S. Dept. of Energy, Washington, DC.

Gifford, R. M. (1980). Carbon storage by the biosphere. *In* "Carbon Dioxide and Climate: Australian Research" (G. I. Pearman, ed.), pp. 167–181. Australian Academy of Sciences, Canberra City.

Gifford, R. M. (1994). The global carbon cycle: A viewpoint on the missing sink. *Aust. J. Plant Physiol.* 21, 1–15.

Goudriaan, J., and Ketner, P. (1984). A simulation study for the global carbon cycle, including man's impact on the biosphere. *Clim. Change* 6, 167–192.

Grulke, N. E., Riechers, G. H., Oechel, W. C., Hjelm, U., and Jasger, C. (1990). Carbon balance in tussock tundra under ambient and elevated atmospheric CO_2. *Oecologia* 83, 485–494.

Harley, P. C., and Tenhunen, J. D. (1991). Modeling the photosynthetic response of C_3 leaves to environmental factors. *In* "Modeling Crop Photosynthesis—from Biochemistry to Canopy" (K. J. Boote and R. S. Loomis, eds.), pp. 17–39. Crop Science Society of America, American Society of Agronomy, Madison, WI.

Harley, P. C., Thomas, R. B., Reynolds, J. F., and Strain, B. R. (1992). Modelling photosynthesis of cotton grown in elevated CO_2. *Plant, Cell Environ.* 15, 271–282.

Harvey, L. D. D. (1989). Effect of model structure on the response of terrestrial biosphere models to CO_2 and temperature increases. *Global Biogeochem. Cycles* 3, 137–153.

Hudson, R. J. M., Gherini, S. A., and Goldstein, R. A. (1994). Modeling the global carbon cycle: Nitrogen fertilization of the terrestrial biosphere and the "missing" CO_2 sink. *Global Biogeochem. Cycles* 8, 307–333.

Johnson, D. W., Henderson, P. H., Ball, J. T., and Walker, R. F. (1996). Effects of CO_2 and N on Growth and N Dynamics in Ponderosa Pine. *In* "Carbon Dioxide and Terrestrial Ecosystems" (G. W. Koch and H. A. Mooney, eds.). Academic Press, San Diego.

Jordan, D. B., and Ogren, W. L. (1984). The CO_2/O_2 specificity of ribulose-1,5-bisphosphate carboxylase/oxygenase: Dependence on ribulose-bisphosphate concentration, pH and temperature. *Planta* 161, 308–313.

Jurick, T. W., Briggs, G. M., and Gates, D. M. (1985). "Carbon Dynamics of Northern Hardwood Forests: Gas Exchange Characteristics," DOE/EV 10091-1. U.S. Dept. of Energy, Washington, DC.

Lawlor, D. W. (1993). "Photosynthesis: Molecular, Physiological and Environmental Process," 2nd ed. 318 pp. Longman Scientific & Technical, Essex, UK.

Luo, Y. (1995). Interspecific variation of photosynthetic responses to elevated CO_2: Implication for ecosystem-level regulations. *Ecology* (in review).

Luo, Y., and Mooney, H. A. (1995). Long-term studies on carbon influx into global terrestrial ecosystems: Issues and approaches. *J. Biogeog.* 22, 2631–2637.

Luo, Y., Field, C. B., and Mooney, H. A. (1994). Predicting responses of photosynthesis and root fraction to elevated CO_2: Interactions among carbon, nitrogen, and growth. *Plant, Cell Environ.* 17, 1194–1205.

Melillo, J. M., McGuire, A. D., Kicklighter, D. W., Moore, B., III, Vorosmarty, C. J., and Schloss, A. L. (1993). Global climate change and terrestrial net primary production. *Nature* 363, 234–240.

Monteith, J. L., and Unsworth, M. H. (1990). "Principles of Environmental Physics," 2nd Ed., 291 pp. Edward Arnold, London.

Mooney, H. A., Vitousek, P. M., and Matson, P. A. (1987). Exchange of materials between terrestrial ecosystems and the atmosphere. *Science* 238, 926–932.

Mooney, H. A., Drake, B. G., Luxmoore, R. J., Oechel, W. C., and Pitelka, L. F. (1991). Predicting ecosystem responses to elevated CO_2 concentrations. *BioScience* 41, 96–104.

Olson, J. S., Watts, J. A., and Allison, L. J. (1983). "Carbon in Live Vegetation of Major World Ecosystems," ORNL-5862. Oak Ridge National Lab., Oak Ridge, TN.

O'Neill, E. G., and Norby, R. J. (1996). Litter Quality and Decomposition Rates of Foliar Litter Produced under CO_2 Enrichment. *In* "Carbon Dioxide and Terrestrial Ecosystems" (G. W. Koch and H. A. Mooney, eds.). Academic Press, San Diego.

Pearcy, R. W., and Ehleringer, J. R. (1984). Comparative ecophysiology of C_3 and C_4 plants. *Plant, Cell Environ.* 7, 1–13.

Potter, C. S., Randerson, J. T., Field, C. B., Matson, P. A., Vitousek, P. M., Mooney, H. A., and Klooster, S. A. (1993). Terrestrial ecosystem production: A process model based on global satellite and surface data. *Global Biogeochem. Cycles* 7, 811–841.

Prentice, K. C., and Fung, I. Y. (1990). The sensitivity of terrestrial carbon storage to climate change. *Nature* 346, 48–51.

Regehr, D. L., Bazzaz, F. A., and Boggess, W. R. (1975). Photosynthesis, transpiration and leaf conductance of *Populus deltoides* in relation to flooding and drought. *Photosynthetica* 9, 52–61.

Ryle, G. J. A., Powell, C. E., and Tewson, V. (1992). Effect of elevated CO_2 on the photosynthesis, respiration, and growth of perennial ryegrass. *J. Exp. Bot.* 43, 811–818.

Sage, R. F. (1994). Acclimation of photosynthesis to increasing atmospheric CO_2: The gas exchange perspective. *Photosynth. Res.* 39, 351–368.

Sage, R. F., Sharkey, T. D., and Seemann, J. R. (1989). Acclimation of photosynthesis to elevated CO_2 in five C_3 species. *Plant Physiol.* 89, 590–596.

Sage, R. F., Sharkey, T. D., and Seemann, J. F. (1990). Regulation of ribulose-1,5-bisphosphate carboxylase activity in response to light intensity and CO_2 in the C3 annuals *Chenopodium album* L. and *Phaseolus vulgaris* L. *Plant Physiol.*, 1735–1742.

Schindler, D. W., and Bayley, S. E. (1993). The biosphere as an increasing sink for atmospheric carbon: Estimates from increased nitrogen deposition. *Global Biogeochem. Cycles* 7, 717–733.

Schlesinger, W. H. (1991). "Biogeochemistry: An Analysis of Global Change," 442 pp. Academic Press, San Diego.

Sharkey, T. D. (1985). Photosynthesis in intact leaves of C3 plants: Physics, physiology, and rate limitations. *Bot. Rev.* 51, 53–105.

Siegenthaler, U., and Sarmiento, J. L. (1993). Atmospheric carbon dioxide and the ocean. *Nature* 365, 119–125.

Smith, T. M., Shugart, G. B., Bonan, G. B., and Smith, J. B. (1992). Modeling the potential response of vegetation to global climate change. *Adv. Ecol. Res.* 22, 93–116.

Stitt, M. (1991). Rising CO_2 levels and their potential significance for carbon flow in photosynthetic cell. *Plant, Cell Environ.* 14, 741–762.

Tans, P. P., Fung, I. Y., and Takahashi, T. (1990). Observational constraints on the global atmospheric CO_2 budget. *Science* 247, 1431–1438.

Taylor, S. E., and Terry, N. (1984). Limiting factors in photosynthesis. V. Photochemical energy supply colimits photosynthesis at low values of intercellular CO_2 concentration. *Plant Physiol.* 75, 82–86.

Thorning, K. W., Tans, P. P., and Komhyr, W. D. (1989). Atmospheric carbon dioxide at Mauna Loa Observatory: 2. Analysis of the NOAA/GMCC data 1974–1985. *J. Geophys. Res.* 94, 8549–8565.

Tissue, D. T., and Oechel, W. C. (1987). Response of *Eriophorum vaginatum* to elevated CO_2 and temperature in the Alaskan tussock tundra. *Ecology* 68, 401–410.

von Caemmerer, S., and Edmonson, D. L. (1986). The relationship between steady-state-gas exchange, in vivo RuP2 carboxylase activity and some carbon reduction cycle intermediates in *Raphanus sativus*. *Aust. J. Plant Physiol.* 13, 669–688.

Wigley, T. M. L., and Raper, S. C. B. (1992). Implications for climate and sea level of revised IPCC emissions scenarios. *Nature* 357, 293–300.

Wong, S. C. (1990). Elevated atmospheric partial pressure of CO_2 and plant growth. II. Nonstructural carbohydrate content in cotton plants and its effect on growth parameters. *Photosynth. Res.* 23, 171–180.

Wullschleger, S. D. (1993). Biochemical limitations to carbon assimilation in C_3 plants—A retrospective analysis of the A/Ci curves from 109 species. *J. Exp. Bot.* 44, 907–920.

21

Biota Growth Factor β: Stimulation of Terrestrial Ecosystem Net Primary Production by Elevated Atmospheric CO_2

Jeffrey S. Amthor and George W. Koch

I. Introduction

Assimilation of atmospheric CO_2 during photosynthesis, less photorespiration, by all the plants in an ecosystem traditionally is called gross primary production (GPP; e.g., mol CO_2 m^{-2} [ground] year^{-1}). The balance of GPP and plant (often called autotrophic) respiration is net primary production (NPP; e.g., mol CO_2 m^{-2} year^{-1}). Net primary production is a measure of plant growth. The difference between NPP and the release of CO_2 by herbivores and decomposers, i.e., heterotrophic respiration, is net ecosystem production (NEP; e.g., mol CO_2 m^{-2} year^{-1}). When considered at the global scale, terrestrial ecosystem NEP represents the net impact of terrestrial ecosystems on the carbon cycle.

Many factors affect ecosystem GPP, NPP, and NEP, including vegetation type and age, temperature, soil texture and nutrient levels, precipitation, solar irradiance, air and soil pollution, and atmospheric CO_2 partial pressure. For example, increasing atmospheric CO_2 can significantly enhance GPP in C_3-dominated ecosystems because (1) present atmospheric CO_2 partial pressure is less than the Michaelis–Menten constant of the primary carboxylation reaction with respect to CO_2 and (2) CO_2 inhibits photorespiration. Also, NPP of C_4-dominated ecosystems might be stimulated by elevated CO_2 because of improved plant water relations (Rogers *et al.*, 1983).

To the extent that plant growth and respiration are related to photosynthesis, NPP also will be sensitive to atmospheric CO_2 partial pressure. For example, if a given increase in atmospheric CO_2 increases GPP by $x\%$ and the ratio of NPP to GPP is unaffected, NPP will also be increased by $x\%$. In controlled-environment experiments with individual plants and small communities, the ratio of NPP to GPP may be unaffected, or even increased, by elevated CO_2 (Amthor, 1996). Few ecosystems have been studied, however, and general effects of elevated CO_2 on the ratio of ecosystem NPP to GPP are unknown. This begs the questions, will source–sink interactions at the level of whole plants and ecosystems facilitate a parallel increase in GPP and NPP in elevated CO_2? When elevated CO_2 stimulates NPP, are "extra" carbon sinks formed or is individual sink activity increased?

With respect to the global carbon cycle, the relationship between NEP and atmospheric CO_2 is more important than the effects of increasing CO_2 on GPP and NPP (Schlesinger, 1991a). Ecosystem NEP in elevated CO_2 has rarely been studied. In the absence of altered temperature or moisture, elevated CO_2 could slow decomposition if it increases litter C/N ratio [Field et al., 1992; O'Neill and Norby, 1996 (this volume)] and might increase herbivory if live tissue quality declines [Lindroth, 1996 (this volume)]. In any case, the magnitudes of the effects of increasing atmospheric CO_2 partial pressure on terrestrial ecosystem GPP, NPP, and NEP are among the most important unanswered questions concerning increasing atmospheric CO_2, the global carbon cycle, and function of the biosphere.

We note that increasing atmospheric CO_2 is not the only environmental change taking place that is likely to affect global terrestrial ecosystem productivity. For example, nitrogen deposition to many terrestrial ecosystems is increasing due to air pollution (Schlesinger, 1991b), and tropospheric levels of the phytotoxic air pollutant ozone are increasing (Chameides et al., 1994). Also, it is possible that Earth's surface temperature will increase by several degrees Celsius during the next decades to centuries because of an enhanced greenhouse effect (Manabe, 1983), and precipitation may also be altered by elevated atmospheric CO_2 (Manabe and Stouffer, 1994). Moreover, human land-use activities such as forest clearing significantly affect local and regional NEP (Houghton et al., 1983). Nonetheless, this chapter will focus on the effects of elevated CO_2 partial pressure on primary production processes, largely in isolation from other environmental changes.

II. Modeling and Measuring Plant and Ecosystem Responses to Elevated CO_2

The primary plant responses to elevated CO_2 are increased photosynthesis and reduced stomatal conductance, although many factors may modulate these responses. Mechanistic models of leaf-level photosynthetic response to CO_2 have been developed (e.g., Farquhar et al., 1980) and extended to the level of the canopy (e.g., Sellers et al., 1992; Baldocchi, 1993; Amthor et al., 1994). Empirical models of stomatal response to CO_2 have also been applied to canopies. Ecosystem-level models (e.g., Running and Hunt, 1993) usually contain less physiological detail than leaf photosynthesis models, but nonetheless may be useful in making predictions of ecosystem responses to elevated CO_2.

Simplicity is important in models of regional and global carbon cycles because the systems are large. Thus, CO_2 interactions with rubisco, stomates, and the growth of individual plants or communities are often reduced to one or a few parameters that represent the net effect of CO_2 on ecosystem productivity in global-scale models. This is a limitation of most global carbon cycle models with respect to predicting global terrestrial ecosystem response to increasing atmospheric CO_2.

One simple approach to including the effects of CO_2 on primary production in regional and global models is the empirical coefficient β, used as follows (e.g., Gifford, 1980; Goudriaan, 1993):

$$\text{NPP} = \text{NPP}_0[1 + \beta \ln(C/C_0)] \quad \text{NPP}_0 \geq 0 \qquad (1)$$

where NPP is net primary production at the atmospheric CO_2 partial pressure C and NPP_0 is net primary production at a reference or baseline atmospheric CO_2 partial pressure C_0. Note that β applies to the primary effects of a change in atmospheric CO_2 on plant metabolism and growth, but not to any secondary effects mediated through climatic change. This β, and similar parameters, has been called the "biota growth factor" and the "biotic growth factor" (e.g., Bacastow and Keeling, 1973; Gifford, 1980). The logarithmic response of NPP to CO_2 used in Eq. (1) was questioned by Gates (1985), who suggested that β might be better calculated with a Michaelis–Menten expression, but Gifford (1980) and Goudriaan et al. (1985) adequately defended the use of Eq. (1) in large-scale models, at least for moderate (tens of pascals) changes in atmospheric CO_2 partial pressure.

A β sensu Eq. (1) of unity represents a 69% increase in NPP with a doubling of atmospheric CO_2, whereas β is about 0.72 if NPP is stimulated 50% by a doubling of CO_2 (see also Fig. 1). If atmospheric CO_2 partial pressure has no effect on NPP, β is zero, and if elevated CO_2 were to reduce NPP (inhibit photosynthesis or growth), β would be less than zero. Because the effects of elevated atmospheric CO_2 partial pressure on all the biology and ecology of the terrestrial biosphere may be reduced to a single parameter such as β in a global carbon cycle model (e.g., Taylor and Lloyd, 1992; Goudriaan, 1993; Rotmans and den Elzen, 1993), it is especially important to understand the basis for chosen values of that parameter. As might be expected, a range of values of β has been used in models (e.g., reviewed

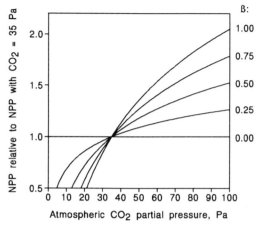

Figure 1 Net primary production (NPP) over a range of atmospheric CO_2 partial pressures relative to NPP at 35 Pa CO_2 for five values of the biota growth factor β.

in Gifford, 1980; Gates, 1985). Goudriaan (1993) surmised that β may be about 0.7 under conditions favorable for plant growth, but that it declines with nutrient shortage; thus, values less than 0.7 are typically used in global carbon cycle models. For example, Rotmans and den Elzen (1993) were able to "balance" the global carbon budget with $\beta = 0.4$ when the effects of temperature on NPP and heterotrophic respiration were also considered. We will return to the use of β as a carbon cycle balancing tool or "knob."

The results of CO_2-enrichment experiments can be used to calculate β as follows:

$$\beta = (NPP_E/NPP_A - 1)/\ln(C_E/C_A) \quad NPP_A > 0 \qquad (2)$$

where NPP_E is measured plant growth in elevated CO_2 partial pressure, NPP_A is measured plant growth in ambient (or a reference) CO_2 partial pressure, C_E is the CO_2 partial pressure in the elevated CO_2 treatment, and C_A is the ambient (or reference) CO_2 partial pressure.

The effect of CO_2 on NEP (β_{NEP}; we relate β without a subscript to NPP) can be summarized analogously to the effect of CO_2 on NPP as follows:

$$\beta_{NEP} = (NEP_E/NEP_A - 1)/\ln(C_E/C_A) \quad NEP_A > 0 \qquad (3)$$

where NEP_E is NEP in elevated CO_2 and NEP_A is NEP in ambient (or reference) CO_2. In cases for which $NEP_A \leq 0$, calculations of β_{NEP} have no meaning. Herein, we use mean experimental values of NPP_A, NPP_E, NEP_A, and NEP_E to calculate β and β_{NEP}, irrespective of statistical differences between mean values in individual studies. We do this in part because most CO_2-enrichment experiments are poorly replicated (two or three replicates is common) and therefore lack power (*sensu* Sokal and Rohlf, 1981) in determining responses to CO_2.

β_{NEP} is a measure of the combined effects of CO_2 on photosynthesis, plant respiration, and heterotrophic respiration, which can all be affected by a number of primary and secondary plant responses to elevated CO_2. For the purposes of large-scale models of carbon cycling, we suggest that NPP and heterotrophic respiration should be separated because they involve fundamentally different processes, i.e., plant photosynthesis and growth as opposed to soil microbial metabolism. Thus, NPP and heterotrophic respiration can be uncoupled in many ecosystems, at least in the short term (1 to a few years), and therefore can be affected differentially by elevated atmospheric CO_2 partial pressure. Likewise, GPP and plant respiration may be affected to different degrees by elevated CO_2, and it may be important to treat them as separate processes in carbon cycle models as well, although photosynthesis and plant respiration are often tightly linked over the course of a year. Thus, our emphasis is on β rather than β_{NEP}, although estimates of β_{NEP} are available from some ecosystem experiments when estimates of β are not.

At the outset we recognize that the experimental database is small. Only a few ecosystems have been exposed to elevated CO_2 in controlled experiments, and NPP has been estimated in only a subset of those few experiments. There are many reasons for this. Measurements of NPP are difficult to make for natural ecosystems; indeed, relatively few true measures if NPP have been made in any CO_2 environment. It is difficult to estimate root growth in the field because of restricted access and because root turnover is not easily measured. This may be a particular limitation to assessing the effects of elevated CO_2 on NPP because an increase in root/shoot ratio may occur with CO_2 enrichment, although this issue is unresolved [Rogers *et al.*, 1994; Curtis *et al.*, 1996 (this volume); Norby *et al.*, 1996 (this volume)]. Often, therefore, only aboveground biomass is measured, and even total aboveground living biomass underestimates aboveground NPP because litter production must be included in NPP. In spite of the limited data, however, we proceed in order to outline the present state of knowledge.

A. Individual-Plant Response to Elevated CO_2

Most elevated-CO_2 experiments have been conducted with individual plants, often grown in pots, rather than plant communities or ecosystems in the field. In hundreds of laboratory and glasshouse experiments, C_3 crop plant yield has been enhanced by 33% on average with a doubling of CO_2 (Kimball, 1983; Rogers *et al.*, 1994), and total biomass accumulation by C_3 crop plants has been enhanced, on average, by 30–50% ($\beta = 0.43$–0.72), although results have been variable (Cure, 1985). Many noncrop herbaceous plants also responded positively to elevated CO_2. For example, from the analysis of Hunt *et al.* (1993), we derive a β value of about 0.62 for herbaceous plants with a competitive strategy (*sensu* Grime *et al.*, 1988), but a value of only 0.14 for herbs with either ruderal or stress-tolerant strategies (*sensu* Grime *et al.*, 1988). In sunlit chambers, estimated aboveground β depended on CO_2 partial pressure (Table I). In such a case, β cannot be considered a constant, which has important implications for the use of β in models over a range of CO_2 partial pressures.

The geometric mean (less than the arithmetic mean) value of β from 398 published observations of growth in elevated-CO_2 experiments with 73 forest tree (all C_3) species in growth chambers, glasshouses, and open-top chambers was ca. 0.43, with a range of ca. -0.6 to 5.8 (Fig. 2; S. D. Wullschleger, personal communication). Needless to say, those experiments involved seedlings and saplings exposed to elevated CO_2 for no more than a few years. The value of β of 0.43 indicates a 30% increase in young tree growth with a doubling of atmospheric CO_2 partial pressure (see also Ceulemans and Mousseau, 1994), which is similar to the 33% increase in crop-plant productivity summarized by Kimball (1983). It is uncertain how well this value of young forest-tree β, i.e., 0.43, applies to whole-forest NPP. Nonetheless, it encapsulates a large amount of available data.

Crassulacean acid metabolism (CAM) plants represent only a few percent of plant species, but are key components of many arid and semiarid ecosystems. Their growth response to elevated CO_2 has been studied infrequently compared with C_3 species. During 1 year of growth at 35 and 65 Pa CO_2 in glasshouses, *Agave deserti* accumulated 3.63 and 4.60 g dry shoot plus root per plant, respectively (Nobel and Hartsock, 1986), which gives 0.43 for β. In the same experiments, *Ferocactus acanthodes* β was about 0.47 for 35 and 65 Pa CO_2 grown plants. When calculated from growth data at 35 and 50 Pa CO_2, however, β was lower. Similarly, for *Opuntia ficus-indica*, β was 0.62 for a CO_2 pressure of 52 Pa compared with 37 Pa, and β was 0.83 for 72 Pa CO_2 compared with 37 Pa (Cui *et al.*, 1993), which is opposite

Table I Rice Aboveground β Estimated from Maximum Biomass (i.e., Excluding Litter Production, Which Was Not Reported) as a Function of CO_2 Partial Pressure Treatments (Based on Baker *et al.*, 1992)

CO_2 partial pressure (Pa)	β
16 vs 33	1.21
25 vs 33	1.82
50 vs 33	0.63
66 vs 33	0.26
90 vs 33	0.25

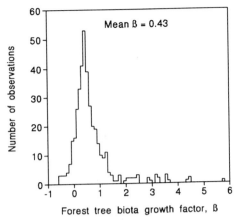

Figure 2 Frequency of observations of β (in intervals of 0.1) from published experiments with 73 forest tree species seedlings and saplings, often grown in pots, involving 398 total observations. Because the distribution is not normally distributed, the geometric mean (ca. 0.43) was calculated as $\log^{-1}(\bar{x}) - 1$, where $\bar{x}$ is the mean of $\log(1 + \beta)$ of the 398 observations (S. D. Wullschleger, personal communication).

to the response of β to CO_2 partial pressure shown in Table I for rice. For two tropical CAM species, elevated CO_2 had insignificant effects on growth, except for enhanced root dry mass accumulation in one species (Ziska et al., 1991).

B. Ecosystem Response to Elevated CO_2

1. An Artificial Ecosystem "Artificial tropical ecosystems" were treated with daytime CO_2 pressures of about 34 and 61 Pa for 3 months in 2.5-m-tall chambers with 0.19 m of soil (replicated once) in Switzerland (Körner and Arnone, 1992). Whole-plant production was ca. 710 g m^{-2} in 34 Pa CO_2 and ca. 860 g m^{-2} in 61 Pa CO_2, assuming 700 g m^{-2} at the start of the experiment (Körner and Arnone, 1992). Thus, β was ca. 0.36, which is on the same order as mean values for crop plants and forest tree seedlings summarized earlier.

2. Crop Ecosystems The growth and productivity of C_3 crops growing in the field in chambers are usually enhanced by elevated CO_2, although there has been considerable variability in responses. In about 15 studies with C_3 crops, β ranged from ca. 0.29 to ca. 1.18 [derived from Lawlor and Mitchell (1991)]. Growth and productivity of C_4 crops can also be stimulated by elevated CO_2; this may be largely related to increased water-use efficiency rather than stimulated photosynthesis per unit leaf area (Rogers et al.,

1983), because C_4 photosynthesis is nearly CO_2-saturated at present atmospheric CO_2 levels.

Although they are useful research tools, open-top chambers alter the microclimate and can themselves affect growth and productivity. The most realistic method of treating crops with elevated CO_2, i.e., free from "chamber effects," is the free-air CO_2 enrichment (FACE) system (Allen *et al.*, 1992). Unfortunately, only a few FACE experiments have been conducted because they are expensive; to date, FACE crop experiments (replicated in space three times) have been conducted in Arizona with cotton in 1989, 1990, and 1991 and with wheat in 1992–1993 and 1993–1994 [Pinter *et al.*, 1996 (this volume)], but only pseudo-control FACE apparatuses have been used for the ambient CO_2 treatments. In those FACE experiments, CO_2 enrichment was at the level of ca. 55 Pa CO_2.

For cotton grown in 1989 at ambient CO_2 (nominally 35 Pa), mean end-of-season total biomass was ca. 1393 g m^{-2}, and end-of-season mean total biomass in the elevated CO_2, i.e., FACE, plots (55 Pa) was *ca.* 1708 g m^{-2} (Kimball *et al.*, 1992), giving a value of *ca.* 0.50 for β. Those biomass values do not include leaf and fruit litter or roots that died before the end of the season for either ambient or elevated CO_2 treatments, so that they underestimate NPP by up to a few hundred grams per square meter. In the 1990 cotton FACE experiments, two irrigation treatments were used: "wet" and "dry" [irrigation amounts are given by Pinter *et al.* (1996, this volume)]. End-of-season total biomass estimates give $\beta = 0.45$ for the dry treatment and $\beta = 0.86$ for the wet treatment (derived from Kimball *et al.*, in preparation). In similar experiments in 1991, cotton β was *ca.* 0.87 in the dry treatment and *ca.* 0.93 in the wet treatment (derived from Kimball *et al.*, in preparation). We note that cotton, a broad-leaved C_3 woody perennial generally grown for less than 1 year, is the most CO_2-responsive field crop studied to date; the mean value of β with respect to cotton biomass accumulation was about 1.21 in glasshouse and chamber experiments [Cure, 1985; see also Radin *et al.*, (1987) and Kimball and Mauney (1993)].

In the 1992–1993 and 1993–1994 wheat FACE experiments, two CO_2 levels (37 and 55 Pa) and two irrigation treatments ("full" and "half") were used [Pinter *et al.*, 1996 (this volume)]. The elevated-CO_2 treatment increased total aboveground biomass production, i.e., aboveground NPP, in both irrigation treatments (Table II). In several chamber studies, wheat β with respect to biomass accumulation was about 0.45 (Cure, 1985) and with respect to grain production was 0.50–0.55 (Kimball, 1983; Cure, 1985).

Crop responses to elevated CO_2 have been positive, but also variable, and generalizations concerning β are difficult to make. Much remains to be learned concerning the CO_2 response of crops when other environmental factors vary. This may be most important with respect to temperature (i.e., global warming), because grain production of both wheat and rice, the

Table II Estimated Wheat Aboveground Biomass Production in the 1992–1993 and 1993–1994 Free-Air CO_2 Enrichment (FACE) Experiments in Arizona

Year	Irrigation treatment	Maximum aboveground living plus dead biomass (g m^{-2})[a]		β
		37 Pa CO_2 ambient plots	55 Pa CO_2 FACE apparatus	
1992–1993	half	1491	1552	0.10
	full	1828	1983	0.21
1993–1994	half	1371	1611	0.44
	full	1874	2053	0.24

[a] Values for biomass are means of four replicates [Pinter et al., 1996 (this volume), and personal communication].

two most important crops on the global scale, can be limited by high temperatures [e.g., Evans, 1993; Baker et al., 1996 (this volume)].

3. Natural Ecosystems A few elevated-CO_2 experiments have been conducted in natural, or near-natural, ecosystems using open-top or closed-top chambers. Unfortunately, estimates of NPP generally are not available from those studies. Moreover, no natural forest or woodland ecosystems have been exposed to elevated CO_2 for periods of months to years, yet forests and woodlands are thought to account for as much as 53–64% of global terrestrial NPP (e.g., Ajtay et al., 1979; Taylor and Lloyd, 1992; Rotmans and den Elzen, 1993).

In a 3-year field experiment using closed-top chambers, twice-ambient compared with ambient CO_2 partial pressure increased tussock tundra NEP during the first summer (NPP was not estimated). Elevated CO_2 also increased NEP during the second summer, but not by as much as during the first year. There was little effect of CO_2 partial pressure on NEP during the third year, presumably due to acclimation processes, although twice-ambient CO_2 in combination with a temperature increase of 4°C resulted in enhanced NEP compared to ambient CO_2 and temperature (Oechel et al., 1994). We cannot, however, calculate β_{NEP} for this ecosystem because NEP_A was negative all 3 years of the experiment (Oechel et al., 1993). In later experiments at a nearby tundra site, similar results were obtained during a single year of CO_2 enrichment: seasonal (summer) NEP was negative at 34 Pa CO_2 but positive at 68 Pa CO_2 (Grulke et al., 1990). Again, because NEP_A was negative, β_{NEP} cannot be given a meaningful value. Nonetheless, a doubling of CO_2 transformed an ecosystem that was a CO_2 source into an ecosystem that was a CO_2 sink during the summer.

Three temperate wetland communities have been exposed to ambient and twice-ambient CO_2 in open-top chambers since 1987: *Scirpus olneyi* (C_3),

Spartina patens (C_4), and *Scirpus olneyi–Spartina patens* mixtures [Drake and Leadley, 1991; Drake, 1992; Drake *et al.*, 1996 (this volume)]. On the basis of maximum aboveground living biomass, we estimate that aboveground β was 0.20–0.43 for the C_3 community during the first 7 years of the experiment, but that aboveground β was negative 4 out of 7 years for the C_4 community (Table III). Belowground and whole-plant β values are unknown. β_{NEP} was greater than zero in all three communities during 1990 (Table III).

Twice-ambient CO_2 significantly increased aboveground biomass production by native tallgrass prairie (C_3–C_4 mixture) in open-top chambers during a dry year, i.e., 1989, but not during a year with closer to average precipitation, i.e., 1990 [Owensby *et al.*, 1993, 1996 (this volume)]. Growth of C_4 but not C_3 species was stimulated by elevated CO_2 in 1989. Plant growth in 1990 was about twice the growth in 1989. Our conservative estimates of prairie aboveground NPP [from Fig. 1 in Owensby *et al.* (1993)] yield an aboveground β of ca. 0.16 for the dry year, i.e., 1989. Root growth was estimated during 1990 with buried ingrowth bags. Elevated CO_2 more than doubled root growth compared with ambient CO_2. Our conservative estimates of prairie whole-plant NPP yield a β of ca. 0.22 for 1990 [from Figs. 3 and 4 in Owensby *et al.* (1993)].

Aboveground growth of California grassland communities enclosed in open-top chambers was sensitive to CO_2 (ca. 38 and 73 Pa were used), but responses depended on grassland type and interannual variations in weather [Field *et al.*, 1996 (this volume)]. A low productivity serpentine-soil grassland (maximum aboveground NPP $\approx$ 160 g dry mass m^{-2} in ambient CO_2 during the 3-year experiment) had β values ranging from 0.55 to 0.80, while a more productive sandstone-soil grassland (maximum aboveground NPP $\approx$ 280 g m^{-2}) had β values ranging from -0.14 to 0.61, with greater NPP in elevated than in ambient CO_2 in only the third of the

Table III Response of Aboveground NPP (β) and NEP (β_{NEP}) in Two Natural Mesohaline Salt Marsh Communities to Twice-Ambient CO_2 Partial Pressure in Open-Top Chambers (Five Replicates of Each Community)

Community	β^a							$\beta_{NEP}{}^b$	
	1987	1988	1989	1990	1991	1992	1993	1988	1990
Scirpus olneyi (C_3)	0.32	0.21	0.20	0.23	0.26	0.43	0.29	1.27	0.79
Spartina patens (C_4)	0.097	-0.039	0.12	-0.23	-0.64	-0.21	0.31	0.58	0.42

[a] Based on aboveground maximum standing biomass during the summer; thus, values do not account for any litter production during the year [from Drake *et al.*, 1996 (this volume)].

[b] From data in Drake and Leadley (1991) and Drake (1992). β_{NEP} applies only to May–November; shoots are usually dormant December–April and CO_2 flux measurements were not made then.

3 years of treatment. Much of the increase in aboveground ecosystem production in elevated CO_2 on serpentine soil was due to increased production by late-season annuals, perhaps because of the extended duration of soil water availability resulting from reduced early-season transpiration in elevated CO_2.

III. Information Needs

Clearly, more measurements of NPP (and NEP) in natural ecosystems in elevated and ambient CO_2 partial pressures are needed. Research with forests has been particularly limited for, in spite of the large carbon stocks and annual carbon fluxes of boreal, temperate, and tropical forests and woodlands, very little is known concerning the responses of large trees and intact forests to elevated CO_2. Establishment of a β for forests, and indeed the global biota, is therefore problematic. All that can be done at present is to extrapolate from the data that do exist, which apply mostly to crops, individual herbaceous plants, short ecosystems, and small—usually isolated—trees.

Better accounting for the carbon that is assimilated within an ecosystem is also needed. How much assimilated carbon is respired by plants and how much is added to new plant structural matter? Of the additional carbon, if any, added to plant structure as a result of elevated CO_2, what fraction is added to perennial tissue such as wood, and what fraction is added to seasonal tissue such as leaves that abscise within 1 or a few years?

Questions of long-term responses to elevated CO_2 are also important. Is the final size or age of perennial plants greater in elevated CO_2? How much of the carbon in litter and exudates is decomposed each year, and how much enters long-term soil carbon pools? Before the role of terrestrial ecosystems in the global carbon cycle can be assessed—in present or elevated CO_2—the carbon cycles of individual terrestrial ecosystems must be better understood. In sum, there are many outstanding questions, but few data.

IV. Where Do We Stand with Respect to β?

Many models of ecosystem, regional, and global carbon cycling include β or a similar parameter (e.g., Taylor and Lloyd, 1992; Goudriaan, 1993; Rotmans and den Elzen, 1993). Model predictions of the effects of elevated CO_2 on ecosystem productivity are important because CO_2 enrichment experiments are expensive and difficult (or impossible) to conduct at the ecosystem level. Moreover, regional and global experiments are impossible;

the present increase in atmospheric CO_2 is not an (good) experiment because CO_2 partial pressure is not being controlled, there is no second CO_2 treatment that allows an assessment of the effects of CO_2 rather than other environmental changes, and there is no replication of the CO_2 treatment. All of this means that caution must be used with β in models. We can say that β is probably greater than zero in many ecosystems, but a different β may be needed for each ecosystem type included in a model, and perhaps even several β's may be needed within each ecosystem type in a global model. This only complicates the issue because so few experimental data concerning β are available, and there are none for many important ecosystems such as boreal, temperate, and tropical forests.

With respect to models, we advocate *a priori* estimates of β, as weak as these may be. On the contrary, values of β often were calculated in order to "balance" carbon budgets. For example, it is not uncommon to calculate CO_2 release by fossil fuel combustion and land-use changes, measure the accumulation of CO_2 in the atmosphere, predict (model) the uptake of CO_2 by oceans, and then assume that the remainder (20–50% of the CO_2 released in fossil fuel combustion in recent years) is taken up by terrestrial ecosystems. The value of β, therefore, is set so that this remainder is equal to predicted NEP under the assumption that preindustrial global NEP was zero. In such a case, β is, in our opinion, actually an error term ε (the error deriving from a combination of measurements and model predictions). This approach may indicate actual values of β at the global scale— that is certainly one hope—although it may also represent no more than cumulative error in the derivation of other terms in the global carbon cycle. Admittedly, the establishment of an *a priori* value of β seems impossible at present because experimental data are so limited, but there is knowledge of CO_2–plant interactions, and this knowledge should be considered in carbon cycle models. In some global models, β is the only parameter that includes CO_2–plant interactions, and its value should not be taken lightly, nor should it be equated with ε.

Irrespective of the structure of carbon cycle models, there is a clear need for many new elevated-CO_2 experiments at the ecosystem level. Forests must be studied, but the potential significance of tropical grasslands should not be overlooked (Gifford, 1980; Fisher *et al.*, 1994). Moreover, stimulation of NPP by elevated atmospheric CO_2 may be time-dependent, so that long-term studies are needed. Natural CO_2 springs provide some opportunity to study long-term responses to elevated CO_2 (e.g., Koch, 1993; Miglietta *et al.*, 1993), but the springs are limited to specific locations, detailed knowledge of site history may be lacking, and there may be large temporal variations in CO_2 partial pressure. Increased NPP due to elevated CO_2 may lead to larger ecosystems, i.e., ecosystems containing more carbon, but the increase in ecosystem size cannot continue indefinitely. Sooner or later, some new limit on ecosystem size will arise and elevated CO_2 will not increase ecosystem size above that new limit. On the other hand, elevated CO_2 may change the time required for an ecosystem to reach its maximum size. Also, carbon cycling within ecosystems may be accelerated in elevated CO_2 (Gifford, 1980). Because of the many unresolved issues, quantitative predictions of the effects of elevated CO_2 partial pressure on global β are problematic, and in view of the likelihood that few ecosystem experiments will be conducted in the near future, reliance on isolated plant and artificial community data to estimate β will continue. This is unfortunate, but preferable to renaming ε as β in carbon cycle models.

References

Ajtay, G. L., Ketner, P., and Duvigneaud, P. (1979). Terrestrial primary production and phytomass. In "The Global Carbon Cycle" (B. Bolin, E. T. Degens, S. Kempe, and P. Ketner, eds.), pp. 129–181. John Wiley & Sons, Chichester, UK.

Allen, L. H., Jr., Drake, B. G., Rogers, H. H., and Shinn, J. H. (1992). Field techniques for exposure of plants and ecosystems to elevated CO_2 and other trace gases. Crit. Rev. Plant Sci. 11, 85–119.

Amthor, J. S. (1996). Plant respiratory responses to elevated CO_2 partial pressure. In "Advances in Carbon Dioxide Effects Research" (L. H. Allen, M. B. Kirkham, D. M. Olszyk, and C. Whitman, eds.). American Society of Agronomy, Madison, WI (in press).

Amthor, J. S., Goulden, M. L., Munger, J. W., and Wofsy, S. C. (1994). Testing a mechanistic model of forest-canopy mass and energy exchange using eddy correlation: carbon dioxide and ozone uptake by a mixed oak–maple stand. Aust. J. Plant Physiol. 21, 623–651.

Bacastow, R., and Keeling, C. D. (1973). Atmospheric carbon dioxide and radiocarbon in the natural carbon cycle: II. Changes from A. D. 1700 to 2070 as deduced from a geochemical model. In "Carbon and the Biosphere" (G. M. Woodwell and E. V. Pecan, eds.), pp. 86–135. U.S. Atomic Energy Commission, National Technical Information Service, U.S. Department of Commerce, Springfield, VA.

Baker, J. T., Laugel, F., Boote, K. J., and Allen, L. H., Jr. (1992). Effects of daytime carbon dioxide concentration on dark respiration in rice. Plant Cell Environ. 15, 231–239.

Baker, J. T., Allen, L. H., Jr., Boote, K. J., and Pickering, N. B. (1996). Assessment of rice responses to global climate change: CO_2 and temperature. In "Carbon Dioxide and Terrestrial Ecosystems" (G. W. Koch and H. A. Mooney, eds.). Academic Press, San Diego.

Baldocchi, D. D. (1993). Scaling water vapor and carbon dioxide exchange from leaves to a canopy: rules and tools. In "Scaling Physiological Processes: Leaf to Globe" (J. R. Ehleringer and C. B. Field, eds.), pp. 77–114. Academic Press, San Diego.

Ceulemans, R., and Mousseau, M. (1994). Effects of elevated atmospheric CO_2 on woody plants. New Phytol. 127, 425–446.

Chameides, W. L., Kasibhatla, P. S., Yienger, J., and Levy, H., II (1994). Growth of continental-scale metro-agro-plexes, regional ozone pollution, and world food production. Science 264, 74–77.

Cui, M., Miller, P. M., and Nobel, P. S. (1993). CO_2 exchange and growth of the crassulacean acid metabolism plant Opuntia ficus-indica under elevated CO_2 in open-top chambers. Plant Physiol. 103, 519–524.

Cure, J. D. (1985). Carbon dioxide doubling responses: a crop survey. In "Direct Effects of Increasing Carbon Dioxide on Vegetation" (B. R. Strain and J. D. Cure, eds.), pp. 99–116. U.S. Department of Energy, National Technical Information Service, U.S. Department of Commerce, Springfield, VA.

Curtis, P. S., Zak, D. R., Pregitzer, K. S., Lussenhop, J., and Teeri, J. A. (1996). Linking above- and belowground responses to rising CO_2 in northern deciduous forest species. In "Carbon Dioxide and Terrestrial Ecosystems" (G. W. Koch and H. A. Mooney, eds.). Academic Press, San Diego.

Drake, B. G. (1992). A field study of the effects of elevated CO_2 on ecosystem processes in a Chesapeake Bay wetland. *Aust. J. Bot.* 40, 579–595.

Drake, B. G., and Leadley, P. W. (1991). Canopy photosynthesis of crops and native plant communities exposed to long-term elevated CO_2. *Plant Cell Environ.* 14, 853–860.

Drake, B. G., Peresta, G., Beugeling, E., and Matamala, R. (1996). Long-term elevated CO_2 exposure in a Chesapeake Bay wetland. In "Carbon Dioxide and Terrestrial Ecosystems" (G. W. Koch and H. A. Mooney, eds.). Academic Press, San Diego.

Evans, L. T. (1993). "Crop Evolution, Adaptation and Yield." Cambridge Univ. Press, Cambridge, UK.

Farquhar, G. D., von Caemmerer, S., and Berry, J. A. (1980). A biochemical model of photosynthetic CO_2 assimilation in leaves of C_3 species. *Planta* 149, 78–90.

Field, C. B., Chapin, F. S., III, Matson, P. A., and Mooney, H. A. (1992). Responses of terrestrial ecosystems to the changing atmosphere: a resource-based approach. *Annu. Rev. Ecol. Syst.* 23, 201–235.

Field, C. B., Chapin, S. F., III, Chiariello, N. R., Holland, E. A., Mooney, H. A. (1996). The Jasper Ridge CO_2 Experiment. In "Carbon Dioxide and Terrestrial Ecosystems" (G. W. Koch and H. A. Mooney, eds.). Academic Press, San Diego.

Fisher, M. J., Rao, I. M., Ayarza, M. A., Lascano, C. E., Sanz, J. I., Thomas, R. J., and Vera, R. R. (1994). Carbon storage by introduced deep-rooted grasses in the South American savannas. *Nature* 371, 236–238.

Gates, D. M. (1985). Global biospheric response to increasing atmospheric carbon dioxide concentration. In "Direct Effects of Increasing Carbon Dioxide on Vegetation" (B. R. Strain and J. D. Cure, eds.), pp. 171–184. U.S. Department of Energy, National Technical Information Service, U.S. Department of Commerce, Springfield, VA.

Gifford, R. M. (1980). Carbon storage by the biosphere. In "Carbon Dioxide and Climate: Australian Research" (G. I. Pearman, ed.), pp. 167–181. Australian Academy of Science, Canberra.

Goudriaan, J. (1993). Interaction of ocean and biosphere in their transient responses to increasing atmospheric CO_2. *Vegetatio* 104/105, 329–337.

Goudriaan, J., van Laar, H. H., van Keulen, H., and Louwerse, W. (1985). Photosynthesis, CO_2 and plant production. In "Wheat Growth and Modeling" (W. Day and R. K. Atkin, eds.), pp. 107–122. Plenum Press, New York.

Grime, J. P., Hodgson, J. G., and Hunt, R. (1988). "Comparative Plant Ecology: A Functional Approach to Common British Species." Unwin Hyman, London.

Grulke, N. E., Riechers, G. H., Oechel, W. C., Hjelm, U., and Jaeger, C. (1990). Carbon balance in tussock tundra under ambient and elevated atmospheric CO_2. *Oecologia* 83, 485–494.

Houghton, R. A., Hobbie, J. E., Melillo, J. M., Moore, B., Peterson, B. J., Shaver, G. R., and Woodwell, G. M. (1983). Changes in the carbon content of terrestrial biota and soils between 1860 and 1980: a net release of CO_2 to the atmosphere. *Ecol. Monogr.* 53, 235–262.

Hunt, R., Hand, D. W., Hannah, M. A., and Neal, A. M. (1993). Further responses to CO_2 enrichment in British herbaceous species. *Funct. Ecol.* 7, 661–668.

Kimball, B. A. (1983). Carbon dioxide and agricultural yield: an assemblage and analysis of 430 prior observations. *Agron. J.* 75, 779–788.

Kimball, B. A., and Mauney, J. R. (1993). Response of cotton to varying CO_2, irrigation, and nitrogen: yield and growth. *Agron. J.* 85, 706–712.

Kimball, B. A., La Morte, R. L., Peresta, G. J., Mauney, J. R., Lewin, K. F., and Hendrey, G. R. (1992). Appendix I: weather, soils, cultural practices, and cotton growth data from the 1989 FACE experiment in IBSNAT format. *Crit. Rev. Plant Sci.* 11, 271–308.

Koch, G. W. (1993). The use of natural situations of CO_2 enrichment in studies of vegetation responses to increasing atmospheric CO_2. In "Design and Execution of Experiments on CO_2 Enrichment" (E.-D. Schulze and H. A. Mooney, eds.), pp. 381–391. Commission of the European Communities, Brussels.

Körner, C., and Arnone, J. A., III (1992). Responses to elevated carbon dioxide in artificial tropical ecosystems. *Science* 257, 1672–1675.

Lawlor, D. W., and Mitchell, R. A. C. (1991). The effects of increasing CO_2 on crop photosynthesis and productivity: a review of field studies. *Plant Cell Environ.* 14, 807–818.

Lindroth, R. L. (1996). CO_2-mediated changes in tree chemistry and tree–Lepidoptera interactions. In "Carbon Dioxide and Terrestrial Ecosystems" (G. W. Koch and H. A. Mooney, eds.). Academic Press, San Diego.

Manabe, S. (1983). Carbon dioxide and climatic change. *Adv. Geophys.* 25, 39–82.

Manabe, S., and Stouffer, R. J. (1994). Multiple-century response of a coupled ocean–atmosphere model to an increase of atmospheric carbon dioxide. *J. Climate* 7, 5–23.

Miglietta, F., Raschi, A., Bettarini, I., Resti, R., and Selvi, F. (1993). Natural CO_2 springs in Italy: a resource for examining long-term response of vegetation to rising atmospheric CO_2 concentrations. *Plant Cell Environ.* 16, 873–878.

Nobel, P. S., and Hartsock, T. L. (1986). Short-term and long-term responses of crassulacean acid metabolism plants to elevated CO_2. *Plant Physiol.* 82, 604–606.

Norby, R. J., Wullschleger, S. D., and Gunderson, C. A. (1996). Tree responses to elevated CO_2 and implications for forests. In "Carbon Dioxide and Terrestrial Ecosystems" (G. W. Koch and H. A. Mooney, eds.). Academic Press, San Diego.

Oechel, W. C., Hastings, S. J., Vourlitis, G., Jenkins, M., Riechers, G., and Grulke, N. (1993). Recent change of Arctic tundra ecosystems from a net carbon dioxide sink to a source. *Nature* 361, 520–523.

Oechel, W. C., Cowles, S., Grulke, N., Hastings, S. J., Lawrence, B., Prudhomme, T., Riechers, G., Strain, B., Tissue, D., and Vourlitis, G. (1994). Transient nature of CO_2 fertilization in Arctic tundra. *Nature* 371, 500–503.

O'Neill, E. G., and Norby, R. J. (1996). Litter quality and decomposition rates of foliar litter produced under CO_2 enrichment. In "Carbon Dioxide and Terrestrial Ecosystems" (G. W. Koch and H. A. Mooney, eds.), Academic Press, San Diego.

Owensby, C. E., Coyne, P. I., Ham, J. M., Auen, L. M., and Knapp, A. K. (1993). Biomass production in a tallgrass prairie ecosystem exposed to ambient and elevated CO_2. *Ecol. Appl.* 3, 644–653.

Owensby, C. E., Ham, J. M., Knapp, A., Rice, C. W., Coyne, P. I., and Auen, L. M. (1996). Ecosystem-level responses of tallgrass prairie to elevated CO_2. In "Carbon Dioxide and Terrestrial Ecosystems" (G. W. Koch and H. A. Mooney, eds.). Academic Press, San Diego.

Pinter, P. J., Jr., Kimball, B. A., Garcia, R. L., Wall, G. W., Hunsaker, D. J., and LaMorte, R. L. (1996). Free-air CO_2 enrichment: responses of cotton and wheat crops. In "Carbon Dioxide and Terrestrial Ecosystems" (G. W. Koch and H. A. Mooney, eds.). Academic Press, San Diego.

Radin, J. W., Kimball, B. A., Hendrix, D. L., and Mauney, J. R. (1987). Photosynthesis of cotton plants exposed to elevated levels of carbon dioxide in the field. *Photosynth. Res.* 12, 191–203.

Rogers, H. H., Bingham, G. E., Cure, J. D., Smith, J. M., and Surano, K. A. (1983). Responses of selected plant species to elevated carbon dioxide in the field. *J. Environ. Qual.* 12, 569–574.

Rogers, H. H., Runion, B. G., and Krupa, S. V. (1994). Plant responses to atmospheric CO_2 enrichment with emphasis on roots and the rhizosphere. *Environ. Pollut.* 83, 155–189.

Rotmans, J., and den Elzen, M. G. J. (1993) Modelling feedback mechanisms in the carbon cycle: balancing the carbon budget. *Tellus* 45B, 301–320.

Running, S. W., and Hunt, E. R., Jr. (1993). Generalization of a forest ecosystem process model for other biomes, BIOME-BGC, and an application for global-scale models. *In*

"Scaling Physiological Processes: Leaf to Globe" (J. R. Ehleringer and C. B. Field, eds.), pp. 141–158. Academic Press, San Diego.

Schlesinger, W. H. (1991a). Climate, environment and ecology. *In* "Climate Change: Science, Impacts and Policy" (J. Jäger and H. L. Ferguson, eds.), pp. 371–378. Cambridge Univ. Press, Cambridge, UK.

Schlesinger, W. H. (1991b). "Biogeochemistry: An Analysis of Global Change." Academic Press, San Diego.

Sellers, P. J., Berry, J. A., Collatz, G. J., Field, C. B., and Hall, F. G. (1992). Canopy reflectance, photosynthesis, and transpiration. III. A reanalysis using improved leaf models and a new canopy integration scheme. *Remote Sens. Environ.* 42, 187–216.

Sokal, R. R., and Rohlf, F. J. (1981). "Biometry," 2nd Ed. Freeman, New York.

Taylor, J. A., and Lloyd, J. (1992). Sources and sinks of atmospheric CO_2. *Aust. J. Bot.* 40, 407–418.

Ziska, L. H., Hogan, K. P., Smith, A. P., and Drake, B. G. (1991). Growth and photosynthetic response of nine tropical species with long-term exposure to elevated carbon dioxide. *Oecologia* 86, 383–389.

22

Response of Terrestrial Ecosystems to Elevated CO_2: A Synthesis and Summary

George W. Koch and Harold A. Mooney

The past decade has seen great progress in understanding the response of terrestrial ecosystems to elevated atmospheric CO_2. At the time of the last major synthesis (Strain and Cure, 1985), there was only a single study of an intact unmanaged ecosystem in elevated CO_2: the Arctic tundra study of Oechel and colleagues (Chapter 10, this volume). Since that time the research focus has shifted from the study of short-term responses of individual species to multiyear field experiments with natural vegetation. There are now more than a dozen major ecosystem-level CO_2 experiments worldwide and more coming on-line each year. The chapters in this volume have summarized findings from many of these studies, emphasizing results on growth and primary productivity, plant and ecosystem CO_2 and water fluxes, and partitioning of carbon and nutrients between plants and soils. This chapter synthesizes the patterns of response seen across different systems and, in so doing, suggests priorities for future ecosystem CO_2 research.

I. Cross-System Comparisons

A. Leaf-Level Responses

Leaf-level data generally show the most consistency in terms of qualitative patterns of response to elevated CO_2 across ecosystems. Although varying in magnitude, a stimulation of leaf net photosynthesis and a reduction in stomatal conductance are observed in all of the herbaceous, woody, and crop system studies reported in this volume, with the exception of the Arctic tundra (Chapter 10) and the C_4-dominated tall grass prairie (Chapter 9). Photosynthesis apparently was unchanged after 3 years of elevated CO_2 in the tundra, while the prairie system showed stimulation of leaf photosynthesis only in years with relatively low precipitation. These exceptions notwithstanding, we conclude that a sustained enhancement in leaf-level photosynthesis, a reduction in stomatal conductance, and the consequent increase in leaf-level water use efficiency are common responses to elevated CO_2 in managed and naturally growing vegetation.

The chemical properties of leaves from nearly all ecosystems also change in predictable way in response to elevated CO_2. A reduction in leaf nitrogen concentration is commonly observed in conjunction with increased concentrations of leaf nonstructural carbohydrates and increased leaf mass to area ratios, although there are exceptions (Chapter 8). These changes have at least two major implications: they may influence rates of herbivory, and if litter chemistry is affected in proportion to green leaf chemistry, rates of litter decomposition may also be altered, but see below. In some temperate deciduous tree species, changes in the mature leaf carbon to nitrogen ratio (C/N ratio) are associated with altered levels of secondary defensive compounds, and together these changes affect the feeding and growth rates of insect herbivores (Chapter 7). Comparable data are lacking for most other systems, but if similar results are obtained, then we expect that changes in leaf chemistry in elevated CO_2 should have the general effect of stimulating insect herbivory due to a reduction in leaf nutritional quality. Herbivory responses may vary greatly among different classes of herbivores, however; in ruminants, reduced forage quality may slow digestion and reduce intake rates (Chapter 9).

How changes in tissue chemistry might influence herbivore population dynamics will require much more extensive *in situ* research. While FACE (free-air CO_2 enrichment) systems allow the treatment of large numbers of plants and herbivores, they have the drawback of being poorly isolated experimental "islands," and many herbivores will move freely between experimental plots and surrounding vegetation. For less mobile herbivores such as scale insects, this may not be a major problem. In addition to the influences of plant tissue chemistry, insect populations may be affected by changes in plant growth and reproductive schedules in elevated CO_2 (see the following).

B. Litter Quality and Decomposition

While the preceding effects of growth CO_2 level on green leaf chemistry have been studied thoroughly, there is much less data on the chemical properties of leaf, root, and other plant tissue litter in elevated CO_2. Evidence from field studies of tallgrass prairie, salt marsh, and temperate deciduous tree species indicates that leaf litter quality (C/N or lignin/N ratios) and mass loss rates of decomposing litter are little affected by CO_2 treatment, in marked contrast to results obtained in studies of plants grown in containers (Chapter 6). If these field results hold across other ecosystems, we would expect minor changes in nutrient cycling due to leaf litter quality effects. Other indirect effects of elevated CO_2, such as stimulation of soil microbial populations by enhanced carbon flow through roots (Chapter 3), may influence decomposition and soil nutrient dynamics in the absence of changes in litter quality. Importantly, because the litter of different plant species may differ chemically, CO_2-induced changes in species composition may indirectly alter decomposition dynamics (Chapter 9).

It should be recognized that soil temperature and moisture exert primary control over decomposition rates; to the extent that increased atmospheric CO_2 drives climatic changes, these effects may greatly outweigh effects on decomposition and nutrient cycling mediated by possible changes in plant litter chemistry or altered carbon supply to microbes. Even in the absence of climate change, however, soil moisture may increase in elevated CO_2 due to reductions in ecosystem evapotranspiration (see the following), and this may alter decomposition or nutrient dynamics.

C. Ecosystem Productivity

In comparing productivity responses to elevated CO_2, we can consider the absolute response, i.e., the change in net primary production in grams of carbon per square meter per year, and the relative response, typically reported as the percentage change relative to the ambient CO_2 treatment. Absolute responses are important in the context of the changing global carbon cycle, while relative responses may be more meaningful when considering the degree to which a given ecosystem is altered by elevated CO_2. As discussed here, the studies in this volume indicate that conclusions regarding the "responsiveness" of different ecosystems to elevated CO_2 depend on whether absolute or relative measures are considered. At this time, cross-system comparisons are problematic because of the different methods (peak biomass, cumulative biomass production, shoot density) used to assess productivity. The adoption of accurate, uniform, and comparable approaches for determining ecosystem carbon budgets should be a primary objective of the CO_2 research community. Nonetheless, preliminary comparisons of the ecosystem experiments summarized in this volume suggest a number of intriguing patterns of response to long-term CO_2 enrichment.

In general, productivity changes in elevated CO_2 are smaller and more varied, both among systems and from year to year within systems, than the widespread and often sizeable stimulation to leaf photosynthesis. The clearest correspondence between photosynthetic and growth responses is for crop systems (Chapters 13–15), where resources other than CO_2 are not strongly limiting and additional carbon fixation translates into additional plant biomass, both above- and belowground. In unmanaged systems there is often little or no significant measurable increase in plant biomass, despite a large increase in photosynthetic carbon uptake in elevated CO_2. Indications are that in such cases much of the increased carbon assimilated is entering the soil via enhanced root growth, turnover, or exudation. Determination of the fate of this extra carbon—incorporation by soil organisms including bacteria and fungi, accumulation in soil organic matter, or release back to the atmosphere as increased soil or root respiration—is a formidable technical challenge, but one that must be met if accurate carbon budgets are to be developed for terrestrial ecosystems.

One of the most striking findings from the studies of unmanaged systems is the large interannual variation in the relative stimulation of productivity by elevated CO_2 (Fig. 1). In the tallgrass prairie, for example, the enhancement of productivity in elevated CO_2 ranged from nearly 80% in one year to negligible in others. There was a similarly high variation among years in the annual grassland and C_3 salt marsh. In each of these systems it appears that CO_2 stimulates productivity more in years in which baseline productivity (at normal CO_2) is low, usually as a result of water and/or high temperature stress. Thus, in the tallgrass prairie, elevated CO_2 enhanced productivity by about 80% in a very dry year, by 40% in a year of intermediate productivity, and only slightly if at all in high productivity years (Fig. 1). A similar trend is apparent for the annual grassland and C_3 salt marsh, while the pattern is weaker for the C_4 salt marsh and absent after 2 years in the Swiss alpine study. Although not reported in this volume, recent analyses of data from the FACE studies of cotton and wheat are consistent with the pattern from unmanaged systems; elevated CO_2 enhances productivity more in drought treatments and in years in which productivity is low due to drought and high temperature stress (P. Pinter, personal communication).

Not included in Fig. 1 are data from the Arctic tundra in which elevated CO_2 stimulated net ecosystem CO_2 uptake in tundra microcosms for 2 years, but not in the third year of the study (Chapter 10; Oechel et al., 1994). Because net ecosystem CO_2 uptake was negative at ambient CO_2 (i.e., the tundra was a net CO_2 source), it is impossible to calculate a relative effect of elevated CO_2 on productivity in this system. The 3-year trend of decreasing stimulation by elevated CO_2 has been interpreted as indicating a progressive and complete homeostatic adjustment in this nutrient-limited ecosystem (Oechel et al., 1994). Given that baseline productivity (at ambient CO_2) showed considerable variation over the course of the study (Oechel et al., 1993, 1994), however, it is not clear to what extent interactions between

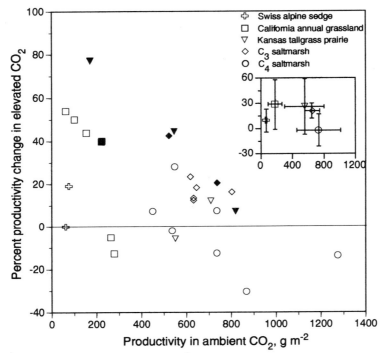

Figure 1 Percentage change in ecosystem productivity in elevated CO_2 as a function of productivity in ambient CO_2 for five unmanaged herbaceous ecosystems subjected to long-term CO_2 enrichment. Percentage change calculated as [(elevated − ambient)/ambient] × 100. Reported measures of productivity for the different systems are as follows: Swiss alpine sedge, peak season aboveground biomass, g m^{-2} (11); California annual grassland, total aboveground biomass, g m^{-2} (8); tallgrass prairie, peak standing biomass, g m^{-2} (9); C_3 salt marsh, shoot density, no. m^{-2} (12); C_4 salt marsh, peak shoot biomass, g m^{-2} (12). Numbers in parentheses refer to data source chapters in this volume. Individual symbols of a type represent different experimental seasons. Annual grassland data include 3 years each for communities on serpentine and sandstone-derived soils. Total aboveground biomass for annual grassland was calculated from the data in Table 3 of Chapter 8 as [peak biomass + (late season annual production/2)]. Solid symbols of a type represent experiments in which elevated CO_2 was reported to have a statistically significant effect on productivity. Inset shows means ± SD for multiple years from each study; axis labels the same as in main figure.

CO_2 and other factors controlling annual productivity (e.g., temperature) may also have contributed to the observed results.

The inverse relationship between the relative change in productivity in elevated CO_2 and baseline productivity can be interpreted in terms of the interaction between the physiological effects of CO_2 and the dominant factor limiting productivity in different ecosystems. In the drought-limited annual grassland and tallgrass prairie, elevated CO_2 improves soil and plant water status by reducing stomatal conductance. This increases water use efficiency and tends to alleviate the water limitation to productivity seen at ambient CO_2 (Chapters 8 and 9). In the salt marsh, factors associated with anaerobiosis and salinity may be the dominant limitations to productivity. Elevated CO_2 does not directly alleviate these stresses, but it does enhance net ecosystem CO_2 uptake in the C_3 community more in years characterized by higher tempratures, conditions that can also reduce annual productivity (Chapter 12). Here, the primary benefit of elevated CO_2 may be the suppression of photorespiration (in C_3 species), but an improvement in plant water relations may also be involved. In the Swiss alpine tundra, low temperatures and, in some years, low light are the major environmental limitations to plant growth. Low temperatures tend to preclude photosynthetic response to elevated CO_2, and so it is expected that alpine tundra would show a minor growth response to elevated CO_2.

The greater relative enhancement in productivity by elevated CO_2 in dry years has the effect of reducing interannual variations in productivity. It is tempting to suggest that elevated CO_2 may act as a productivity buffer against drought; further experiments and modeling studies should test this idea. In any case, understanding of the interaction of elevated CO_2 and other environmental factors in controlling interannual variations in productivity in different ecosystems should be a top priority of future research.

The variation in average productivity responses to elevated CO_2 observed across different unmanaged ecosystems does not appear to conform to the interannual pattern within systems (Fig. 1, inset). Although on average the lowest relative enhancement of productivity does occur in the high productivity salt marsh, there is no clear trend toward greater stimulation of productivity by elevated CO_2 in the less productive systems. (In fact, many of the apparent productivity responses are not statistically significant, and it is difficult to draw any conclusions). The two least productive ecosystems, alpine sedge and California annual grassland (particularly that on serpentine-derived soils, Chapter 8), are characterized by species with inherently low growth rates and by limitations (low temperature in the alpine community and highly infertile soils in the serpentine grasslands) that elevated CO_2 cannot directly reduce in the short term, unlike water stress. In such systems, changes in species composition may be required for significant productivity increases to be realized.

Consistent with other studies, the crop studies in this volume report large percentage increases in productivity under elevated CO_2, averaging about 33% for wheat, rice, cotton, and beans (including the "wet" treatments at ambient and 550 ppm CO_2 for FACE wheat and cotton and at ambient and approximately twice-ambient CO_2 in the other studies). The results for crop and unmanaged systems support the idea that elevated CO_2 can produce large percentage increases in productivity over the short-term in resource-rich ecosystems and in systems where it leads to an increased availability or efficiency of use of a limiting resource (Field *et al.*, 1992).

Absolute changes in productivity give a different impression of comparative ecosystem response to elevated CO_2 than do relative productivity responses (Fig. 2). Not surprisingly, the tendency is for more productive systems to show greater increases in primary production, with crop systems posting the largest gains and the low productivity alpine sedge and annual grassland showing small absolute increases. Productivity gains in crop sys-

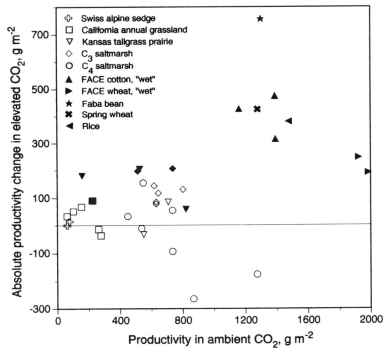

Figure 2 Absolute change in ecosystem productivity in elevated CO_2 as a function of productivity in ambient CO_2 for different unmanaged and crop ecosystems. Absolute change calculated as (productivity in elevated CO_2) − (productivity in ambient CO_2). Units are g m^{-2} for all unmanaged systems except C_3 saltmarsh (shoot density, see Fig. 1 legend). Reported measures of seasonal productivity for crop systems are as follows: FACE cotton, final harvest aboveground + belowground dry matter, g m^{-2} (Chapter 13; Kimball *et al.*, 1992; P. Pinter, personal communcation); FACE wheat, peak aboveground biomass, g m^{-2} (Chapter 13; P. Pinter, unpublished data); faba bean, total aboveground dry matter, g m^{-2} (Chapter 14); spring wheat, total aboveground production, g m^{-2} (Chapter 14); rice, total aboveground dry matter, g m^{-2} (Chapter 15; Baker *et al.*, 1992). Solid symbols of a type represent experiments in which elevated CO_2 was reported to have a statistically significant effect on production.

tems range from ca. 250 to 750 g m^{-2}. Again, unmanaged systems show considerable interannual variability; productivity in the tallgrass prairie was enhanced by ca. 60–200 g dry matter m^{-2} year^{-1} in 3 out of 5 years, the largest increase occurring in a year of intermediate productivity in ambient CO_2. In the California annual grassland on serpentine soils (the three lowest productivity points in this class), there is a strong tendency toward an inverse relationship between relative and absolute responses, such that high percentage increases are associated with small absolute increases in productivity (compare Figs. 1 and 2). In the C_4 salt marsh community, it appears that elevated CO_2 has no effect on phytomass production. In some years the C_3 salt marsh community shows large increases in shoot density in elevated CO_2, although it is not clear how this translates into a biomass response.

Absolute increases in net ecosystem CO_2 uptake reported for the Arctic tussock tundra during the first 2 years of a 3-year study (Oechel *et al.,* 1994), if representative of annual biomass productivity, would greatly exceed the CO_2 enhancements for the other unmanaged eocsystems shown in Fig. 2 (ca. 400 g C m^{-2}). During the third year, however, this system responded similarly to the alpine sedge community, with the ambient and elevated CO_2 treatments showing no significant differences in net ecosystem CO_2 exchange. Conclusions regarding the potential for biomass responses to elevated CO_2 in Arctic tundra may depend on the nature of the relationship of measured seasonal net ecosystem CO_2 exchange and actual annual biomass production. The degree of nutrient limitation, and thus potential responses to increased CO_2, is strongly affected by temperature in the tundra, and $CO_2 \times$ temperature interactions must be included in any analysis of CO_2 effects in tundra ecosystems (Chapter 10).

The pattern of more productive systems showing larger absolute productivity increases in elevated CO_2 emphasizes the critical need for CO_2 experiments in temperate forests and tropical forests and grasslands. These highly productive systems already play a major role in the global carbon cycle and should show a strong primary production response (at least in the short-term) to elevated CO_2.

D. Evapotranspiration and Soil Moisture

On the basis of stomatal responses to elevated CO_2, ecosystem evapotranspiration (ET) generally would be expected to decrease in elevated CO_2 in the absence of compensatory changes in leaf area. Reductions in ET should be most pronounced where soil evaporation is low compared to transpiration (e.g., in high LAI systems) and where transpiration is closely coupled to leaf surface conductance—the conductance most affected by CO_2 concentration. Although few systems have been examined, it appears that preliminary results are not inconsistent with these expectations.

In the three unmanaged systems in which it was estimated (Chapters 8, 9, and 12), ET apparently was lower in elevated CO_2, suggesting that CO_2-induced reductions in stomatal conductance tend to outweigh the increases in leaf area index seen in some studies (Chapters 9 and 12). In the FACE study of cotton and wheat (Chapter 13), energy balance and modeling approaches indicated a reduction in ET of 5–10% in elevated compared to ambient CO_2. In both the annual grassland and tallgrass prairie (Chapters 8 and 9, respectively), soil moisture was higher in elevated CO_2; in the annual grasslands this was true only where vegetation cover was high.

A reduction in ecosystem ET has important implications for the duration of the growing season in seasonally droughted systems, and there are indications that a lengthening of the period of moisture availability in elevated CO_2 may result in shifts in species composition (Chapters 8 and 9). Continuous measurements of ET and soil moisture throughout the entire growing season and subsequent drought would indicate whether the reduction in ET seen during the growing season in elevated CO_2 results in a carrying over of "unused" moisture from year to year or whether available moisture is fully exploited by the lengthening of the growing season.

E. Growth Patterns and Phenology

Where comparisons have been made for trees and crops, indeterminant species show a greater growth response to elevated CO_2 than do determinant species (Chapters 5 and 13), presumably because they have greater flexibility in adjusting to altered resource availability than do determinant species. The seasonal timing of plant activity apparently is affected by elevated CO_2 in some systems, although there are no clear patterns as yet. In some crop and tree species, as well in unmanaged ecosystems, senescence was delayed and carbon fixation was extended in elevated CO_2 (Chapters 5, 8, 10, 13). In the Arctic tundra and annual grassland, this resulted from an enhancement in growth and photosynthesis in subdominant species that remained active late in the growing season (Chapters 8 and 12), rather than from a strong phenological response of the dominant species. In Sitka spruce, a shortening of the growing season apparently was related to the altered nutritional status in elevated CO_2 (Chapter 4). Clearly, much additional research is needed on the influence of CO_2 level on phenological patterns in different ecosystems. In beginning to address more complex ecological interactions, it will be important to examine plant phenological responses to CO_2 in relation to annual activity patterns of herbivores and pollinators.

II. Future Research Needs

The studies summarized in this volume reveal many new complexities and questions about the response of terrestrial ecosystems to elevated CO_2. In addition to the suggestions for future work already made, we discuss topics that we believe should be addressed by future ecosystem CO_2 studies.

A. Additional Ecosystem Experiments

The ecosystem productivity responses summarized herein, while a valuable starting point, are insufficient for developing accurate predictions of the response of the terrestrial biosphere to rising atmospheric CO_2. What we can say at this time is that different terrestrial ecosystems respond differently to elevated CO_2. The range of ecosystems encompassed by this general statement is still quite small, however, and does not include the major forest types of the world, which are exactly those systems that are presently responsible for most terrestrial primary production.

Differential ecosystem response to CO_2 implies that the use of a single value of β, the biota growth factor, in global carbon cycle models is likely to be highly inaccurate (Chapter 21), unless that value is developed with knowledge of the response to elevated CO_2 of Earth's major vegetation types. On the basis of our analysis, it is highly productive systems (e.g., temperate and tropical forests) that are potentially the most responsive to elevated CO_2 in terms of absolute changes in primary production, and it is these systems about which we know the least.

It is no accident that CO_2 experiments have yet to be conducted in forests. These are expensive (for ecological research) and technically challenging undertakings. Yet there is no doubt that these experiments must be conducted. Experiments with tree saplings in open-top chambers give an important glimpse of the response of certain forest species under specific conditions, but cannot incorporate the host of biotic and abiotic interactions that may alter the magnitude, and even the direction, of forest carbon sequestration in elevated CO_2 (Chapter 16). Fortunately, there is now a FACE experiment beginning in a temperate coniferous forest. Hopefully this single experiment will not be used to represent the responses of all of the world's forests to elevated CO_2. At the very least we would hope for a study in a humid tropical forest, in which interactions with biogeochemical processes may differ greatly from temperate systems.

A second critical ecosystem, in terms of potential response to rising CO_2, is the tropical grassland. Experiments here should be technically quite manageable and yet vitally important if, as recent evidence suggests, these systems are currently and may continue to be globally important carbon sinks (Fisher et al., 1994). A third class of systems that should be studied is those near transitions between biomes where elevated CO_2 might be expected to alter the environmental gradient along which the vegetation transition is arranged. An example of this would be grassland–shrubland ecotones where a CO_2-induced change in water balance may cause biome boundaries to shift over time, which in turn might alter biogeochemical cycling and biophysical properties of the land surface.

There are arguments to be made in support of CO_2 experiments in many other communities and ecosystems, but the scarcity of resources dictates that large-scale studies will be few. Although many systems will not be studied experimentally, there may well be low cost, long-term monitoring approaches that could be initiated now that would greatly enhance future understanding of the effects of rising CO_2 and other changing environmental factors on these systems.

B. CO_2–Environment Interactions

We have seen that ecosystem response to elevated CO_2 depends on average resource levels and plant characteristics, which is consistent with past predictions (e.g., Field *et al.,* 1992). From multiyear data we have also learned that CO_2 responses strongly depend on the level of factors, such as precipitation, that may vary considerably on an interannual basis. Moreover, there is now evidence that ecosystem responses to elevated CO_2 can themselves alter levels of soil moisture and nutrients. Improvement of our understanding of the direct and indirect CO_2–environment interactions is critically important to developing better predictions of long-term ecosystem responses to rising CO_2 and should be a central focus of future research.

Several studies presented here indicate that belowground processes and the nitrogen cycle in particular are key regulators of productivity responses in elevated CO_2, yet there is at present a weak understanding of carbon flows to different belowground pools—root production and exudation, bacteria, fungi, and various soils organic matter fractions—and of the interaction of these flows with soil nitrogen transformations. There are contrasting hypotheses for the expected sign of the feedback linking altered carbon and nitrogen cycles in elevated CO_2 (Chapter 16). Will mineral nitrogen be immobilized as microbial populations and activity increase due to increased rhizodeposition (i.e., negative feedback), or will microbes receiving increased carbon input rely more heavily on older soil organic matter, thereby making more nitrogen available to plants (positive feedback; Chapter 3)? What occurs may depend on initial soil fertility and, to the extent that it may influence plant–microbe competition for mineral nitrogen, the degree of seasonal synchrony of microbial and root activity. Detection of the pathways of altered carbon flow and their relationship to microbial processes should be a top priority of future research.

Given that it is likely that vegetation response to increased CO_2 will depend on the extent to which the nitrogen cycle can be altered, it will be important to examine the various ways in which this might occur. These include changes in the C/N ratios of major ecosystem pools and fluxes, altered distributions of nitrogen between plants and soils, and changes in the magnitude of nitrogen inputs and outputs (Shaver *et al.*, 1992), effects that may occur to varying degrees in different systems. Thus, new information must be obtained on the flexibility of C/N ratios in different organic matter pools, on the control of organic matter partitioning among pools, and on the "openness" of the nitrogen cycle in different ecosystems.

The strong CO_2 × environment interactions challenge experimental detection of changes in productivity responses due to hypothesized CO_2-induced feedbacks via, for example, litter decomposition or nutrient cycling. If elevated CO_2 gradually changes ecosystem processes, then ambient CO_2 treatments are not adequate controls for sorting out CO_2 × environment interactions associated with interannual weather variations because these interactions may operate differently in a system subjected to, say, 5 years vs 1 year of elevated CO_2 or any number of years of ambient CO_2. Separation of cumulative effects of CO_2 treatments from year to year environmental interactions may require the addition of new elevated CO_2 treatments during each year of a multiyear study in order to provide a "year one" baseline against which to compare the responses of multiyear experimental plots. At the very least, these problems emphasize that CO_2 experiments in natural systems should be of sufficient duration to capture most of the characteristic interannual variation in major climatic variables.

C. Population, Community, and Evolutionary Responses

The studies presented in this volume primarily focus on relationships between CO_2 concentration and fundamental ecosystem processes: primary productivity, water and energy exchange, and carbon and nutrient cycling between ecosystem pools. Responses at the level of population genetics, species composition, and community structure may require very long periods to develop, and it is doubtful whether the duration of most CO_2 experiments is sufficient to reveal more than the direction of initial change in many systems. The studies presented here provide an early glimpse of responses that deserve a great deal more experimental and modeling effort.

What is clear from the limited data in this volume is that simple predictions of species interactions based on physiological responses of isolated plants will not reliably predict population- or community-level responses. This is seen in comparing the two experiments involving both C_3 and C_4 species. Although results from the salt marsh (Chapter 12) support the dogma that photosynthesis and growth of C_3 plants are more responsive to elevated CO_2 than C_4 plants, the opposite is true in the tallgrass prairie (Chapter 9) where elevated CO_2 results in increases in leaf area, biomass, and population sizes of C_4 grasses and decreases for C_3 species. In the latter case, the authors point out that the results may have been related to the absence of large herbivores in the normally grazed tallgrass system. Grazing and fire are essential influences on species composition in many ecosystems, and it will be very challenging yet important to incorporate these factors into analyses of CO_2 effects on ecosystem structure and composition.

In seasonally droughted system, the CO_2-induced increase in water use efficiency apparently may alter system-level water availability and, in turn, the relative performance of early and late season species (Chapter 8). This response, like many others, is highly variable among years and an understanding of its significance for long-term changes in community composition will likely require population models incorporating competition and the physiological and phenological characteristics of dominant species or functional types. Many studies have already seen highly significant effects of open-top chambers on primary ecosystem processes. It is likely that chambers strongly impact processes such as pollination, dispersal, and germination. Chamber effects represent a problem that must be solved, whether through greater reliance on FACE systems or otherwise, in order to derive meaningful conclusions regarding population-level phenomena from experimental studies.

More detailed information on population, community, and evolutionary responses to elevated CO_2, based on continuing work from the studies in this volume and from other research groups, will be presented in a companion volume (Körner et al., 1996).

III. Conclusions

At this time, we can conclude that carbon input to ecosystems is generally enhanced by elevated CO_2. From the available data it appears that the degree to which elevated CO_2 increases primary production does not vary in a predictable way across different ecosystems, although there is a tendency toward larger absolute increases in productivity in more productive systems. Forested ecosystems have not been examined.

Productivity responses to elevated CO_2 are highly variable from year to year within systems, and CO_2 interactions with the water cycle appear to be particularly important in mediating this variation. In seasonally droughted systems, elevated CO_2 may stimulate productivity to a greater degree in relatively dry, low productivity years by partially alleviating water limitation to productivity. In some situations, elevated CO_2 reduces evapotranspiration and increases soil moisture availability so that the growing season is lengthened, and this may promote changes in species composition.

Elevated CO_2 generally increases plant biomass production less than it increases ecosystem CO_2 uptake. The location and chemical fate of the extra carbon assimilated in elevated CO_2 are poorly known in most experiments and are critical to understanding long-term carbon dynamics and the potential for increased carbon sequestration in elevated CO_2.

The nitrogen cycle may be altered in numerous ways by elevated CO_2, none of which is adequately understood at this time. The scant field data indicate little if any change in leaf litter quality and decomposition due to elevated CO_2. Elevated CO_2 may have important indirect influences on decomposition, for example, by increasing soil moisture in some systems. There is evidence for both positive and negative feedbacks on plant productivity via elevated CO_2-induced increases in belowground carbon flows and microbial activity. In forest systems, the nitrogen cycle becomes progressively more closed with stand age, and this may significantly impact the potential for elevated CO_2 to alter plant nitrogen availability.

Changes in leaf chemical properties in elevated CO_2 alter rates of herbivory by insects and ruminants. Understanding of the population or ecosystem implications of altered plant–herbivore interactions in elevated CO_2 will require new experimental approaches at relevant spatial and temporal scales.

Ecosystem CO_2 experiments are lacking in systems that are extremely important to the global carbon cycle and human economies, most notably temperate and tropical forests and tropical grasslands. It is unlikely that accurate predictions of the role of the terrestrial biosphere in the future global carbon cycle will be developed without knowledge of how these systems respond to increasing atmospheric CO_2.

Acknowledgments

We thank Chris Field and Geeske Joel for their input to the analyses and ideas presented here.

References

Baker, J. R., Laugel, F., Boote, K. J., and Allen, L. H. (1992). Effects of daytime carbon dioxide concentration on dark respiration in rice. *Plant Cell Environ.* 15, 231–239.

Field, C. B., Chapin, F. S., III, Matson, P. A., and Mooney, H. A. (1992). Responses of terrestrial ecosystems to the changing atmosphere: A resource-based approach. *Annu. Rev. Ecol. Systematics* 23, 201–235.

Fisher, M. J., Rao, I. M., Ayarza, M. A., Lascano, C. E., Sanz, J. I., Thomas, R. J., and Vera, R. R. (1994). Carbon storage by introduced deep-rooted grasses in the South American Savannas. *Nature* 371, 236–238.

Kimball, B. A., La Morte, R. L., Peresta, G. J., Mauney, J. R., Lewin, K. F., and Hendrey, G. R. (1992). Appendix I: Weather, soils, cultural practices, and cotton growth data from the 1989 FACE experiment in IBSNAT format. *Crit. Rev. Plant Sci.* 11, 271–308.

Körner, C., and Bazzaz, F. A., eds. (1996). "Biological Diversity in a CO_2-Rich World." Physiological Ecology Series. Academic Press, San Diego.

Oechel, W. C., Hastings, S. J., Vourlitis, G., Jenkins, M., Riechers, G., and Grulke, N. (1993). Recent change of Arctic tundra ecosystems from a net carbon dioxide sink to a source. *Nature* 361, 520–523.

Oechel, W. C., Cowles, S., Grulke, N., Hastings, S. J., Lawrence, B., Prudomme, T., Riechers, G., Strain, B., Tissue, D., and Vourlitis, G. (1994). Transient nature of CO_2 fertilization in arctic tundra. *Nature* 371, 500–503.

Shaver, G. R., Billings, W. D, Chapin, F. S., III, Giblin, A. E., Nadelhoffer, K. J., Oechel, W. C., and Rastetter, E. B. (1992). Global change and the carbon balance of arctic ecosystems. *BioScience* 42, 433–441.

Strain, B. R., and Cure, J. D. (1985). "Direct Effects of Increasing Carbon Dioxide on Vegetation." United States Department of Energy, DOE/ER-0238.

Index

Aboveground
  and belowground, nutrient cycles, 87–89
  grassland growth, 408–409
  live biomass in annual grassland, 139–140
  respiration, 256
  rice, biomass, 269
  study methods at UMBS, 45
  total biomass, 261–262
Acclimation
  crops, inclusion in models, 326–327
  photosynthesis, to elevated CO_2, 349–351
Accuracy, in hierarchic and single-level models, 322–323
Acer saccarum, see Sugar maple
Acid detergent fiber, in biomass of grasses, 157
Adaptation
  crops, inclusion in models, 326–327
  long-term processes at canopy level, 260
Agave deserti, growth response to elevated CO_2, 404
Aiken clay loam, at Placerville site, 24–25
Air distribution system
  for alpine vegetation studies, 183
  in microcosms, 136–137
Allocation pattern
  biomass, depending on nutrient conditions, 73–74, 83–85
  controlled by plant genetics, 369–371
  leaf-level model, 350
  long-term, in elevated CO_2, 353–354
Alps, European Central, closed vegetation studies, 179–195
Andropogon gerardii, see Big bluestem
Antioxidants, *see also specific antioxidants*
  subcellular localization, 300–301
Arctic ecosystems
  direct effects of elevated CO_2, 163–174
  tundra microcosms, 418, 420, 422

  tussock tundra exposed to elevated CO_2, 167–171
  NEP, 407
Artifactual restrictions
  chamber-enclosed trees, 54
  pot studies, 1, 97–100, 291–294
  test environments, 215–216
Artificial forest floor, as artifact, 37
Ascorbate
  reduction of H_2O_2, 301–303
  regeneration, 307
  tobacco, ozone sensitivity, 306–307
Ascorbate peroxidase, detoxification of reactive oxygen species, 301–303
Beech
  biomass and partitioning, 73–74
  carbohydrates, 81–82
  chlorophyll and nitrogen content, 79–81
  growth strategy and response to elevated CO_2, 71–85
  leaf area and leaf number, 74–77
  phenology, 77
  photosynthesis and respiration, 78–79
Belowground
  and aboveground, nutrient cycles, 87–89
  carbon flows, 428
  increased C flux, 43–44, 50
  spatial deployment changes, 84
Betula papyrifera, see Paper birch
Big bluestem
  aboveground biomass, total N, 155–157
  photosynthetic capacity, 154
  total biomass, 151–152
Bigtooth aspen, UMBS 1991 study, 45–49
Biomass
  aboveground
    in annual grassland, 139–140
    rice, 269
    spring wheat and faba bean, 261–262
  alpine
    above- and belowground, 183–184
    distribution, 180–182

432   Index

Biomass, *continued*
  annual increment, 11
  cotton and wheat crops, 220, 227–231
  in ecosystem models, 360–361
  fine root
    Douglas fir stands, 33–34
    response to CO_2, 49
  and phenology responses, 191–192
  production
    C_3 and C_4 species, 151–152
    and nutrient limitations, 155, 159
    roots and shoots of grasses, 199
  seedlings
    ponderosa pine, 26–27
    Sitka spruce, 64–68
  soil microbial, 44
  sweet chestnut and beech, 73–74
Biota growth factor β, *see also*
    Photosynthetic β factor
  and CO_2-stimulated terrestrial ecosystem NPP, 399–411
  in leaf photosynthesis model, 382–393
  single value for, 424
Botta's pocket gopher
  disturbance of soils at Jasper Ridge, 128–129
  elimination from experimental area, 133
Boundary layer resistance, and canopy density, 209–210
Branches, Sitka spruce
  in branch bags, 58–59
  photosynthesis, 61–63

Canopy
  air temperature, 187–188
  assimilation model, 277–279
  C balance, semifield study, 252–262
  closed, responses to CO_2 concentration, 15–17
  density, and boundary layer resistance, 209–210
  energy balance, 226
  gas exchange rates, 221
  long-term adaptation processes, 260
  photosynthesis, 222–225
  photosynthesis and transpiration, plant-level model, 352–353
  rice, photosynthetic rates, 269–274
  senescence, 234–235
Carbohydrates
  availability to tissues, 61–63
  deciduous trees, CO_2 effects, 7

  increased allocation to roots and mycorrhizae, 24
  sinks, 166–168, 170–171
  supply, and growth, inclusion in crop modeling, 329–331
  sweet chestnut and beech, 81–82
  temperature effects, 335
  total nonstructural, 185, 192–194
Carbon, *see also* Net carbon exchange
  allocation
    among vegetative organs, 353–354
    to reproductive structures, 354–355
    to roots, 84–85
  assimilation
    ecosystem, 209
    measurement, 220–221
    and water use efficiency, 148
  balance
    and ecosystem gas exchange, 189–191
    in forest stands, 53–54
  budget
    global, balancing, 402, 410
    soil, 238
  flow
    in alpine ecosystem, 185
    belowground, 428
    dependent on allocation pattern, 369–371
  input to root zone, 43–44
  oxidation and reduction cycles, 387–388
  photosynthetic influx, and atmospheric CO_2 increase, 381–394
  stocks, in arctic soils, 164–165
Carbon cycling
  in ecosystem models, 360–361
  large-scale models, 403
  in yellow poplar and white oak system, 13–15
Carbon dioxide, *see also* Elevated carbon dioxide; Net ecosystem CO_2 exchange
  above- and belowground responses, 41–50
  atmospheric increase, and photosynthetic C influx, 381–394
  doubling, interspecific differences, 257–259
  effect on
    arctic plant and ecosystem function, 163–174
    cotton and wheat, 235–236
    ponderosa pine, 23–38

stomata, 351
water relations, 225–227
efficient uptake in arctic plants, 171–172
interactions with
  environment, 425–426
  nitrogen, 283–295
  nutrients, 2–3, 75–77, 339
  temperature, 17–19, 317–340
Jasper Ridge experiment, 121–142
regime in open tops, 185–186
response modeling, 260–261
rice responses, 265–281
treatment of cotton and wheat, 219
twice-ambient, 407–408
uptake by spring wheat and faba bean, 251–262
Carbon dioxide enrichment
alpine plants responsive to, 177–178
effects on
  biomass partitioning, 76
  litter quality and decomposition, 92–96
  increased rice grain yield, 280
  tree seedlings, 3–19
Carbon mineralization, *see also* Nitrogen mineralization
and soil fauna, 96
Carbon : nitrogen ratio
litter, 88, 94–96, 363
Sitka spruce, 65
soils and vegetation, 35–37, 167
Carbon–nutrient balance theory, 106–107
Castanea sativa, see Sweet chestnut
Caterpillar, herbivory, elevated CO_2 effects, 112–115
CENTURY model, 131
Chaparral, shrub response surface, 359–360
Chesapeake Bay wetlands, long-term exposure to elevated CO_2, 197–212
Chlorophyll, sweet chestnut and beech, 79
Chloroplasts, production site of reactive oxygen species, 300–303
Climate change
ecosystem hydrologic responses, 364–365
global, rice responses, 265–281
regional and global responses, forecasting, 368–369
and species distribution, modeling, 357
Clones
loblolly pine, 66–67
Sitka spruce, 58–61, 67–68

Closed-top chamber, *see also* Open-top chamber
tussock tundra NEP study, 407
Comminution, *see* Fragmentation
Community
alpine plant, sedge-dominated, 179–180
and ecosystem, responses to tree chemistry changes, 115–117
Community models, 355–359
Computer programs, *see also* Models
ROOTS, 46
SYSTAT, 26
Constraints
on alpine plant response to elevated CO_2, 178–179
CO_2 fertilization effect by N availability, 283–284
on measuring CO_2 response in deciduous forests, 1–3
Correlative models, climate, 368–369
Cotton
biomass, leaf area, and fruit production, 227–230
lint yields, 231–233
test crops at FACE facility, 218–220
Crops
C_3, ecosystem, 405–406
cotton and wheat, FACE studies, 215–244
mechanistic simulation models, 317–340
spring wheat and faba bean, growth and CO_2 uptake, 251–262
Cyclophysis, and CO_2 responses of young trees, 66

Dark respiration rate
leaves, 6
rice canopy, 272–277
sweet chestnut and beech, 78–79
wetlands C_4 grasses, 209–210
Decomposition
CO_2 enrichment effect, 92–96
and litter quality, 89–92, 416–417
and nutrient cycling, 363–364
Detoxification, reactive oxygen species, 300–303
Development
alpine plant, 184
crop, temperature effects, 333–335
insect, effects of foliar chemistry changes, 114–117
Dispersal, plant, CO_2 effects, 362–363

Douglas fir
  fine root biomass, 33–34
  N translocation, 288
  tannin content, 90
Drought stress, combined with elevated CO_2, 308–309, 427
Dry mass, white oak and yellow poplar
  final harvest, 7–9
  whole-tree, 11

Ecosys model, 226–227
Ecosystem
  alpine, gas exchange studies, 183
  arctic
    CO_2 loss to atmosphere, 173–174
    dynamics, 163–174
  artificial, crop, and natural, 405–409
  artificial tropical, 290
  and community, responses to tree chemistry changes, 115–117
  forest, responses to elevated CO_2, 3–19
  hydrology, modeling, 364–365
  mature forest, applicability of results with seedlings, 32–38, 291–294
  model, Jasper Ridge CO_2 experiment, 121–142
  modeling effects of elevated CO_2, 347–371
  net CO_2 exchange, 199–202, 209–212
  physiognomy, and species composition, 361–363
  productivity, 417–422
  species composition changes, 87–89
  tallgrass prairie, response to elevated CO_2, 147–160
  terrestrial
    NPP, 399–411
    response to elevated CO_2, 415–428
Edinburgh, open-top chamber studies with Sitka spruce, 56–68
Elevated carbon dioxide
  crop responses, 317–340
  and environmental stress, 299–311
  and growth strategy, deciduous trees, 71–85
  long-term, Chesapeake Bay wetland, 197–212
  plant- and ecosystem-level modeling, 347–371
  Sitka spruce responses, 53–68
  tallgrass prairie responses, 147–160
  terrestrial ecosystem responses, 415–428
  and tree chemistry, 107–112
Emergence, crops in The Netherlands, 257–258
Empiricism
  versus mechanism, in crop models, 320–323
  in predicting ecosystem responses, 347–348
Energy balance
  canopy, evapotranspiration estimates from, 226
  crops, inclusion in crop modeling, 331–332
Environmental stress, and elevated CO_2, trees affected by, 299–311
European Central Alps, vegetation and elevated CO_2, 179–195
Evapotranspiration
  estimates from
    canopy energy balance, 226
    soil moisture depletion, 226
  potential, for test crops, 219–220
  reduction under CO_2 enrichment, 153–154
  and soil moisture, 422–423
  stomatal resistance effects, 364–365
Extrapolation
  with mechanistic and empirical models, 321–323
  spatial, $\mathscr{L}$ factor to global scale, 388–390

Faba bean
  crop growth modeling, 256–259
  harvestable yield, 254–256
  model description and validation, 253–254
  semifield experimental setup, 252–253
  temperature and light effects, 259–260
FACE, see Free-air CO_2 enrichment
Fagus sylvatica, see Beech
Farquhar model, C_3 photosynthesis, 327–328, 383, 386
Fauna, soil, and C mineralization, 96
Fertilization
  CO_2
    cotton, 228–229
    N cycling constraints, 283–284
    nitrogen, Placerville site, 27–32, 37–38

Fine roots
  allocation of additional C, 210
  biomass
    in Douglas fir stands, 33–34
    response to CO_2, 49
  dynamics, 286
  yellow poplar and white oak, 5, 14–15
Foliage
  biomass, steady state after crown closure, 35
  high-CO_2, insects fed on, 114–115
  leaching, and forest N cycling, 284–289
  litter
    dead and green, 156–157
    quality and decomposition rates, 87–101
  N concentration, 7, 26–28
Forage quality, changes under elevated CO_2, 157
Forbs
  C_3, increase under elevated CO_2, 93–94
  in tallgrass prairie experiment, 150–160
Forecasting, with top-down approaches, 368–369
Forests
  carbon cycling, 13–15
  deciduous
    above- and belowground responses to CO_2, 41–50
    constraints on measuring CO_2 response, 1–3
    interactions between CO_2 and N, 283–295
  Lepidoptera species, 106
  mature, applicability of results with seedlings, 32–38, 291–294
  nitrogen cycling, 284–289
Fragmentation, in litter decomposition, 90–92
Free-air CO_2 enrichment
  capital costs, 293
  cotton and wheat test crops, 218–244
  crop ecosystem, 406
  facilities, 17, 100, 217
  history, 216–217
  performance and reliability, 217–218
Fruit production, cotton, 227–230

Gap models, 356–357
Gas exchange
  canopy, 221
  ecosystem
    and C balance, 189–191
    net, 199–202, 209–212
  ecosystem-scale, 130
  whole-plant closed system, 73
  yellow poplar and white oak, 5–6
G'DAY model, 13–14, 360
Generality, and tractability, model ecosystems, 124–125
GePSi model, 131
Global carbon cycle models, β values, 401–402
Global change research
  forest role, 54
  model ecosystem role, 124–125
Global climate change, rice responses, 265–281
Gopher, see Botta's pocket gopher
Gossypium hirsutum, see Cotton
Grain filling rates, temperature effects, 334–335
Grain yield, wheat, 233–235, 323
Grasses, C_3 and C_4
  Chesapeake Bay wetlands, 198–212
  elevated CO_2 effects, 93–94
  tallgrass prairie, 150–160
Grasslands
  California, aboveground growth, 408–409
  serpentine and sandstone, Jasper Ridge CO_2 experiment, 121–142
  tropical, carbon sinks, 424
Green leaf area index, see also Leaf area index
  canopy, FACE studies, 223–224
  CO_2 enhancement, 235–236
  under DRY treatment, 227, 231
Gross primary production, atmospheric CO_2 effects, 399–400
Growth
  and carbohydrate supply, inclusion in crop modeling, 329–331
  crop, temperature effects, 333–335
  leaf, phenology, 60–61
  limited, and C assimilation, 209
  patterns, and phenology, 423
  and photosynthesis, feedback, 2–3
  root
    and rooting volume, 55
    study methods at UMBS, 45–46
  seedling, and physiological responses, 72–73
  spring wheat and faba bean, and CO_2 uptake, 251–262

Growth, *continued*
and yield responses
cotton and wheat, 227–236
rice, 267–269
Growth dynamics
ponderosa pine, 23–38
white oak and yellow poplar, 9–13
Growth efficiency index, 11–13
Growth strategy, and tree responses to elevated CO_2, 71–85

Head production, wheat, 230–231
Herbivores
insect, elevated CO_2 effects, 112–115
interactions with plants, modeling, 365–366
Herbivory
effect of leaf chemical changes, 416
in natural ecosystems, 149
Hierarchic models, crop plants, 320–323
HYBRID model, 359, 362
Hydrocarbons, volatile, potential reaction partners for ozone, 307

Insects
increased herbivory, 149
phytophagous, interactions with trees, 105–118
Integrated Forest Study sites, 35–36
Interference models, individual-based, 357–358
Interspecific competition, and phenology, 261
Irrigation, WET and DRY treatments, 219–236

JASPER model, 131
Jasper Ridge Biological Preserve, ecosystem-level CO_2 experiment, 121–142
Juvenility, and water use efficiency, 54

Kansas State University, tallgrass prairie experimental site, 150–160
Knowledge needs
in crop modeling, 337–339
NPP and NEP in natural ecosystems, 409

Lateral root mass, estimation, 5
Leaching
foliar, and forest N cycling, 284–289
in litter decomposition, 90–92

Leaf area index, *see also* Green leaf area index
calculation from dry matter growth, 254
closed stand, 16–17
increase from high tillering rates, 166
Leaf area production, in yellow poplar and white oak, 13–14
Leaf-level responses, pattern consistency, 415–416
Leaf-level $\mathcal{L}$ factor
application to long-term studies, 390–392
biochemical basis, 386–388
in leaf photosynthesis model, 382–393
spatial extrapolation to global scale, 388–390
Leaves
ascorbate levels, 308–309
CO_2 assimilation, 328
formation under low temperatures, 178
green and senesced, N concentration, 90
gross photosynthesis, maximum rate, 253–254
growth, phenology, 60–61
LF_{max}, in canopy assimilation model, 277–279
nitrogen content, 194
photosynthesis, 5–6, 221–222
photosynthesis models, 349–351, 382–393
senescence, 231
stomatal conductance models, 351
total area and number, 74–77
water potentials, 225
Lepidoptera, interactions with trees, 105–118
Light, effects on faba bean and wheat, 259–260
Light energy, environmental stress effects, 309–311
Light use efficiency, *see also* Water use efficiency
CO_2 effect, 253
cotton and wheat, 236–238
Lignin
concentration in C_4 grasses, 157
increased amounts in elevated CO_2, 88
Lignin:N ratio
CO_2 enrichment effect, 96
litter, 363–364
LINKAGES model, 16–17, 359, 362
Lint yield, cotton, 231–233, 237–238

Liquidambar styraciflua, see Sweet gum
Liriodendron tulipifera, see Yellow poplar
Litter
　belowground production and turnover, 364
　dead and green foliage, 156–157
Litter quality
　CO_2 enrichment effect, 92–96
　and decomposition, 89–92, 416–417
Loblolly pine
　clonal plants, 66–67
　and sweet gum, litter quality differences, 88

MAESTRO model, 57
Maricopa Agricultural Center, FACE facility, 217
Markov models, 358
Maturity, tree, effect on responses to elevated CO_2, 53–68
MECCAs, see also Microcosms
　artificial ecosystem units, 130
　design, 134–138
Mechanism
　versus empiricism, in crop models, 320–323
　in predicting ecosystem responses, 348
Memory
　ozone memory effect, 306
　seedling–sapling level of forest, 293
Methane, flooded rice fields as source, 280
Microbial catabolism, in litter decomposition, 90–92
Microclimate
　alteration by open-top chamber, 325, 406
　and CO_2 concentration, 185–188
Microcosms, see also MECCAs
　arctic tundra, 418, 420, 422
　for different experiments, 134–138
　at Jasper Ridge site, 129–131
Migration, plant
　CO_2 effects, 362–363
　speed, 371
Mineralization, see Carbon mineralization; Nitrogen mineralization
Modeling
　elevated CO_2 effects, challenges, 347–371
　plant and ecosystem responses to elevated CO_2, 400–409
Models, see also Computer programs
　canopy assimilation, 277–279
　crop, with state and rate variables, 319–320
　decomposition and nutrient cycling, CENTURY, 131
　ecosystem, Jasper Ridge CO_2 experiment, 121–142
　global carbon cycle, 401–402
　leaf-level, 349–351
　leaf photosynthesis, 383–393
　mathematical, ecosys, 226–227
　mechanistic simulation, 317–340
　plant-level, 351–355
　plant photosynthesis and growth, GePSi, 131
　population, community, and stand, 355–359
　population dynamics, JASPER, 131
　process-based, MAESTRO, 57
　process-based plant–soil simulation, G'DAY, 13–14, 360
　simulating potential crop production, 253–262
　stand competition, LINKAGES, 16–17, 359, 362
　two-phase negative exponential, 91
　UMBS conceptual, 42–44
Moth, herbivory, elevated CO_2 effects, 112–115
Mycorrhizae
　carbohydrate allocation, 24
　turnover, C input via, 43
　vesicular arbuscular, 46

Needles
　age, and growing season, 63–65
　biomass increase in elevated CO_2, 59
　N and CO_2 concentrations, 26–28
　Norway spruce, superoxide dismutase, 301, 304–306
NEP, see Net ecosystem production
Net carbon exchange, in tallgrass prairie experiment, 153–154
Net ecosystem CO_2 exchange, Chesapeake Bay wetlands, 199–202, 209–212
Net ecosystem production
　affecting factors, 399–400
　tussock tundra, 407
Netherlands, Wageningen Rhizolab facility, 252–262
Net primary production, and β factor, 382

Nitrogen
  allocation
    among vegetative organs, 353–354
    to reproductive structures, 354–355
  availability
    increase with elevated CO_2, 16–17
    and litter decomposition, 155
  content of leaves, 194
  distribution between plants and soils, 425–426
  foliar concentration, 7, 26–28
  interactions with CO_2 in forests, 283–295
  limitation, as continuum, 35, 294–295
  retranslocated, 90–91
  sweet chestnut and beech, under elevated CO_2, 79–81
Nitrogen cycling
  elevated CO_2 effects, 289–291
  in forests, 284–289
Nitrogen mineralization, *see also* Carbon mineralization
  and decomposition rates, 91–92
  and immobilization, 48–49
  increased with organic matter turnover, 43–44
  in rhizosphere soils, 30, 37–38
  at tree seedling site, 4
North America, coocurrence of white oak and yellow poplar, 3–5
Norway spruce, combined effects of CO_2 and ozone, 303–307
NPP, *see* Net primary production
Nutrients, *see also* Requirement
  arctic plants limited by, 172–174
  cycling
    and decomposition, 363–364
    stages, 286
  interactions with CO_2, 2–3, 75–77, 339
  internal recycling, 56
  limitations, and biomass production, 155
  mineral, role in CO_2 effects, 182–183
  reallocation within canopy, 352–353
  retranslocation in coniferous species, 90–92
  variations, microcosm study at Jasper Ridge, 129–131

Open-top chamber, *see also* Closed-top chamber
  alpine vegetation experiments, 182–195
  alteration of microclimate, 325
  Chesapeake Bay wetlands study, 198
  design, 131–133
  Edinburgh studies with Sitka spruce, 56–68
  field experiments with tree saplings, 3–5
  Jasper Ridge CO_2 experiments, 129–142
  modified, in studies with deciduous trees, 72–85
  Placerville site study with ponderosa pine, 24–38
  wheat agrosystem, 234
Oryza sativa, *see* Rice
Oxidative stress, in trees, protection from, 299–311
Ozone
  with CO_2, effects on plants, 303–307
  as environmental stress, 299–300

Paper birch
  insect herbivory, 112–115
  tree chemistry, elevated CO_2 effects, 107–112
Partitioning
  biomass, sweet chestnut and beech, 73–74
  models, 329–331
Phase change, in trees, 54
Phenology
  accelerated rates, 234–235
  and biomass, 191–192
  and growth patterns, 423
  interaction with CO_2, 168
  in interspecific competition, 261
  leaf growth, 60–61
  root growth, 55
  sweet chestnut and beech, 77
  tree, CO_2 enrichment effect, 110
Photorespiration
  crops, inclusion in crop models, 327–329
  H_2O_2 formed in, 309–310
  inhibition, 300
  and photosynthetic apparatus, 303–304
Photorespiratory carbon oxidation cycle, 387–388
Photosynthate
  excess, sink strength alteration, 9
  expanding sinks, 239
  sink demand, 207–209
Photosynthesis
  acclimation, 210–211
  C_3 and C_4, 148, 327–329
  canopy, plant-level model, 352–353
  cotton and wheat crops, measurement, 220–221

doubling in CO_2-enriched plants,
    61–63
  gross, maximum rate, 253–254
  and growth, feedback, 2–3
  interaction with growth and
    development, 334–335
  leaf and canopy, 221–225
  leaf-level model, 349–351, 382–393
  light-saturated, leaves, 5–6
  prolonged, in arctic plants, 169–170
  responses to elevated CO_2, 49
  rice canopy, 269–274
  sweet chestnut and beech, 78–79
Photosynthetic capacity
  C_3 species, 154
  elevated CO_2 effects, 206–210
  loss with increased CO_2 concentration,
    279–280
Photosynthetic carbon reduction cycle,
  387–388
Photosynthetic β factor, defined from $\mathscr{L}$
  factor, 392–393
Photosynthetic photon flux density
  cotton and wheat, 221
  in crop models, 319–320
  rice, 273, 277–278
Physiognomy, ecosystem, and species
  composition, 361–363
Picea sitchensis, see Sitka spruce
Pinus ponderosa, see Ponderosa pine
Pinus taeda, see Loblolly pine
Placerville site
  N fertilization, 294
  ponderosa pine growth and N dynamics
    study, 23–38
  soil solution NO_3^-, 291
Plant-level models, 351–355
Plants
  alpine, CO_2 enrichment effects, 177–178
  arctic, direct effects of elevated CO_2,
    163–174
  C_3, photosynthesis model, 383–393
  C_3 and C_4, elevated CO_2 effects, 426
  community structure, incorporation into
    ecosystem models, 362–363
  interactions with herbivores, modeling,
    365–366
  response to elevated CO_2, modeling,
    400–409
Plant–soil simulation models, process-
  based, 13–14

Ponderosa pine
  growth and N dynamics, 23–38
  seedlings, N cycling, 289
Population models, 355–358
Populus euramericana, UMBS 1992 study,
    45–49
Populus grandidentata, see Bigtooth aspen
Populus tremuloides, see Quaking aspen
Pot studies, artifactual restrictions, 1,
    97–100, 291–294
Prairie, tallgrass
  ecosystem-level responses to elevated
    CO_2, 147–160
  twice-ambient CO_2, 408
Primary production
  in ecosystem models, 360–361
  gross, atmospheric CO_2 effects, 399–400
  increases in more productive systems,
    421–422
  net, and β factor, 382
Productivity, ecosystem, 417–422

Quaking aspen
  insect herbivory, 112–115
  tree chemistry, elevated CO_2 effects,
    107–112
  UMBS research, 46
Quantum efficiency, temperature and CO_2
  effects, 277–278
Quantum flux density, in alpine ecosystem
  studies, 185–187
Quercus alba, see White oak
Quercus rubra, see Red oak

Radish, combined effects of CO_2 and
  ozone, 303–307
Rainfall distributors, in microcosms, 137
Reactive oxygen species, detoxification,
  300–303
Reciprocal exposure experiments, with
  arctic plants, 168–169
Red oak
  insect herbivory, 112–115
  tree chemistry, elevated CO_2 effects,
    107–112
Regeneration rate, ribulose 1,5-
  bisphosphate, 387–388
Regional and global models
  bottom-up approaches, 366–368
  top-down approaches, 368–369
Regrowth cores, in study of elevated CO_2,
  203

Replication, experiment, limitations, 122–124
Reproductive structures, C and N allocation, 354–355
Requirement, *see also* Nutrients calculation, 284–289
Resource use models, 356
Respiration, *see also* Dark respiration rate
  aboveground and soil–root, 256
  inclusion in crop modeling, 336–337
  rice canopy, seasonal trends, 274–277
Response surface, chaparral shrubs, 359–360
Responsiveness, ecosystems to elevated CO_2, 417–422
Rhizomes
  biomass production, 210
  elevated CO_2 effect, 203
Rhizosphere
  labile C input, 43
  N availability, CO_2 effect, 24–32, 290
Ribulose 1,5-bisphosphate, carboxylation, 386–387
Ribulose-1,5-bisphosphate-carboxylase/oxygenase, *see* Rubisco
Rice
  aboveground β, estimate, 404
  SPAR experiments
    canopy assimilation model, 277–279
    growth and yield, 267–269
    photosynthesis, 269–274
    respiration, 274–277
Root closure
  greater root biomass with, 38
  and N uptake, 32–34
Roots, *see also* Fine roots
  C and N allocation, 353–354
  grasses, biomass, 199
  growth
    phenology, 55
    study methods at UMBS, 45–46
  ingrowth bag biomass, 156
  N and CO_2 concentrations, 26–30
  restriction factor, 83–84, 98–99
  sedges and grasses, production, 202–203
Root–shoot partitioning, carbon, 330–331
Root–shoot ratio
  CO_2 and N effects, 30–31
  elevated CO_2 effects, 335
  maturation state effects, 65–66
  model-predicted response to drought, 324–325

ROOTS software, 46
Rubisco
  arctic plants, content reduction, 171
  coactivated by CO_2, 327
  CO_2-binding, 388
  elevated CO_2 effect, 24
  kinetics, 321–322
  limiting leaf photosynthesis, 383
  N reallocated away from, 43
  oxygen-binding, 303
  reduction under high CO_2, 194
  rice canopy, decreased activity, 273

Salt marsh
  Chesapeake Bay wetlands study, 197–212
  productivity, 420, 422
Sap flow, transpiration estimates, 226
Scaling
  and nested modeling approach, 359–360
  plant responses to population and community, 358–359
Scirpus olneyi, *see* Sedges
Scotland, Edinburgh, open-top chamber studies with Sitka spruce, 56–68
Seasonal trends, rice canopy dark respiration, 274–277
Sedges
  arctic, responses to elevated CO_2, 165–166
  C_3
    Chesapeake Bay wetlands, 198–212
    C:N ratio, 92–94
    dense mats, and C sinks, 194–195
    dominating alpine plant community, 179–180
Seedlings
  and ecosystem-level response, 291–294
  forest tree, CO_2 enrichment, 3–19
  ponderosa pine
    applicability of results to mature forests, 32–38
    biomass, 26–27, 30
    N cycling, 289
  Sitka spruce
    Edinburgh open-top chamber study, 58–59
    growth and biomass, 64–68
    soil-planted beech and chestnut, 73–77
Shoots
  C and N allocation, 353–354
  grasses, biomass, 199
  sedges and grasses, production, 202–203

Shrubs
  arctic, responses to elevated CO_2, 165–166
  chaparral, response surface, 359–360
Single-level models, 321–323
Sinks
  for additional carbohydrates, 166–168, 170–171
  atmospheric CO_2, in Arctic, 164
  carbon
    and dense sedge mats, 194–195
    in tropical grasslands, 424
  photoassimilates, 271–272
  photosynthate, 207–209, 239
Sitka spruce, responses to elevated CO_2, 53–68
Size
  plants, and experimental tractability, 128
  trees, and maturity, 55–56
Soil–plant–atmosphere research units, rice studies, 266–281
Soils
  Aiken clay loam, Placerville site, 24–25
  alpine, composition, 180
  arctic, carbon stocks, 164–165
  carbon budget and sequestration, 238
  disturbance during potting, 291
  ion uptake, 336–337
  Jasper Ridge, C content and disturbance, 128–129
  as major N pool in mature forests, 35–38
  moisture, and evapotranspiration, 226, 422–423
  NO_3^-, affecting insect performance, 114–115
  rhizosphere
    NH_4^+ and NO_3^-, 26
    N mineralization, 30, 48
  serpentine, 135–136, 420
Source–sink balance
  and growth strategies, 71–72
  models predicting, 329–330
  pot-grown sweet chestnut and beech, 83
  in tallgrass prairie, 159–160
Sour orange trees, growth efficiency, 15–16
SPAR, see Soil–plant–atmosphere research units
Species composition changes
  among arctic plants, 173–174
  annual plants at Jasper Ridge site, 127–128
  ecosystem, 87–89
  and ecosystem physiognomy, 361–363
  at elevated CO_2, 93–94, 163, 290–291
Species distribution, as function of climate, modeling, 357
Springs, cold CO_2, 171–172
Spring wheat
  biomass, leaf area, and head production, 230–231
  crop growth modeling, 256–259
  grain yield, 233–235
  harvestable yield, 254–256
  model description and validation, 253–254
  semifield experimental setup, 252–253
  temperature and light effects, 259–260
  test crops at FACE facility, 218–220
Stand competition models, 16–17
Stand models, 355–359
Stem elongation, elevated CO_2 effect, 77
Stem mass, CO_2 enrichment effect, 9–13
Stems, faba bean, increased yield, 255–256
Stomatal conductance
  decrease at elevated CO_2, 239
  inclusion in crop modeling, 331–332
  leaf-level model, 351
  response to CO_2 enrichment, 5–6
  and stomatal density, 225–226
  sugar maple seedlings, 18–19
  and water use efficiency, 148, 157–158
Sugar maple
  insect herbivory, 112–115
  stomatal conductance, 18–19
  tree chemistry, elevated CO_2 effects, 107–112
Superoxide dismutase
  ozone effects, 304–305
  in spruce needles, 301
Sweet chestnut
  biomass and partitioning, 73–74
  carbohydrates, 81–82
  chlorophyll and nitrogen content, 79–81
  growth strategy and response to elevated CO_2, 71–85
  leaf area and leaf number, 74–77
  litter, 96
  phenology, 77
  photosynthesis and respiration, 78–79
Sweet gum, and loblolly pine, litter quality differences, 88
SYSTAT software, 26

Tallgrass prairie
  ecosystem-level responses to elevated CO_2, 147–160
  productivity, 418
  twice-ambient CO_2, 408
Tannin content
  Douglas fir, 90
  hardwoods, 110
Temperature
  arctic soil, 173
  canopy air, 187–188
  and decomposition rates, 88–89
  effects on
    C_3 photosynthesis, 328
    faba bean and wheat, 259–260
    global change, rice responses, 265–281
    globally averaged, 385
    increased, crop responses, 332–335
    interaction with CO_2, in forest stand dynamics, 17–19
    night, constraint on cell differentiation, 179
Testing, crop models, 323–326
Thomomys bottae, *see* Botta's pocket gopher
Time of day, and net ecosystem CO_2 exchange, 199–202
Tissues
  belowground, C allocation, 172–173
  chemical composition, alteration, 116–117
  composition of Jasper Ridge annuals, 140
  foliar, suitability for insects, 106
  N concentration
    perennials, 288
    plant, 155–157
    wetlands grasses and sedges, 197–212
  types, comparative analysis, 63–68
  woody
    N concentrations, 28–30, 34–35
    N use efficiency, 289
Tobacco, ascorbate concentrations, 306–307
Tractability, and generality, of model ecosystems, 124–125
Transpiration, *see also* Evapotranspiration
  canopy, plant-level model, 352–353
  cotton and wheat crops, measurement, 220–221
  inclusion in crop modeling, 331–332
  sap flow, estimates, 226

Trees, *see also specific types*
  chemistry, CO_2-mediated changes, 105–118
  oxidative stress, protection from, 299–311
  responses to elevated CO_2
    and growth strategy, 71–85
    maturity effects, 53–68
    UMBS conceptual model, 42–44
  shade-tolerant, growth response to elevated CO_2, 290–291
Triticum aestivum, *see* Spring wheat
Tulip tree, *see* Yellow poplar
Tussock tundra
  exposed to elevated CO_2, 167–171
  NEP, 407

UMBS, *see* University of Michigan Biological Station
University of Arizona, Maricopa Agricultural Center, 217
University of Michigan Biological Station, research site, 42–50
University of Orsay, modified open-top chamber studies with deciduous trees, 72–85
University of Wisconsin, biotron studies with hardwoods, 108–112
Upscaling, with MAESTRO process-based model, 57

Vicia faba, *see* Faba bean
Volatile hydrocarbons, potential reaction partners for ozone, 307

Wageningen Rhizolab, semifield studies with wheat and faba bean, 252–262
Water potential, midday, in high-CO_2 field plots, 138–139
Water stress
  effect on photosynthetic capacity, 158–159
  interaction with CO_2, 224–225
  in tallgrass prairie, 153–154
Water use efficiency, *see also* Light use efficiency
  cotton and wheat, 236–238
  ecosystem-level, elevated CO_2 effects, 152–154
  in juvenile and mature trees, 54
  and reduced stomatal conductance, 148, 419–420

Wetlands
    Chesapeake Bay, long-term elevated CO_2 exposure, 197–212
    communities, β values, 407–408
White oak
    carbon cycling, 13–15
    gas exchange, 5–6
    growth dynamics, 9–13
    metabolites and dry mass, 7–9
    seedlings, senesced leaves from, 289–290

Yellow poplar
    carbon cycling, 13–15
    gas exchange, 5–6
    growth dynamics, 9–13
    litter quality and mass loss rates, 94–96
    metabolites and dry mass, 7–9
Yield
    and growth responses
        cotton and wheat, 227–236
        rice, 267–269
    harvestable, for faba bean and wheat, 254–256

Zeaxanthin, ascorbate role, 308–309

Physiological Ecology
A Series of Monographs, Texts, and Treatises

Continued from page ii

F. S. CHAPIN III, R. L. JEFFERIES, J. F. REYNOLDS, G. R. SHAVER, and J. SVOBODA (Eds.). Arctic Ecosystems in a Changing Climate: An Ecophysiological Perspective, 1991

T. D. SHARKEY, E. A. HOLLAND, and H. A. MOONEY (Eds.). Trace Gas Emissions by Plants, 1991

U. SEELIGER (Ed.). Coastal Plant Communities of Latin America, 1992

JAMES R. EHLERINGER and CHRISTOPHER B. FIELD (Eds.). Scaling Physiological Processes: Leaf to Globe, 1993

JAMES R. EHLERINGER, ANTHONY E. HALL, and GRAHAM D. FARQUHAR (Eds.). Stable Isotopes and Plant Carbon–Water Relations, 1993

E.-D. SCHULZE (Ed.). Flux Control in Biological Systems, 1993

MARTYN M. CALDWELL and ROBERT W. PEARCY (Eds.). Exploitation of Environmental Heterogeneity by Plants: Ecophysiological Processes Above- and Belowground, 1994

WILLIAM K. SMITH and THOMAS M. HINCKLEY (Eds.). Resource Physiology of Conifers: Acquisition, Allocation, and Utilization, 1995

WILLIAM K. SMITH and THOMAS M. HINCKLEY (Eds.). Ecophysiology of Coniferous Forests, 1995

MARGARET D. LOWMAN and NALINI M. NADKARNI (Eds.). Forest Canopies, 1995

BARBARA L. GARTNER (Ed.). Plant Stems: Physiology and Functional Morphology, 1995

GEORGE W. KOCH and HAROLD A. MOONEY (Eds.). Carbon Dioxide and Terrestrial Ecosystems, 1996